Student's Companion

Pharmaceutical Chemistry

Student's Companion

Pharmaceutical Chemistry

Dr. Kishor S Jain

M.Pharm., Ph.D., FIC.

Principal
Rajmata Jijau Shikshan Prasarak Mandal's College of Pharmacy,
Dudulgaon, Moshi-Alandi Road, Pune-412105, Maharashtra.

Deepali K Kadam

M.Pharm.

Assistant Professor, Department of Pharmaceutical Chemistry
K. K. Wagh College of Pharmacy
Hirabai Haridas Vidyanagari, Amrut Dham, Panchavati
Nashik 422003, Maharashtra

PharmaMed Press

An imprint of BSP Books Pvt. Ltd.
4-4-309/316, Giriraj Lane,
Sultan Bazar, Hyderabad - 500 095.

Pharmaceutical Chemistry
by **Kishor S Jain and Deepali K Kadam**

Disclaimer: The authors and the publishers have taken due care to provide the authentic, reliable and up to date information related to the subject. However, neither the authors nor the publisher shall be responsible for any liability for any damage caused as a result of use of this book. The respective user must check the accuracy from other sources too.

Published by

PharmaMed Press
An imprint of BSP Books Pvt. Ltd.
4-4-309/316, Giriraj Lane, Sultan Bazar, Hyderabad - 500 095.
Phone: 040-23445688; Fax: 91+40-23445611
E-mail: info@pharmamedpress.net
www.bspbooks.net/www.pharmamedpress.net

ISBN: 978-93-90211-77-7 (Hardback)

Foreword

Pharmaceutical Chemistry is a backbone subject with sound knowledge of various other subjects like inorganic chemistry, organic chemistry, biochemistry, pharmacology, physical chemistry and analytical chemistry.

After going through the contents of the book entitled, **"Pharmaceutical Chemistry"**, authored by an eminent teacher, Dr. Kishor. S. Jain and co-author, Ms. Deepali K. Kadam, I felt that the authors have justified their zealous efforts. Creating interest in Pharmaceutical Chemistry among the Diploma in Pharmacy students through a book on the subject is always a challenge for any author. A variety of books by Indian authors are available on the subject, but a majority of them do not come up to the expectations of the learners. This book besides being comprehensive, is also with very simple and in very lucid language to understand. It will surely help the students and teachers in gaining better clarity of concepts for better preparations for Board as well as entrance exams.

This book stresses through simple and easy to understand language the mechanisms of actions, branded formulations, multiple examples along with structures and IUPAC nomenclatures, diagrams/figures, synthetic schemes for some important drugs as well as, question banks which include multiple choice questions, short answer questions and long answer questions on various included chapters. Various topics covered include- Introduction to Pharmaceutical chemistry- including impurities and limit tests, Volumetric & Gravimetric Analysis, Inorganic Pharmaceuticals, Nomenclature of Organic & Heterocyclic Compounds, Medicinal Chemistry (Study of the various categories of medicinal compounds with respect to classification, chemical name, chemical structure of some compounds, uses, stability and storage conditions, different types of formulations and their popular brand names).

This simple, systematic and thought-provoking presentation of the contents in this book would be of immense help to the students and teachers at diploma level.

I extend my best wishes to the authors for showing this book the light in the need, which would facilitate enhanced learning and teaching with pleasurable experiences.

Dr. Sanjay S. Toshniwal
(M.Pharm, Ph.D, LLB, IDDP, Indian Patent Attorney).
Central Council Member, Pharmacy Council of India, New Delhi.
Founder and General Secretary- Dr S K Toshniwal Educational & Research Trust.
Principal & Director - Vidarbha Institute of Pharmacy, Washim (M.S.).
IPR-Patent Professional.
Director, Marathwada Pharma Cluster, Aurangabad (M.S.).
Independent Director- KPM, Aurangabad (M.S.) (Past)
Technical Expert-BPPI, Dept of Pharmaceuticals, Govt of India. (Past)
All India Topper-Indian Patent Agent Exam (Govt of India).
Recipient of four National Awards including Best Pharmacy Teacher-2014 (IPC Trust)
Granted Four National and International Patents.

Preface

Great satisfaction and gratitude we express to all students and teachers all over the state as well as out of state pharmacy colleges for such a great response and acceptance to our earlier books on PHARMACEUTICAL ANALYSIS; PHARMACEUTICAL ORGANIC CHEMISTRY & PHARMACEUTICAL INORGANIC CHEMISTRY (both Theory & Practicals) for B.Pharm. degree students from past 16 years. We wish to state that their very positive feedbacks have encouraged us in continuing our endeavours and zealous efforts to now write a book for the Diploma in Pharmacy students in the subject- Pharmaceutical Chemistry. This book is comprehensive yet very simple and lucid to understand. Our aim is to provide the students and teachers not only what they need as per the syllabus, but also help them in gaining better clarity of concepts for better preparations for Board as well as entrance exams.

Therefore, we have stressed on simplicity, easy to understand language, mechanisms of actions, branded formulation, multiple examples along with structures and IUPAC nomenclatures, diagrams/figures, synthetic schemes for some important drugs as well as, question banks which include multiple choice questions, short answer questions and long answer questions after each of the total 31 chapters. Various topics covered are those specified in the syllabus of Pharmacy Council of India, and these include, different therapeutic classes of drugs and their classifications.

The contents as per the syllabus are broadly classified into

Chapter 1. Introduction to Pharmaceutical chemistry- including impurities and limit tests

Chapter 2. Volumetric & Gravimetric Analysis

Chapter 3. Inorganic Pharmaceuticals

Chapter 4. Nomenclature of Organic & Heterocyclic Compounds

Chapters 5-31: Medicinal Chemistry- Study of the various categories of medicinal compounds with respect to classification, chemical name, chemical structure (compounds marked with*) uses, stability and storage conditions, different types of formulations and their popular brand names

Thanks to the wonderful team of PharmaMed Press of BSP-Books Pvt. Ltd. especially, Hon. Anilbhai Shah and his team, and excellent co-ordination, co-operation and communication by Mr Naresh Davar to meet the timeline and maintain the quality of this book. We once again wish an enjoyable learning experience for the students and satisfactory teaching for the teachers of the subject – Pharmaceutical Chemistry and a Happy New Academic Year!

- Authors

CONTENTS

CHAPTER 1

Introduction to Pharmaceutical Chemistry

CHAPTER 2

Volumetric Analysis

CHAPTER 3.1

Haematinics

CHAPTER 3.2

Gastro-Intestinal Agents

CHAPTER 3.3

Topical Agents

CHAPTER 3.4

Dental Products

CHAPTER 3.5

Medicinal Gases

CHAPTER 4

Introduction to Nomenclature

CHAPTER 5

Drugs Acting on Central Nervous System: Anaesthetics

CHAPTER 5.2

Drugs Acting on Central Nervous System: Sedatives and Hypnotics

CHAPTER 5.3

Drugs Acting on the Central Nervous System: Antipsychotics

CHAPTER 5.4

Drugs Acting on the Central Nervous System: Anticonvulsants

CHAPTER 5.5

Drugs Acting on the Central Nervous System: Antidepressants

CHAPTER 6.1

Drugs Acting on Autonomic Nervous System: Sympathomimetic Agents

CHAPTER 6.2

Drugs Acting on Autonomic Nervous System: Adrenergic Antagonists

CHAPTER 6.3

Drugs Acting on Autonomic Nervous System: Cholinergic Drugs and Related Agents

CHAPTER 6.4

Drugs Acting on Autonomic Nervous System: Cholinergic Blocking Agents: Natural & Synthetic

CHAPTER 7.1

Drugs Acting on Cardiovascular System: Anti-Arrhythmic Drugs

CHAPTER 7.2

Drugs Acting on Cardiovascular System: Anti-Hypertensive Agents

CHAPTER 7.3

Drugs Acting on Cardiovascular System: Antianginal Agents

CHAPTER 8

Diuretics

CHAPTER 9

Hypoglycemic Agents

CHAPTER 10

Analgesic and Anti-Inflammatory Agents

CHAPTER 11.1

Anti-Infective Agents: Antifungal Agents

CHAPTER 11.2

Anti-Infective Agents: Urinary Tract Anti-Infective Agents

CHAPTER 11.3

Anti-Infective Agents: Antitubercular Agents

CHAPTER 11.4

Anti-Infective Agents: Antiviral Agents

CHAPTER 11.5

Anti-Infective Agents: Antimalarials

CHAPTER 11.6

Anti-Infective Agents: Sulfonamides

CHAPTER 12

Antibiotics

CHAPTER 13

Anti-Neoplastic Agents

Chapter 1

INTRODUCTION TO PHARMACEUTICAL CHEMISTRY

CONTENTS

- Introduction to Pharmaceutical chemistry: Scope and objectives
- Sources and types of errors: Accuracy, precision, significant figures
- Impurities in Pharmaceuticals: Sources and effect of impurities in Pharmacopoeial substances, the importance of the limit test, Principle and procedures of Limit tests for chlorides, sulfates, iron, heavy metals and arsenic.

♦ LEARNING OBJECTIVES ♦

After completing this chapter, the student should be able to understand:

1. Scope and objectives of Pharmaceutical Chemistry.
2. Sources and types of errors.
3. Accuracy, precision, significant figures.
4. Impurities in Pharmaceuticals
5. Different sources of impurities
6. Principles and procedures of various Limit tests.

1.1 INTRODUCTION TO PHARMACEUTICAL CHEMISTRY

Pharmaceutical chemistry involves drug chemistry, quality assurance and analysis of drugs and study of various analytical techniques used, drug metabolism, pharmacology, as well as cures and remedies for diseases. **[Fig.1.1].**

Under the broad scope of Pharmaceutical Chemistry, various branches studied are – Pharmaceutical Inorganic chemistry, Pharmaceutical Organic Chemistry, Pharmaceutical Biochemistry, Pharmaceutical Analysis, Pharmaceutical Medicinal Chemistry, etc. Even new drug discovery, pharmacokinetics, and quality assurance as part of pharmaceutical chemistry. Thus, the scope of Pharmaceutical Chemistry is very broad, and these components are taught to more or less extent in various courses of Pharmacy.

To Summarize- "Pharmaceutical Chemistry deals with the Drugs' Structure, Synthesis, Analysis, Medicinal Effect, Mechanism of Action, Metabolism and Quality Assurance".

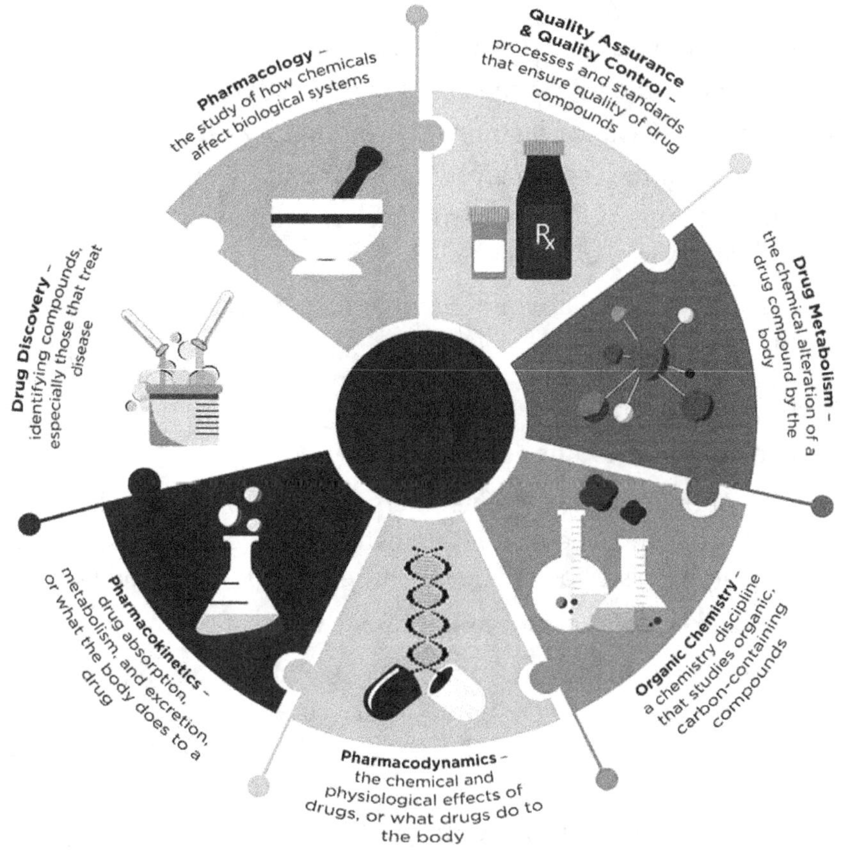

Fig. 1.1 Scope of Pharmaceutical Chemistry

1.2 SOURCES AND TYPES OF ERRORS, ACCURACY, PRECISION, SIGNIFICANT FIGURES

1.2.1 Common sources of errors

These include instrumental, environmental, procedural, and human errors. An error is the major aspect of laboratory science in physical and chemical testing, and its test findings are the primary scientific basis for assessing product quality. Physical and chemical laboratory experiments include three primary sources of errors: **systematic *error, random error*** and ***human error***. These sources of errors in the laboratory should be studied well before any further action.

1.2.1.1 System Errors in laboratory experiments- It applies to the repeated measuring of the same object under repeated conditions of measurement. The amount of the error value is either positive or negative, which is called the **fixed system error** in laboratory experiments and laboratory tests.

The error is caused primarily by:

- The incorrect method of measurement in laboratory experiments
- The incorrect method of using the instrument in laboratory experiments
- The failure of the measuring instrument in laboratory experiments
- The performance of the testing tool itself in laboratory experiments
- The inappropriate use of the standard material and the changing environmental conditions in laboratory experiments

With certain steps and proper laboratory equipment, these sources of errors can be minimised and corrected.

Different types of system errors are:

1.2.1.1.1 Method error in laboratory experiments

The method error in laboratory experiments refers to the error created by the **very process** or **procedure** of physical and chemical examination. This error is inevitable, so the test result is often low or high. This error can be minimised through method validation.

1.2.1.1.2 Instrument error in laboratory experiments

The instrument error in test labs is caused primarily by **laboratory instrument inaccuracy**. For instance, if the meter dial or the zero point is inaccurate, the measurement result would be too small or too big. This error can be minimised through instrument calibration at regular intervals.

1.2.1.1.3 Reagent error in laboratory experiments

The reagent error in lab tests is caused primarily by the **impure reagent** or the **inability to meet the experimental provisions**; such as the existence of impurities in the reagent used in the physical and chemical testing phase; or the existence of contaminated water or reagent contamination that may influence the results of the examination; or the storage or operating climate.

1.2.1.2 Random error in laboratory experiments

Error caused by various **unknown factors** is known as random error. This error poses erratic changes at random, primarily due to a variety of small, independent, and accidental factors. The random error is atypical from the surface. Since it is accidental, the random error is often called an **unmeasurable** or **accidental error**.

1.2.1.3 Human Error in laboratory experiments

Human error in laboratory experiments and laboratory tests primarily refers to the mistake in the **physical and chemical inspection phase** caused by the factors of the **inspector**; particularly in the following three aspects:

a. **Operational error in laboratory experiments:** This is due to a lack of operation precautions. e.g., weighing a hygroscopic substance without effective moisture or humidity control can cause an error in weight.

b. **Subjective error in laboratory experiments:** It is caused by differences in subjective observations and perceptions from person to person regarding physical and chemical tests.

    e.g., different perceptions of sharpness in colour change at the endpoint

c. **Negligence error in laboratory experiments:** refers to the mistake caused during the physical and chemical examination by the analyst, e.g., reading mistake, operation error, measurement error etc.

1.2.2 Accuracy

Accuracy is the closeness of the measurement to its true or accepted value; any deviation from this value is expressed by the **error**. It measures agreement between a result and its actual true value. We can never determine accuracy precisely because the true value of a measured quantity can never be known exactly. We need to use an accepted value. Accuracy is expressed in terms of either *absolute* or *relative* error.

Precision, on the other hand, describes the agreement (or closeness or less deviation in values) among several results that have been obtained similarly. **(Fig. 1.2)**

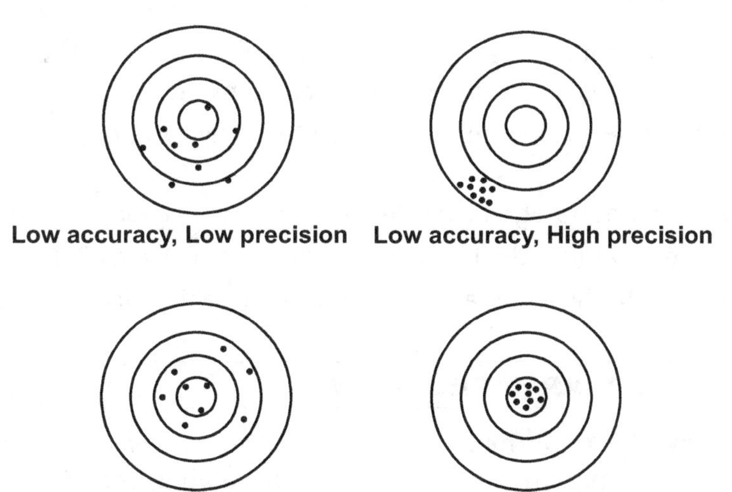

Fig. 1.2 Illustration of Accuracy and Precision showing their distribution in and around the Bull's Eye of a Dart Board

To determine accuracy, we have to know the true value, and this value is exactly what we are trying to obtain through the analysis. Do we know the answer precisely (precision), also, do we know it accurately (accuracy)?

1.2.3 Precision

Precision describes the reproducibility of measurements, i.e., the closeness of results to each other. Precision is determined by repeating the measurement on replicate samples.

Precision is a function of the deviation from the mean d_i, or just the deviation, which is defined as:

$$d_i = \left| X_i - \bar{X} \right|$$

The lesser the deviation, the higher the precision. In other words, precision is the closeness of results to others that have been obtained in precisely the same way.

Three terms to describe the precision of a set of replicate data are standard deviation, variance and coefficient of variation.

1.2.4 Significant Figures

One of the best ways of indicating reliability is to give a confidence interval at the 90% or 95% confidence level. Another method is to report the absolute standard deviation or the coefficient of variation of the data.

   A less satisfactory but more common indicator of the quality of data is the **significant figure** convention:

- A simple way of indicating the probable uncertainty associated with an experimental measurement is to round the result so that it contains only significant figures.
- The significant figures in a number are all the certain digits plus the first uncertain digit.
- A zero may or may not be significant depending on its location in a number.
- A zero surrounded by other digits is always significant (such as 30.24 ml).
- Zeros that only locate the decimal point for us are not significant.

1.3 IMPURITIES IN PHARMACEUTICALS

Pharmaceutical chemicals and formulations must maintain a very high degree of purity. A compound is considered impure if it contains foreign matter/impurities. These impurities affect its potency. Although it is almost impossible to attain 100% purity, technically as well as cost-wise, it is still possible to achieve a high degree of purity.

Although, Purity can be achieved through the process of purification, it is often economically less viable. Alternatively, a reasonably acceptable purity can be achieved by controlling various sources or reasons that add to the impure nature of an active pharmaceutical ingredient or drug and excipients used in pharmaceutical formulations. Pharmacopoeias have fixed the limit for these impurities.

1.3.1 Sources and Effects of Impurities in Pharmacopoeial Substances

Impurities may enter or are formed in a drug substance during any of the following three stages during;

1. Manufacturing.
2. Purification and processing.
3. Storage and packaging.

1. During Manufacturing:

(a) **Raw materials employed:** Impurities present in raw materials may be carried through the manufacturing process to contaminate the final product. Impurities such as; As, Pb, heavy metals, chlorides associated with sodium compounds, sulfuric acid with copper sulfate and hydrochloric acid with iron chloride, are some common examples. Likewise, many elements accompany others in traces.

(b) **Reagents used in the manufacturing process:** The quality and purity of reagents used for manufacturing the drug substances are very important. If reagents used in the manufacturing process contain some impurities, these may find entry into the final product. For example, Sulphuric acid is used in many chemical processes. This acid often has lead present in it. Anions like Cl^- and SO_4^{-2} are common impurities in many substances because hydrochloric acid and sulphuric acid are used in processing.

(c) **Solvents used in the manufacturing process:** The manufacturing processes may involve single or multiple steps (unit operations). Naturally, solvents play an important role next to the main reagents as most of the chemical reactions involved in these processes are solvent based. If proper quality/purity of solvents is not assured, they may add to the impurities. Solvents like toluene and *n*-butanol contain water as an azeotrope. Alcoholic solvents also may be contaminated with water, and ethyl acetate can contain acetic acid in small amounts. Thus, the quality of solvents needs to be assured and controlled.

(d) **Reaction vessels:** The reaction vessels employed in the manufacturing process may be metallic (cast iron, mild steel, stainless steel) or mild steel with a glass lining. Nowadays, wooden or other metallic vessels are not used in the pharma industry. Some solvents and reagents employed in the process may react with the metals of the reaction vessels, leading to their corrosion and passing traces of metal impurities into the solution, contaminating the final product. Similarly, glass vessels may leach traces of alkali into the solvent. Even if acids like HCl, if by chance, contain a small amount of fluoride, it can itch the glass lining and begin the metallic contamination. Lead antimony, bismuth etc., can crop up as impurities from the vessels.

(e) **Intermediate products in the manufacturing process:** Some intermediates which are produced during the manufacture may be carried out through the final product as impurities. Intermediates are products of (i) incomplete conversion of reactants to final products or (ii) side or competing reactions, or (iii) decomposition of products formed due to poor process control. In the manufacturing process of KI, the intermediate iodate is the main impurity. Similarly, sodium bromate is the impurity in NaBr.

(f) **Defects in manufacturing process:** Poor mixing and non-adherence to optimum reaction conditions (proper temperature, pressure and pH) may lead to impurities. E.g., Improper heating (failing to achieve bright red temperatures) in the process of manufacturing Zinc Oxide can lead to unoxidised metallic Zn as an impurity.

$$2\ Zn + O_2 \rightarrow 2\ ZnO + Zn\ \textbf{(impurity)}$$

(g) **Manufacturing hazards:** In industrial areas, the atmosphere is contaminated with dust particles (Al_2O_3, silica glass, carbon, gases like; H_2S, SO_2, CO_2, CO, etc.). While manufacturing pharmaceutical products, these impurities may enter the final product. Accidental inclusion of dirt or glass or porcelain or silica or carbon or fibre particles due to poor manufacturing practices and facilities unable to check atmospheric and cross contaminations can lead to unwanted particulate matters in the product in many ways. These need to be checked and controlled. Wear and tear of machinery may shed metallic particles.

2. During Purification and Processing:

If not properly controlled, impurities are often added during the purification processes, mainly through the purifying reagents, solvents or vessels used.

(a) Reagents used to remove other impurities: Some chemicals are sometimes added to remove or precipitate another substance. This may also give rise to a source of impurity. For example, $BaCl_3$ is added to remove excess sulphate in $AlCl_3$. Hence, $AlCl_3$ is likely to contain Ba as an impurity.

(b) Solvents used in the process of purifications: Often, the solvents used for purification can be sources of impurities. These solvents range from organic solvents to acids (organic as well as mineral) and, of course, water. Water is the cheapest solvent and most widely used. Therefore, it is known as the universal solvent.

Types of water used are:

(i) **Tap water:** It contains impurities of Na^+, Ca^{2+}, Mg^{2+}, CO_2^{-3}, and SO_4^{-2}, which, when used, appear as impurities in the final product.

(ii) **Softened water:** It contains Na^+ and Cl^- ions as impurities that may appear as impurities in the final product when used.

(iii) **Demineralised water:** Though it is free from all the above inorganic ion impurities, it still contains organic impurities like; salts of carboxylic acids, N and S, etc.

(iv) **Distilled water:** Considered to be the best. It is pure water and free from all inorganic and organic impurities, but its production cost is very high.

(c) **Contamination due to vessels and equipment used for purification:** During the purification processes, if the vessels are defective or not perfectly cleaned and dried, they may add impurities like; metallic ions, rust, glass particles, moisture, etc. The other equipment, mainly the filters, centrifuges, dryers, etc., also need to be cleaned and dry.

3. **During Storage and Packaging:**

(a) **Errors in the packaging or use of substandard packaging material:** During the process of packaging or filling and sealing, whether applicable for solid dosage forms or liquid dosage forms or API, proper material which can ensure complete foolproof packaging without access to the atmosphere and light will ensure the stability of the product. Thus, the quality and strength of packaging material is very important. For example, if the aluminium foil for tablet strip or cap for a liquid formulation bottle is of substandard quality, it can add to impurities. This may lead to recalls of entire batches from the market. This is very critical for parenteral formulations.

(b) **Faulty packaging processes:** Most pharmaceutical packaging processes are assembly lined automated, generally involving pressing and sealing with heat. If the process parameters are not optimised or are tampered with, then it may lead to contaminations. For example: Nowadays, most parenteral products are in polymer containers using FFS (Form-Fill-Seal) processes which involve proper heating, filling, sealing and congealing cycles. Any changes in process parameters can be hazardous.

(c) **Microbial contamination:** Microbial contaminations, mainly in the form of fungal and bacterial growth, may be due to the result of improper storage conditions as well as faulty packaging. The products for parenteral administration and ophthalmic preparations have to undergo sterility tests.

1.3.2 Effects of Impurities on Pharmaceuticals

1. Some impurities, if present beyond certain tolerance limits, can cause untoward side effects that can lead to unpleasant reactions—for example, Heavy metals like; Pb, Fe and As salts.

2. Some impurities which are otherwise harmless in nature and without any therapeutic effect, if present in considerable proportions, dilute the active strength or potency of the drug substance—for example, Na, K, Cl, SO_4, CO_3 salts.

3. Some impurities may be able to catalyse the degradation, thereby shortening the shelf life of the drug substance.

4. Some impurities, by their chemical nature, can interact with the drug substance to affect its purity and potency. Such impurities are said to be incompatible with the drug substance/s.

5. Some impurities, by virtue of their unstable nature like; hygroscopic nature, oxidisable nature, etc., can bring about change in the physical properties like; change in appearance, taste, odour, stability, etc., of the drug substance causing technical difficulties in its use as well as formulation.

1.4 LIMIT TESTS

Limit Tests are quantitative or semi-quantitative tests designed to identify or control small quantities of impurities. These tests should be specific and sensitive.

Limit = A value or amount that is likely to be present in a substance.

Test = To examine or to investigate.

Impurity = A foreign matter present in a compound.

Definition:

A limit test is defined as "a quantitative or semi-quantitative test designed to identify and control small quantities of impurities which are likely to be present in the substance".

1.4.1 Importance of Limit Tests

1. To find out the harmful amount of impurities.
2. To find out avoidable/unavoidable amounts of impurities.

Types of Limit Tests:

1. Comparison method.
2. Quantitative determination.
3. Test in which there is no visible reaction.

General Principles:

1. If the sample is lighter (in colour/turbidity/opalescence) than the standard solution, then it is within the pharmacopoeial limit (accepted).

2. If the sample is darker/heavier than the standard solution, it is above the pharmacopoeial limit (rejected).

3. **Specificity of a Limit Test:** A given limit test for a trace impurity should involve some selective reaction of the reagent with the trace impurity under consideration/detection specifically characteristic only to it.

4. **Sensitivity of a Limit Test:** As most limit tests involve dilute solutions and results are based on the concentration of the trace impurity, the results may take longer to become observable or appreciable. Thus, consideration of the duration of the test needs to be of prime consideration in designing the limit test.

Nesslers' Cylinder (IP appendix VII A127):

It is a clear glass cylinder with a normal capacity of 50 ml. However, some of Nessler's cylinders are of 100 ml capacity. The overall height is about 15 cm, the external height to the 50 ml mark is 11.0 to 12.4 cm, and the thickness of the wall is around 1.0 to 1.5 mm, while the thickness of the base is about 1.0 to 3.0 mm. The external height to the 50 mark of cylinders used for the test must not differ by more than 1 mm in the given pair.

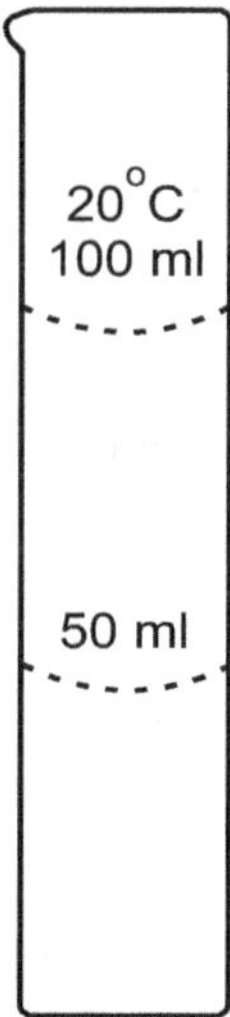

Fig. 1.3 Nessler's Cylinder

General Precautions:

1. The liquid used must be clean and filtered, if necessary.

2. The Nessler's cylinder must be made of colourless glass and have the same inner diameter in the given pair.

3. Detecting opalescence or colour development must be performed in daylight.

4. Comparison of turbidity should be done against a black background.

5. Comparison of colour should be done against a white background.

1.4.2 Principle and Procedure of Limit Test for Chloride

Aim: To perform the limit test for Chloride.

Apparatus: Pair of Nessler's cylinders, Glass rods, Stand, Measuring cylinder, Spatula, Pipette.

Chemicals: Dilute Nitric acid (10% v/v), 0.1 M Silver nitrate, Test sample, Distilled water.

Preparations:

1. **0.1 M Silver Nitrate:** Dissolve 1.7 gm of silver nitrate in 10 ml and make up the volume of 100 ml with water.

2. **Chloride Standard Solution (25 ppm Cl⁻):** Dilute 5 volumes of a 0.0824% w/v sodium chloride solution to 100 volumes with water.

Principle: The limit test for chloride involves the chemical reaction between Silver nitrate and soluble chlorides in the presence of dilute nitric acid to form precipitation of Silver chloride, which is insoluble in dilute nitric acid. The turbidity (opalescence) formed in the test solution is compared with the standard solution in Nessler's glass cylinder by viewing them transversely against a dark background. If the opalescence in the sample is less than the standard, it passes the test. If it is more than the standard, it fails the test.

$$NaCl + AgNO_3 \xrightarrow{\text{dil.HNO}_3} AgCl \downarrow + NO_3^-$$

Sodium Chloride   Silver Nitrate   Silver Chloride   Soluble Nitrate

Procedure: Take two 50 ml Nessler's cylinders. Label one as "Test" and the other as 'Standard'.

(A) Test Solution:

1. Dissolve the specified quantity of substance under examination in distilled water or prepare a solution as directed in an individual monograph and transfer it to a Nessler's cylinder.

2. Add 10 ml of dil. HNO_3 (However, when nitric acid is used in the preparation of the solution, then dil. HNO_3 should not be added).

3. Dilute to 50 ml with distilled water and add 1 ml of 0.1 M Silver nitrate solution. Stir the mixture immediately with a glass rod and allow it to stand for 5 minutes.

(B) Standard Solution:

1. Take the mixture of 10 ml of Chloride Standard solution (25 ppm Cl⁻) and 5 ml of distilled water and transfer to a Nessler's cylinder.

2. Add 10 ml of dil. HNO_3.

3. Dilute to 50 ml with water and add 1 ml of 0.1 M Silver nitrate solution. Stir the mixture immediately with a glass rod and allow it to stand for 5 minutes.

Table 1.1 Procedure for Preparation of Standard and Test Solution

Test Solution	Standard Solution
• Dissolve the specified quantity of substance in distilled water or prepare a solution as directed in individual monograph, in a Nessler's cylinder.	• Take mixture of 10 ml of Chloride Standard solution (25 ppm Cl^-) and 5 ml of distilled water and transfer to a Nessler's cylinder.
• Add 10 ml of dil. HNO_3.	• Add 10 ml of dil. HNO_3.
• Dilute to 50 ml with distilled water.	• Dilute to 50 ml with distilled water.
• Add 1 ml of 0.1 M Silver nitrate solution.	• Add 1 ml of 0.1 M Silver nitrate solution.
• Stir immediately with a glass rod.	• Stir immediately with a glass rod.
• Allow to stand for 5 minutes.	• Allow to stand for 5 minutes.

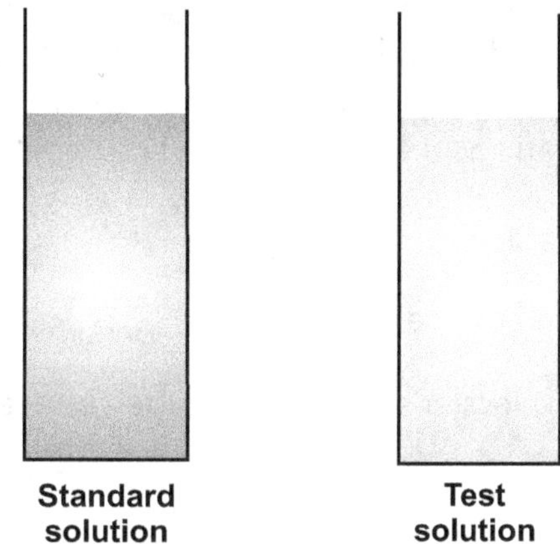

Fig. 1.4 Comparison of Standard and Test Solutions

Observation and Conclusion: The opalescence in STD is more than that of the test. Thus, the sample passes the limit test for Chloride.

Note: *Sometimes, the solution to be tested has to be prepared by a special method and instruction to this effect, if given, must be followed for preparing the test solution. The opalescence in the sample and the standard solution is compared by keeping the Nessler's cylinder against the proper background and observing side by side.*

1.4.3 Principle and Procedure of Limit Test for Sulphate

Aim: To perform the limit test for Sulphate.

Apparatus: Pair of Nessler's cylinder, Glass rod, Stand, Measuring cylinder, Spatula, and Pipette.

Chemicals: Dilute hydrochloric acid, Barium chloride, Ethanolic sulphate standard solution (10 ppm), 5 M Acetic acid, Sulphate Standard solution (10 ppm), Distilled water.

Preparation:

1. **Ethanolic Sulphate Standard solution (10 ppm):** Dilute 1 volume of a 0.181% w/v solution of potassium sulphate in ethanol (30%) to 100 volumes with ethanol (30%).

2. **5M Acetic acid:** Dilute 285 ml of glacial acetic acid to 1000 ml with distilled water.

3. **Sulphate Standard Solution (10 ppm):** Dilute 1 volume of a 0.181% w/v solution of Potassium sulphate in distilled water to 100 volumes with the same solvent.

4. **Barium chloride (25%w/v):** Dissolve 25 gm of Barium chloride in 100 ml of distilled water.

Principle: It is based upon the chemical reaction between Barium chloride and soluble Sulphate. The turbidity produced is compared with the standard solution. The small quantity of Potassium sulphate increases the sensitivity of the test. Alcohol prevents supersaturation, and more uniform turbidity develops. If the turbidity produced in the test is more intense than the standard turbidity, it fails the test. Acetic acid is added to prevent the precipitation of other anions like; phosphate, oxalate, borate, etc., with a solution of Barium chloride. Therefore, Acetic acid precipitates only sulphate ions.

$$SO_4^{2-} \quad + \quad BaCl_2 \quad \longrightarrow \quad BaSO_4\downarrow \quad\quad 2Cl^-$$

Soluble sulphate $\qquad$ Barium chloride $\qquad\qquad$ Barium sulphate $\quad$ Soluble chloride

Procedure: Take two 50 ml Nessler's cylinders. Label one as "Test" and the other as 'Standard'.

(A) Test Solution:

1. Take 10 ml of 25% w/v solution of Barium chloride in a Nessler's cylinder.

2. Add 1.5 ml of Ethanolic sulphate standard solution (10 ppm SO_4^{-2}), mix and allow to stand for 1 minute.

3. Add 15 ml of the solution prepared as directed in the monograph or solution of the specified quantity of the substance under examination in 15 ml of water.

4. Add 0.15 ml of 5M Acetic acid.

5. Add a sufficient quantity of distilled water to produce 50 ml. Stir immediately with a glass rod and allow it to stand for 5 minutes.

(B) Standard Solution:

1. Take 10 ml of 25% w/v solution of Barium chloride in a Nessler's cylinder.

2. Add 1.5 ml of Ethanolic sulphate standard solution (10 ppm SO_4^{-2}), mix and allow to stand for 1 minute.

3. Add 15 ml of Sulphate standard solution (10 ppm SO_4^{-2}).

4. Add 0.15 ml of 5M Acetic acid.

5. Add a sufficient quantity of distilled water to produce 50 ml. Stir immediately with a glass rod and allow it to stand for 5 minutes.

Table 1.2 Procedures for Preparation of Test and Standard Solutions

Test Solution	Standard Solution
• Transfer 1 ml of 25% w/v solution of Barium chloride in a Nessler's cylinder.	• Transfer 1 ml of 25% w/v solution of Barium chloride in a Nessler's cylinder.
• Add 1.5 ml of Ethanolic sulphate standard solution (10 ppm SO_4^{-2}).	• Add 1.5 ml of Ethanolic sulphate Standard solution (10 ppm SO_4^{-2}).
• Add 15 ml of Test solution.	• Add 15 ml of Sulphate Standard solution (10 ppm SO_4^{-2}).
• Add 0.15 ml of 5M Acetic acid.	• Add 0.15 ml of 5M Acetic acid.
• Dilute to 50 ml with distilled water and allow to stand for 5 minutes.	• Dilute to 50 ml with distilled water and allow to stand for 5 minutes.
• Observe the turbidity developed and compare with that of the standard.	• Observe the turbidity developed and compare with that of the sample.

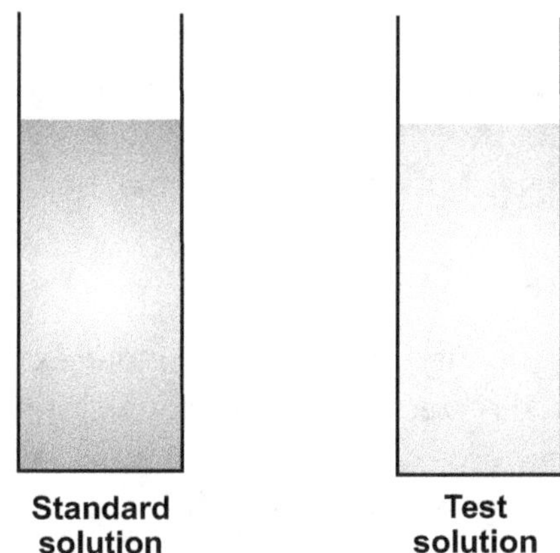

Fig. 1.5 Comparison of Standard and Test Solution

Observation and Conclusion: The turbidity in STD is seen more than that of the Test. Thus, the sample

1.4.4 Principle and Procedure of Limit Test for Iron

Aim: To perform the limit test for Iron.

Apparatus: Pair of Nessler's cylinders, Glass rods, Stand, Measuring cylinder, Spatula, Pipette.

Chemicals: 20% Iron free Citric acid, Thioglycollic acid, Iron Standard solution (20 ppm), Iron-free ammonia solution.

Preparation: Iron Standard Solution (20 ppm): Dilute 1 volume of 0.172% w/v solution of Ferric ammonium sulphate in 0.05 M Sulphuric acid to 10 volumes with water. Contains iron in the ferric state.

Principle: The test involves the chemical reaction between ferrous iron and thioglycollic acid in the presence of citric acid and ammoniacal alkaline medium. Wherein a pale pink to deep reddish purple colour is obtained. Ferric iron is reduced to ferrous iron by the thioglycollic acid, and ferrous thioglycollate is produced. Citric acid forms a soluble complex with iron and prevents its precipitation by ammonia as ferrous hydroxide. Ferrous thioglycollate is colourless in neutral or acid solutions. The colour due to the ferrous compound gets destroyed by oxidising agents. To avoid the interference of other ions, 20% iron-free citric acid is used. Citric acid forms a complex with other metal cations. The colour produced from the test substance is compared by viewing vertically with a standard solution (Ferric ammonium sulphate). If the colour of the test solution is less dark than the standard solution, then the test sample passes the test.

$$Fe^{2+} + 2 \begin{array}{c} CH_2SH \\ | \\ | \\ COOH \end{array} \longrightarrow \begin{array}{c} CH_2SH \quad\quad OOC \\ | \quad\quad\quad\quad\quad | \\ Fe \\ | \quad\quad\quad\quad\quad | \\ COO \quad\quad HSH_2C \end{array} + 2H^+$$

Thioglycolic acid   Ferrous thioglycolate

Procedure: Take two 50 ml Nessler's cylinders. Label one as 'Test' and the other's as 'Standard'.

(A) Test Solution:

1. Dissolve the specified quantity of substance under examination in distilled water or prepare a solution as directed in an individual monograph and transfer it to a Nessler's cylinder.
2. Add 2 ml of 20% w/v solution of iron-free citric acid and 0.1 ml of Thioglycollic acid. Mix, and make alkaline with iron-free ammonia solution.
3. Dilute to 50 ml with distilled water and allow it to stand for 5 minutes.

(B) Standard Solution:

1. Take 2 ml of Iron Standard solution (20 ppm) in 20 ml distilled water and transfer to a Nessler's cylinder.
2. Add 2 ml of 20% w/v solution of iron-free Citric acid and 0.1 ml of Thioglycollic acid. Mix, and make alkaline with iron-free ammonia solution.
3. Dilute to 50 ml with distilled water and allow it to stand for 5 minutes.

Table 1.3 Preparation of Test and Standard Solutions

Test Solution	Standard Solution
• Dissolve the specified quantity of substance under examination in distilled water and transfer in a Nessler's cylinder.	• Take 2 ml of Iron Standard Solution (20 ppm) in 20 ml water and transfer to a Nessler's cylinder.
• Add 2 ml of 20% w/v solution of iron free Citric acid and 0.1 ml of Thioglycollic acid.	• Add 2 ml of 20% w/v solution of iron free Citric acid and 0.1 ml of Thioglycollic acid.
• Make alkaline with iron free ammonia solution.	• Make alkaline with iron free ammonia solution.
• Dilute to 50 ml with distilled water and allow to stand for 5 minutes.	• Dilute to 50 ml with distilled water and allow to stand for 5 minutes.
• Observe the intensity of the purple, colour developed by viewing vertically and compare with that of the standard.	• Observe the intensity of the purple, colour developed by viewing vertically and compare with that of the sample

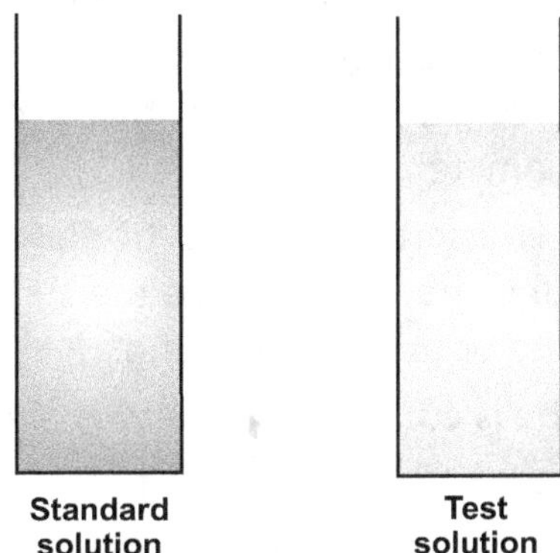

Fig. 1.6 Comparison of Standard and Test Solution

Observation and Conclusion: The intensity of the colour in STD is seen more than that of the Test. Thus, the sample passes the limit test for iron.

Note: *All the reagents used in the limit test for iron should themselves be iron-free. Hence, they themselves should conform to the limit.*

Table 1.4 Examples of Test Samples

Sample	Preparation
Sodium chloride	Dissolve 1 gm of Sodium chloride in distilled water. Use resulting solution for the limit test.
Sodium acetate	10 gm of Sodium acetate + 100 ml of distilled water. Use 20 ml of resulting solution for the limit test.
Magnesium sulphate	5.0 gm of Magnesium sulphate + 50 ml of distilled water. Use the 2 ml of resulting solution for the limit test.

Observation: The purple colour produced in the sample solution should not be greater than the standard solution. If the purple colour produced in the sample solution is less than the standard solution, then the sample will pass the limit test for iron and *vice versa*.

Reasons: Citric acid helps in the precipitation of iron by ammonia by forming a complex with it. Iron-free citric acid is used for complex metal cations other than iron, if any. Thioglycolic acid helps to oxidise Iron (II) to Iron (III). Ammonia is used to make the solution alkaline.

Result: Given sample passes/fails the limit test for iron.

1.4.5 Principle and Procedure of Limit Test for Arsenic

Aim: To perform the limit test for Arsenic.

Apparatus: Lead acetate, Cotton, Glass rod, Stand, Spatula, Measuring cylinder, Pipette (1 ml and 5 ml).

Apparatus for Limit Test for Arsenic:

1. A 120 ml capacity, wide-mouthed bottle fitted with a rubber bung through which passes a glass tube of height approximately 20 cm and diameter 6-8 mm is used. One end of this tube is constricted like that of a pipette with a 1 mm diameter having a hole of 2 mm diameter.

2. When the rubber bung is inserted into the bottle containing 70 ml of liquid, the constricted end of the tube should be above the surface of the liquid, and the hole inside should be below the bottom of the bung.

3. The upper end of the tube is cut off square and is either slightly rounded or ground smooth.

4. The rubber bungs (about 25 mm × 25 mm), each with a hole bored centrally and through, exactly 6.5 mm in diameter, are fitted with a rubber band or spring clip for holding them tightly in place.

5. The glass tube is lightly packed with cotton wool, previously moistened with Lead acetate solution and dried so that the upper surface of the cotton wool is not less than 25 mm below the top of the tube.

6. The upper end of the tube is then inserted into the narrow end of one of the pair of rubber bungs; to a depth of 10 mm (the tube must have a rounded-off end).

7. A piece of Mercuric chloride paper is placed flat on the top of the bung, and the other bung placed over it and secured by means of the spring clip in such a manner that the holes of the two bungs meet to form a true tube of 6.5 mm diameter interrupted by a diaphragm of Mercuric chloride paper.

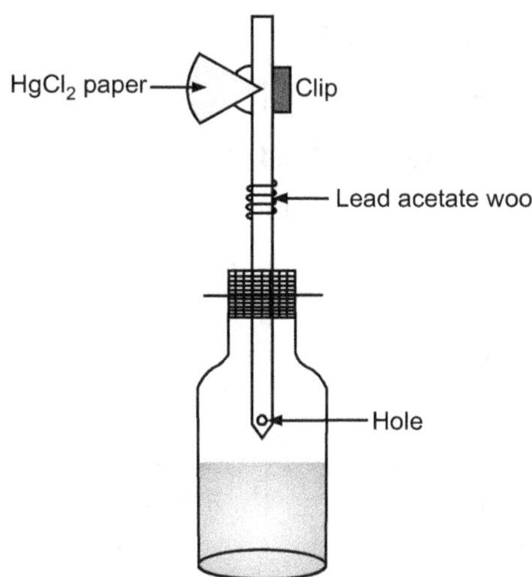

Fig. 1.7 Apparatus for Limit Test for Arsenic

Chemicals: Arsenic Standard Solution (10 ppm As), 1M Potassium Iodide, 10 g of Zinc AsT, Mercuric chloride test paper, Distilled water.

Preparation:

1. **Arsenic Standard Solution (10 ppm As):** Dissolve 0.33 gm of Arsenic Trioxide in 5 ml of 2M Sodium hydroxide and dilute to 250 ml with distilled water. Dilute 1 volume of this solution to 100 volumes with distilled water.

2. **1 M Potassium Iodide (AsT):** Dissolve 166 gm of Potassium Iodide in sufficient distilled water to produce 1000 ml.

3. **2 M Sodium Hydroxide Solution (AsT):** Dissolve 40 gm of Sodium hydroxide in distilled water to make the volume to 1000 ml with the same solvent.

4. **Lead Acetate Cotton Wool (AsT):** Immerse absorbent cotton in a mixture of 10 volumes of Lead Acetate solution and 1 volume of 2M Acetic acid. Drain off the liquids by absorbing them with filter paper and dry them at the R.T. store in a tightly-closed container.

5. **Mercuric Chloride Paper:** Prepared by impregnating filter paper in Saturated Mercury Chloride solution and drying at 60°C, avoiding contact with metal and in the dark. The paper used should be smooth, white and not less than 25 mm.

6. **Standard Hydrochloride Acid (AsT):** Add 1 ml of Stannous Chloride (AsT) solution to 100 ml of Hydrochloride Acid (AsT).

Principle: This test is based on the **Gutzeit test**. The limit test for Arsenic involves the reduction of all Arsenic into Arsine gas due to the combined action of Zinc and Potassium iodide. When Arsine gas comes in contact with dry Mercuric chloride paper, it produces a yellow to brown stain. The intensity and length of the stain is directly proportional to the amount of Arsenic present in the test substance.

$$Zn + 2HCl \rightarrow ZnCl_2 + 2(H)$$

$$2As + 6(H) \rightarrow 2AsH_3^{-}\uparrow$$

$$HgCl_2 + AsH_3 \rightarrow HgCl_2 \cdot AsH_3$$

Yellow complex

Procedure: Take two sets of Arsenic Test Apparatus. Label one as 'Test' and the other as 'Standard'.

Table 1.5 Preparation of Standard and Test Solution

Test Solution	Standard Solution
• Dissolve given sample in 50 ml distilled water and add 10 ml of Stannated Hydrochloric Acid and transfer into the Arsenic Test apparatus bottle.	• Transfer 10 ml Arsenic Standard solution into the Arsenic Test apparatus bottle. Add 10 ml of Stannated Hydrochloric Acid.
• Add 5 ml of 1 M Potassium Iodide (AsT) and 10 gm of Zinc (AsT).	• Add 5 ml of 1 M Potassium Iodide (AsT) and 10 gm of Zinc (AsT).
• Immediately assemble the apparatus and keep the solution aside for 40 minutes.	• Immediately assemble the apparatus and keep the solution aside for 40 minutes.
• Compare the stain obtained on the Mercuric Chloride paper with that in the apparatus containing Standard Solution.	• Compare the stain obtained on the Mercuric Chloride paper with that in the apparatus containing Test Solution.

Observation and Conclusion: If the intensity of colour of the stain in the STD is seen more than that of the test, then the sample passes the limit test for arsenic.

Notes:

1. Lead acetate pledgers or papers are used to trap any Hydrogen sulphide which may be evolved along with Arsine gas.

2. Stannous chloride is essential for the complete evolution of Arsine. In the Arsenic Limit test, preference is given to Stannous salts because they reduce Arsenic to an Arsenious state and sometimes to the metallic state, whereas, Cadmium salts in themselves are not reducing agents.

3. Care must be taken that the Mercuric chloride paper remains quite dry during the test.

4. The most suitable temperature for running the test is generally about 40°C.

5. The tube must be washed with Hydrochloric acid (AST), rinsed with distilled water and dried between succeeding tests.

1.4.6 Principle and Procedure of Limit Test for Heavy Metals

Aim: To perform the limit test for Heavy Metals.

Apparatus: Nessler's cylinders, Glass rods, Stand, Measuring cylinder, Spatula, Pipette.

Principle: It is based on the reaction between the solution of heavy metals (i.e., Lead, Mercury, Bismuth, Tin, Cobalt, Manganese, etc.) and a saturated solution of Hydrogen sulphide. In acidic media, it produces metal sulphides, which are distributed in a colloidal state and produce reddish/black colour. The Colour intensity of the test solution is compared with the Standard Lead nitrate solution.

$$\textbf{Heavy Metals} + \textbf{H}_2\textbf{S} \longrightarrow \textbf{Heavy Metal Sulphide}$$

$$\textbf{Pb(NO}_3)_2 \xrightarrow{\ \textbf{H}_2\textbf{S (excess)}\ } (\textbf{Pb}^+) + \textbf{H}_2\textbf{S}\cdot\textbf{PbS} + \textbf{2HNO}_3$$

$$\textbf{PbCl}_2 + \textbf{Na}_2\textbf{S} \longrightarrow \textbf{PbS}^- + \textbf{2NaCl}$$

Chemicals: Lead Standard Solution (20 ppm Pb), Dilute Acetic Acid, Dilute Ammonia Solution, Hydrogen Sulphide solution, Distilled water.

Preparation:

1. **Lead Standard Solution (20 ppm Pb):** Dilute 1 volume of lead standard solution to 5 volumes with distilled water.
2. **Hydrogen Sulphide Solution:** Freshly prepared saturated solution of hydrogen sulphide in distilled water.
3. **Dilute Acetic Acid:** Dilute 57 ml of Glacial Acetic acid to 1000 ml with distilled water.
4. **Dilute Ammonia Solution:** Dilute 425 ml of Strong Ammonia solution to 1000 ml with distilled water.

Procedure:

Table 1.6 Preparation of Standard and Test Solutions

Standard Solution	Test Solution
• Pipette out 1 ml of Lead Standard Solution (20 ppm Pb) and transfer to 50 ml Nessler's cylinder. Dilute with distilled water to 25 ml.	• Place 25 ml of Test solution which is prepared as directed in individual monograph or dissolves the specified quantity of the substance in distilled water to produce 25 ml, in a Nessler's cylinder.
• Adjust a pH between 3 to 4 by using dilute acetic acid or dilute ammonia solution.	• Adjust a pH between 3 to 4 by using dilute acetic acid or dilute ammonia solution.
• Dilute with distilled water to about 35 ml and mix.	• Dilute with distilled water to about 35 ml and mix.
• Add 10 ml of freshly prepared hydrogen sulphide solution and mix.	• Add 10 ml of freshly prepared hydrogen sulphide solution and mix.
• Dilute to 50 ml with distilled water, allow standing for 5 minutes and viewing downwards over a white surface.	• Dilute to 50 ml with distilled water, allow standing for 5 minutes and viewing downwards over a white surface.

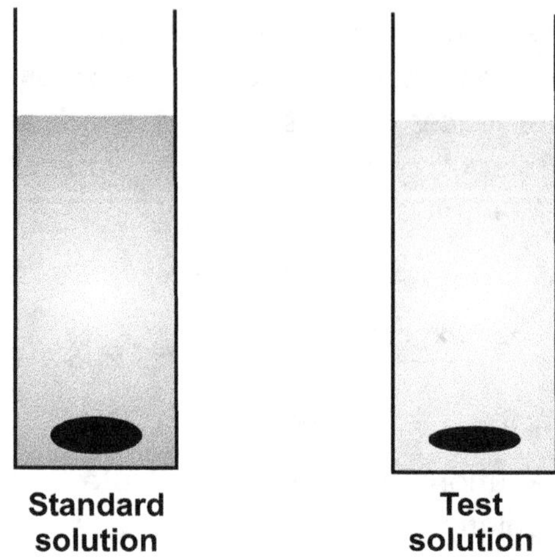

Fig. 1.8 Comparison of Standard and Test Solutions

Observation and Conclusion: The quantity of the black ppt of Lead Sulphide in STD solution is seen more than that of the Test solution. Thus, the sample passes the limit test for heavy metals.

QUESTION BANK

A. MULTIPLE CHOICE QUESTIONS

1. Which of the following error is caused by poor calibration of the instrument?
   A. Random error
   B. Gross error
   C. System error
   D. Precision error

Answer: C

2. Random errors in a measurement system are due to ....
   A. Environmental changes
   B. Use of un calibrated instrument
   C. Poor cabling practices
   D. Unpredictable effects

Answer: D

3. If the quantity to be measured remains constant while taking the repeated measurements, then the random errors can be eliminated by ....
   A. Calculating the mean of the number of repeated measurements
   B. Calculating the median of the number of repeated measurements

    C. Calculating the sum of the numbers of repeated measurements

    D. Either (a) or (b)

Answer: A

4. The error between the mean of finite data set and the mean of infinite data set is known as ....

    A. True error of the mean

    B. Standard error of the mean

    C. Finite error

    D. Infinite error

Answer: B

5. Which of the following statements is NOT true?

    A. Accuracy expresses the correctness of measurement

    B. Precision represents reproducibility of measurement

    C. High degree of precision implies a high degree of accuracy also

    D. High degree of accuracy implies a high degree of precision also

Answer: C

6. Which error affects the precision of measurement?

    A. Human error

    B. Systematic (or determinate) errors

    C. Random error

    E. Gross error

Answer: C

7. The closeness of the measurement to its true or accepted value is expressed by the error is known as ....

    A. Accuracy

    B. Precision

    C. Mean

    D. Median

Answer: A

8. The middle result when replicate data are arranged according to increasing or decreasing value is called as ....

    A. Median

    B. Mean

    C. Precision

    D. Accuracy

Answer: A

9. PPM means ....
    A. percent per million,
    B. percent parts per million,
    C. percent purity in millions,
    D. parts per million

Answer: D

10. Impurities in pharmaceutical preparations may be due to the following sources ....
    A. Raw materials,
    B. Manufacturing process,
    C. Chemical instability,
    D. All of the above

Answer: D

11. Limit test for Chloride has been based upon the reaction between .... and .... to obtain turbidity.
    A. Chloride and silver nitrate
    B. Iron and silver nitrate,
    C. Chloride and sulphuric acid
    E. None of these

Answer: A

12. The principle behind the limit test of Sulphate is ....
    A. To compare Opalescence of test solution with standard,
    B. To compare the Purple colour of the test solution with the standard,
    C. To compare the Pink colour of the test solution with the standard,
    D. None of the above

Answer: A

13. If the sample is lighter (in colour/ turbidity/ opalescence) than the standard solution, then it is ....
    A. Accepted
    B. Rejected,
    C. Rechecked
    D. None of above

Answer: A

14. What is the role of alcohol in the limit test of Sulphate?
    A. Prevents supersaturation and develops uniform turbidity
    B. Increases the sensitivity of the test

    C. Prevents precipitation

    D. All of these

Answer: A

15. In the limit test of Arsenic, Mercuric Chloride used produces a .... stain.
    - A. Pink
    - B. Black
    - C. Red
    - D. Yellow

Answer: D

16. Limit test of Iron, ferric iron is reduced to ferrous iron by ....
    - A. Ammonia
    - B. Thioglycollic acid
    - C. Chloroform
    - D. Acetic acid

Answer: B

B. SHORT ANSWER QUESTIONS

1. Define the following terms:

(a) Error	(b) Accuracy
(c) Precision	(d) Significant figures
(e) Mean	(f) Median
(g) Random error	(h) Systematic error

2. Enlist the different types of errors.

3. What do you mean by Impurity?

4. Enlist various sources of impurities.

5. What are limit tests, and what is their significance?

6. Write specifications of Nessler's cylinder as per Pharmacopoeia.

7. Why Nessler's cylinder is required for the limit test?

8. Why only distilled water is used in the limit test?

9. Write the principle behind the limit test for Chlorides.

10. What is the modified limit test for Chlorides?

11. Write the reaction involved in the limit test for Chlorides.

12. What is the difference between a normal Chloride limit test and a modified limit test for Chloride?

13. How 0.1 M Silver Nitrate solution is prepared?

14. Discuss the principle of the limit test for Chloride.

15. Discuss limit test of Chloride for Potassium permanganate.

16. How standard Chloride solution is prepared?

17. What is the concentration of Chloride standard solution?

18. Why is nitric acid added to the Chloride limit test?

19. Why is silver nitrate used in limit tests for Chlorides?

20. What is the importance of the Chloride limit test?

21. Write the principle behind the limit test for Sulphate.

22. Write the reaction involved in the limit test for Sulphate.

23. Discuss the limit test for Sulphate in brief.

24. Why Potassium Sulphate in a small concentration is required in the limit test of Sulphate?

25. What is the role of Ethanolic Potassium sulphate in the limit test for Sulphate?

26. How standard Ethanolic Potassium sulphate solution is prepared?

27. What is the concentration of Sulphate in standard solution?

28. How Barium chloride solution is prepared?

C. LONG ANSWER QUESTIONS

1. Classify and discuss various types of errors with pertinent examples.

2. What are errors? Enlist sources of errors and explain types of errors with suitable examples.

3. What steps should be taken to minimise the following errors?
   (a) Instrumental errors   (b) Method errors   (c) Personal errors

4. Give the principle involved in the limit test for Iron as per I.P and describe it.

5. Write in detail raw materials as a source of impurity.

6. Enlist various sources of impurities. Discuss manufacturing hazards in detail.

7. Write in detail the Limit Test of Arsenic with its modifications.

8. Enlist sources of impurities in pharmaceutical substances and explain their effects on pharmaceutical substances.

9. Describe the limit test for Iron.

10. Discuss various sources of impurities in pharmaceutical substances.

11. Discuss the limit test for sulphate.

12. Explain the principle involved in the limit test for Lead I.P.

13. Illustrate the sources of impurities in pharmaceutical substances.

14. What is the effect of impurities on the quality of pharmaceuticals? Explain.

∎∎∎

Chapter 2

VOLUMETRIC ANALYSIS

CONTENTS

Volumetric analysis:
- Fundamentals of volumetric analysis
- Acid-base titration
- Non-aqueous titration
- Precipitation titration
- Complexometric titration
- Redox titration
- Gravimetric analysis: Principle and method.

♦ LEARNING OBJECTIVES ♦

After completing this chapter, the student should be able to understand:

1. Fundamental concepts of volumetric analysis
2. Definition of an Acid & base, types & applications of Acid-Base titration
3. Definition of a Non-aqueous titration, solvents used in Non-aqueous titrations, applications of Non-aqueous titrations.
4. Definition of a Precipitation titration, applications of Precipitation titrations.
5. Definition of a Complexometric titration, applications of Complexometric titrations.
6. Definition of a Redox titration, applications of Redox titrations.
7. Definition of Gravimetric analysis, Principle of Gravimetric analysis, method of Gravimetric analysis, steps involved in Gravimetric analysis.

2.1 FUNDAMENTALS OF VOLUMETRIC ANALYSIS

2.1.1 Introduction to Pharmaceutical Analysis

2.1.1.1 Definition

Pharmaceutical analysis is a branch of practical chemistry that involves a series of processes for identification, determination, quantification and purification of a substance, separation of the components of a solution or mixture, or determination of the structure of chemical compounds. The substance may be a single compound or a mixture of compounds, and it may be isolated or in any of the dosage forms. The substances used as pharmaceuticals are from various synthetic or natural (animal, plant, marine, microbial or mineral) sources.

2.1.1.2 Scope

The process of analysis can be broadly categorised as; a) qualitative (identification) and b) quantitative (estimation). The sample to be analysed is called an **analyte,** and based on the size of the analyte, quantitative analysis can be termed as; macro (0.1 gm or more), semi-micro (0.01 gm to 0.1 gm), micro (0.001 gm to 0.01 gm), sub-micro (0.0001 gm to 0.001 gm), ultra-micro (below 10–4 gm) and trace analysis (100 to 10000 ppm).

1. **Qualitative analysis** is performed to establish the composition of natural/synthetic substances. These tests are performed to indicate whether the substance or compound is present in the sample or not. Various qualitative tests involve; the detection of evolved gas, formation of precipitates, limit tests, colour change reactions, determination of melting point and boiling point etc.

2. **Quantitative analysis** is mainly used to quantify any compound or substance in the sample. These techniques are based on (a) the quantitative performance of a suitable chemical reaction and either ensuring the amount of reagent added to complete the reaction or measuring the amount of reaction product obtained, (b) the characteristic movement of a substance through a defined medium under controlled conditions, (c) electrical measurement, (d) measurement of some spectroscopic properties of the compound or substance (analyte).

2.1.1.3 Different Techniques of Quantitative Analysis

Based on the above, the different techniques of quantitative analysis can be categorised as under:

(i)   Chemical methods/techniques:

    (a)   Volumetric or titrimetric methods

    (b)   Gravimetric methods

    (c)   Gasometric methods

(ii)   Electrical methods/techniques

(iii) Instrumental methods/techniques

(iv) Biological and microbiological techniques

Let us see the above techniques/methods in more detail.

(i) Chemical Methods:

    (a)   Titrimetric or Volumetric Methods:

        These involve the reaction of a substance to be determined with an appropriate reagent as a standard solution and the determination of the volume of solution required to complete the reaction. Volumetric methods require simple and lesser apparatus, and they are susceptible to having lower accuracy.

Various types of titrimetric methods are:

(i) Acid-base titrations (neutralization reactions)

(ii) Complexometric titrations

(iii) Precipitation titrations

(iv) Oxidation-reduction titrations

(v) Non-aqueous titrations

(b) Gravimetric Methods:

In gravimetric analysis, a substance to be determined is converted into an insoluble precipitate in the purest form, which is then collected and weighed. It is a rather time-consuming process. In electrogravimetry, electrolysis of the sample is carried out on the electrodes, and the deposited product is weighed after drying.

Thermogravimetry (TG) records the change in weight, differential thermal analysis (DTA) records the difference in temperature between a test substance and an inert reference material, and differential scanning calorimetry (DSC) records the energy needed to establish a zero-temperature difference between a test substance and reference material.

(c) Gasometric Methods:

Gasometry involves the measurement of the volume of gas evolved or absorbed in a chemical reaction. Some of the gases which are analysed by Gasometry are CO_2, N_2O, N_2, cyclopropane, amyl nitrate, ethylene, helium etc.

(ii) Electrical Methods:

Electrical methods of analysis involve the measurement of electric current, voltage or resistance in relation to the concentration of some species in the solution.

Electrical methods of analysis include:

(a) Potentiometry

(b) Conductometry

(c) Polarography

(d) Voltammetry

(e) Amperometry

Potentiometry measures the electrical potential of an electrode in equilibrium with an ion to be determined. Conductometry measures the electrical conductivity of an electrode with a reference electrode. While, Polarography, Voltammetry and Amperometry measure electrical current at a micro-electrode.

(iii) Instrumental Methods of Analysis:

Instrumental methods involve measuring some physical properties of the compound or a substance. These methods are employed for the determination of minor or trace concentrations of elements in the sample. Instrumental methods are preferred due to their selectivity, high speed, accuracy and simplicity of analysis. Any change in the properties of the system is detected by measurement of absorbance, specific rotation, refractive index, migration difference, charge to mass ratio etc.

Spectroscopic methods of analysis depend upon the measurement of the amount of radiant energy of a particular wavelength emitted by the sample.

Methods that include; absorption of radiation are ultra-violet, visible, infrared, atomic absorption, nuclear magnetic resonance spectroscopy etc.

Emission methods involve heating or electrical treatment of the sample so that the atoms are raised to the excited state to emit the energy, and the intensity of this energy is measured. Emission methods include; emission spectroscopy, flame photometry, fluorimetry, etc.

Chromatographic techniques and electrophoretic methods are separation methods for the mixture of compounds but are also applied for the identification of compounds of mixtures. Various chromatographic techniques are GC, HPLC, TLC, HPTLC, PC etc.

Mass spectrometry involves the vaporisation of material using a high vacuum, and the vapour is bombarded by a high-energy electron beam. Vapour molecules undergo fragmentation to produce ions of varying sizes. These ions are differentiated by accelerating them in an electrical field and then deflecting them in a magnetic field. Each kind of ion gives a peak in the mass spectrum.

(iv) Biological and Microbiological Methods:

Biological methods are used when the potency of a drug or its derivative cannot be properly determined by any available physical or chemical methods. These estimations are called bioassays. Microbiological methods are used to observe the potency of antibiotic or antimicrobial agents. In the antimicrobial assay, inhibition of growth of bacteria in the sample is compared with that of the standard antibiotic. These methods include the cup plate method and turbidimetric analysis.

2.1.1.4 Methods of Expressing Concentration

Solutions may be described as **single-phase systems** composed of two or more chemical substances representing **homogeneous molecular dispersion**. In general, the components of a solution retain their individual identities. Thus, a solution is properly termed a homogeneous mixture on the basis of the variability of composition. The properties of a solution are uniform throughout the mixture because the dispersion of the solute molecules in the solvent is on a molecular scale, making the molecules indistinguishable by usual observation procedures. Colloidal solutions, in contrast to true solutions, contain very small particles, but these are not of molecular dimensions and may be observed by various techniques.

I. Concentration Expressions:

The concentration of solute in a solution may be expressed in many ways, depending upon the convenience of those concerned with its use. Chemists more frequently prefer to work with the number of moles or equivalents of a particular solute. These quantities are also important to pharmacists, as been seen in the use of milliequivalents per litre (mEq/Ltr) for electrolyte solutions. Pharmacists also use percentage concentrations or some similar expressions of the constituents by parts. The commonly employed concentration expressions are reviewed in the following on a physical basis (weight or volume).

Expression of Strengths:

Percent Concentration: The Physical Methods of Expressing the Concentration. The term "Percent" or, more usually, the symbol "%" is used with one of the four different meanings in the expression of concentrations according to circumstances. In order for the meaning to be attached to the expression in each instance to be clear, one of the following notations is used.

(a) **Percent w/w (%w/w):** It (Percentage weight in weight) expresses the number of grams of solute in 100 gm of product. The concentrations of strong acids, as available commercially, are expressed in this way.

For example: H_2SO_4 98.0% w/w; CH_3COOH 33% w/w etc.

Also, expressing the percentage purity of the solid dosage forms such as tablets, capsules etc., as percent weight/weight (%w/w) is another example.

(b) **Percent w/v (%w/v):** It (Percentage weight in volume) expresses the number of grams of solute in 100 ml of product (i.e. 100 ml of solution). This is a common way of specifying the solution composition of mixtures of miscible liquids or solids in liquids.

For example: H_2O_2 solution 5-7% w/v; $BaCl_2$ solution 10% w/v etc.

(c) **Percent v/v (%v/v):** It (Percentage volume in volume) expresses the number of millilitres of solute in 100 ml of product. e.g. Alcohol 95% v/v.

(d) **Percent v/w (%v/w):** It (Percentage volume in weight) expresses the number of millilitres of solute in 100 gm of product. Usually, the strength of solutions of solids in liquids is expressed as percentage weight in volume, of liquids as percentage volume in volume and of gases in liquids as percentage weight in weight. This last %v/w is not commonly practised these days.

(e) **Parts per million (ppm):** When the concentration of a solution is expressed as parts per million (ppm), it means weight in weight unless otherwise specified.

For example, In the limit test of chloride, 25 ppm of Cl^-.

II. The Chemical Methods of Expressing Concentration:

The Chemical methods of expressing concentration are based upon a chemical formula or combining power. The word "Concentration" is frequently used as a general term referring to a quantity of substance in a defined volume of solution. But for quantitative titrimetric analysis, use is made of standard solutions in which the base unit of quantity employed is the "mole". This follows the definition given by the **International Union of Pure and Applied Chemistry (IUPAC)**. A solution containing very small amounts of solute may be expressed in millimolar (mM) concentration, defined as "the number of millimole/ml of the solution" (1 mM = 1×10^{-3} M).

1. **Molarity (M):** The molarity of a solution expresses the number of moles (gram molecular weights) of solute contained in 1000 ml (1 Litre) of the solution. A solution containing 1 mole of solute in each litre of the total solution is said to be one molar (M) solution.

2. **Molality (m):** The molality of a solution expresses as the number of moles of a solute contained in 1000 gm of a solvent. This method of denoting concentration is used in many equations to express the thermodynamic properties of solutions.

3. **Normality (N):** The normality of a solution expresses the number of equivalents (gram equivalent weight) of the solute in one litre of solution. This is generally a much more useful expression since it is directly related to reactive concentrations of various species in the solution.

4. **Equivalent weight of acid:** Equivalent weight of the substance is that weight which contains 1 gm of replaceable hydrogen ions (1.008 gm).

5. **Equivalent weight of base:** Equivalent weight of the base is that weight which contains one gram of replaceable hydroxyl groups, i.e., 17.008 gm of hydroxyl ion. Equivalent weight is generally defined as "the mass of a substance, especially in grams that displace or combines with or are equivalent to eight grams of oxygen or one gram of hydrogen". It is used in expressing the combining powers of elements.

6. **Saturated solution:** A saturated solution is one which has dissolved all the solute it is capable of holding at a given temperature. The temperature is a very crucial aspect of saturated solutions, and unless otherwise specified, the temperature is assumed to be 25°C. An example of a saturated solution is the Boric acid solution used as an Eyewash. The concentration of this solution is near enough to saturation (4.5-5 %w/v) that a drop to below usual temperature will cause the boric acid to crystallise from the solution. This represents a caution to its use in the eye. If crystals are present, the solution should be warmed to dissolve them.

7. **Formality (F):** Formality may be defined as "the number of gram formula weight (GFW) of the solute dissolved per litre of the solution".

$$\text{Formality (F)} = \frac{\text{GFW}}{\text{Litres of solution}}$$

2.1.1.5 Primary and Secondary Standards

(a) Primary Standards

A few volumetric reagents are obtainable in such a high state of purity that they can be directly used for the preparation of standard solutions by the use of accurately weighed quantities. Many others, including some of the most useful, cannot be so used, either because they are hydrated and subject to small variations in the content of water of crystallisation; or because, for some other reason, they cannot be guaranteed to have the degree of necessary purity. A reagent of this kind is made up to approximately the correct concentration, the solution so obtained is standardised against a pure compound, and a factor is calculated.

Primary standards are defined as "the substances of known high purity (99.95-100.05%) with respect to the active component which should be stable at temperatures for drying used to standardise the volumetric reagents", or Primary standards are "the substances which are available in pure form with the definite chemical composition". Several primary standards are available from the National Bureau of Standards and other chemical suppliers. Several substances of analytical reagent grade are sufficiently pure to serve as primary standards for ordinary work, or they may be made so by recrystallisation.

Requirements for Primary Standards:

One of the requirements needed for volumetric analysis is a substance having a known purity. The solution of this substance can be employed as a titrant. There are few known substances which can be used as a primary standard. The requirements for primary standard substances are as follows:

A primary standard should satisfy the following requirements:

1. It must be easy to obtain, purify, dry (preferably at 110°C-120°C), and preserve in a pure state. (This requirement is not usually met by hydrated substances since it is difficult to remove surface moisture completely without effecting partial decomposition.)

2. The substance should be unaltered in the air during weighing. This condition implies that it should not be hygroscopic, oxidised by air, or affected by carbon dioxide. The primary standard should maintain an unchanged composition during storage.

3. The substance should be capable of being tested for impurities by qualitative and other tests of known sensitivity. (The total amount of impurities should not, in general, exceed 0.01-0.02 %.)

4. It should have a high relative molecular mass, so the weighing errors may be negligible. (The precision in weighing is ordinarily 0.1-0.2 mg; for an accuracy of 1 part in 1000. It is necessary to employ samples weighing at least about 0.2 gm).

5. The substance should be readily soluble under the conditions in which it is employed.

6. The reaction with the standard solution should be stoichiometric and practically instantaneous. The titration error should be negligible or easy to determine accurately by experiment.

In practice, an ideal primary standard is difficult to obtain, and a compromise between the above ideal requirements is usually necessary. The substances commonly employed as primary standards are indicated below.

Only substances which satisfy these requirements can be used as primary standards, and their solutions may be prepared by accurate weighing of the required quantity of the substance and dissolving in water to produce the required volume of the solution accurately.

Examples of Primary Standards:

These are materials which, after drying under the specified conditions, are recommended for use as primary standards in the standardisation of volumetric solutions.

The following are recommended for use as primary standards, as per the Indian-Pharmacopoeial committee for various titrations.

Table 2.1 Types of Titrations and their Primary Standards

Type of Titration	Primary Standard
I. Acid - Base Titrations	
1. Alkalimetry (Sodium hydroxide)	Potassium hydrogen phthalate*
2. Acidimetry (Hydrochloric acid Sulphuric acid)	Anhydrous sodium carbonate*
II. Oxidation-Reduction titrations (Redox titrations)	
1. Permanganometry (Potassium permanganate)	Arsenic trioxide, Sodium oxalate, Potassium dichromate
2. Iodometry (Sodium thiosulphate	Potassium bromate*, Potassium dichromate, Potassium iodate
3. Iodimetry (Iodine)	Arsenic trioxide*
4. Ceriometry (Cerric ammonium sulphate)	Arsenic trioxide*
III. Precipitation titrations	
Mohr's method (Silver nitrate)	Sodium chloride*
IV. Complexometric titrations	
Disodium edetate	Granulated zinc* Calcium carbonate Magnesium sulphate
V. Non-Aqueous titration (N.A.T)	
1. Perchloric acid	Potassium hydrogen phthalite
2. Sodium methoxide	Benzoic acid*
3. Lithium methoxide	Benzoic acid*
4. Tetrabutyl ammonium hydroxide	Benzoic acid*

Used as Ideal Primary Standard for standardising the volumetric reagents given in Indian Pharmacopoeia.

Normally hydrated salts are not used as primary standards as they lose varying amounts of water on dying. Because of the difficulty of efficient drying, the primary standard substance is generally dried between 105°C and 110°C before use.

(b) Secondary Standards

A secondary standard is a substance which may be used for standardisations and whose content of the active substance has been found by comparison against a primary standard. A secondary standard solution contains an exactly known amount of the substance in a unit volume of the solution and is expressed as normality or molarity and can be determined by titrating against a primary standard. It follows that a secondary standard solution is a solution in which the concentration of dissolved solute has not been determined from the weight of the compound dissolved but by reaction (titration) of a volume of a primary standard solution. Thus, a sodium hydroxide solution may be standardised by titrating against a standard solution of Potassium Hydrogen Phthalate or a standard solution of Hydrochloric Acid (Secondary standard). A secondary standard is a substance which for one or more of the reasons, cannot be used as a primary standard. e.g., Sodium Hydroxide cannot be used as a primary standard for the reason that it absorbs water and carbon dioxide from the atmosphere and the composition of its solution is subject to wide variations at different periods.

Similarly, sodium thiosulphate absorbs CO_2 from the atmosphere and gets decomposed, and a deposit of sulphur settles at the bottom. Similarly, the compositions of solution of various other substances like mineral acids, potassium permanganate, Iodine etc., are also variable at different times. Therefore, these cannot be used as primary standards. The normality or molarity of the solution of such substances can be found by titrating their solutions against an appropriate standard solution of a primary standard; in other words, the solution may be standardised by titrating against the standard solution of a primary standard.

Volumetric solutions, also known as Standard Solutions, are solutions of reagents of known concentration intended primarily for use in quantitative determinations. Concentrations are usually expressed in terms of molarity (M). It is not always possible or essential to prepare volumetric solutions of a accurate desired theoretical molarity. A solution of approximately the desired molarity is prepared and standardised by titration against a solution of a primary standard. The molarity factors so obtained are used in all calculations where such standardised solutions are employed. As the strength of a standard solution may change upon standing, the molarity factor should be re-determined frequently. Volumetric solutions should not differ from the prescribed strength by more than 10%, and the molarity should be determined with a precision of 0.2%. When solutions of a reagent are used in several molarities, the details of the preparation and standardisation are usually given for the most commonly used strength. Stronger or weaker solutions are prepared and standardised using the proportionate amount of the reagent or by making an exact dilution of a stronger solution.

Volumetric solutions prepared by dilution should be re-standardized either or directed for the stronger solution or by comparison with another volumetric solution having a known ratio to the stronger solution. The water used in preparing volumetric solutions complies with the requirements of the monograph on purified water unless otherwise specified. When used for the preparation of unstable solutions such as potassium permanganate or sodium thiosulphate, it should be freshly boiled and cooled. When a solution is to be used in an assay in which the end-point is determined by an electrochemical process (e.g., potentiometric), the solution must be standardised in the same way.

2.2 ACID-BASE TITRATIONS

2.2.1 Concepts of Acid-base

2.2.1.1 Arrhenius Theory

Arrhenius's definition of an acid is - a substance which yields hydrogen ion (H^+) in an aqueous medium', and that of a base is - a substance which yields hydroxy ions (OH^-) in an aqueous medium'.

However, these definitions have *two* serious short-comings:

(a) They lack an explanation of the behaviour of acids and bases in non-aqueous media, and

(b) Acidity is associated with hydrogen ion - a relatively simple particle; but basicity is associated with hydroxyl ion - a relatively complex entity.

2.2.1.2 Lowry and Bronsted's Theory

Just after the First World War in 1923, Bronsted and Bjerrum in Denmark and Lowry in Great Britain jointly put forward a more acceptable and satisfactory theory of acids and bases, which is devoid of objections earlier raised in Arrhenius' definition.

According to Lowry and Bronsted's theory, 'an acid is a substance capable of yielding a proton (hydrogen ion), while a base is a substance capable of accepting a proton'. Thus, a complementary relationship exists between an acid and a base that may be expressed in a generalised fashion as below:

$$A_{acid} \rightleftharpoons H^+ + B_{base}$$

Conjugate Acid-Base Pair:

The pair of substances, which by virtue of their mutual ability to either gain or lose a proton, is called a conjugate acid-base pair. A few typical examples of such pairs are:

Acid-Base:

$$HNO_3 \rightleftharpoons H^+ + NO_3^-$$

$$HCl \rightleftharpoons H^+ + Cl^-$$

$$CH_3COOH \rightleftharpoons H^+ + CH_3COO^-$$

$$NH_4^+ \rightleftharpoons H^+ + NH_3$$

$$HPO_4^{2-} \rightleftharpoons H^+ + PO_4^{3-}$$

$$H_3O^+ \rightleftharpoons H^+ + H_2O$$

$$HCO_3^- \rightleftharpoons H^+ + CO_3^{2-}$$

$$Al(H_2O)_6^{+++} \rightleftharpoons H^+ + Al(H_2O)_5\,OH^{2+}$$

It is quite evident from the above examples that not only molecules but also anions and cations can act as acids and bases.

In an acid-base titration, the acid will not release a proton unless the base capable of accepting it is simultaneously present. In other words, in a situation where actual acid-base behaviour exists, then an interaction should involve two sets of conjugate acid-base pairs, represented as:

$$A_1 + B_2 \rightleftharpoons B_1 + A_2$$
$$\text{acid}_1 \quad \text{base}_2 \qquad \text{base}_1 \quad \text{acid}_2$$

$$HCl + NH_3 \rightleftharpoons Cl^- + NH_4^+$$
$$\text{acid}_1 \quad \text{base}_2 \qquad \text{conjugate} \quad \text{conjugate}$$
$$\text{base}_1 \qquad \text{acid}_2$$

Some other examples include:

$$CH_3COOH + H_2O \rightleftharpoons CH_3COO^- + H_3O^+$$

$$H_2SO_4 + H_2O \rightleftharpoons HSO_4^- + H_3O^+$$

$$H_2O + CN^- \rightleftharpoons OH^- + HCN$$

$$NH_4^+ + S_2^- \rightleftharpoons NH_3 + HS^-$$

In short, the species that essentially differ from each other by possesing or not possessing one proton are known as conjugate base and acid, respectively. Sometimes, such a reaction is termed a protolytic reaction or protolysis, where A_1 and B_1 make the first conjugate acid-base pair and A_2 and B_2 the other pair.

Merits of Lowry-Bronsted Theory:

It has two points of merit, which are:

(a) Hydrochloric acid on being dissolved in water undergoes a protolytic reaction, thus:

$$HCl \ + \ H_2O \ \rightleftharpoons \ Cl^- \ + \ H_3O^-$$
$$acid_1 \qquad base_2 \qquad base_1 \qquad acid_2$$

It may be observed that H_3O^+, known as hydronium or oxonium ion, is invariably formed when an acid is dissolved in water.

Likewise, ammonia, on being dissolved in water, is also subjected to protolysis, thus:

$$NH_3 \ + \ H_2O \ \rightleftharpoons \ NH_4^+ \ + \ OH^-$$
$$base_1 \qquad acid_2 \qquad acid_1 \qquad base_2$$

(b) All proton-transfer reactions may be handled, thus:

$$NH_4^+ \ + \ S^{-2} \ \rightleftharpoons \ NH_3 \ + \ HS^-$$
$$acid_1 \qquad base_2 \qquad base_1 \qquad acid_2$$

Demerits of Lowry-Bronsted Theory:

It does not hold good for non-protonic solvents, for instance: BF_3, $POCl_3$ and SO_2.

2.2.1.3 Lewis's Theory

Lewis (1923) put forward another definition of acids and bases solely dependent on the giving or taking of an electron pair. According to Lewis: 'an acid is an electron pair acceptor, whereas a base is an electron pair donor'. Therefore, it is evident that whenever any neutralisation occurs, the formation of an altogether new co-ordinate covalent bond between the electron pair donor and acceptor atoms takes place.

Thus, Lewis's definition is a much broader definition that includes coordination compound formation as acid-base reactions, besides Arrhenius and Lowry-Bronsted acids and bases. Examples:

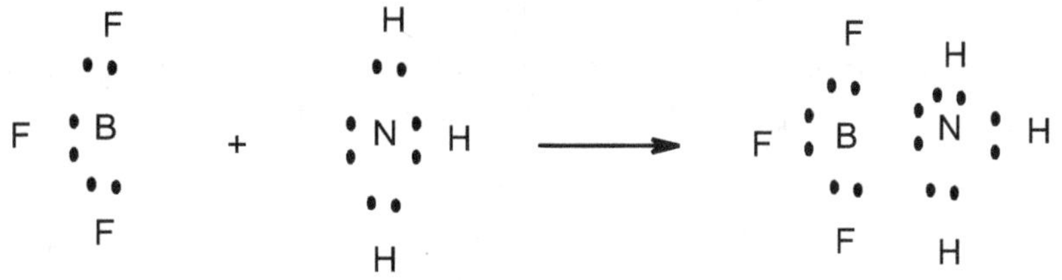

Boron trifluoride
(acid)

Ammonia
(base)

The reaction of boron trifluoride (acid) with ammonia (base) results in a stable octet configuration between by the mutual sharing of a pair of electrons between the latter (donor) and the former (acceptor).

$$Ag^+ \; + \; 2\; \overset{..}{\underset{..}{:}}N\overset{..}{:}H \longrightarrow \left[Ag \; :N:H \right]^{+}_{2}$$

<table>
<tr><td align="center">Electron
Acceptor
(acid)</td><td align="center">Electron
Donor
(base)</td></tr>
</table>

The reaction of ammonia (base) with Ag^+ (acid) results in a stable configuration due to the mutual sharing of a pair of electrons between the latter (donor) and former (acceptor).

2.2.1.4 Usanovich Theory

Usanovich (1934) modified the Lewis concept of acid and base by removing the restriction of either donation or acceptance of the electron pair in a more generalised fashion. According to him:

Acid: It is a chemical species that reacts with a base, thereby giving up cations or accepting anions or electrons.

Base: It is a chemical species that reacts with an acid, thereby giving up anions or electrons or combining with cations.

Unlike Arrhenius, Lowry-Bronsted and Lewis acids and bases, Usanovich's concept in a much broader sense includes all the oxidising agents as acids and the reducing agents as bases,

For example:

$$Fe^{2+} \rightleftharpoons Fe^{3+} + e^{-}$$
$$\text{Base} \qquad\qquad \text{Acid}$$

In the Iron (II)-Iron (III) system, the ferric ion (III) acts as an oxidising agent and is an acid, while the ferrous ion (II) acts as a reducing agent and is a base.

$$Ce^{3+} \rightleftharpoons Ce^{4+} + e^{-}$$
$$\text{Base} \qquad\qquad \text{Acid}$$

Similarly, in the Cerous (III)-Ceric (IV) system, the ceric ion (IV) behaves as an oxidising agent and acts as an acid, while the cerous ion (III) behaves as a reducing agent and acts as a base.

2.2.1.5 Lux-Flood Concept

The concept of acid-base reactions concerning the oxide ion was first introduced by Lux (1929) and supported by Flood (1947). According to the Lux-Flood concept - 'an acid is the oxide-ion acceptor while a base is the oxide donor'.

Examples:

$$MgO \;+\; SiO_2 \;\rightarrow\; MgSiO_3$$
$$CaO \;+\; SO_3 \;\rightarrow\; CaSO_4$$

In the above reactions, both MgO and CaO are the oxide ion donors and hence act as bases, whereas SiO_2 and SO_3 are the oxide-ion acceptor and hence act as acids. Ultimately, the Lux-Flood acid and base react to form magnesium silicate ($MgSiO_3$) and calcium sulphate ($CaSO_4$) salts, respectively.

2.2.2 Properties of Acids

For the properties of acids and bases, we will use the Arrhenius definition.
1. Acids release a hydrogen ion into a water (aqueous) solution.
2. Acids neutralise bases in a neutralisation reaction.
3. Acids corrode active metals.
4. Acids turn blue litmus to red.
5. Acids taste sour.

2.2.3 Properties of Bases
1. Bases release a hydroxide ion into the water solution.
2. Bases denature the protein.
3. Bases turn red litmus to blue.
4. Bases taste bitter.
5. Bases neutralise acids in a neutralisation reaction.

2.2.4 Indicators used in Acid Base Titration

An indicator is a substance which is used to determine the endpoint of a titration. In acid-base titrations, organic substances (weak acids or weak bases) are generally used as indicators. They change their colour within a certain pH range.

The colour change and the pH range of some common indicators are tabulated below:

Table 2.2 Characteristics of Indicators

Indicator	pH Range	Colour change
Methyl orange	3.2-4.5	Pink to yellow
Methyl red	4.4-6.5	Red to yellow
Litmus	5.5-7.5	Red to blue
Phenol red	6.8-8.4	Yellow to red
Phenolphthalein	8.3-10.5	Colourless to pink

2.2.5 Concept of Strong and Weak Acids/Bases

A strong acid is an acid which dissociates completely in water. That is, all the acid molecules break up into ions and solvate (attach) to water molecules. Therefore, the concentration of hydronium ions in a strong acid solution is equal to the concentration of the acid.

Examples of strong acids: Perchloric acid ($HClO_4$), Nitric acid (HNO_3), and Sulfuric acid (H_2SO_4).

The majority of acids exist as weak acids, an acid which dissociates only partially. On average, only 1% of a weak acid solution dissociates in water in a 0.1 mol/L solution. Therefore, the concentration of hydronium ions in a weak acid solution is always less than the concentration of the dissolved acid.

Examples of strong bases: Lithium hydroxide (LiOH), Potassium hydroxide (KOH), Barium hydroxide [$Ba(OH)_2$], and Calcium hydroxide [$Ca(OH)_2$].

2.2.6 Buffers

A buffer is a solution that resists changes in pH. A buffer is made with a weak acid and a soluble salt containing the conjugate base of the weak acid or a weak base and a soluble salt containing the conjugate acid of the weak base.

Some examples of buffer material pairs are:

1.   acetic acid (CH_3COOH) and sodium acetate (CH_3COONa);
2.   hydrofluoric acid (HF) and potassium fluoride (KF);
3.   carbonic acid (H_2CO_3) and sodium bicarbonate ($NaHCO_3$);
4.   ammonium hydroxide (NH_4OH) and ammonium nitrate (NH_4NO_3),
5.   nitrous acid (HNO_2) and lithium nitrite ($LiNO_2$).

2.2.6.1 Buffer Capacity

Buffer capacity is a measure of the efficiency of a buffer in resisting changes in pH. Conventionally, the buffer capacity is expressed as the amount of strong acid or base, in gram-equivalents, that must be added to 1 litre of the solution to change its pH by one unit. Calculate the buffer capacity (β) as:

$$\beta = \frac{\Delta B}{\Delta pH}$$

where, ΔB =   Gram equivalent of strong acid/base to change pH of 1 litre of buffer solution

ΔpH = The pH change caused by the addition of strong acid/base.

2.2.7 Neutralization Curves

If a pH meter is used, pH can be recorded on a regular basis as a titrant is added. A plot of the pH (as the vertical axis) against the volume of titrant would produce a slopping curve that is particularly steep around the equivalence point. pH is a measure of the H_3O^+ concentration in a solution.

Adding one or two drops to a neutral solution greatly changes the H_3O^+ concentration by a factor of 10 or more. This is what makes the titration curve so steep in that one region and therefore makes the equivalence point easy to identify in the graph. Therefore to find the amount of titrant needed to neutralise titrate is therefore easy to accurately quantify.

2.2.7.1 Neutralization Curve for Strong Acid and Strong Base Titration

Consider, e.g. 100 ml 1M HCl titrated with 1M NaOH. After observing changes in pH value, we will get a graph as follows.

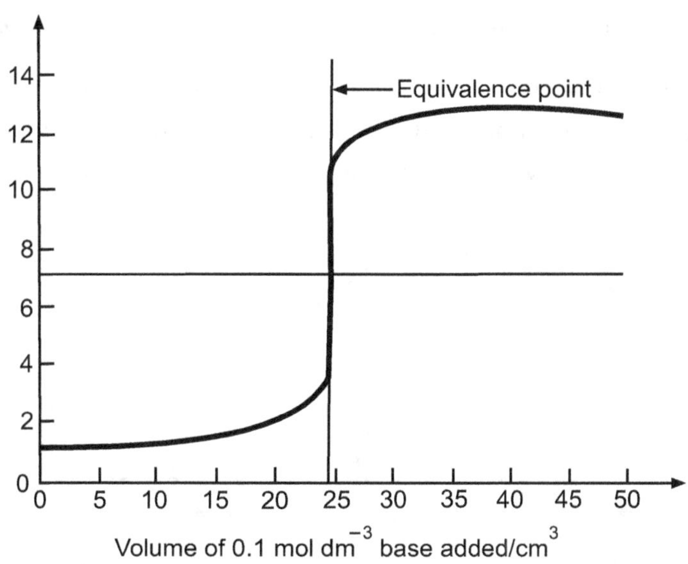

Fig. *2.1 Neutralization Curve for Strong Acid and Strong Base Titration*

2.2.7.2 Neutralization Curve for Weak Acid and Strong Base Titration

Consider a titration between 100 ml of 0.1N CH_3COOH and 0.1N NaOH. After observing changes in pH value, we will get a graph as follows.

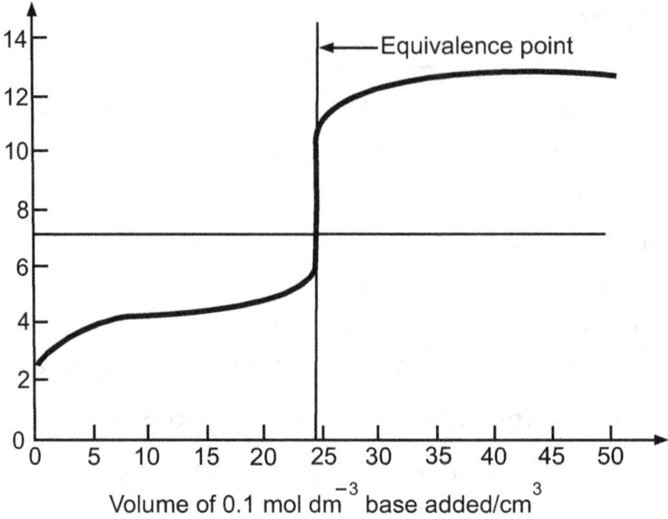

Fig. 2.2 Neutralization Curve for Weak Acid and Strong Base

2.2.7.3 Titration Curve between Weak Base and Strong Acid Titration

Consider a titration between 100 ml of 0.1N aqueous ammonia solution with 0.1N HCl. After observing changes in pH value, we will get a graph as follows.

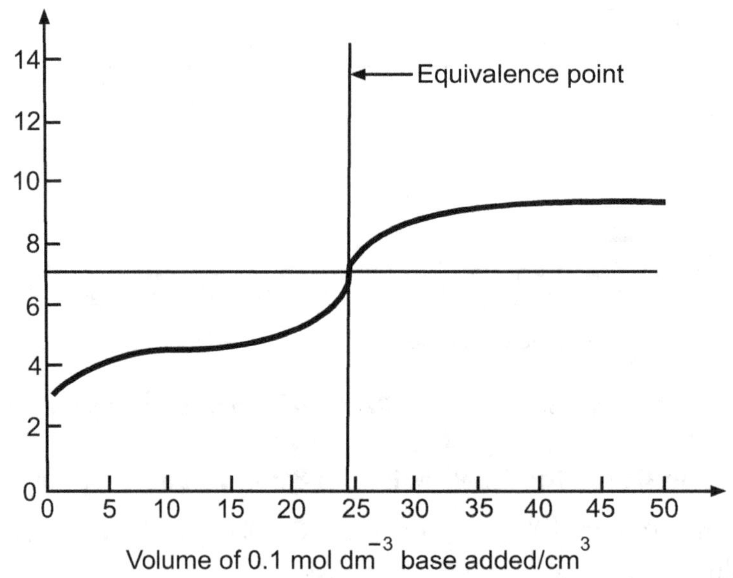

Fig. 2.3 Titration Curve between Weak Base and Strong Acid

2.2.7.4 Titration Curve between Weak Acid and Weak Base

Consider the titration of 100 ml of 0.1 N CH_3COOH with 0.1 N ammonia. After observing changes in pH value, we will get a graph as follows.

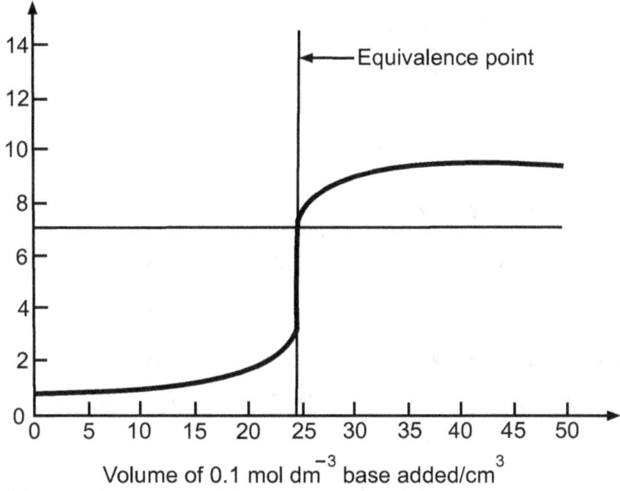

Fig. 2.4 Titration Curve between Weak Acid and Weak Base

2.3 NON-AQUEOUS TITRATIONS

2.3.1 Introduction

During the past four decades, a plethora of newer complex organic medicinal compounds have taken cognisance in the therapeutic armamentarium. Evidently, these compounds posed two vital problems of quality control, both in pure and dosage forms, by virtue of their inherent characteristics, namely:

    (a)   Poor solubility, and

    (b)   Weak reactivity in an aqueous medium.

2.3.2 Advantages of Non-Aqueous Titrations

1. Elimination of poor solubility of substances.
2. Enhancement of weak reactivity of substances.
3. Selective titration by using suitable solvent and titrant of acidic/basic components of physiologically active moiety of a salt.
4. Maintenance of speed, precision, accuracy and simplicity at par with classical methods of analysis.
5. Weak bases with Kb values less than 10^{-6} can be titrated satisfactorily by non-aqueous titrations. The reason being that in an aqueous medium and at higher Kb values ($> 10^{-6}$), the solvent water competes progressively with the basic species in the solution for the proton of the solvent.

2.3.3 Properties of Non-aqueous Solvent

1. Should be liquid at room temperature.
2. Should not be highly toxic.

3. Should have the capability of self-dissociation.
4. Should have acid-base character.
5. Should have dielectric constant.
6. Should exert levelling and differentiating effects.

2.3.4 Different Types of Solvents

There are three types of solvents used in non-aqueous titration:

(a) Protophilic Solvents:

They are essentially basic in nature and normally react with acids to form solvated protons. These solvents are either dissociating or non-dissociating in nature. These solvents are generally used for titration of weak acids.

Weakly acidic substances react with protophilic solvents and increase the concentration of cations A+, therefore called as lyonium ions.

$$HA^+ + OH^- \longrightarrow H^+ + A^-$$

For example, these include pyridine, acetic anhydride, liquid ammonia, ethylene diamine, and dimethylformamide.

(b) Protogenic Solvents:

They are acidic in nature and character, e.g., sulphuric acid. They exert a 'levelling effect' on bases, i.e., they become indistinguishable in strength when dissolved in strongly basic solvents due to their enhanced affinity of strong bases for protons. These solvents are generally used for titration of weak bases.

Weakly basic substances react with protogenic solvents and increase the concentration of anion (OH^-) lyate ions. A lyate ion is considered the basic character of a substance.

$$BOH + H^+ \longrightarrow BH^+ + OH^-$$

For example: Acetic acid and sulphuric acid.

(c) Amphiprotic Solvents:

They possess both protophillic and protogenic characteristics. They undergo dissociation to a very less extent. Acetic acid is employed chiefly as a solvent for the titration of basic substances, and its dissociation can be depicted as shown below:

$$CH_3COOH \longrightarrow H^+ + CH_3COO^-$$

In the above instance, acetic acid behaves as an acid.

For example, Water, alcohol, and acetic acid.

Perchloric Acid is a very strong acid. When it is made to dissolve in acetic acid, the latter can behave as a base and form an 'onium ion' after combining with protons donated by the perchloric acid.

Thus, we have:

$$HClO_4 \rightleftharpoons H^+ + ClO_4$$

$$CH_3COOH + H^+ \rightleftharpoons CH_3COOH_2^+$$

Onium ion

As the $CH_3COOH_2^+$ ion can instantly donate its proton to a base, therefore, a solution of perchloric acid in glacial acetic acid behaves as a strongly acidic solution.

Pyridine is a weak base, and when dissolved in acetic acid, the latter exerts its levelling effect and subsequently increases the basic characteristics of the pyridine. Therefore, it is practically feasible to titrate a solution of a weak base in acetic acid against a mixture of perchloric acid in acetic acid. Thus, a sharpened point is achieved, which otherwise cannot be obtained when the titration is performed in an aqueous medium.

The various reactions with perchloric acid, acetic acid and pyridine are summarised below:

$$HClO_4 + CH_3COOH \rightleftharpoons CH_3COOH_2^+ + ClO_4^-$$

$$C_6H_5N + CH_3COOH \rightleftharpoons C_6H_5NH^+ + CH_3COO^-$$

$$CH_3COOH_2^+ + CH_3COO^- \rightleftharpoons 2CH_3COOH$$

Summing up: $\quad HClO_4 + C_6H_5N \rightleftharpoons C_6H_5NH^+ + ClO_4^-$

Acetonitrile, acetone and dimethylformamide:

These non-aqueous solvents exert a greater differential in the protophillic properties of many substances than in the corresponding aqueous solutions due to the levelling effect of water in the latter solutions.

Hence, the most acidic substance in aqueous solutions of a number of acids is the formation of the hydronium ion, as shown below:

$$HB + H_2O \longrightarrow H_3O^+ + B^-$$

Acid $\qquad\qquad\qquad\qquad$ Hydronium ion

It is pertinent to observe here that the following inorganic acids almost exhibit equal strength in aqueous solutions, whereas in non-aqueous solvents, their 'acidity' retards in the following order:

$$HClO_4 > HBr > H_2SO_4 > HCl > HNO_3$$

In glacial acetic acid (an acidic solvent) and dioxane (a neutral solvent), the perchloric acid ($HClO_4$) behaves as more acidic (i.e., less protophyllic) than HCl; and, therefore, many base-hydrochlorides (i.e., chlorides) may be titrated with standard $HClO_4$, just as carbonates may be titrated in aqueous solution with standard HCl.

In short, it is possible to titrate mixtures of two or three components selectively with a single titration by the wisdom of the right choice of solvent for non-aqueous titrations.

2.3.5 Levelling Effect, Leveling and Differentiating Solvents

The apparent strength of a protonic acid depends on the solvent in which the acid is dissolved. When all the acids are stronger than H_3O^+ ion are added to H_2O, they donate as a proton to H_2O to H_3O^+ ion and appear to have equal strength since all these acids are levelled to the strength of H_3O^+ ion which is left in solution and is common to all such solutions.

The strength of all the acids becomes equal to that of the H_3O^+ ion, which is called the levelling effect of the solvent, water, and here water is called a levelling solvent for all these acids. In an aqueous solution, all very strong bases like Na^+, H^-, $Na^+NH_2{}^-$, and $Na^+OC_2H_5{}^-$ are levelled to the strength of OH^- ion, for they react completely with H_2O to produce OH^- ions. The solvent in which complete proton transfer occurs is called levelling solvents.

In other words, the solvent in which the solute is ~100% ionised is called levelling solvents.

Since HF and HCl are both ~100% ionised in liquid NH_3 to give ~100% NH_4^+ ions, these appear to be of equal strength and Liq. NH_3 acts as a levelling solvent for HF and HCl. In H_2O, HF is only partially ionised, whereas HCl and HBr are ~ 100% ionised. Thus, H_2O is a differentiating solvent for HF, but for HCl and HBr, it is a levelling solvent.

Several mineral acids are partially ionised in glacial CH_3COOH medium because CH_3COOH is a poor proton-acceptor but rather a better proton donor. CH_3COOH, therefore, acts as a differentiating solvent towards the mineral acids. But, for bases, CH_3COOH act as a levelling solvent.

2.3.6 Methodology for Non-Aqueous Titrations

For non-aqueous titrations, the following four steps are usually taken into consideration, namely:

(i)   Preparation of 0.1 N Perchloric acid.

(ii)   Standardization of 0.1 N Perchloric acid.

(iii) Choice of Indicators.

(iv) Effect of Temperature on Assays.

Preparation of 0.1 N Perchloric Acid

Materials Required: 8.5 ml of perchloric acid (70.0 to 72.0%); 1 litre of glacial acetic acid; 30 ml of acetic anhydride.

Procedure: Gradually mix 8.5 ml of perchloric acid to 900 ml of glacial acetic acid with vigorous and continuous stirring. Now add 30 ml acetic anhydride and make up the volume to 1 litre with glacial acetic acid and allow to stand for 24 hours before use. The acetic anhydride reacts with the water (approx. 30%) in perchloric acid and some traces in glacial acetic acid, thereby making the resulting mixture practically anhydrous. Thus, we have:

$$H_2O \ + \ (CH_3CO)_2O \longrightarrow 2CH_3COOH$$

Acetic anhydride   Acetic acid

Precautions:

The following precautions must be observed:

(a) Perchloric acid is usually available as a 70 to 72% mixture with water (sp. gr. 1.6). It usually undergoes a spontaneous explosive decomposition and, therefore, is always available in the form of a solution.

(b) Conversion of acetic anhydride to acetic acid requires 40-45 minutes for its completion. It being an exothermic reaction, the solution must be allowed to cool to room temperature before adding glacial acetic acid to the volume.

(c) Avoid adding an excess of acetic anhydride, especially when primary and secondary amines are to be assayed, because these may be converted rapidly to their corresponding acetylated non-basic products:

$$R\!-\!NH_2 \ + \ (CH_3CO)_2O \longrightarrow R\,NH\,(CH_3CO) \ + \ CH_3COOH$$

Primary amine   Acetylated product

(d) Perchloric acid is not only a powerful oxidising agent but also a strong acid. Hence, it must be handled very carefully. Perchloric acid has a molecular weight of 100.46, and 1 L of 0.1 N solution shall contain $1/10^{th}$ the equivalent weight or 10.046 gm. To prepare 1 L of the standard perchloric acid solution requires 8.5 ml (sp. gr. 1.6) volume and a purity of 72%, which will calculate as 9.792 gm of $HClO_4$.

Standardisation of 0.1 N Perchloric Acid

Standardisation of acid titrant may be carried out with any substance with a strong base and known purity. Potassium hydrogen phthalate (KHP) is used as a primary standard because of its high purity.

KHP is amphoteric in nature. It acts as an acid with a base and as a base with an acid. It contains COOH and COOK groups in the structure. When it reacts with a base, the COOH group donates its H^+ and and when it reacts with acid ($HClO_4$), the COOK group accepts a proton and forms acid and thus, acts as a base.

Perchloric acid

During titration, a crystalline precipitate of $KClO_4$ appears, but this does not interfere with the location of the endpoint.

Therefore,   204.14 gm $C_8H_5O_4K \equiv$ 1000 ml of 1N $HClO_4$

or   0.02041 gm of $C_8H_5O_4K \equiv$ 1 ml of 0.1 N $HClO_4$

Procedure: Weigh accurately about 0.5 gm of potassium hydrogen phthalate in a 100 ml conical flask. Add 25 ml of glacial acetic acid and attach a reflux condenser fitted with a silica-gel drying tube. Warm until the salt gets dissolved completely. Cool and titrate with 0.1 N perchloric acid by making use of either of the following two indicators:

(a) acetous crystal violet - 2 drops, endpoint Blue to Blue-Green (0.5% w/v).

(b) acetous oracet blue B - 2 drops, endpoint Blue to Pink.

Choice of Indicators

A number of indicators stated below are commonly used in non-aqueous titrations. It is, however, necessary to mention that the same indicator must be used throughout for carrying out the standardisation, titration and neutralisation of mercuric acetate solution.

Table 2.3 Information of Indicator

Sr. No.	Name of Indicator	Colour Change Observed		
		Basic	**Neutral**	**Acidic**
1.	Crystal violet (0.5% w/v in glacial acetic acid)	Violet	Blue-Green	Yellowish Green
2.	Oracet blue B (0.5% w/v in glacial acetic acid)	Blue	Purple	Pink
3.	α-Naphtholbenzein (0.2% w/v in glacial acetic acid)	Blue or Blue Green	Orange	Dark-Green
4.	Quinalidine red (0.5% w/v in methanol)	Magenta	-	Almost colourlesss

Effect of Temperature on Assays

Generally, most of non-aqueous solvents possess greater coefficients of expansion as compared to water, which is why small differences in temperature may afford significant and appreciable errors that can be eliminated by the application of appropriate correction factors. Hence, it is always advisable to carry out standardisation and titration, preferably at the same temperature. In a situation where these temperature parameters cannot be achieved, the volume of titrant may be corrected by the application of the following formula:

$$V_c = V[1 + 0.001 (t_1 + t_2)]$$

where, V_c = Corrected volume of titrant,

    V = Volume of titrant measured,

    t_1 = Temperature at which titrant was standardised, and

    t_2 = Temperature at which titration was performed.

2.3.7 Applications of Non-aqueous Titration

1. Determination of weak bases. For example, weak base antipyrine in water does not give a sharp end point, but acetic acid provides an endpoint break.
2. Very weak base e. g. Caffeine gives a good endpoint in acetic anhydride.
3. Determination of alkali salts of weak acids.
4. A mixture of bases of different strengths can be analysed by selecting a differentiating solvent for bases. For example (acetonitrile)
5. Determination of substances that are not themselves basic. For e. g., Amine salts dissolved in acetic acid are treated with an excess of mercurous acetate, and the amine is freed and can be titrated as a base with perchloric acid.
6. Imide is a moderately strong acid, and phenols are weak acids which can be titrated in non-aqueous media.
7. Drugs can be titrated in their dosage form without the interference of pharmaceutical adjuncts. Sometimes stearic acid, in tablet formulation, interferes in endpoint detection by consuming base in direct titration, and this substance must be removed before an acidic active ingredient is titrated.
8. Percentage of purity is determined by the assays.

   Example: Sulphanilamide is dissolved in 50 ml of dimethylformamide, and five drops of thymol blue indicator are added. The resulting solution is titrated with sodium methoxide, and the endpoint is detected as blue.
9. **Used in the determination of the concentration expressions.**

   Example: Isoprenaline solutions are mixed with glacial acetic acid and titrated with 0.1N perchloric acid using crystal violet as an indicator.

    1 ml of 0.1 N perchloric acid $\equiv$ 0.5206 gm of isoprenaline

Example: 0.2 g of ethambutol is dissolved in the mixture of 100 ml of acetic acid and 5 ml of mercuric acetate solution, and then it is titrated with 0.1 M of perchloric acid ($HClO_4$) using crystal violet as an indicator.

1 ml of 0.1M $HClO_4 \equiv 0.01386$ gm of ethambutol.

10. **Used in the determination of hydrophobic compounds.**

Example: Used in the determination of Amantadine HCl Barbiturates alkaloids etc.

Method: Weigh 0.1 gm of sample dissolved in 5 ml of pyridine and 0.25 ml of thymolphthalein solution, and 10 ml of silver nitrate-pyridine isoprenaline.

Used in the determination of diuretics.

Example: Small quantity of the drug is dissolved in anhydrous pyridine, which is heated and then cooled. The resulting solution is titrated with 0.1M of tetrabutyl ammonium hydroxide solution.

1ml of 0.1M tetrabutyl ammonium hydroxide $\equiv 0.01488$ gm of hydrochlorothiazide

11. **Used in the determination of steroids.**

Examples: Methyl testosterone, Estradiol, etc.

Method: Sample solution is mixed with 2 ml of dimethylformamide and 25 ml of chloroform. 5 ml of the resulting solution is taken, and then two drops of thymol blue indicator solution is added and titrated with methanolic potassium hydroxide solution. Simultaneously blank is carried out.

12. **Used in the determination of antitubercular drugs.**

Example: 0.2 gm of the drug is dissolved in the mixture of 100 ml of acetic acid and 5 ml of mercuric acetate solution. Then the resulting solution is titrated with 0.1M perchloric acid ($HClO_4$) using crystal violet as an indicator.

1 ml 0.1M $HClO_4 \equiv 0.01386$ gm of ethambutol

13. **Used in the determination of adrenergic drugs.**

Method: Drug solutions are mixed with glacial acetic acid and titrated with 0.1N perchloric acid using crystal violet as an indicator.

1 ml of 0.1N perchloric acid $\equiv 0.5206$ gm of isoprenaline
$\equiv 0.3193$ gm of noradrenaline
$\equiv 0.05767$ gm of salbutamol
$\equiv 28.08$ gm of xylometazolin

2.4 PRECIPITATION TITRATIONS

2.4.1 Introduction

It is a type of volumetric analysis which is based on the formation of precipitates, i.e. insoluble products, by reaction between two species.

Precipitation titration is based on reactions between ionic species that yield ionic compounds with limited solubility, thus formed reaction product is in the precipitated form. Hence, they are known as precipitation titrations.

For example: When NaCl is titrated with Silver nitrate, insoluble Silver chloride is formed.

$$Ag^+ + Cl^- \longrightarrow AgCl$$

The requirements for precipitation titrations are:

- Precipitate must be practically insoluble.
- Precipitation reaction must be rapid and quantitative.
- Co-precipitation must not occur.
- It must be possible to detect equivalence points during titration.

2.4.2 Endpoint detection in precipitation titrations

Endpoint detection in precipitation titrations can be done by the following methods:

1. Mohr's method
2. Volhard's method
3. Fajan's method

2.4.2.1 Mohr's Method

This method was described by K. F. Mohr in 1865 for the estimation of halides (mostly chloride) using silver ions (silver nitrate) as titrant. The endpoint detection depends on the formation of coloured precipitates.

This method uses potassium chromate as an indicator. Initially, the titrant (silver nitrate) forms precipitates with the analyte (e.g., chloride). After completion of precipitation reaction between the analyte and titrant, titrant forms precipitate with the indicator-potassium chromate.

$$Ag^+ + Cl^- \longrightarrow AgCl$$
$$CrO_4^{2-} + 2Ag^+ \longrightarrow Ag_2CrO_4$$

The precipitates of silver chromate are of brick red colour. Hence, the endpoint is marked by the formation of red brick precipitates.

The concentration of the indicator is important. Silver chromate should start precipitating at the equivalence point, i.e. only after the analyte has been consumed.

The concentration of the indicator can be determined as follows:

Consider the titration of 0.1M NaCl with 0.1 M AgNO$_3$ in the presence of dilute K$_2$CrO$_4$ solution.

$$K_{sp(AgCl)} = 1.2 \times 10^{-10} = [Ag^+][Cl^-]$$
$$K_{sp(Ag_2CrO_4)} = 1.7 \times 10^{-12} = [Ag^+][CrO4^{-2}]$$

As the $AgNO_3$ solution is added, AgCl will precipitate first as the concentration of chloride ions is greater as compared to chromate ions.

At the first point where the red silver chromate is just precipitated, both the salts AgCl and Ag_2CrO_4 will be in equilibrium.

$$K_{sp(AgCl)} = [Ag^+][Cl^-]$$

$$[Ag^+] = Ksp_{(AgCl)}/[Cl^-] \qquad ..... (1)$$

$$(K_{sp(Ag_2CrO_4)}) = [Ag^+]^2 [CrO_4^{-2}]$$

$$[Ag^+]^2 = [K_{sp(Ag_2CrO_4)}]/[CrO_4^{-2}]$$

$$[Ag^+] = [K_{sp(Ag_2CrO_4)}]^{1/2}/[CrO_4^{-2}]^{1/2} \qquad ..... (2)$$

Comparing equations (1) and (2),

$$[Ag^+] = K_{sp(AgCl)} = (K_{sp\,(Ag_2CrO_4)})^{1/2}$$

$$[Cl^-] = [CrO_4^{-2}]^{1/2}$$

$$\frac{[Cl^-]}{[CrO4^{-2}]^{1/2}} = \frac{1.2 \times 10^{-10}}{(1.7 \times 10^{-12})^{1/2}} = 9.2 \times 10^{-5}$$

At the equivalence point,

$$[Cl^-] = 1.1 \times 10^{-5}$$

Keeping the value of Cl^-, we get the following value for indicator concentration.

$$[CrO_4^{-2}] = 1.4 \times 10^{-2}\,M = 0.014\,M$$

Thus, the concentration of potassium chromate solution should be 0.014 M so that it will precipitate out only after all the chloride has been precipitated.

Mohr's method should be applied only to a neutral or slightly alkaline solution, i.e. pH 6.5-9, because chromate ions form dichromate ions in an acidic solution.

$$2CrO_4^{-2} + 2H^+ \rightleftharpoons HCrO_4^- \rightleftharpoons Cr_2O_7^{2-} + H_2O_5$$

In an alkaline solution, silver hydroxide may be precipitated.

Acidic solution can be made neutral by adding calcium carbonate or sodium hydrogen carbonate. Alkaline solution can be made neutral by adding acetic acid and then calcium carbonate.

Titration of iodide and thiocyanate is not successful because AgI and AgSCN adsorb chromate ions strongly, hence a false, indistinct endpoint results.

2.4.2.2 Volhard's Method

This method was discovered by Jacob Volhard. In this method, the endpoint is marked by the formation of a soluble-coloured compound.

It is an indirect method for the determination of chloride, in which excess silver nitrate solution is added to the flask along with the sample (For example, sodium chloride). Ferric ammonium sulphate is used as an indicator. The remaining silver nitrate is titrated with thiocyanate in an acid solution. It is also called Thiocyanometry.

Principle: It is a back type of titration. The analyte is quantitatively precipitated by the addition of standard silver nitrate. The excess silver nitrate remaining is back titrated with ammonium thiocyanate (NH_4SCN) using ferric ammonium sulphate as an indicator. Initially, silver thiocyanate is precipitated. After the equivalence point, when no silver ion is remaining, added thiocyanate reacts with ferric ion to give reddish brown ferric thiocyanate.

$$AgNO_3 + X^- \longrightarrow AgX + NO_3^-$$

$$AgNO_3 + SCN^- \longrightarrow AgSCN + NO_3^-$$

$$Fe^{+3} + SCN^- \longrightarrow [FeSCN]^{+2}$$

$$K_{sp} \, AgCl = 1.2 \times 10^{-10}$$

$$K_{sp} \, AgSCN = 7.1 \times 10^{-13}$$

When all excess silver ion has reacted, thiocyanate may react with AgCl precipitate since silver chloride has more solubility than silver thiocyanate. So, near the end point of the back titration of excess silver ion, silver chloride reacts slowly with ammonium thiocyanate, which in turn leads to overconsumption of thiocyanate ions, incorrect endpoint and too low values of chloride analysis.

$$AgCl + SCN^- \longrightarrow AgSCN + Cl^-$$

This reaction takes place before a reaction occurs with a ferric ion; hence a titration error is introduced.

To overcome this, Silver chloride precipitate may be filtered or modified Volhard's method may be used.

Modified Volhard's Method:
It is used for the determination of Sodium Chloride or Potassium Chloride. In this method, organic liquid like nitrobenzene or dibutyl phthalate is added to coat silver chloride precipitates so that they do not interfere with the titration of excess of silver chloride.

With bromides, titration error is small; with iodides, the titration error is even less due to the low solubility of silver iodide and bromide than silver thiocyanate.

Volhard titration must be carried out in acidic conditions to prevent precipitation of ferric ions of indicator as the hydrated oxide and to prevent precipitation of silver ion as hydroxide.

The concentration of ferric ammonium sulphate indicator of about 0.01 M is employed because at a higher concentration yellow colour of ferric ion interferes in endpoint detection.

2.4.2.3 Fajan's Method

It is also called as indicator adsorption method. It was discovered by Kazimierz Fajan. This method involves the use of adsorption indicators.

Such indicators are adsorbed on the surface of the precipitate at the equivalence point, and this adsorption is accompanied by a colour change.

For example, Acidic dyes - Fluorescein, Eosin, Diiodomethylfluorescein, and Dichloro-fluorescein.

Basic dyes: Rhodamine series.

Principle: Adsorption of the indicator at the endpoint:

When a sodium chloride precipitate is titrated with silver nitrate, silver chloride precipitate will adsorb chloride ions which are initially in excess. This layer will, in turn, hold the secondary adsorbed layer of sodium ions, as indicated in Fig. 2.5 (a). After the equivalence point, silver ions are in excess, hence silver chloride ions now adsorb silver ions which in turn will adsorb nitrate ions, as indicated in Fig. 2.5 (b).

But if sodium salt of fluorescein is present, fluorescein ions would be adsorbed instead of nitrate ions as a secondary adsorbed layer. This adsorption leads to the formation of pink colour as the structure of fluorescein gets modified by the adsorption of silver ions, as indicated in Fig. 2.5 (c).

$$Ag^+ + Fl^- \longrightarrow AgFl$$

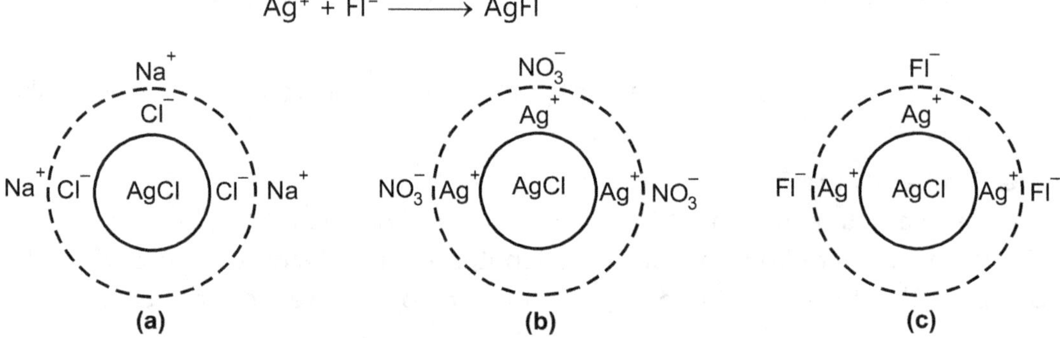

Fig. 2.5 Adsorption of Silver Ions

Conditions which govern the choice of adsorption indicators in Fajan's method:
- Indicator ion should have a charge opposite to that of precipitating agent.
- Solution should be concentrated enough to give a sharp colour change.
- Indicator should be adsorbed only after the equivalence point.

- Precipitate particles should be of colloidal dimension to maximise the quantity of indicator adsorbed.
- Silver halides are sensitised to the action of light by the layer of an adsorbed indicator like fluorescein, hence precipitation titrations of halides using argentometry and adsorption indicators should be carried out with minimum exposure to sunlight.

Limitations of Fajan's method:

- If the concentration of species is very low, there will not be enough precipitates to allow colour change to be observed.
- Method is pH dependent, as the indicator must be ionised.

Some Adsorption Indicators:

Fluorescein:

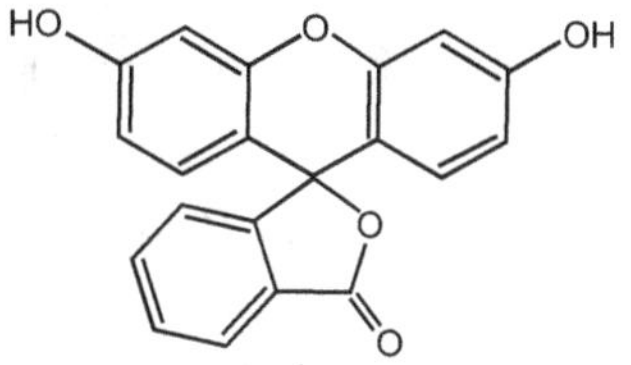

0.2 % solution of the sodium salt of fluorescein in water/alcohol is used.

Eosin:

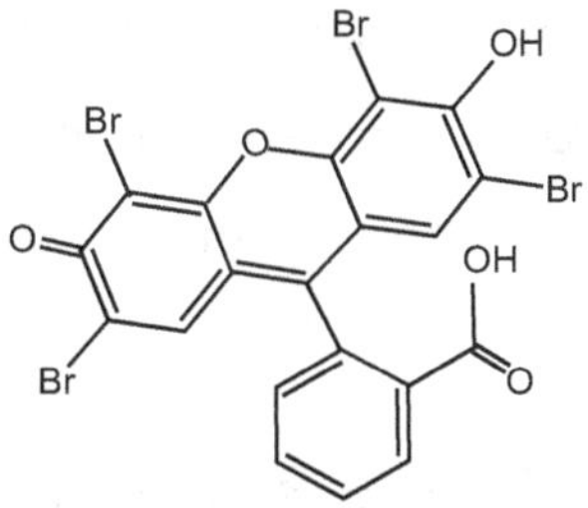

It is orange coloured powder dye, sparingly soluble in alcohol. Bromides, iodides and thiocyanate, even in dilute solution, maybe titrated accurately.

Dichlorofluorescein:

It is a weakly acidic solution. 0.1 % solution is used as an indicator. Some other adsorption indicators are; diiodomethylfluorescein, thorin, bromocresol green, bromophenol blue, etc.

2.4.3 Application of Precipitation Titrations

- Determination of anions: Halides (Bromide, chloride, fluoride).
- Determination of divalent anions (S^{2-}).
- Determination of mercaptans (CH_3SH).
- Determination of fatty acids.
- Determination of the presence of metals in some pharmaceutical dosage forms.

- In the plating bath industry.
- In gravimetric analysis.
- Assay of various pharmaceutical dosage forms.

2.5 COMPLEXOMETRIC TITRATIONS

2.5.1 Introduction

Complexometric titration is a form of volumetric analysis in which the formation of a coloured complex is used to indicate the end point of a titration. Complex is formed by a reaction of a metal ion (a cation) with an anion or neutral molecule.

$$\textbf{Metal + Ligand} \longrightarrow \textbf{ML}$$

M = Metal ion, L = Ligand, ML = Complex

An indicator capable of producing an unambiguous colour change is usually used to detect the end point of the titration.

The accuracy of complexometric titrations is high and they are useful for detecting the concentration of metal ions at millimole level.

Some Important Terms:

Central ion: Metal ion is called a central ion.

Ligand: Group attached to the central ion is called a ligand.

Counter ion: The ion is necessary for electroneutrality if the complex is charged.

Coordination number: Number of bonds formed by the metal ion is known as the coordination number of the metal.

For example:

$$Ag^{+2} + 2CN^- \rightleftharpoons Ag(CN)^-_2$$

Ag^{+2} is the central ion, CN^- is the ligand, and the Coordination number is 2.

2.5.2 Ligand (Complexing Agents)

A ligand is an ion or molecule with a lone pair of electrons that can be donated to a central metal atom to form a coordination complex.

The nature of metal-ligand bonding can range from covalent to ionic. Ligands are usually considered electron donors attracted to the metal at the centre of the complex. Metals are electron acceptors. Thus, the reaction by which a complex is formed can be regarded as a Lewis acid-base reaction.

Classification of Ligands

Ligands are classified based on the number of sites available for bonding.

Unidentate (Single toothed): Ligands that have a single donor group.

For example, NH_3 (ammonia) and CN^- (Cyanide).

Ligands which contain more than one donor group are known as multidentate ligands. They may be classified as:

Bidentate: Ligands that have two donor groups. Examples: Glycine (NH_2-CH_2-COOH), ethylenediamine, $C_2O_4^{2-}$ (oxalate ion), etc.

Tridentate: Ligands which have three donor groups. For example, diethyl triamine.

Tetradentate: Ligands which have four donor groups. For example, triethylenetetramine.

Hexadentate ligands: Ligands that have six donor groups. For example, EDTA (Ethylene Diamine Tetracetic Acid).

$$HOOC-CH_2 \diagdown \qquad \qquad \diagup CH_2-COOH$$
$$:N-CH_2-CH_2-N:$$
$$HOOC-CH_2 \diagup \qquad \qquad \diagdown CH_2-COOH$$

Structure of EDTA

Chelating Agents

These are chemical compounds which can bind with metal ions through multiple coordination bonds to form stable, water-soluble complexes. The complexes so formed are called chelates. Examples of chelating agents are EDTA (Ethylene Diamine Tetracetic Acid), DTPA (Diethylene Triamine Pentaacetic Acid), NTA (Nitrilo 2, 2, 2-Triacetic Acid), etc.

2.5.3 EDTA

It is the most widely used titrant in complexometric titrations. It is a versatile chelating agent. It can form four or six bonds with a metal ion, and it forms chelates with both transition-metal ions and main-group ions.

The EDTA molecule has 6 potential sites for bonding a metal ion: the four carboxyl groups (Hydrogen atoms can be replaced by metal ions) and the two amino groups (each amino group has unshared pair of electrons). Thus, EDTA is a hexadentate ligand.

Disodium salt of EDTA is generally preferred as it is a water-soluble chelating agent. It is non-hygroscopic and a very stable sequestering agent.

Advantages of EDTA:

1. Low cost.
2. It is readily available.
3. It is a hexadentate ligand.
4. It forms a strainless 5-membered ring.
5. It forms water-soluble stable complexes.

6. It has a high molecular weight, hence less weighing errors.
7. It forms fairly stable 1:1 complexes with all metals except alkali metals (For example, Na, K).

Preparation and Standardization of 0.05 M Disodium EDTA:

Preparation of 0.05 M Disodium EDTA: Dissolve 1.86 gm of disodium edetate in sufficient water to produce 100 ml.

Standardisation of 0.05 M Disodium EDTA:

Principle: Disodium edetate is standardised using zinc granules. Zinc granules are dissolved by using dil HCl. Bromine water is added to ensure oxidation of trace iron impurity to ferric ion, forming a less stable complex with EDTA than a ferrous form of iron. Sodium hydroxide is added to make the solution neutral, resulting in the precipitation of zinc hydroxide. To dissolve the precipitate, ammonia buffer pH 10 is added. When precipitates get completely dissolved, 5 ml excess is added, which acts as a buffer and maintains the pH of the solution at 10 required for the formation of zinc edetate complex.

Procedure: Weigh accurately 0.8 gm of granulated pure zinc. Dissolve in 12 ml dilute hydrochloric acid by warming, add a few drops of bromine water, and boil to remove excess bromine. Cool and add sufficient distilled water to produce 250 ml in a volumetric flask. Pipette 25 ml of solution in a conical flask, neutralise by 2 N sodium hydroxide solution. Dilute to about 150 ml with water. Add sufficient ammonia buffer pH 10 to dissolve the precipitate. Add slight excess and titrate with disodium edetate solution by using Mordant Black II as an indicator till the pink colour turns blue.

$$\text{Molarity of Disodium EDTA} = \frac{\text{Weight of zinc} \times 1000}{\text{Mol. Wt. of Zinc} \times \text{Volume of Disodium EDTA}}$$

Some of the indicators commonly used and official in Indian Pharmacopoeia are as follows:

1. **Calcon (Mordant black 17, Solochrome Dark blue):** Sodium 2-hydroxyl (2-hydroxyl-1-naphthylazo) naphthalene-4-sulphonate.

   It gives purple-red colour with calcium ions in an alkaline solution. In the absence of metal ions, it gives blue colour. It is employed as a calcon mixture which is one-part calcon and 99 parts of sodium chloride.

2. **Catechol Violet:** 1% solution is used as an indicator. It forms highly coloured complexes with a wide range of metal ions. Complexes with metals are blue in colour in acidic

solutions. Thorium (Th^{4+}) complex at pH 3 and bismuth at pH 1.5 is stable. Other metals like; magnesium, manganese, cobalt, zinc, calcium, cadmium, etc. can be titrated in the pH range of 4-7. A 0.1 % solution in water is employed as an indicator.

3. **Eriochrome Black T (Mordant Black - 2):** Chemically, it is sodium 1-(1-hydroxy-2-naphthylazo)-5-nitro-2-naphthol-4-sulphonate. A 0.2 gm of dye with 2 gm of hydroxylamine hydrochloride in 50 ml methylamine hydrochloride or 50 ml methyl-alcohol is employed in titration. A mixture of 0.2 percent mordant black-2 with 100 parts of sodium chloride is also employed as an indicator. It has blue colour at pH 10, and the complex has red colour at pH 10. Below pH 6.3 and above pH 11.5, the dye has a reddish colour. Use of a buffer of pH 10 is essential. Metals like magnesium, calcium, cadmium, zinc, etc., give excellent results when directly titrated.

4. **Mordant Blue 3 (Solochromecyanne R):** It is an indicator specific to aluminium. In the presence of aluminium, it gives purple and pink colours when EDTA is in excess.

5. **Murexide (Ammonium purpurate):** A mixture of 0.2 per cent dispersed sodium chloride is used to determine calcium at pH 12.0. Complex of calcium murexide gives blue-violet colour while indicator solution has red violet colour. This indicator gives better results when 0.5% of naphthol green is incorporated in it. Calcium can be estimated in the presence of magnesium (for which it forms a less stable complex than calcium). This indicator was official in I.P. 1966.

6. **Xylenol orange:** One part of the indicator in 99 parts of potassium nitrate is prepared for use. Alternatively, 0.1% solution in 50% alcohol is used. It gives red-violet colour with mercury, lead, and zinc metals in an acidic solution. For bismuth, pH 1 to 3, for lead, pH 4 to 5, zinc pH 8 to 9 and for cadmium and mercury, pH 5 to 6 is used. In the absence of metal ions, it gives yellow colour.

7. **Pyridylazonaphthol (PAN):** It gives red colour in the presence of copper ions and yellow in the absence of copper ions. A 0.1% solution in alcohol is employed in titrations.

2.5.4 Applications of Complexometric Titrations

Complexometric Titrations are used in the determination of various elements by different titration methods as follows:-

Direct titration: Al, Ca, Mg, Zn, Cd, Cu, Ni, Co, Pb, Ba, Mn, Hg, etc.

Back titration: Pb, Hg, Ni, Al.

Replacement method: Ca, Hg, Pb, Fe

Determination of Hardness of Water: Water hardness due to Ca and Mg is expressed as the amount of Ca and Mg ions in ppm. The total Ca and Mg is titrated with standard EDTA solution using Eriochrome black T as an indicator.

2.6 REDOX TITRATIONS

2.6.1 Introduction

Redox titration is one of the methods of volumetric analysis, which involves chemical reactions based on reduction and oxidation (redox reactions). In an oxidation-reduction titration, oxidising agent reacts with the reducing agent, and there is the transfer of electrons from one component to another. This leads to a change in the valency of the element. In an oxidation-reduction titration, oxidising agents gain electrons from reducing agents and reducing agents donate their electrons to oxidising agents. Both oxidation and reduction reactions take place simultaneously, so titration is called redox titration.

2.6.2 Concepts of Oxidation and Reduction

Oxidation may be defined as the "Loss of electrons or hydrogen or Gain of oxygen by atom, molecule or ion". Oxidation is also known as de-electronation.

Look at the reactions below:

$$C + O_2 \longrightarrow CO_2$$

Here carbon has gained oxygen, and it has been oxidised to carbon dioxide.

$$Fe^{+2} - e^- \longrightarrow Fe^{+3}$$
$$Sn^{+2} - 2e^- \longrightarrow Sn^{+4}$$

In the above reactions, there is a loss of electrons, hence the concerned ions are said to be oxidised.

Reduction may be defined as "gain of electrons or gain of hydrogen or loss of oxygen by atom, molecule or ion".

$$CuO + H_2 \longrightarrow Cu + H_2O$$

In the above reaction, copper oxide has been reduced to copper by loss of oxygen.

Now, look at the reaction below:

$$Cu^{+3} + e^- \longrightarrow Cu^{+2}$$

Here there is a gain of 1 electron, so a cupric ion is said to be reduced to a cuprous ion. In a chemical reaction, oxidation and reduction occur simultaneously. If one atom is oxidised, the other is reduced.

$$2Fe^{+2} + 2Cl_2 \longrightarrow 2(FeCl_2)$$

In the above reaction, the ferrous ion is oxidised to ferric ion by loss of 1 electron, and chlorine has been reduced to a chloride ion by a gain of 1 electron. Therefore, these reactions are called as redox reactions. Redox reactions are also known as electron transfer reactions. During such reactions change in the valency of atoms occurs.

Oxidising agents: These are the agents that themselves get reduced, i.e. they accept electrons and oxidise other elements. They are also called oxidisers or oxidants. Commonly used oxidising agents are potassium dichromate, potassium permanganate, ceric sulphate, iodine, etc.

Reducing agents: These are the agents that themselves get oxidized, i.e. they donate electrons and reduce other elements. They are also called reductants or antioxidants. Examples of reducing agents are oxalic acid, ferrous sulphate, titanious sulphate, sodium thiosulphate, etc.

Let us see the reaction below:

$$2KMnO_4 + 10FeSO_4 + 2H_2SO_4 \longrightarrow K_2SO_4 + 2MnSO_4 + 5Fe_2(SO_4)_3 + 8H_2O$$

Here, electrons are transferred from $FeSO_4$ to $KMnO_4$.

Mn (+7) in $KMnO_4$ has been reduced to Mn (+2) in $MnSO_4$. Fe(+2) in $FeSO_4$ has been oxidised to Fe (+3) in $Fe_2(SO_4)_3$. Thus, $KMnO_4$ acts as an oxidising agent and $FeSO_4$ acts as a reducing agent in an acidic medium.

Half Reaction:

In redox titrations, two reactions take place simultaneously, i.e., oxidation and reduction. Oxidation and reduction reactions are therefore known as half-reactions.

Consider the following reaction:

$$Ce^{+4} + Fe^{+2} \rightleftharpoons Ce^{+3} + Fe^{+3}$$

Oxidation half-reaction may be written as:

$$Fe^{+2} \rightleftharpoons Fe^{+3} + e^-$$

Reduction half-reaction may be written as:

$$Ce^{+4} + e^- \rightleftharpoons Ce^{+3}$$

2.6.3 Redox Potential

Redox titrations can also be carried out potentiometrically, like acid-base titration. Redox potential can be measured by measuring the potential difference of a cell in which redox half reaction is coupled with a standard reference cell, i.e. standard hydrogen electrode. Electrode at which the redox half-reaction occurs is called the indicator electrode.

Oxidation Reduction Potential (ORP or Redox Potential) is a system's capacity to either release or accept electrons from chemical reactions. Potential can be calculated by the Nernst equation.

Nernst Equation:

Potential at indicator is calculated by the Nernst equation.

$$E = E° - \frac{2.303RT}{nf} \log \frac{[\text{Reducing agent}]}{[\text{oxidising agent}]}$$

Where: E = Redox potential at a specific concentration

E° = Standard electrode potential

N = Number of electrons transferred

F = Faraday's constant = 96500 coulombs

R = Universal gas constant = 8.314 J/mol-K

T = Absolute temperature

2.6.4 Common Indicators used for Detection of Endpoint in Redox Titration:

Following indicators can be used in redox titrations to determine the endpoint.
- Internal or redox indicators
- Self-indicators
- External indicators
- Specific indicators
- Potentiometric methods

a. Internal Indicators

These redox indicators (also called Oxidation-Reduction indicators) undergo an oxidation-reduction reaction at the equivalence point. The indicator has different colours in oxidised and reduced forms. Thus, the endpoint is marked by a colours change. Oxidation-reduction equilibrium needs to be established very quickly. Therefore, only a few classes of organic redox systems can be used for indicator purposes.

There are two common types of redox indicators:

1. Metal-organic complexes (E.g. Phenanthroline)

2. True organic redox systems (E.g. Methylene blue)

Sometimes coloured inorganic oxidants or reductants (E.g., Potassium permanganate, Potassium dichromate) are also called redox indicators. They cannot be classified as true redox indicators because of their irreversibility. Redox indicators with true organic redox systems involve a proton as a participant in their electrochemical reaction. Therefore, sometimes redox indicators are also divided as independent or dependent on pH.

Diphenylamine

Diphenylamine was the first redox indicator to be used in the titrimetric analysis. But as it is not easily soluble in water and ions like mercury, tungsten interfere with its action, the barium or sodium salt of diphenylamine sulfonic acid is used. In reduced form, it is colourless, and in the oxidized form, it is violet.

In the oxidised form presence of a long conjugated system leads to the absorption of light in the visible region.

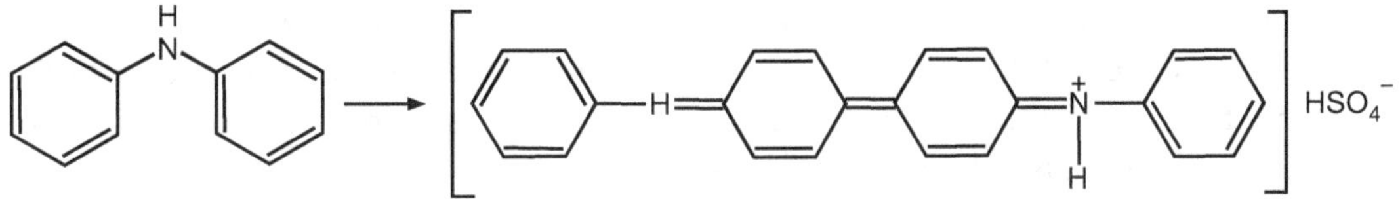

Diphenylbenzidine violet undergoes further oxidation if it is allowed to stand with an excess of dichromate solution. Excessive oxidation is irreversible, and red or yellow products are produced.

Ferroin: It is a complex of iron and 1,10 phenanthroline. The complex is of blood red colour.

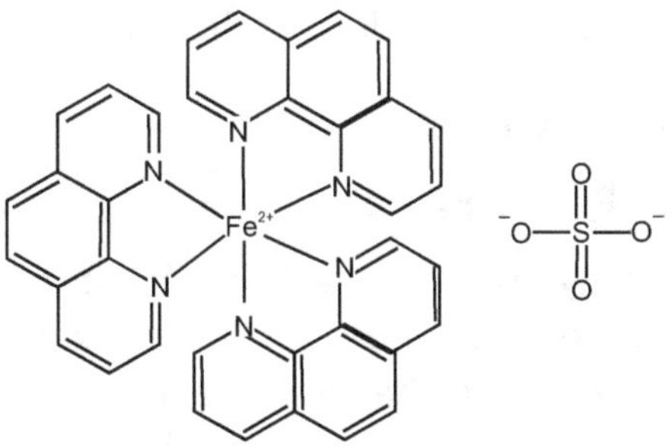

Fe (II) can be oxidised to Fe (III), and the complex (ferric) is of pale blue colour.

b. Self-Indicators

A coloured substance may act as its own indicator.

Examples include potassium permanganate, ceric ammonium sulphate, and iodine.

KMnO$_4$ gets reduced in redox titration as follows:

$$MnO_4^- + 4H^+ \rightleftharpoons Mn^{2+} + 4H_2O$$

KMnO$_4$ has dark purple colour due to MnO$_4^-$ which on reduction gives Mn^{2+}, which is colourless. At the endpoint excess drop of KMnO$_4$ remains in the solution, which imparts pink colour to the solution. In very dilute solutions, KMnO$_4$ cannot be used as a self-indicator as colour cannot be detected in very dilute solutions.

c. External Indicators

These are also called spot test indicators. These were used earlier when no indicators were available. Near the equivalence point, drops of solution are removed and brought into contact with a dilute freshly prepared solution on a spot plate. The endpoint is reached when the first drop fails to give the given colour.

For example: Ferricyanide ion was used to detect iron (II) ion by the formation of iron (II) ferricyanide (Turnbull's blue) on a spot plate outside the titration vessel.

d. Specific Indicators

It is a substance which reacts in a specific manner with one of the reagents in a titration to produce colour. For example: Starch gives deep blue colour with iodine. 1% solution of starch is used. It should be prepared in warm water and must be used freshly as it is prone to bacterial degradation. Starch is used as starch mucilage or starch paste. Thiocyanate gives a red colour with iron (III).

e. Potentiometric Method

It is a physicochemical method where the equivalence point is detected by large changes in potential.

This method can be used when suitable indicators are not available, and the visual method is of limited accuracy (coloured solutions or very dilute solutions). Potential is measured by a potentiometer. At the endpoint, there is a sudden rise in potential.

2.6.5 Some important Redox Titrations

a. Titrations using Potassium Permanganate

Potassium permanganate is widely used as an oxidising agent in redox titrations. It is readily available and inexpensive. It requires no indicators except in very dilute solutions. Permanganate undergoes a variety of chemical reactions since manganese can exist in oxidation states of +2, +3, +4, +6, and +7. It reacts rapidly with reducing agents as follows:

$$MnO_4^- + 8H^+ + 5e^- \rightleftharpoons Mn^{+2} + 4H_2O$$

Some substances require heating or the use of a catalyst to speed up the reaction.

Preparation of 1 M KMnO₄

Dissolve 158 gm of KMnO$_4$ in water to make up a volume to 1000 ml. Heat on a water bath for 1 hour and allow standing for 2 days. Filter through glass wool.

Following precautions must be taken during its preparation:

1. Weigh KMnO$_4$ on a watch glass and not paper since cellulose fibres of the paper are attacked by KMnO$_4$.
2. Filter through non-reducing filters like glass wool or asbestos.
3. Heating is done in a clean glass vessel to destroy reducing substances.
4. All the apparatus must be clean so that traces of grease on glass do not catalyse the decomposition of KMnO$_4$.
5. Store the solution in the dark or in amber-coloured bottles.
6. Acidic solutions of KMnO$_4$ are not stable.

$$MnO_4^- + 4H^+ \longrightarrow MnO_2 + 2H_2O$$

Standardisation of KMnO₄:

Primary standards like Resinous oxide, Sodium oxalate/Oxalic acid, Sodium thiosulphate, and Iron can also be used to standardise potassium permanganate.

Standardisation with Sodium Oxalate:

Dissolve 6.7 gm of sodium oxalate previously dried at 110°C in water and then makeup volume to 1 litre. Pipette out 20 ml of this solution to a conical flask, add 5 ml of concentrated sulphuric acid and then warm to 70°C. Titrate against KMnO$_4$ until pink colour persists for 30 s.

$$5Na_2C_2O_4 + 2KMnO_4 + 8H_2SO_4 \longrightarrow K_2SO_4 + 2MnSO_4 + 10CO_2 + 5Na_2SO_4 + 8H_2O$$

Reaction is slow at room temperature. Hence, it is heated to about 60°C. Even at elevated temperatures, the reaction starts slowly. But the rate increases as Mn(II) is formed. Hence, the reaction is termed autocatalytic.

Precautions:

1. Clean the flask with concentrated H$_2$SO$_4$ or hydrogen peroxide after each titration.
2. Too high temperature leads to the formation of brown colour solution due to decomposition of KMnO$_4$ to MnO$_2$.
3. Insufficient acid leads to the formation of MnO$_2$, which gives brown colour to the solution.

Oxalic acid can also be used for standardisation, but sodium oxalate is purer than oxalic acid. Also, oxalic acid is hygroscopic.

Determination with $KMnO_4$:

In titrations involving $KMnO_4$, it is used as an oxidising agent. In acidic solutions, it reduces to Mn^{+2}; in neutral or alkaline solutions, it reduces to Mn^{+4} (MnO_2). In an acidic solution, it acts as a self-indicator. In alkaline or neutral solutions, diphenylamine or phenanthroline is used as an indicator. It is used for the determination of Iron in iron ores, as well as the oxides of metals like lead and manganese.

Estimation of Ferrous Sulphate:

Weigh the Ferrous sulphate sample. Transfer it to a conical flask. Titrate with standard potassium permanganate solution till pink colour persists for 30 sec.

b. Cerimetry

Titrations involving determinations with cerium compounds are called Cerimetry. Atomic weight of Cerium is 140 gm/mol. It exists in 2 oxidation states: +4 and +3.

In a tetravalent (Quadrivalent) state, it acts as a powerful oxidising agent. It is of dark yellow colour in concentrated solution. On reduction, trivalent cerium ion is obtained, which is colourless. Hence, it can be used as a self-indicator. In dilute solutions, ferroin is used as an indicator.

Some of the compounds of Cerium, such as Ceric ammonium nitrate $(NH_4)_2Ce(NO_3)_6$, Ceric ammonium sulphate $(NH_4)_4Ce(SO_4)_4$, Ceric hydroxide $Ce(OH)_4$, can be used for determination.

Preparation of 0.1 M Ceric Ammonium Nitrate Solution:

Dissolve 65 gm of ceric ammonium sulphate with help of gentle heat in a mixture of 30 ml sulphuric acid and 500 ml water. Cool it. Filter if turbid and dilute with water to 1000 ml.

Standardisation of 0.1 M Ceric Ammonium Nitrate Solution:

Weigh accurately about 0.2 gm of arsenic trioxide, previously dried at 105°C for 1 hour and transfer to a 500 ml conical flask. Wash with 25 ml of 8% w/v solution of sodium hydroxide, swirl to dissolve, add 100 ml of water and mix. Add 30 ml of dilute sulphuric acid, 0.15 ml of osmic acid solution, 0.1 ml of ferroin solution and titrate with ceric ammonium nitrate until pink colour is changed to very pale blue.

Factor: Each ml of 0.1 M ceric ammonium nitrate is equivalent to 0.004946 gm of As_2O_3.

Advantages of Cerimetry over Pemanganometry:

- Cerium can be reduced to only one oxidation state.
- It is a very strong oxidising agent, and its intensity can be varied using different acids.
- Sulfuric acid solutions are extremely stable.

- No effect of light or heating for a short time.
- Cerium nitrate is available in sufficiently pure form to directly prepare a standard solution.

Applications of Cerimetry:

- Titration of hydrogen peroxide.
- Ferrous compounds can be analysed.
- Oxalates can be analysed using ceric sulphate.
- Analysis of tartaric acid, phthalic acid, and salicylic acid.
- Nitrogen compounds like hydroxylamines and nitrites can also be analysed.

Estimation of Ferrous Sulphate:

Dissolve 2.5 gm of sodium bicarbonate in a mixture of 150 ml of water and 10 ml of sulphuric acid. When effervescence ceases, add about 0.5 gm of the substance under examination, accurately weighed, shake gently to dissolve and titrate with 0.1 M ceric ammonium nitrate, using 0.1 ml of ferroin solution as an indicator, until the red colour disappears.

Factor: 1 ml of 0.1 M ceric ammonium nitrate is equivalent to 0.02780 gm of $FeSO_4.7H_2O$.

c. Iodine Titrations

Iodine titrations may be classified as Iodimetry and Iodometry.

Iodimetry:

In iodimetry, a standard solution of iodine is used as a titrant. It is based on the following reaction:

$$2I^- \rightleftharpoons I_2 + 2\ e^-$$

It is used for the estimation of reducing agents like arsenates (H_3AsO_3) and thiosulphates ($Na_2S_2O_3$).

Direct Titration with Iodine:

Principle:

In iodimetry titration, a standard solution of iodine is used for quantitative oxidation of reducing agents.

For example: Estimation of arsenious acid (H_2AsO_3)

$$H_2AsO_3 + H_2O + I_2H_3 \rightleftharpoons AsO_4 + 2H^+ + 2I^-$$

This method is known as direct titration.

Back Titration with Iodine:

Principle:

In this method, a known excess quantity of standard solution of iodine is added to the reducing agent to be titrated. The remaining excess quantity of iodine is then determined by using sodium thiosulphate as a titrant.

For example: Estimation of sodium bisulphate

$$NaHSO_3 + I_2 + H_2O \longrightarrow NaHSO_4 + 2HI$$

$$I_2 + 2Na_2S_2O_3 \longrightarrow 2NaI + Na_2S_4O_6$$

Sodium thiosulphate   Sodium tetrathionate

Preparation of Iodine Solution:

Dissolve 14 gm of iodine in a solution of 36 gm of potassium iodide in 100 ml of water, add 3 drops of hydrochloric acid and dilute with water to 1000 ml.

Iodine is slightly soluble in water (0.00134 mol/lit). Therefore, potassium iodide is added to increase the solubility and to decrease the volatility of iodine.

$$I_2 + I^- \longrightarrow I_3^- \text{ (triodide)}$$

$$I_3^- + I_2 \longrightarrow I_5^- \text{ (pentaiodide)}$$

$$I_5^- + I_2 \longrightarrow I_7^- \text{ (heptaiodide)}$$

Standardisation of Iodine Solution with Arsenic Trioxide:

Weigh 0.15 gm of arsenic trioxide previously dried at 105°C for 1 hour and dissolve in 20 ml of 1 M sodium hydroxide solution by warming if necessary. Dilute with 40 ml of water, add 2 drops of methyl orange solution and add dilute HCl solution until the yellow colour changes to pink. Then add 2 gm of sodium bicarbonate, dilute with 50 ml of water and add 3 ml of starch mucilage as an indicator. Slowly titrate with iodine solution until the permanent blue colour is obtained.

Standardisation of Iodine Solution with Sodium Thiosulphate:

Pipette out iodine solution and titrate it against standard sodium thiosulphate solution until the solution has a pale yellow colour. Add starch solution and continue the titration until the solution is colourless.

Indicators in Iodimetric Titrations:

Starch is used as an indicator in iodometric titrations. Iodine forms blue coloured complex with β amylose constituent of starch. Starch solutions are often decomposed by bacteria. So fresh solutions of starch should be used. Usually, boric acid is used as a preservative. Starch is used in the form of mucilage or paste. 1% solution of soluble starch is used.

Determinations with Iodine:

Generally used for estimation of reducing substances.

HCN:

$$HCN + I_2 \rightleftharpoons ICN + HI$$

Sulphur (sulfide):

$$H_2S + I_2 \rightleftharpoons 2H^+ + 2I^- + S$$

Reducing the power of these substances depends upon the hydrogen ion concentration, and only by proper adjustment of pH these reactions can be made quantitative.

Conditions for Iodimetric Titrations:

1. As iodine is volatile, titration must be conducted in cold conditions.
2. As the solubility of iodine is low in the water, excess of KI must be used to dissolve it.
3. Sufficient time should be given before titration, as iodine ionises too slowly.
4. The reaction mixture should be kept in the dark, as light accelerates the side reactions. Iodide ions are oxidised to iodine by atmospheric oxygen.

$$4I^- + 4H^+ O_2 \longrightarrow 2I_2 + 2H_2O$$

Iodometry:

In this type of titration, iodine is generated in situ. To a known volume of sample (oxidising agent), an excess but known amount of iodide (e.g. potassium iodide) is added. The oxidising agent oxidises iodide to iodine. Iodine dissolves in the iodide-containing solution to give triiodide ions, which have a dark brown colour.

The triiodide ion solution is then titrated against standard thiosulfate solution to give iodide using starch indicator till the blue colour changes to colourless.

Reactions which occur are given below:

$$2I^- + 2e \longrightarrow I_2$$
$$I_2 + I^- \longrightarrow I_3^-$$
$$Na_2S_2O_3 + I_3^- \longrightarrow NaI + Na_2S_4O_6$$

(Sodium thiosulphate)   (Sodium tetrathionate)

Preparation of 0.1 M Sodium Thiosulphate Solution:

Dissolve 25 g of sodium thiosulphate and 0.2 g of sodium carbonate in carbon dioxide-free water and dilute to 1 litre with water.

Standardisation of Sodium Thiosulphate:

(a) Dissolve accurately weighed 0.200 gm of potassium bromate in water to produce 250 ml to 50 ml of this solution. Add 2 gm of potassium iodide and 3 ml of 2M hydrochloric acid and titrate with sodium thiosulphate solution using starch indicator until the blue colour disappears.

Factor: Each ml of 0.1 M sodium thiosulphate is equivalent to 0.002784 gm of $KBrO_3$.

(b) With potassium dichromate: Sodium thiosulphate can be standardised against primary standards like potassium dichromate (AR). Potassium dichromate acts as an oxidising agent.

A known quantity of potassium dichromate is dissolved in water, acidified with hydrochloric acid, and an excess of potassium iodide is added.

Potassium dichromate oxidises KI to iodine, and the liberated iodine is then titrated with sodium thiosulphate using starch as an indicator.

$$K_2Cr_2O_7 + 6KI + 14HCl \longrightarrow 3I_2 + 8KCl + 2CrCl_3 + 7H_2O$$

Pot. dichromate Pot. Iodide   Chromium chlorate

$$2Na_2S_2O_3 + I_2 \longrightarrow Na_2S_4O_6 + 2NaI$$

Sod. thiosulphate   Sod. tetrathionate

End Point Detection:

Starch is used as an indicator in iodometric titrations. Starch is added towards the end of the titration.

Reason: When the concentration of iodine is high, it gets bonded with starch relatively strong, and desorption becomes slow. This makes detection of the endpoint relatively difficult.

So, the solution is titrated till the solution gets pale and then starch is added.

Conditions for Iodometric Titrations:

- Titration must be carried out in acidic conditions.
- In highly alkaline conditions, NaOI (Sodium hypoiodide) is formed, which is a strong oxidising agent compared to iodine.

$$2NaOH + I_2 \longrightarrow NaOI + NaI + H_2O$$

- Reaction mixture must be kept in the dark to decrease the volatilisation of iodine.
- Titration must be carried out in a closed flask and in cold conditions to prevent the volatilisation of iodine.
- Iodine is very slightly soluble in water, so there is a need to add an excess of iodine to dissolve Iodine.

Applications of iodometry:

- Determination of oxygen dissolved in water.

Method of Winkler:

- To the water sample, Mn(II) salt, NaI and NaOH is added.

- White $Mn(OH)_2$ is precipitated and is quickly oxidised to brown $Mn(OH)_3$. The solution is acidified, and $Mn(OH)_3$ oxidises iodide to iodine, titrated with a standard sodium thiosulphate s.

$$O_2 + Mn(OH)_2 + 2H_2O \longrightarrow Mn(OH)_3$$

$$Mn(OH)_3 + I^- + H^+ \longrightarrow Mn^{2+} + I_2 + H_2O$$

2.7 GRAVIMETRIC ANALYSIS: PRINCIPLE AND METHOD

2.7.1 Introduction

Gravimetric method is one of the methods of quantitative analysis, which is concerned with the process of producing and weighing a compound or element in as pure form as possible after some chemical treatment has been carried out on the substance to be examined.

The measurement step in the gravimetric analysis is weighing. Analytes can be separated from the sample matrix/interferences by various methods.

2.7.2 Types of Gravimetry

(i) Precipitation gravimetry:

This method involves the conversion of the analyte to a sparingly soluble precipitate. The precipitates are then filtered, washed free of impurities and converted to a product of known composition by suitable heat treatment. The compound is weighed. That can be weighed.

(ii) Volatilisation gravimetry:

In this method, the sample solution is volatilised at a specific temperature. The product is then collected and weighed. If the analyte is volatile, its mass is determined indirectly from the loss in mass of the sample.

(iii) Electrogravimetry:

In electrogravimetry, the analyte is separated by deposition on an electrode by an electric current. Mass of this compound provides a measure of analyte concentration. It is based on Second faraday's law, i.e. when a given current is passed in series through the solution containing various ions, the amount of substances that are deposited will be in the ratio of their chemical equivalent.

2.7.3 Steps in Precipitation Gravimetry

1. Sample being analysed is weighed accurately.
2. Weighed sample is dissolved in a suitable solvent.
3. Interfering species are removed by using a suitable separation method.
4. Experimental environment is adjusted, i.e. pH adjustment by adding a suitable buffer, change of oxidation state, concentrating or diluting the solvent, etc.
5. Suitable precipitating reagent is added.

6. Precipitation is usually carried out in hot dilute solutions.
7. Precipitate is then separated from the mother liquor by filtration.
8. Precipitate is washed using a suitable solvent.
9. Precipitates are dried or ignited.
10. Percent of substance in the sample is then calculated.

2.7.4 Advantages of Gravimetric Analysis

- It is accurate and precise when using a modern analytical balance.
- Possible sources of error are readily checked since filtrates can be tested for completeness of precipitation, and precipitates may be examined for the presence of impurities.
- Absolute method since it involves direct measurement without any form of calibration being required.
- Relatively inexpensive apparatus.

2.7.5 Unit operation in Gravimetric Analysis

1. Sampling
2. Preparation of solution
3. Precipitation
4. Testing completeness of precipitation
5. Digestion
6. Filtration
7. Washing
8. Drying or Ignition

1. **Sampling:** Representative samples must be taken correctly to ensure that the results are accurate. The sample should be homogenous and in powder form.

2. **Preparation of solution:** The sample is dissolved in an appropriate solvent, and the sample solution is prepared. To minimise sample loss due to solution preparation, funnels should be used; solutions for dissolution must be added through the funnel. Some form of preliminary separation may be required sometimes. Proper adjustments like the volume of solution, pH, and temperature must be done. If the substance is soluble in acids and alkalies, and if the reaction is likely to evolve gases, then allow gas evolution to cease before proceeding. If heating is required for dissolution, a water bath or asbestos gauze is used.

3. **Precipitation:** Specific or at least selective precipitant should be used. Precipitant must be such that it should react with the analyte to form the precipitates, which are easy to filter, wash, sufficiently stable, sparingly soluble and should have constant known composition after drying. Precipitation is generally carried out in a resistant glass beaker with slow addition of a dilute solution of precipitating reagent with

constant stirring. This reaction must be carried out in a hot solution. Boiling of the solution must be avoided to prevent loss due to spattering. Only moderate excess of precipitant must be used. If too much excess of precipitant is used, it may lead to an increase in solubility.

4. **Testing the completeness of precipitation:** The completeness of precipitation is checked by carefully adding a few drops of precipitant from the sides of the beaker wall to the solution with the settled precipitate. If no turbidity is seen at the edge, precipitation is complete.

5. **Digestion:** In this step, the precipitates are allowed to age in the mother liquor. It is carried out at a higher temperature to speed up the process. It improves the purity of precipitates. It also helps to minimise surface adsorption. However, amorphous precipitates should not be kept over as they can be easily contaminated by the adsorption.

6. **Filtration:** In this operation, precipitates are separated from the mother liquor by filtration. The choice of filter media depends on the nature of the precipitate, the cost of filter media, and the heating temperature required for drying. Various filter media used in the gravimetric analysis are,

   (i)   Filter paper

   (ii) Filter pulp

   (iii) Filter mats

   (iv) Permanent porous filter discs

   (i) **Filter paper:** Quantitative filter papers used in the gravimetric analysis should have low ash content. Lower values of ash content are usually achieved by washing with hydrochloric and hydrofluoric acid during its manufacture. Circular filter papers with 07.0, 09.0, 11.0, 12.5 cm are available. They are available in different degrees of porosities. Generally, three degrees of porosities are there: one for gelatinous and coarse particles, second for medium size particles and third for very fine precipitates. The speed of filtration also varies for these papers. It is fast for first, medium for second and slow for third. Filter papers for quantitative work are treated with nitric acid to improve their mechanical strength.

   (ii) **Filter pulp:** Fox filtration, filter pulp is used for gelatinous precipitates as pores of quantitative filter paper get clogged by the gelatinous precipitates. Filter pulp can be prepared by macerating the filter paper torn into pieces with hot distilled water. Whitman filter clipping is ashless and used for his purpose. Filter pulp tablets are also available and are easily disintegrable. Addition of filter pulp is made immediately before filtration and after the precipitation. The precipitate, along with filter pulp, is transferred into the crucible. For very fine particles which are difficult to coagulate can also be filtered by using filter pulp.

(iii) Filter mats:

    (a) Gooch crucible: Gooch first employed a filter mat of purified asbestos supported inside the platinum crucible. The bottom of the crucible was perforated with numerous small holes. Late porcelain and silica gooch crucibles were also invented. Porcelain variety is used by beginners. Asbestos used is generally long fibered, white, silky and anhydrous. It is prepared by boiling it with hydrochloric acid. The crucibles can be dried in an oven and muffle furnace to constant weight.

    (b) Munroe crucibles: These are platinum gooch crucibles. These are resistant to chemicals. Finest particles can be filtered, filtration is rapid, and high temperatures can be used for heating. But due to the high cost of platinum, they are not used for routine analysis.

    (c) Glass fibre discs: Inexpensive glass fibre discs are available for use instead of an asbestos mat. A circle of glass fibre filter paper may be placed in a coarse porosity filter crucible for rapid filtration. Glass fibres are made from fine borosilicate glass fibres. They have excellent retention [properties with rapid filtration it remains unaffected by chemical reagents and heating up to 500°C].

(iv) Permanent porous filter discs: Sintered glass crucibles are an example of this type of filter. There is no preparation of filter mat as in the case of Gooch crucibles.

    (a) Sintered glass crucibles: These are made up of resistant glass like pyrex glass and have the porous disc of sintered ground glass fused in the body of the crucible. These filter discs have various porosities (G1, G2, G3, G4). Their pore sizes (average diameter) are 100-120 microns, 40-50 microns, 20-30 microns and 5-10 microns, respectively. The crucible G1 type is used for coarse particles, while the G4 type is for very fine particles.

    (b) Silica crucibles: These are used when drying of precipitates is to be done above 200°C. They can withstand temperatures up to 1000°C. These are resistant to chemicals but affected by phosphates, strong alkalies and hydrofluoric acid.

    (c) Porcelain crucibles: These are glazed inside and outside. The bottom consists of a plate of porous porcelain. They should not be heated directly over the flame.

7. Washing of precipitates: The impurities on the surface of precipitates can be removed by washing precipitates. Also, washing helps in removing the entrapped mother liquor. Many precipitates are washed with a washing solution instead of water. Ideal washing solvent should have no solvent action on precipitate but should dissolve foreign matter easily.

The washing solutions are of the following types:

(a) **Washing with the solution of precipitant:** Precipitates are washed with dilute solutions of precipitant. For example: Lead sulphate precipitate is washed with dilute sulphuric acid.

(b) **Washing with the solution of electrolyte:** Washing is done with electrolytes sometimes to avoid peptization. For example: Silver Chloride precipitates are washed with nitric acid.

(c) **Washing with substances which suppress the hydrolysis of precipitate:** Sometimes precipitates undergo hydrolysis leading to an increase in solubility. It can be prevented by washing with the solution which suppresses hydrolysis, For example, washing of $MgNH_4PO_4$ precipitate of $BaSO_4$ with distilled water.

8. **Drying and ignition of precipitates:** After filtration, a gravimetric precipitate is heated until its mass becomes constant. Heating removes the solvent and any volatile species carried down with the precipitate. Some precipitates are also ignited to decompose the solid and form a compound of known composition. This new compound is often called the weighing form.

The temperature required to produce a suitable weighing form varies from precipitate to precipitate.

Air drying (Ambient temperature):

$MgNH_4PO_4 \cdot 6H_2O$ is dried by washing with a mixture of alcohol and ether and drawing air over the precipitate for a few minutes.

But there may be a danger of incomplete removal of water by washing, so this procedure is not usually recommended.

Air drying (Low temperature):

The precipitates are dried at 100°C–130°C.

For example: Silver chloride precipitates can be dried at this temperature.

Ignition: Ignition at high temperatures is required for the complete removal of water that is occluded or very strongly adsorbed and for the complete conversion of some precipitates to the desired form.

Gelatinous precipitates such as hydrous oxides adsorb water strongly and must be heated to very high temperatures to remove water completely.

For example: Ignition of calcium oxalate to calcium oxide is an example of a chemical change that requires a high temperature for a complete reaction.

Errors during Ignition:

o There may be incomplete removal of water or solvent. If filter paper is employed, the carbon present in it may cause a reduction of precipitates. Substances like AgCl are never filtered on paper. Filtering crucibles are employed.

o Over-ignition may lead to decomposition.

o Reabsorption of water or carbon dioxide by an ignited precipitate on cooling may occur. To avoid this, the crucible should be properly covered and kept in a desiccator as they cool.

Determination of Optimum Drying and Ignition Temperatures: Studies on the ignition temperatures required for different precipitates can be done using thermogravimetric analysis (TGA).

In this method, thermobalance is used, which allows a sample to be weighed while it is in a furnace. The data is recorded in the form of a graph of the weight of the precipitate vs temperature, called a pyrolysis curve. From this curve, the optimum drying temperature can be decided.

9. **Weighing and calculations:** The residue after drying is cooled to room temperature in the desiccator and then weighed accurately on the analytical balance to find out the weighted form.

$$\% \text{ A} = \frac{gA}{g} \text{ sample} \times 100$$

Where

gA = grams of analyte

g sample = grams of sample taken for the analysis. The percent of the sample can be calculated first by knowing the weight, and then percentage calculations are made.

2.7.6 Applications of Gravimetric Analysis

Estimation of Barium Sulphate

Procedure:

Weigh about 0.6 gm in a platinum crucible; add 5 gm of sodium carbonate and 5 gm of potassium carbonate and mix. Heat to 1000°C and maintain at this temperature for 15 minutes. Allow to cool and suspend the residue in 150 ml of water. Wash the crucible with 2 ml of acetic acid and add to the suspension. Cool in ice and filter by decantation, transferring as little of the solid matter as possible to the filter. Wash the residue with successive quantities of a 2 per cent w/v solution of sodium carbonate until the washings are free from sulphate, and discard the washings. Add 5 ml of dilute hydrochloric acid to the filter and wash through into the vessel containing the bulk of the solid matter with water. Add 5 ml of hydrochloric acid and dilute to 100 ml with water. Add 10 ml of a 40 per cent w/v solution of ammonium acetate, 25 ml of a 10 per cent w/v solution of potassium dichromate and 10 gm of urea. Cover, digest in an oven at 80°C to 85°C for 16 hours, and filter while still hot through a sintered-glass filter (porosity No. 4), washing the precipitate initially with a 0.5 per cent w/v solution of potassium dichromate and finally with 2 ml of water. Dry to constant weight at 105°C.

1 gm of the residue is equivalent to 0.9213 gm of $BaSO_4$.

QUESTION BANK

A. MULTIPLE CHOICE QUESTIONS

1. The number of moles of solute dissolved in 1 kg of a solvent is known as ....
   A. Molarity
   B. Molality
   C. Formality
   D. Normality

   Answer: B

2. In volumetric analysis, the substance being titrated is known as ....
   A. Titrant
   B. Titrate
   C. Titer
   D. Filtrate

   Answer: B

3. Gram .... of solute dissolved in one litter of solvent gives 1 Normal solution.
   A. Molecular weight
   B. Formula weight
   C. Equivalent weight
   D. None of the above

   Answer: C

4. One molar sodium hydroxide solution is prepared by dissolving .... gm of sodium hydroxide in 1 lit. of solvent.
   A. 40
   B. 36
   C. 45
   D. 30

   Answer: A

5. % v/v represents ....
   A. Number of ml of solute in 100 gm of solvent
   B. Number of ml of solute in 100 ml of solvent
   C. Number of gm of solute in 100 ml of solvent
   D. Number of gm of solute in 1000 ml of solvent

   Answer: B

6. .... is also known as volumetric analysis.
    A. Gravimetric analysis
    B. Titrimetric analysis
    C. Both (a) and (b)
    D. None of the above

Answer: B

7. .... is prepared in the laboratory for specific analysis.
    A. Primary standard
    B. Secondary standard
    C. Standard substance
    D. All of the above

Answer: C

8. .... records the difference in temperature between a test substance and an inert reference material.
    A. Thermogravimetric analysis
    B. Differential scanning calorimetry
    C. Differential thermal analysis
    D. Gravimetric analysis

Answer: C

9. Methyl orange has a pH range of _____.
    A. 2.2 to 4.0
    B. 3.2 to 4.4
    C. 4.4 to 5.4
    D. 4.2 to 6.2

Answer: B

10. What is the pH range of phenolphthalein?
    A. 3.1–4.4
    B. 8.3–11.0
    C. 4.5–6.3
    D. 6.3–8.3

Answer: B

11. Which of the following is a property of acid?
    A. Sour
    B. Corrosive
    C. Change litmus blue to red
    D. All of the above

Answer: D

12. According to ...., theory acid is an electron acceptor
    A. Arrhenius theory
    B. Lewis theory
    C. Ostwalds theory
    D. Lowry Bronsted theory

Answer: B

13. An Arrhenius acid is defined as a chemical species that.......
    A. Is a proton donor
    B. Is a proton acceptor
    C. Produces hydroxide ions in a solution
    D. Produces hydrogen ions in a solution

Answer: D

14. A solution that resists changes in pH is called as
    A. Buffer solution
    B. Standard solution
    C. Neutral solution
    D. Saturated solution

Answer: A

15. Which of the following is the general property of the base
    A. Taste sour
    B. Turn litmus red
    C. Conduct electric current in solution
    D. The concentration of H_3O^+ is greater than the Concentration of OH

Answer: C

16. PH is given as ....
    A. $pH = Log\ [H^+]$
    B. $pH = -Log_{10}\ [H^+]$
    C. $pH = 1/[H^+]$
    D. $pH = -Log_{10}\ [OH^-]$

Answer: B

17. Which is valid reason/s for non-aqueous titrations
    A. The reactant is insoluble in water
    B. The reactant is reactive with water
    C. The sample is too weak acid or too weak base
    D. All of the above

Answer: D

18. The following are protogenic solvents except for _____.
    A. Acetic acid
    B. Formic acid
    C. Pyridine
    D. Propionic acid

Answer: C

19. Glacial acetic acid is an example of _____.
    A. Protogenic solvent
    B. Protophillic solvent
    C. Amphiprotic solvent
    D. Aprotic solvent

Answer: C

20. The assay of _____ is performed by non-aqueous titration.
    A. Magnesium sulphate
    B. Barium sulphate
    C. Ephedrine hydrochloride
    D. Benzoic acid

Answer: C

21. Primary standard used for the Standardization of 0.1 N Perchloric Acid is
    A. Potassium hydrogen phthalate
    B. Granulated zinc
    C. Calcium carbonate
    D. Magnesium sulphate

Answer: A

22. Solvent as an acid on titration with base and act as a base on titration with acid is known as
    A. Protogenic
    B. Protophilic
    C. Aprotic
    D. Amphoteric

Answer: D

23. Which solvent can be used in non-aqueous titration
    A. Acetic acid
    B. Glacial acetic acid
    C. Semi-normal acetic acid
    D. Water

Answer: B

24. Which of the following is used as an indicator in Mohr's method?
    A. Starch
    B. Ferric Ammonium Sulphate
    C. Phenolphthalein
    D. Potassium Chromate

Answer: D

25. .... gm of $AgNO_3$ in 1 litre distilled water gives 0. 1 M $AgNO_3$ solution.
    A. 40
    B. 17
    C. 1.7
    D. 4.0

Answer: B

26. Which of the following is used as an indicator in Volhard's method?
    A. Ferric Ammonium Sulphate
    B. Eosin
    C. Fluorescin
    D. Potassium Chromate

Answer: A

27. Adsorption indicators are used in.............
    A. Mohr's method
    B. Volhard's method
    C. Fajan's method
    D. Modified Volhard's method

Answer: C

28. EDTA is a .... ligand.
    A. Tetradentate
    B. Bidentate
    C. Hexadentate
    D. Tridentate

Answer: C

29. EDTA is standardised using...........
    A. Magnesium
    B. Zinc
    C. Calcium
    D. Manganese

Answer: B

30. Erichrome Black-T in complex with metal gives .... colour
    A. Red
    B. Blue
    C. Violet
    D. Green

Answer: A

31. Reduction means ....
    A. Removal of oxygen
    B. Removal of hydrogen
    C. Addition of oxygen
    D. Loss of electrons

Answer: A

32. Equivalent weight of $KMnO_4$ in the acidic condition is ....
    A. 52.66
    B. 158
    C. 31.6
    D. 108

Answer: C

33. In the titration of oxalic acid using $KMnO_4$ in an acidic medium, the best indicator is...........
    A. Starch
    B. Phenolphthalein
    C. Diphenylamine
    D. $KMnO_4$

Answer: D

34. Iodine can be standardised using..............
    A. Arsenic trioxide
    B. Sodium thiosulphate
    C. Both a and b
    D. None of the above

Answer: C

35. For gravimetry________ filter papers are used
    A. Ashless
    B. Ash
    C. Normal
    D. None of the above

Answer: A

B. SHORT ANSWER QUESTIONS

1. Define non-aqueous titration.
2. Enlist the different types of solvents used in the non-aqueous titration.
3. Give examples of Protophilic Solvents.
4. Give examples of Protogenic Solvents.
5. Give the Properties of a Non-aqueous Solvent.
6. What are precipitation titrations?
7. Which indicators are used in Fajan's method?
8. What is the concentration of potassium chromate indicator required for Mohr's method?
9. Name some adsorption indicators.
10. How will you prepare a 0.1 M $AgNO_3$ solution?
11. What are complexometric titrations?
12. What are ligands?
13. Which ligand is widely used in complexometric titrations?
14. Name indicators used in complexometric titrations.
15. Which primary standard is used to standardise disodium edetate?
16. Give the difference between iodometry and iodimetry.
17. How will you prepare and standardise 0.1 N sodium thiosulphate?
18. How will you prepare and standardise the 0.1 N Iodine solution?
19. Give applications of gravimetric analysis.
20. What are the advantages of the gravimetric method of analysis?

C. LONG ANSWER QUESTIONS

1. Draw neutralisation curves for:
   Strong Acid vs Strong Base.
   Strong Acid vs Weak Base.
   Weak Acid vs Strong Base.
2. What is acid base titration? Draw neutralisation curves with suitable examples.
3. Write a note on Non-aqueous Titration.
4. Give the preparation and standardisation of 0.1 N Perchloric acid.
5. Write a note on Mohr's method.
6. Write a note on Volhard's method.
7. What is Fajan's method of precipitation titration? Give its limitations.

8. Give applications of precipitation titration.
9. Explain the concepts of oxidising agents and reducing agents in redox titrations.
10. Write a note on Permanganometry.
11. Write a note on Common Indicators for Detection Endpoint in Redox Titration.
12. Write a note on Iodometry
13. Write a note on Iodimetry
14. Write a note on unit operations in gravimetry.
15. Write a note on Gravimetry.

■■■

HAEMATINICS

CONTENTS

Pharmaceutical formulations, market preparations, storage conditions and uses of inorganic haematinic products:

- Ferrous sulphate,
- Ferrous fumarate,
- Ferric ammonium citrate,
- Ferrous ascorbate,
- Carbonyl iron

♦ LEARNING OBJECTIVES ♦

After completing this chapter, the student should be able to understand:

1. The term haematinic; the role of haematinic products
2. Chemical Names of various inorganic haematinic products
3. Chemical Structures of different inorganic haematinic products
4. Uses, Stability and storage conditions of inorganic haematinic products

3.1.1 INTRODUCTION

A haematinic is a nutrient required for the formation of blood cells in the process of hematopoiesis [the formation of blood cellular components]. In other words, haematinics are substances that are essential to the proper formation of the components of blood. Examples of hematinics include folic acid, vitamin B_{12}, and iron. In addition, vitamin D, which helps maintain the health of bones the reservoirs of new blood cells may also have a role in protecting haemoglobin and stimulating the formation of new blood cells. The deficiency of haematinics can lead to anaemia [deficiency of red cells or haemoglobin in the blood]. In cases of anaemia, haematinics can be administered as medicines in order to increase the haemoglobin content of the blood.

In short, haematinics are the agents used for the formation of blood to treat various types of anaemia as an iron supplement used to treat or prevent low blood levels of iron (such as those caused by anaemia or pregnancy). Iron is an important mineral that the body needs to produce red blood cells and keep one in good health. Iron is mostly absorbed from the duodenum and upper jejunum.

Erythropoietin (EPO) is a hormone that stimulates erythropoiesis, which can also be given as a medicine to increase the haemoglobin content of the blood, but EPO is not classified as a hematinic.

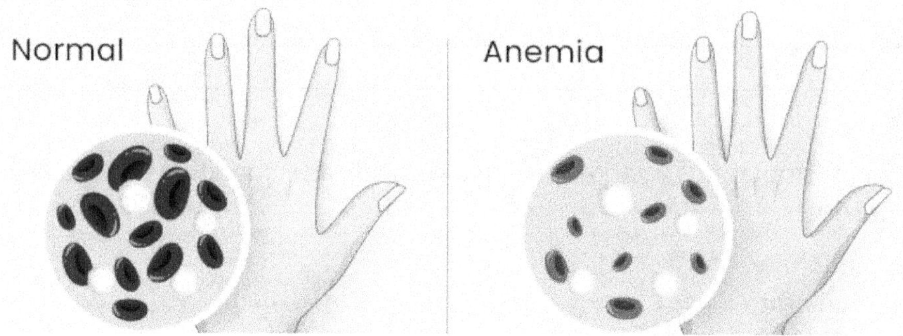

Fig. 3.1.1 Anemia

Anaemia is a condition in which one lacks enough healthy red blood cells to carry adequate oxygen to body's tissues. Having anaemia, also referred to as low haemoglobin, can make a person feel tired and weak. There are many forms of anaemia, having their own causes.

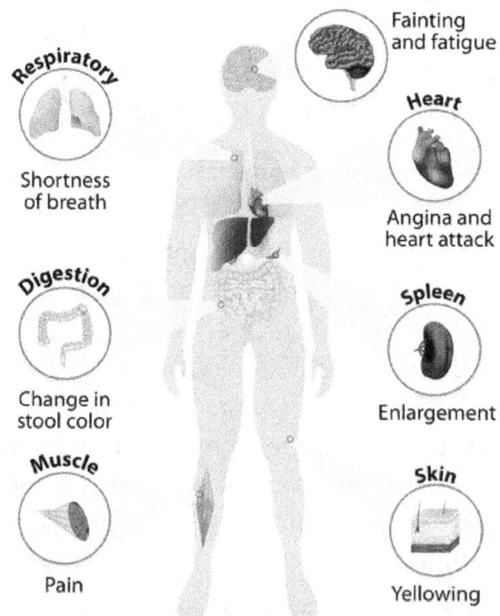

Fig. 3.1.2 Symptoms of anaemia

Symptoms: Following are the commonly associated symptoms with anaemia

- Fatigue.
- Weakness.
- Pale or yellowish skin.
- Irregular heartbeats.
- Shortness of breath.
- Dizziness or lightheadedness.
- Chest pain.
- Cold hands and feet.

Types of Anemia: The different types of anaemia are linked to various diseases and conditions. Actually, many types of anaemia exist, but the four main types are

- Iron-deficiency anaemia: a condition in which blood lacks adequate healthy red blood cells,
- Pernicious anaemia: the inability of the body to properly utilise vitamin B_{12}, which is essential for the development of red blood cells.
- Aplastic anaemia: a condition that occurs when the body stops producing enough new blood cells.
- Hemolytic anaemia: a disorder in which red blood cells are destroyed faster than they can be made

3.1.2 HAEMATINICS

Herein, we are to discuss the properties, preparation, applications, purity, formulations and market preparations of mainly some inorganic iron-containing haematinic products.

3.1.2.1 Ferrous Sulphate (I.P.): (Green Vitriol, Iron Vitriol): Molecular formula: $FeSO_4, 7H_2O$;

Molecular Weight: 278.02.

It should contain NLT 98.0% and NMT 103.35% of $FeSO_4, 7H_2O$.; Percentage iron content = 36.8

Preparation

Prepared by dissolving iron dust or filings in excess of dilute H_2SO_4. The iron dissolves with effervescence. When the reaction subsides, the liquid is boiled to concentrate, filtered and cooled after that to obtain the crystals, which are filtered and dried at room temperature.

$$Fe + H_2SO_4 + nH_2O \longrightarrow FeSO_4, 7H_2O + H_2O + H_2$$

Properties

It is a pale, bluish green, crystalline or granular solid, which is odourless with a saline metallic, astringent taste. It is efflorescent in dry air and may get oxidised to ferric salt

(Fe4(OH)2(SO4)3), having brownish colour. It is acidic to litmus with a pH of about 3.7. It is soluble in water but practically insoluble in alcohol. When heated, it decomposes to ferric oxide, sulphur dioxide and sulphuric acid.

$$2(FeSO_4, 7H_2O) \longrightarrow Fe_2O_3 + SO_2 + H_2SO_4 + 13H_2O$$

Applications

Used most widely in the form of oral iron preparations like tablets, capsules, syrups, reconstitutable dry powders, etc., as a hematinic (promoting the formation of haemoglobin) to treat iron deficiency, particularly in uncomplicated anaemias.

Dose: 200 mg 2-3 times/day

Brand: FEROBID-Z Capsule (Sterkem Pharma Pvt. Ltd)

Tests for Purity

1. Identification (reactions of Fe and SO_4).

2. Acidity (1% w/v solution in water requires NMT 1.0 ml of 0.1 N NaOH, Indicator: Methyl orange),

3. Tested for heavy metals, As, Cu etc.

4. Test for basic sulphates (Insoluble sulphates): May be detected by the presence of turbidity in an aqueous solution.

Assay

By Redox Titration: Aqueous sample solution is titrated against 0.1 N KMnO4 solution in the presence of dil. H_2SO_4 to a coloured endpoint.

$$2\ FeSO_4 + 2KMnO_4 + 4H_2SO_4 \longrightarrow K_2SO_4$$

3.1.2.2 Ferrous Fumarate: (Iron(II) fumarate; Feostat); IUPAC Name: Iron (2^+) $(2E)$-but-2-enedioate; Molecular formula: $C_4H_2FeO_4$; Mol. wt: 169.90.

Pure ferrous fumarate has an iron content of 32.87%; therefore, one tablet of 300 mg of iron fumarate will contain 98.6 mg of iron (548% Daily Value based on 18 mg RDI).

Preparation

Anhydrous ferrous fumarate is prepared by mixing a water-soluble ferrous salt and a water-soluble salt of fumaric acid in an aqueous medium at a temperature over about 70 °C

(a)

+ Ferrous $\longrightarrow$

(b)

Fe^2

Properties: It occurs as a reddish-orange odourless powder, slightly soluble in water and has a melting point: 280°C.

Fe^{2+}

Applications

1. Ferrous fumarate is used to treat iron deficiency anaemia (a lack of red blood cells caused by having too little iron in the body). Used most widely as in the form of oral iron preparations like tablets, capsules, syrups, reconstitutable dry powders, etc., as a hematinic (promoting the formation of haemoglobin) to treat iron deficiency, particularly in uncomplicated anaemias. It replaces iron in the body when the body does not produce enough on its own.

2. It is recommended for use in the fortification of foods for infants and young children.

3. It is also used in ink and dye manufacturing, as well as agriculture.

Dose:

Adult: ferrous fumurate 325 mg and folic acid 0.35 mg, 2-3 times/day

Premature neonates: 2 to 4 mg elemental iron/kg/day divided every 12 to 24 hours (maximum daily dose = 15 mg).

Infants and children <12 years: Prophylaxis: 1 to 2 mg elemental iron/kg/day (maximum 15 mg) in 1 to 2 divided doses.

Brand: AUTRIN Capsule (Wyeth Limited)

Tests for Purity

1. Identification To 1.5 g, add 25 mL of dilute hydrochloric acid (1 in 2). Dilute with water to 50 mL, heat to dissolve, then cool, filter on a fine-porosity, sintered-glass crucible, wash the precipitate with dilute hydrochloric acid (3 in 100), saving the filtrate for Identification test B, and dry the precipitate at 105 °C the IR absorption of a potassium bromide dispersion of the dried precipitate so obtained exhibits maxima only at the same wavelengths as that of a similar preparation of USP Fumaric Acid RS

2. A portion of the filtrate obtained in the preceding test responds to the tests for Iron

3. Tested for heavy metals, As, Cu etc.

Assay

Transfer 500 mg of Ferrous Fumarate, accurately weighed, to a 500-mL conical flask, and add 25 mL of dilute hydrochloric acid (2 in 5). Heat to boiling, add a solution of 5.6 g of stannous chloride in 50 mL of dilute hydrochloric acid (3 in 10) dropwise until the yellow colour disappears, then add 2 drops in excess. Cool the solution in an ice bath to room temperature, add 10 mL of mercuric chloride solution (1 in 20), and allow it to stand for 5 minutes. Add 200 mL of water, 25 mL of dilute sulfuric acid (1 in 2), and 4 mL of phosphoric acid, then add 2 drops of orthophenanthroline TS, and titrate with 0.1 N ceric sulfate VS. Perform a blank determination and make any necessary correction.

Each mL of 0.1 N ceric sulfate is equivalent to 16.99 mg of $C_4H_2FeO_4$.

3.1.2.3 Ferric Ammonium Citrate: (Ammonium ferric citrate; Ammonium iron(III) citrate; FerriSeltz); IUPAC Name: 2-Hydroxypropane-1,2,3-tricarboxylate, ammonium iron(3⁺) salt; Molecular formula: $(NH_4)_5[Fe(C_6H_4O_7)_2]$; Molecular Weight: 261.98. Pure ferric ammonium citrate is a complex salt of an undetermined structure composed of 16.5 to 18.5 percent iron, approximately 9 percent ammonia, and 65 percent citric acid.

Preparation: Ferric ammonium citrate (iron (III) ammonium citrate) is prepared by the reaction of ferric hydroxide with citric acid, followed by treatment with ammonium hydroxide, evaporating, and drying. The resulting product occurs in two forms depending on the stoichiometry of the initial reactants.

(1) Ferric ammonium citrate (iron (III) ammonium citrate, CAS Reg. No. 1332-98-5) is a complex salt of an undetermined structure composed of 16.5 to 18.5 percent iron, approximately 9 percent ammonia, and 65 percent citric acid and occurs as reddish brown or garnet red scales or granules or as a brownish-yellowish powder.

(2) Ferric ammonium citrate (iron (III) ammonium citrate, CAS Reg. No. 1333-00-2) is a complex salt of an undetermined structure composed of 14.5 to 16 percent iron, approximately 7.5 percent ammonia, and 75 percent citric acid and occurs as thin transparent green scales, as granules, as a powder, or as transparent green crystals.

Properties: It occurs as reddish brown or garnet red scales or granules or as a brownish-yellowish powder, with a faint ammonia odour. It is soluble in water.

Applications:
1. As a hematinic
2. As a food ingredient and as an acidity regulator

Dose: Adult: 210 mg (1 tablet) *p, o, t. i. d.,* with meals for Iron Deficiency Anemia;

Brand: HAEM-UP Syrup (Cadila Pharmaceuticals Ltd.)

Tests for Purity:

Ferric Ammonium Citrate contains not less than 16.5 percent and not more than 18.5 percent of iron(Fe^{+3}).

Test A: Ignite about 0.5 g: it chars and leaves a residue of iron oxide.

Test B: To 10 mL of a solution of Ferric Ammonium Citrate (1 in 100), add 6 N ammonium hydroxide dropwise: the solution darkens, but no precipitate forms.

Test C: To 5 mL of a solution of Ferric Ammonium Citrate (1 in 100), add 0.3 mL of potassium permanganate TS and 4 mL of mercuric sulfate TS, and heat the mixture to boiling: a white precipitate forms.

Ferric citrate: To a solution of Ferric Ammonium Citrate (1 in 100), add potassium ferrocyanide TS: no blue precipitate is formed.

Assay: Transfer about 1 g of Ferric Ammonium Citrate, accurately weighed, to a 250-mL conical flask and dissolved in 25mL of water and 5 mL of hydrochloric acid. Add 4 g of potassium iodide, insert the stopper, and allow to stand protected from light for 15 minutes. Add 100 mL of water, and titrate the liberated iodine with 0.1 N sodium thiosulfate

VS, using starch TS as the indicator. Perform a blank determination, and make any necessary corrections.

Each mL of 0.1 N sodium thiosulfate is equivalent to 5.585 mg of iron (Fe).

3.1.2.4 Ferrous ascorbate: (Ferrous cevitamate, Iron(2$^+$) L-ascorbate; L-Ascorbic acid, iron complex; (+)-Iron(II) L-ascorbate), Molecular formula: $C_{12}H_{14}FeO_{12}$ Molecular Weight: 406.08, IUPAC Name: (2*R*)-2-[(1*S*)-1,2-dihydroxyethyl]-3,4-dihydroxy-2*H*-furan-5-one;(2*R*)-2-[(1*S*)-1,2-dihydroxyethyl]-5-oxo-2*H*-furan-3,4-diolate; iron(2$^+$); Percentage iron content = 13.75

Ferrous ascorbate functions as a catalyst in various metabolic activities in our bodies, accelerating oxygen delivery and use and aiding cell development and proliferation. This promotes the synthesis of red blood cells and haemoglobin.

Ferrous Ascorbate is a combination of vitamin C and iron. Ferrum or iron helps to replenish the iron stores in the human body. Ascorbate contains vitamin C, which improves iron levels in the body by increasing its absorption from different sources. Ferrous ascorbate also helps to produce an increased number of red blood cells and haemoglobin. It increases oxygen supply to various parts of the body and promotes cell growth and proliferation. A six-carbon compound related to glucose. It is found naturally in citrus fruits and many vegetables. Ascorbic acid is an essential nutrient in human diets and necessary to maintain connective tissue and bone. Its biologically active form, vitamin C, functions as a reducing agent and coenzyme in several metabolic pathways. Vitamin C is considered an antioxidant.

Properties: It is a grey solid melting > 400°C.

Preparations: Earth metal salts or their hydroxides are reacted with ferrous sulfate to get corresponding ferrous salts, which are reacted with ascorbic acid in an aqueous medium at slightly acidic or neutral conditions, followed by filtration to get ferrous ascorbate in the mother liquor. Alternatively, ascorbic acid salt can be treated with ferrous salt followed by the filtration of the same to get ferrous ascorbate in the mother liquor. The said mother liquor, obtained by any means, can be used for spray drying to get ferrous ascorbate.

Optionally, ferrous salts are isolated by filtration and then subjected to reaction with ascorbic acid in an aqueous medium at a slightly acidic or neutral medium, followed by spray drying to get ferrous ascorbate.

Applications

1. Ferrous Ascorbate is used to treat iron deficiencies in the body. Thus, it acts as a health supplement for those with low iron content in their diet. Iron increases the production of red blood cells and haemoglobin. Vitamin C present in Ferrous Ascorbate aids in the effective and maximum absorption of dietary iron. Ferrous Ascorbate is the main treatment agent for haemoglobin disorders (anaemia) related to iron deficiency and underlying chronic kidney disorders.

2. Can be used in infant formula

3. As food supplement

4. Ferrous Ascorbate is better when compared to other iron supplements, as it causes minimal gastrointestinal tract problems. Also, it is easily absorbed in the human body with less troublesome interactions with dietary products and other medicines.

5. Ferrous Ascorbate is also suitable for long-term use as compared to other iron supplements.

Test for identity: Dissolve a quantity of the powdered tablets containing 10 mg of Iron in 2 ml of water, and add 1 ml of potassium ferricyanide solution; a blue precipitate is formed that does not dissolve on the addition of 5 ml of dilute hydrochloric acid.

Assay: Ferrous ascorbate equivalent to elemental iron is determined by UV Spectroscopy

Brands: LIVOGEN-CZ tablets (Merck); FERIKIND tablets (Mankind), CheriXT (Dhani)

FEROCAD injection (Cadila Pharmaceuticals).

Dose: In India, ferrous ascorbate tablets (each tablet contains 100 mg elemental iron as ferrous ascorbate, with 1 mg folic acid) in the recommended dose of 1 tablet daily is followed.

3.1.2.5 Carbonyl iron: (Pentacarbonyl iron); Molecular formula: $Fe(CO)_5$; Molecular Weight: 195.90;

Simply put, elemental iron is defined as the amount of iron available for absorption by your body. Not all irons are 100% absorbable. For example, ferrous gluconate is 12% elemental iron, while carbonyl is 100% elemental iron. That's why carbonyl is often referred to as "pure iron."

Carbonyl iron is a highly pure (97.5% for grade S, 99.5+% for grade R) iron, prepared by chemical decomposition of purified iron pentacarbonyl. It usually has the appearance of a grey powder composed of spherical microparticles. It is not a salt. It absorbs at 69%, compared to only 10 to 15%, with most other forms of iron. Most of the impurities are carbon, oxygen, and nitrogen.

Properties: It is a straw-yellow to brilliant orange liquid or a greyish powder with a musty odour.

Preparation: $Fe(CO)_5$ is produced by the reaction of fine iron particles with carbon monoxide.

The industrial production of this compound is somewhat similar to the Mond process in that the metal is treated with carbon monoxide to give a volatile gas. In the case of iron pentacarbonyl, the reaction is more sluggish. It is necessary to use an iron sponge as the starting material and harsher reaction conditions of 5–30 MPa of carbon monoxide and 150–200 °C. Similar to the Mond process, sulfur acts as a catalyst. The crude iron pentacarbonyl is purified by distillation.

1. $FeI_2 + 4\ CO \rightarrow Fe(CO)_4I_2$

2. $5\ Fe(CO)_4I_2 + 10\ Cu \rightarrow 10\ CuI + 4\ Fe(CO)_5 + Fe$

Applications: Carbonyl iron powder is used to treat iron deficiency and as an iron dietary supplement.

Brand: Cofol-Z capsules (CIPLA); Fefol-z capsules (GSK)

Dose: As a dietary supplement:

Male: 8 mg orally once daily; Female: 18 mg orally once daily; Pregnant female: 27 mg orally once daily; Lactating female 9 mg orally once daily

For Iron Deficiency Anemia

Adult: 300 mg orally every 12 hours; may increase to 300 mg every 6 hours or 250 mg ER orally every 12 hours in case of severe iron deficiency anaemia,

Pediatric: 4-6 mg/kg orally divided every 8 hours.

QUESTION BANK

A. MULTIPLE CHOICE QUESTIONS

1. _______________ increases the absorption of iron
    A. Vitamin A
    B. Vitamin C
    C. Folic acid
    D. Pyridoxin

    Answer: C

2. _____________ can be given intravenously
    A. Ferrous sulfate
    B. Ferric ammonium citrate
    C. Ferrous fumarate
    D. Ferrous ascorbate

    Answer: D

3. Iron is mainly absorbed from _______________
    A. duodenum and upper jejunum
    B. lower jejunum
    C. stomach
    D. ileum

    Answer: A

4. Match the following drugs and their common names
    A. Ferrous Sulphate   (i)   Ferrous cevitamate
    B. Ferrous fumarate   (ii)   FerriSeltz
    C. Ferric ammonium citrate   (iii) Green Vitriol
    D. Ferrous ascorbate   (iv)   Feostat

    Answer:
    (a) (iii)
    (b) (iv)
    (c) (ii)
    (d) (i)

5. Match the following drugs and their famous brands
    A. Ferrous Sulphate   (i)   FEROBID-Z Capsule
    B. Ferrous fumarate   (ii)   AUTRIN Capsule
    C. Ferric ammonium citrate   (iii) HAEM-UP Syrup

D. Ferrous ascorbate   (iv)   LIVOGEN-CZ tablets

Answer:

(a) (i)

(b) (ii)

(c) (iii)

(d) (iv)

6. Match the following drugs and their appearance

    A. Ferrous Sulphate   (i)   reddish-brown scales or granules

    B. Ferrous fumarate   (ii)   grey solid

    C. Ferric ammonium citrate   (iii) reddish-orange powder

    D. Ferrous ascorbate   (iv) bluish green solid

Answer:

(a) (iv)

(b) (iii)

(c) (i)

(d) (ii)

7. Ferrous iron (Fe^{2+}) is recommended over ferric iron (Fe^{3+}) for hematinic purposes due to

    A. better absorption

    B. efficacy

    C. lesser side effects

    D. all of above

Answer: A

8. Match the following iron salts and their % iron content

    A. Ferrous Sulphate   (i)   36.8

    B. Ferrous fumarate   (ii)   32.87

    C. Ferric ammonium citrate   (iii) 16-18.5

    D. Ferrous ascorbate   (iv) 13.75

Answer:

(a) (i)

(b) (ii)

(c) (iii)

(d) (iv)

9. The molecular formula of Carbonyl Iron is___
    A. $Fe(CO)_3$
    B. $Fe(CO)_4$
    C. $Fe(CO)_6$
    D. $Fe(CO)_5$

Answer: D

10. Match the following

A. Iron-deficiency anaemia	(i)	lack of adequate healthy red blood cells,
B. Pernicious anaemia	(ii)	Body is unable to utilise vitamin B_{12}, essential for the development of RBCs.
C. Aplastic anaemia	(iii)	the body stops producing enough new blood cells.
D. Hemolytic anemia	(iv)	a disorder in which red blood cells are destroyed faster than they can be made.

Answer:
(a) (i)
(b) (ii)
(c) (iii)
(d) (iv)

B. SHORT ANSWER QUESTIONS

1. Write applications of any three hematinic agents
    (a) Ferrous sulphate
    (b) Ferrous fumarate.
    (c) Ferric ammonium citrate
    (d) Ferrous ascorbate.
    (e) Carbonyl Iron

2. Write preparation and properties of any two hematinic agents
    (a) Ferrous sulphate.
    (b) Ferrous fumarate.
    (c) Ferric ammonium citrate
    (d) Ferrous ascorbate.

3. Discuss in detail the Pentcarbonyl iron

C. LONG ANSWER QUESTIONS

1. Discuss the term anaemia and its types and give an account of various inorganic iron-containing haematinic compounds.

■■■

GASTRO-INTESTINAL AGENTS

CONTENTS

Pharmaceutical formulations, market preparations, storage conditions and uses of inorganic

- Gastro-intestinal Agents: Introduction
- Acidifying agents
- Antacids:
  - Aluminium hydroxide gel,
  - Magnesium hydroxide,
  - Magaldrate,
  - Sodium bicarbonate,
  - Calcium Carbonate;
- Adsorbents;
- Protectives;
- Cathartics

♦ LEARNING OBJECTIVES ♦

After completing this chapter, the student should be able to understand:

1. Classification of various inorganic Gastrointestinal Agents
2. Chemical Names of various inorganic Gastrointestinal Agents
3. Chemical Structures of different inorganic Gastrointestinal Agents
4. Uses, Stability and storage conditions of inorganic Gastrointestinal Agents
5. Different types of formulations and popular brand names of inorganic Gastrointestinal Agents

3.2.1 GASTRO-INTESTINAL AGENTS: INTRODUCTION

The Gastro-Intestinal Tract (G.I.T.) is one of the important body systems comprising of many organs. These include the oral cavity of the mouth, oesophagus, stomach, small intestine (duodenum, jejunum, and ileum), large intestine (cecum, colon, and rectum) and anus. Also present are accessory organs contributing to the digestive process, such as the salivary glands, pancreas, gallbladder, liver etc.

Each of these organs plays an important role in the uptake as well as digestion of food. The system comprising these organs has a variety of enzymes involved in the breakdown and digestion of foodstuff. The main function of this system is the digestion of food particles and absorption of digestive contents (nutrients, electrolytes, minerals, and fluids) - into the circulatory system for cellular use. Undigested material passes through the lower intestinal tract with the aid of peristalsis to the rectum and anus and is excreted as faeces or stool.

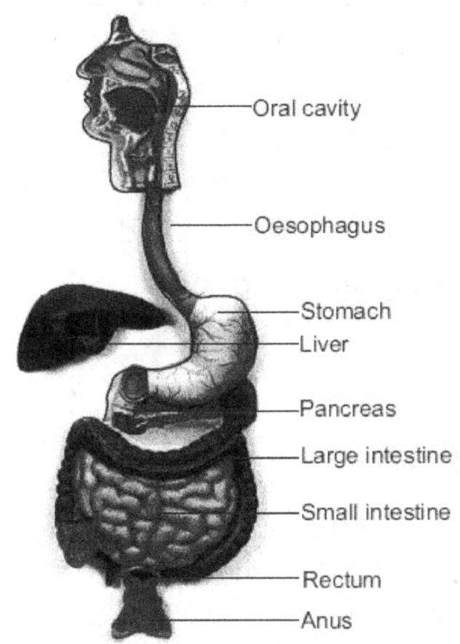

Fig. 3.2.1 *The General Structure of the Digestive System*

The digestive system comprises of two types of organs:

   (a)   Digestive organs

   (b)   Accessary digestive organs

Functions of the Digestive System

1. To ingest food
2. To transport food
3. To digest food into simpler components that can be absorbed and utilised by the body.
4. To absorb necessary nutrients into the bloodstream.
5. To expel waste from the body

Any disorder or dysfunction of any of these component organs leads to a variety of symptoms and ailments. It is well said and established that a healthy GIT system contributes greatly to the well-being of a person. There are a variety of disorders and ailments associated with the GIT, ranging from minor to severe, and there are many therapies, including drug therapy, to cure them. Some minor symptomatic ailments are curable by the use of some drugs of inorganic compositions. Most of these drugs are available over the counter (OTC) without prescription.

Therefore, they need to be dealt with care by the pharmacist in particular. The pharmacist should be able to judge between a minor ailment and a serious ailment and accordingly decide on the usage of these drugs.

These drugs of inorganic origins are mainly divided into five main categories (thus, in the scope of this chapter), namely; Inorganic agents used to treat gastrointestinal disorders include-

1. Products for altering gastric pH: Acidifying Agents: Used to treat achlorhydria (absence of HCl in the gastric secretion), e.g. Dilute HCl.

2. Antacids: Used to treat hyperchlorhydria and peptic ulcers.

3. Protectives for intestinal inflammation.

4. Adsorbents for intestinal toxins.

5. Cathartics or laxatives for constipation.

3.2.2 ACIDIFYING AGENTS

The process of gastric acid secretion occurs at the level of the parietal cells of oxyntic glands in the gastric mucosa of the stomach through the stimulation of any of the three receptors; muscarinic -M_1 receptors, gastrin receptors, histamine-H_2 receptors, producing 2-3 litres of gastric juice per day (HCl of pH 1).

Acid secretion is a physiologically important process of the stomach as;

1. Gastric acid induces activation of pepsinogen, converting it into pepsin (protolytic enzyme). This enzyme is formed at the acidic pH and is responsible to initiate the digestive process.

2. Gastric acid kills bacteria and other microbes, ensuring a stable intragastric environment.

3. Gastric acid softens the fibrous food and converts it into forms suitable for digestion.

Thus, if there is a lack of sufficient gastric acid in the stomach, it can cause GIT disturbances. Achlorhydria or anacidity is the condition which occurs due to the absence of HCl in gastric secretion.

There are two types of achlorhydrias:

1. Gastric secretion is devoid of HCl even after the stimulation with histamine phosphate. This condition occurs in gastritis or gastrectomy.

2. In the second case, there is a lack of HCl, but it may be secreted upon the administration of histamine phosphate. This second condition occurs in chronic nephritis, alcoholism etc.

Symptoms of achlorhydria range from mild diarrhoea, epigastric pain, sensitivity to spicy/pungent food, to even pernicious anaemia due to lack of intrinsic factor, leading to poor absorption of B_{12}.

3.2.2.1 Dilute Hydrochloric Acid I.P. (HCl: M. Wt. 36.46)

It is given in the treatment of achlorhydria. It contains 10% w/w HCl. This dilute HCl is prepared by mixing 274 g of HCl and 726 g of purified water. It occurs as a clear, colourless, and odourless liquid strongly acidic to litmus. The acid should be diluted with 50 volumes of water or juice and sipped through a glass tube to prevent its direct action on the dental enamel.

Assay: An accurate amount of HCl is transferred to a conical flask containing 40 ml of water, and the solution is titrated with 1 N NaOH using methyl orange as an indicator.

Each one ml of 1N NaOH is equivalent to 0.036 g of HCl.

Use: Dilute HCl is used as a gastric acidifier when the level of HCl in the gastric juice is low. Since it is not clear whether the decrease of gastric HCl content is due to any specific cause or is a symptom associated with many physiological conditions, it is doubtful whether the administration of dilute hydrochloric acid will serve any useful clinical purpose.

3.2.3 ANTACIDS

The walls of the stomach are lined with cells that secrete mucus, pepsinogen and hydrochloric acid. The hydrochloric acid concentration of the stomach ranges from 0.03 M to 0.003 M, corresponding to a pH range of about 1.5 to 2.5. The mucus lining of the stomach protects the stomach walls from the action of stomach acid. When excess acid is produced, a condition known as acid indigestion results. If excess acid is forced into the oesophagus, acid reflux or "heartburn" results. High acid concentrations can damage the stomach lining resulting in ulcers. Excess stomach acid results in a state of discomfort known as acid indigestion or heartburn. If the condition remains neglected, it may lead to peptic ulcers of various types. Today hyperacidity is linked mainly to "Hurry, Worry & Curry", though there can also be other pathological reasons.

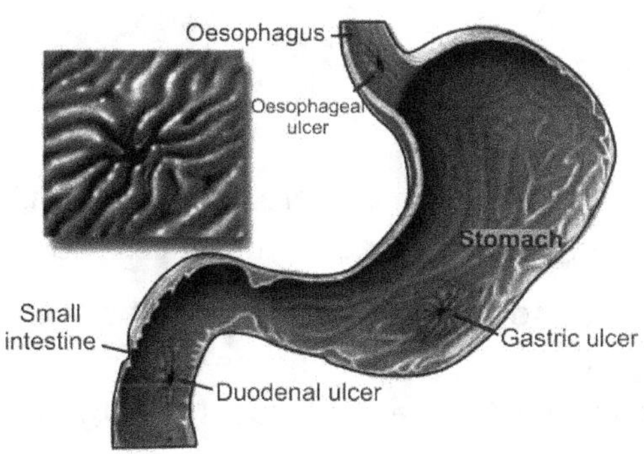

Fig. 3.2.2 Different types of ulcers due to hyperacidity

Acid indigestion may result from a variety of factors, including -

- Overeating
- Alcohol consumption
- Eating certain foods
- Anxiety
- Smoking
- Certain drugs, like NSAIDs, i.e., Aspirin, Ibuprofen, Diclofenac etc.

Compounds employed in therapy to treat these conditions are known as antacids. Antacids give relief from pain due to hyperacidity. The beneficial effects of antacids are due to two mechanisms, i.e., neutralisation of gastric juice and reduction in protolytical activity of pepsin. Elevation of the pH of the gastric content to more than pH 4-5 results in the inactivation of the enzyme pepsin, and at this pH, the damaging effect of the gastric acid on the mucosa is minimum.

Characteristics of an ideal antacid:

1. It should not be absorbed in the systemic circulation
2. It should not cause systemic alkalosis.
3. It should neither be a constipation nor laxative.
4. It should produce a rapid and long-lasting effect.
5. Its reaction with gastric acid should not cause the evolution of large quantities of CO_2 gas.
6. It should be palatable and inexpensive.
7. It should inhibit pepsin

Classification of Antacids: These are broadly classified into three categories;

- Inorganic Antacids
- H_2 - Receptor Antagonists
- Proton Pump Inhibitors (PPI)

Alternatively, antacids are classified as follows:

(1) On a pharmacological basis:

    (a) Systemic or absorbable: These are soluble, readily absorbable and capable of producing systemic electrolyte disturbances. E.g., $NaHCO_3$.

    (b) Non-systemic or non-absorbable: They are not absorbed and act exclusively in the GIT. E.g., Calcium containing (e.g., $CaCO_3$), Aluminium containing (e.g., $Al(OH)_3$), Magnesium containing (e.g., $MgCO_3$).

(2) On the basis of inorganic elements:

(a) Aluminium-containing antacids: E.g., Aluminium hydroxide gel, Aluminium phosphate gel etc.

(b) Calcium containing antacids: $CaCO_3$, tribasic Calcium phosphate.

(c) Mg containing antacids: Mg trisilicate, $MgCO_3$, milk of magnesia ($Mg(OH)_2$), MgO etc.

(d) Na-containing antacids: $NaHCO_3$.

(e) Combined antacids –

(i) Aluminium hydroxide and $Mg(OH)_2$ gel.

(ii) Aluminium hydroxide gel and Mg trisilicate.

The inorganic compounds have been used as antacids for a long time in therapy, and they still enjoy the faith of physicians and patients for symptomatic relief. These are weak bases that are used to neutralise excess stomach acid.

Reactions of Inorganic Antacids

Antacids react with HCl in the stomach. Some common antacid reactions include –

$$CaCO_3 + 2HCl \rightarrow CaCl_2 + H_2O + CO_2$$

$$NaHCO_3 + HCl \rightarrow NaCl + H_2O + CO_2$$

$$Al(OH)_3 + 3HCl \rightarrow AlCl_3 + 3H_2O$$

$$Mg(OH)_2 + 2HCl \rightarrow MgCl_2 + 2H_2O$$

$$MgO + 2HCl \rightarrow MgCl_2 + H_2O$$

Some of the common antacids are –

3.2.3.1 Aluminum Hydroxide Gel (IP): Molecular formula: $Al(OH)3$, M. Wt: 78.

It is a hydrated suspension of aluminium oxide with variable amounts of basic aluminium carbonate. It contains not less than 3.5 % and not more than 4.4% w/w Al_2O_3.

Preparation

It is prepared by the addition of a hot solution of potash alum to a hot solution of sodium carbonate.

$$3Na_2CO_3 + KAl(SO_4)_2 + 3H_2O \rightarrow 3Na_2SO_4 + K_2SO_4 + 3CO_2 + 2Al(OH)_3$$

potash alum

Properties

Aluminium hydroxide gel is a white viscous suspension. It reacts with gastric acid to give $AlCl_3$, which causes nausea and vomiting.

$$3HCl + Al(OH)_3 \rightarrow AlCl_3 + 3H_2O$$

It's a very popular antacid and causes the neutralisation of gastric acid (H_3O^+) by the following mechanism;

$$Al(OH)_3 + 3H_2O \rightarrow Al(OH)_3.(H_2O)_3]$$

$$[Al(OH)_3.(H_2O)_3] + H_3O \rightarrow [Al(OH)_2.(H_2O)_4]^+ + H_2O$$

$$[Al(OH)_2.(H_2O)_4]^+ + H_3O^+ \rightarrow [Al(OH).(H_2O)_5]^{++} + H_2O$$

$$[Al(OH).(H_2O)_5]^{++} + H_3O^+ \rightarrow [Al(H_2O)_6]^{+++} + H_2O$$

Applications

1. Used as a slow-acting antacid to treat hyperacidity.

2. Effective for use in treatments of ulcers and hyperchlorhydria.

Dose: Around 7.5 ml - 15 ml, 4-6 times a day. This 7.5 ml to 15 ml amounts to approximately 150-300 mg of aluminium hydroxide.

Tests for Purity

The following tests are carried out to test for purity:

1. pH (Its pH should be between 5.5 and 8.),
2. Arsenic,
3. Heavy metals,
4. Chloride,
5. Sulphate,
6. Neutralising capacity or acid consuming capacity,
7. Microbial limits.

Assay

An accurately weighed quantity of the gel is dissolved in concentrated hydrochloric acid by warming and diluted with water. To an aliquot of 10ml, a known excess of 0.05M disodium edetate is added, and the mixture is neutralised with 1M sodium hydroxide using methyl red as an indicator. This neutral mixture is warmed in a water bath for half an hour to ensure the complexation of aluminium by sodium edetate. Hexamine is added, and it is back titrated with 0.05M lead nitrate using xylenol orange as an indicator.

Each ml of 0.05M disodium edetate = 0.002549g of Al_2O_3

Notes: *This is a complexometric titration, and the sodium edetate is allowed to complex aluminium under conditions where metals such as calcium and magnesium do not interfere. The excess sodium edetate left after complexation with aluminium is over and is back titrated.*

Brands (Market Preparations): *Diegene (Abott); Gelusil (Pfizer)*

3.2.3.2 Magnesium Hydroxide: (Milk of Magnesia)- Molecular formula: $Mg(OH)_2$, Mol. wt: 68

It contains 95.0 to 100.5% of $Mg(OH)_2$.

Preparation

1. It is either obtained from light magnesium oxide or water in the presence of NaOH.

$$MgO + H_2O \rightarrow Mg(OH)_2$$

or

2. Through the action of magnesium sulphate and sodium hydroxide.

$$MgSO_4 + 2NaOH \rightarrow Mg(OH)_2 + Na_2SO_4$$

Properties

It is a bulky white, odourless powder insoluble in water or alcohol but soluble in dilute acids. Milk of Magnesia is a gelatinous translucent suspension with creamy white consistency. Water may separate out on standing.

Applications

1. Used as an antacid.

2. It is also used as a laxative. It is preferred over magnesium oxide.

Tests for Purity

1. Characteristic tests for Mg,

2. Test for purity: Magnesium hydroxide is tested for (i) soluble alkalis, (ii) soluble carbonates, (iii) acid insoluble matter, and (iv) Tests for As, Ca, and heavy metals. (v) Alkalinity, (vi) Soluble salts, (vii) Microbial count.

Assay

Weighed amount of the sample is dissolved in 25 ml of 1N H2SO4. Excess of acid is back titrated against 1N NaOH using methyl red indicator.

$$\text{Each ml of 1N } H_2SO_4 \cong 0.02917 \text{ g of } Mg(OH)_2$$

Brands (Market Preparations): Deys' Milk of Magnesia (Deys Drugs & Chemicals); Philips Mom (Win Medicare Limited).

3.2.3.3 Magaldrate (hydrated magnesium aluminate)

Molecular Formula: $AlMg_2(OH)_7 \cdot H_2O$; Mol. wt: 164.

Magaldrate is a common aluminium and magnesium-containing antacid drug that is used for the treatment of duodenal and gastric ulcers and esophagitis from gastroesophageal reflux.

Magaldrate is a hydroxymagnesium aluminate complex that is converted rapidly by gastric acid into $Mg(OH)_2$ and $Al(OH)_3$, which are absorbed poorly and thus provide a sustained antacid effect. It is available in the form of oral suspension or tablets.

Magaldrate may negatively influence drugs like tetracyclines, benzodiazepines and indomethacin. High doses or prolonged usage may lead to an increment of defecation and a reduction in faeces consistency. In some cases, it can alter the functionality of the gastrointestinal tract, occasionally provoking constipation or diarrhoea.

Brands (Market Preparations): Megs (Kentreck Labs); Merigel (Ind Swift); Nilcid MPS (Abbott)

3.2.3.4 Sodium Bicarbonate (Baking soda): Molecular Formula: $NaHCO_3$; Mol. wt: 84

Sodium bicarbonate is commonly known as baking soda, bread soda, cooking soda, or bicarbonate of soda. It is a salt composed of a sodium cation and a bicarbonate anion. Sodium bicarbonate is a white solid that is crystalline but often appears as a fine powder.

Preparation: Sodium bicarbonate can be prepared by carbon dioxide reacts with the sodium hydroxide aqueous solution. This reaction initially produces sodium carbonate.

$$2NaOH + CO_2 \rightarrow Na_2CO_3 + H_2O$$

Then, adding carbon dioxide to this reaction produces sodium bicarbonate.

$$Na_2CO_3 + CO_2 + H_2O \rightarrow 2NaHCO_3$$

Properties: It is a white solid crystalline chemical compound, usually in its powder form. It is insoluble in ethanol and slightly soluble in methanol and acetone. At a temperature of 20 degrees Celsius, its solubility in water is 96 grams per litre.

Applications

- Sodium bicarbonate mixed with water can be used as an antacid to treat acid indigestion and heartburn.

- A mixture of sodium bicarbonate and polyethene, such as PegLyte, dissolved in water and taken orally, is an effective gastrointestinal lavage preparation and laxative prior to gastrointestinal surgery, gastroscopy, etc.

- Intravenous sodium bicarbonate in an aqueous solution is used to treat or when insufficient sodium or bicarbonate ions are in the blood. In cases of respiratory acidosis, the infused bicarbonate ion drives the carbonic acid/bicarbonate buffer of plasma to the left and thus raises the pH. For this reason, sodium bicarbonate is used in medically supervised cardiopulmonary resuscitation. Infusion of bicarbonate is indicated only when the blood pH is markedly low ($< 7.1–7.0$).

- It is used for the treatment of hyperkalemia, as it will drive K+ back into cells during periods of acidosis. Since sodium bicarbonate can cause alkalosis, it is sometimes used to treat aspirin overdoses. Aspirin requires an acidic environment for proper absorption, and a basic environment will diminish aspirin absorption in cases of overdose. Sodium bicarbonate has also been used in the treatment of tricyclic antidepressant overdose. It can also be applied topically as a paste, with three parts baking soda to one part water, to relieve some kinds of insect bites and stings (as well as accompanying swelling).

- Sodium bicarbonate can be added to speed up the onset of their effects and make their injection less painful. It is also a component of Moffett's solution, used in nasal surgery.
- It is used to have clean teeth and mouth

Other Applications:

- It is used as pest control to kill cockroaches and to control fungal growth as a disinfectant. Also, to protect armpits from bad smell and irritation
- It is used in cooking, especially to bake food items
- It is used in medicine to be injected intravenously to the prevention of chemotherapy side effects
- It is used to wash kitchen products due to its antibacterial properties

Dose: Around 325 milligrams (mg) to 2 grams one to four times a day.

Brands (Market Preparations): Sobisis Forte (La Renon Healthcare); Nodosis (Steadfast MediShield).

Test for Purity: Compare the colour of the pH paper with the pH scale typically printed on the pH paper pack and assign the pH of the solution accordingly. If the pH is around 8, the sample is sodium bicarbonate. If pH is in the range of 9.5 to 10, it is sodium carbonate.

Assay: Weigh accurately about 3 g of Sodium Bicarbonate, mix with 100 mL of water, add methyl red TS, and titrate with 1 N hydrochloric acid VS. Add the acid slowly, with constant stirring, until the solution becomes faintly pink.

Heat the solution to boiling, cool, and continue the titration until the faint pink colour no longer fades after boiling.

Each mL of 1 N hydrochloric acid is equivalent to 84.01 mg of $NaHCO_3$.

3.2.3.5 Calcium Carbonate (Precipitated Chalk): Molecular Formula: $CaCO_3$; Mol. wt: 100

Calcium Carbonate is a calcium salt used as an antacid to relieve the symptoms of indigestion and heartburn. It is also used to prevent osteoporosis, as a calcium supplement, and to treat high phosphate levels in patients with kidney disease.

Preparation: calcium carbonate is prepared from calcium oxide. Water is added to give calcium hydroxide then carbon dioxide is passed through this solution to precipitate the desired calcium carbonate, referred to in the industry as precipitated calcium carbonate (PCC).

$$CaO + H_2O \rightarrow Ca(OH)_2$$

$$Ca(OH)_2 + CO_2 \rightarrow CaCO_3\downarrow + H_2O$$

Properties: Calcium carbonate is an odourless, fine white powder. It has a density of 2.71 g/mL and a melting point of 1,339°C in its calcite form. Calcium carbonate is insoluble in water and stable at normal temperatures. When heated to high temperatures, it decomposes to form calcium oxide with the release of carbon dioxide.

Applications: Medicinally, it is used as;

- an antacid
- a calcium supplement.
- Filler in cosmetics.

Tests for Purity

To 1 g of Calcium Carbonate II, add 10 mL of water and 7 mL of diluted acetic acid (1 in 4). It effervesces and dissolves. When boiled and neutralised with ammonia TS, this solution responds to all the tests for Calcium Salt in the Qualitative Tests.

Assay: Transfer about 200 mg of Calcium Carbonate previously dried at 200°C for 4 hrs to an accurately weighed 250 mL beaker. Moisten thoroughly with a few mL waters and add dropwise sufficient 3N HCl to dissolve. Add 100 mL of water, 15 mL 1N NaOH and 300 mg hydroxy naphthol blue, and titrate against 0.05 M disodium edetate VS until the solution is distinct blue in colour.

Each mL of 0.05 M disodium edetate is equivalent to 5.004 mg of $CaCO_3$

Brands (Market Preparations) : Calcium Sandoz (Novartis); Intacal 500 (Intas).

3.2.3.6 Combination Antacids

1. None of the antacids has all the properties of an ideal antacid.
2. Calcium and Aluminium antacids cause constipation.
3. Magnesium antacids reduce constipation but have laxative action.
4. Most of the market preparations contain combinations of antacids to approach an ideal.
5. These combinations balance both the constipation (Al and Ca) and laxative (Mg) side effects of antacids.
6. In some mixtures of two antacids, one has a rapid onset of action, while the other has prolonged action.

Some examples of Combination Antacids

1. Aluminum hydroxide gel + magnesium hydroxide (CremalinR) - A suspension containing both agents in 2-4% w/v.
2. A very popular combination of aluminium and magnesium hydroxides, known as Magaldrate (hydrated magnesium aluminate).

3. Aluminum hydroxide gel + magnesium trisilicate combination (Gelusil[R]). This is often combined with simethicone, a defoaming agent, to get relief from flatulence.

4. Aluminum hydroxide gel + calcium carbonate. This shows both rapid and prolonged action.

5. Multiple Antacid Combinations. E.g. Algicon Tablets/Suspension have 4 antacids combination. It contains,

   (a) Aluminum hydroxide gel - 360 mg

   (b) Magnesium alginate - 500 mg

   (c) Magnesium carbonate - 320 mg

   (d) Potassium bicarbonate - 100mg

3.2.4 ADSORBENTS & PROTECTIVES

These are chemically inert agents commonly used for treating mild diarrhoea, dysentery, or other GIT disturbances due to their ability to absorb gases, chemical and drug poisons, and bacterial toxins. These agents mainly include bismuth compounds, kaolin, activated charcoal and pectin.

3.2.4.1 Bismuth Compounds: These are traditionally used for their mild astringent, antacid, antiseptic actions and protective effects on the mucous membranes. These include bismuth subcarbonate, bismuth subnitrate, bismuth subsalicylate, bismuth subgallate, etc.

(a) Bismuth subcarbonate: Molecular formula: $[(BiO_2)_2(CO_3)]_2.H_2O$

It is a basic carbonate of bismuth which on ignition yields NLT 90% and NMT 92% of BiO_3 calculated on a dry basis.

Preparation: By the reaction of bismuth nitrate and sodium carbonate in an aqueous medium

$$4Bi(NO_3)_3 + 6Na_2CO_3 + H_2O \rightarrow [(BiO_2)_2(CO_3)]_2.H_2O + 2NaNO_3 + 4CO_2 \uparrow$$

Properties: White or pale yellow, odourless and tasteless powder, insoluble in water and alcohol but

dissolves in HCl. Decomposes into yellow Bi_2O_3 when ignited.

Tests for Purity

1. Identification (reactions of Bi and CO_3),

2. Tests for purity: Loss on drying (NMT 2% w/w),

3. Tests for chloride, sulphate and nitrate,

4. Tests for heavy metals, Pb, As, Ag, Cu etc.

Dose: 1 to 3 gm in divided doses.

Assay: By gravimetric method. Accurately weighed sample is ignited in a tared crucible to constant weight. The bismuth trioxide (Bi2O3) so obtained is cooled and weighed.

Applications:

1. As astringent and absorbent in the treatment of diarrhoea, dysentery, ulcerative colitis
2. As a mild antacid
3. As a topical protective

(b) Kaolin: It is hydrated aluminium silicate occurring as soft white or off-white powder or lumps. It has a clay-like taste and odour, particularly when moistened. It is insoluble in water, alcohols, acids and alkali. It occurs in two forms, heavy and light.

Applications

1. It is normally used as an absorbent in a combination of pectin.
2. It is used to reduce GIT and mucal inflammations.
3. It is used in topically applied dusting powders

(c) Pectin: It is produced commercially as a white to light brown powder, mainly extracted from citrus fruits, and is used in food as a gelling agent, particularly in jams and jellies. It is also used in fillings, medicines, and sweets, as a stabiliser in fruit juices and milk drinks, and as a source of dietary fibre. It is mainly used for treating diarrhoea in combination with kaolin as an adsorbent and protectant combination. It works by absorbing excess fluids and reducing intestinal movement.

(d) Activated Charcoal: It is a residue from the destructive distillation of various organic materials treated specially to increase its absorptive power. It is a fine black, odourless, tasteless powder free from gritty matter. It is used as an absorbent in the treatment of diarrhoea. It is also used as an antidote for drug poisoning.

3.2.5 CATHARTICS

Cathartics are agents that quicken, increase and facilitate the evacuation (defecation) from the bowel. The milder forms are called **Purgatives,** and still milder forms are available, which are for frequent use, and they are **Laxatives.**

They are the most widely used OTC medications (Over The Counter, which do not require a medical prescription).

Constipation: It is a condition where there is infrequent bowel movement.

Causes: Diet (lack of fibre and liquids), lack of exercise, age, irregular bowel habits, drugs induced, or disease state.

Signs and Symptoms: Infrequent defecation, nausea, vomiting, anorexia, rectal bleeding, weight loss in, chronic constipation.

Treatment: Drugs that promote evacuations of the bowel.

A. Laxatives: Elimination of soft but formed stools. They are generally used to relieve constipation. It has mild action.

B. Purgatives: It has stronger action, resulting in more fluid evacuation.

Mechanism of action:

- Inhibiting $Na^+K^+ATPase$ of villous cells, impairing electrolyte and water imbalance.
- Stimulating adenylyl cyclase in crypt cells increasing water and electrolyte secretion.
- Enhancing PG synthesis in the mucosa, which increases secretion.
- Increasing NO synthesis, which enhances secretion and inhibits non-propulsive contractions in the colon
- Structural injury to the absorbing intestinal mucosal cells.

All three are generally administered by oral route and sometimes by rectal route (enema or suppository). They all act by retaining the body fluid and do not cause excessive dehydration.

Cathartics act by 4 mechanisms;

1. Through local irritation of G.I.T. directly and thereby stimulating peristalsis. These are called as stimulants. E.g., Rhubarb, senna, podophyllum, castor oil, bisacodyl etc.

2. Through increasing the bulk of intestinal contents and thereby stimulating peristalsis. E.g., Bulk purgatives like ispagol, CMC, gums etc.

3. By acting as lubricants of the GIT and thereby facilitating the smooth evacuation of faeces. E.g., Liquid paraffin, glycerine, mineral oils etc.

4. By increasing the osmotic load of the intestine by absorbing water and thereby stimulating peristalsis. These are saline cathartics. E.g., sodium phosphate, magnesium sulphate, sodium potassium tartarate, magnesium carbonate etc.

(a) **Sodium Phosphate:** (Disodium phosphate, dibasic sodium phosphate, disodium hydrogen phosphate) – Molecular formula: $Na_2HPO_4. 7H_2O$ ($12H_2O$), M.Wt.: 268.07 (358.14).

   Contains NLT 98.5 % and NMT 101.0% Na_2HPO_4 on dried basis.

Preparation: It is prepared by the action of phosphoric acid on sodium carbonate— neutralisation and concentration of the solution yield colourless granular salt.

$$H_2PO_4 + Na_2CO_3 \rightarrow Na_2HPO_4 + H_2O + CO_2$$

Properties

It is colourless white granular salt. It effervesces in dry air. It is freely soluble in water and slightly soluble in alcohol. Its aqueous solution is basic in nature.

Purity

1. Identification Tests: 10 % w/v solution should give confirmatory tests for sodium and phosphate.

2. pH: 2% aqueous solution should be 9-12,

3. Tests for purity: Heavy metals as well as Ca, Mg and ions like chloride, sulphate and loss on drying.

Assay

Sample solution in water is titrated against 0.5N H_2SO_4 using a bromocresol green indicator to a green colour endpoint.

$$2Na_2HPO_4 + H_2SO_4 \rightarrow 2NaH_2PO_4 + Na_2SO_4$$

$$\text{Each ml of } 0.\ 5N\,H_2SO_4 \cong 0.0071 \text{ g of } Na_2HPO_4$$

1. Because of its poor absorption in the GIT, it is considered safe and is widely used as saline cathartic (laxative).

2. It is an important component of phosphate buffer

(b) Sodium Potassium Tartrate (Rochelle salt): Molecular formula: $C_4H_4O_6NaK.4H_2O$
   Mol. wt: 282.2

It has NLT 99.0% and NMT 104.0% of $C_4H_4O_6NaK.4H_2O$

Preparation

Sodium potassium tartrate is prepared by the action of potassium bitartrate on sodium carbonate in boiling water. Neutralisation by the evolution of CO2 followed by concentration of the solution yields colourless white salt.

$$2KHC4H_4O_6 + Na_2CO_3 + 6H_2O \rightarrow 2C_4H_4O_6NaK.4H_2O + CO_2$$

Properties

Sodium potassium tartarate appears as colourless white crystals often coated with white powder. It effervesces in dry air. It has a cooling, saline taste. It is freely soluble in water and practically insoluble in alcohol. Its aqueous solution is basic in nature.

Purity

1. **Identification tests**: 10 % w/v solution should give confirmatory tests for sodium and potassium; on heating emits the odour of burn sugar, and the residue is alkaline to litmus, giving effervescence with acids.

2. **Tests for purity**: Heavy metals such as As, Fe as well as chloride, sulphate and loss on drying,

3. **Acidity and alkalinity** by titrating against standard NaOH and HCl solutions using phenolphthalein as an indicator.

Assay-Gravimetric – Volumetric: Accurately weighed sample is carbonised in a silica crucible, the residue boiled with 15 ml 0.5N H_2SO_4 and filtered. Filtrate is diluted to 50 ml with water and excess acid titrated with 0.5 N NaOH using a methyl orange indicator.

$$\text{Each ml of 0. 5NH}_2\text{SO}_4 \equiv 0.071 \text{ g of C}_4\text{H}_4\text{O}_6\text{NaK.4H}_2\text{O}.$$

Applications

1. Because of its poor absorption in the GIT, it is considered safe and is the widely used saline cathartic (laxative).

2. It is an important component of effervescent powders and is an important food additive.

(c) Magnesium Carbonate: (Magnesite) – Molecular formula: $MgCO_3$, Mol. wt: 84.31

Properties: It is an inorganic salt that is a white solid. Several hydrated and basic forms of magnesium carbonate also exist as minerals. Magnesite consists of white trigonal crystals.

The anhydrous salt is practically insoluble in water, acetone, and ammonia. All forms of magnesium carbonate react with acids to release CO_2.

Preparation

Magnesium carbonate is ordinarily obtained by mining the mineral magnesite. Magnesium carbonate can be prepared in the laboratory by reaction between any soluble magnesium salt and sodium bicarbonate.

$$MgCl_2 + 2NaHCO_3 + H_2O \longrightarrow MgCO_3(s) + 2NaCl + H_2O + CO_2$$

High-purity industrial routes include a path through magnesium bicarbonate -combining magnesium hydroxide and carbon dioxide. A slurry of magnesium hydroxide is treated with 3.5 to 5 atm. of carbon dioxide below 50 °C, giving the soluble bicarbonate, then vacuum drying the filtrate, which returns half of the carbon dioxide and water.

$$Mg(OH)_2 + 2CO_2 \longrightarrow Mg(HCO_3)_2$$

$$Mg(HCO_3)_2 \longrightarrow MgCO_3 + CO_2 + H_2O$$

Applications

1. A laxative to loosen the bowels and colour retention in foods.

2. High purity magnesium carbonate is used as an antacid,

3. As an additive in table salt to keep it free flowing.

4. It is an important food additive known as E504.

(d) Magnesium Sulphate : (Epsom Salt): $MgSO_4.7H_2O$ Mol. wt: 246.5

It has NLT 99% and NMT 100% of $MgSO_4$ calculated on a dried basis.

Preparation: The heptahydrate can be prepared by neutralising sulphuric acid with magnesium carbonate or oxide, but it is usually obtained directly from natural sources. Anhydrous magnesium sulphate is prepared only by the dehydration of a hydrate.

$$MgCO_3 + H_2SO_4 \longrightarrow MgSO_4 + CO_2 + H_2O$$

Properties:

It is a colourless white powder highly soluble in water but insoluble in alcohol. It has a cool, saline taste. The anhydrous form is strongly hygroscopic and can be used as a desiccant.

Purity

1. **Identification Tests**: 10 % w/v solution should give confirmatory tests for magnesium and sulphate;

2. **Tests for purity**: Heavy metals As, Fe as well as chloride, sulphate and loss on drying.

3. **Acidity and alkalinity**: 1 g sample dissolved in 10 ml water should yield a clear solution neutral to litmus.

Assay

An aqueous solution of the accurately weighed sample made alkaline with ammonia-ammonium chloride buffer is titrated against 0.05 M disodium edetate to the blue endpoint using the Mordant Black 11 indicator.

Each ml of 0. 05M disodiumedetate $\equiv$ 0.00602 g of $MgSO_4$.

Applications

1. Magnesium sulphate is a common pharmaceutical preparation of magnesium, commonly known as Epsom salt, used both externally and internally. Epsom salt is used in bath salts and for isolation tanks.

2. Oral magnesium sulphate is commonly used as a saline laxative or osmotic purgative.

3. Magnesium sulphate is the main preparation of intravenous magnesium and is also used as an anticonvulsant.

QUESTION BANK

A. MULTIPLE CHOICE QUESTIONS

1. All of the following statements about antacid are true, Except:
   A. Weak bases that neutralise gastric pH
   B. Inhibits the formation of pepsin
   C. Aluminium antacids cause diarrhoea, and magnesium antacids cause constipation
   D. Aluminium antacids cause constipation, and magnesium antacids cause diarrhoea

Answer: C

2. Inorganic agent used to treat GIT agent
   A. products for altering gastric pH
   B. productive for intestinal inflammation
   C. adsorbents for intestinal toxins.
   D. all of the above

Answer: A

3. The goal of antacid therapy.
   A. $\downarrow$ Concentration of acid in gastric juice
   B. Maintain Gastric pH between 3.5 and 7
   C. $\uparrow$ Concentration of acid
   D. Both (a) and (b)

Answer: B

4. $Al(OH)_3$ gel contains
   A. basic aluminium carbonate
   B. basic aluminium chloride
   C. basic aluminium bicarbonate
   D. all of the above

Answer: A

5. Milk of Magnesia is
   A. Magnesium carbonate
   B. Magnesium chloride
   C. Magnesium hydroxide
   D. Mixture of all

Answer: C

6. Calcium-containing antacids differ from aluminium-containing antacids
   A. depend upon their basic property
   B. do not have any amphoteric effect
   C. do not cause systemic alkalosis
   D. all of the above

Answer: A

7. Magaldrate is a combination of
   A. Aluminium and Calcium hydroxides
   B. Aluminium and Magnesium hydroxides
   C. Aluminium and Sodium hydroxides
   D. Aluminium and Calcium carbonates

Answer: B

8. Gelusil[R] is a combination of
    A. Aluminium hydroxide gel and magnesium trisilicate
    B. Aluminium hydroxide gel and Magnesium hydroxide
    C. Aluminium hydroxide gel and Sodium carbonate
    D. Aluminium and Calcium carbonates

Answer: B

9. Which of the following come under the category of Adsorbents and Protectives
    A. Bismuth subcarbonate
    B. Kaolin
    C. Activated charcoal
    D. All of the above

Answer: D

10. Which of the following statements is true
    A. Laxatives are milder in action than purgatives
    B. Purgatives are milder in action than laxatives
    C. Both have the same degree of action
    D. None of the above

Answer: A

B. SHORT ANSWER QUESTIONS

1. Justify the role of acidifying agents in medical treatment and physiology.

2. Enumerate the roles of gastric acid in the body

3. Classify and briefly discuss the important categories of gastrointestinal agents.

4. What are antacids? Classify them and discuss the characteristics of an ideal antacid.

5. What are combination antacids? and discuss some important combinations.

6. What is the role of simethicone in antacid preparations?

7. Name the inorganic compound or combination involved in the following;

    Synonym or Brand Name Inorganic compound or combination involved

    1. Baking Soda

    2. Precipitated Chalk

    3. Gelusil

    4. Magnesite

    5. Magaldrate

    6. Milk of Magnesia

C. LONG ANSWER QUESTIONS

1. Write a note on any one class of inorganic antacids:
   - (i) Aluminium
   - (ii) Magnesium
   - (iii) Sodium.

2. Discuss saline cathartics with respect to their definition, role, mechanism of action and classification in detail with suitable examples.

3. Give a brief monograph of any one of the following important inorganic GIT agents
   - (i) Aluminium hydroxide gel
   - (ii) Calcium Carbonate
   - (iii) Magnesium Hydroxide,
   - (iv) Sodium Bicarbonate.

4. Discuss in detail the GIT agents categorised as protectives and adsorbents

5. Write short notes on -
   - (i) Bismuth compounds,
   - (ii) Bismuth subcarbonate,
   - (iii) Kaolin,
   - (iv) Activated charcoal,
   - (v) Pectin.

6. Discuss any one of the following saline cathartics in detail:
   - (i) Sodium phosphate,
   - (ii) Sodium potassium tartarate,
   - (iii) Magnesium carbonate,
   - (iv) Magnesium sulphate

■■■

TOPICAL AGENTS

CONTENTS

Inorganic Pharmaceuticals: Topical Agents:

Pharmaceutical formulations, market preparations, storage conditions and uses of the Topical agents:

- Silver Nitrate,
- Ionic Silver,
- Chlorhexidine Gluconate,
- Hydrogen peroxide,
- Boric acid,
- Bleaching powder,
- Potassium permanganate

♦ LEARNING OBJECTIVES ♦

After completing this chapter, the student should be able to understand:

1. Classification of various inorganic Topical Agents
2. Chemical Names of various inorganic Topical Agents
3. Chemical Structures of different inorganic Topical Agents
4. Uses, Stability and storage conditions of inorganic Topical Agents
5. Different types of formulations and popular brand names of inorganic Topical Agents

3.3.1 TOPICAL AGENTS

3.3.1.1 Definition: The "topical agents" are substances applied to the body surface, including applications within the body cavities that open outside. (e.g. the oral, vaginal). These compounds act locally with skin or mucous membranes, mainly in a mechanical or physical manner. Topical agents do not absorb directly into the circulation. These compounds have a little pharmacological effect. These compounds are that, when they are applied, produce a variety of effects like adsorbent, astringent, demulcent, emollient or protective. Some compounds also exhibit antimicrobial, astringent activity topically. Some inorganic compounds are having topical local activity.

These agents can be categorised broadly into three groups

1. Protectives & Adsorbents
2. Antimicrobials
3. Astringents.

Herein, the syllabus focuses more on Antimicrobial drugs, still, a brief on the other two categories is presented.

1. **Protectives and Adsorbents:** Protectives are the agents that cover the skin or mucous membrane from possible irritants. Adsorbents are chemically inert substances that can adsorb dissolved or suspended particles or gases, toxins, bacteria etc., and give a protective coating to the inflamed mucus membranes and skin. E.g., Talc, Zinc oxide, Zinc stearate, Calamine, Titanium dioxide, Dimethicone, etc.

2. **Astringents:** These are protein precipitants with limited penetration power. An astringent coagulates the protein on the surface of the cell and brings out a hardening effect. It constricts the tissue as well as Small Blood vessels. Astringents are mild Antimicrobial Agents. E.g., Alum, Zinc sulphate, Silver nitrate, and Boric Acid.

Uses of Astringents:

- Styptic to arrest minor bleeding by coagulation of blood
- Antiperspirant to reduce perspiration by constricting pores of the skin.
- Anti-inflammatory action
- At high concentration to remove unwanted tissue growth
- Internally they can be used to treat diarrhoea
- As cosmetics to skin tone and bring out the hardening effect
- dental products can promote the hardening of the gums.
- Can reduce the cell permeability

3. **Antimicrobial agents** are chemicals whose preparations help reduce or prevent infection due to microbes. Several terms employed in describing or classifying antimicrobial agents are as given below.

   a. **Antiseptics:** These are substances that can kill or prevent the growth of microorganisms. This term is specific for preparations which are to be applied to living tissues.

   b. **Disinfectants:** These are the substances that prevent infection by the destruction of pathogenic microorganisms.

   c. **Germicides:** These are substances which kill micro-organisms.

   d. **Bacteriostatics:** These are substances which primarily function by inhibiting the growth of bacteria. Thus, bacteriostatic drugs or agents do not kill but arrest the growth of bacteria.

   e. Bactericidal: These are the substances which prevent infection by completely killing (cidal) the bacteria causing it.

   f. Sterilizers: This refers to the agents used to render an object completely free of micro-organisms.

3.3.1.2 Mechanism of Action

Antimicrobial actions involve any of the following three mechanisms:

1. Oxidation.

2. Halogenation.

3. Protein binding or precipitation.

(1) Oxidation Mechanism: Compounds acting by this mechanism belong to a class of oxygen-liberating compounds like peroxides and peroxyacids, permanganate and certain oxo-halogen anions.

These anti-infective agents bring about oxidation of active functional groups present in proteins or enzymes vital to the growth or survival of micro-organisms. This causes a change in the conformation of the proteins, thereby altering their function. e.g., the free sulphydryl group has been essential for the functioning of a variety of proteins and enzymes. This free nature of the sulphydryl group gets destroyed by oxidation resulting in the formation of a disulfide bond.

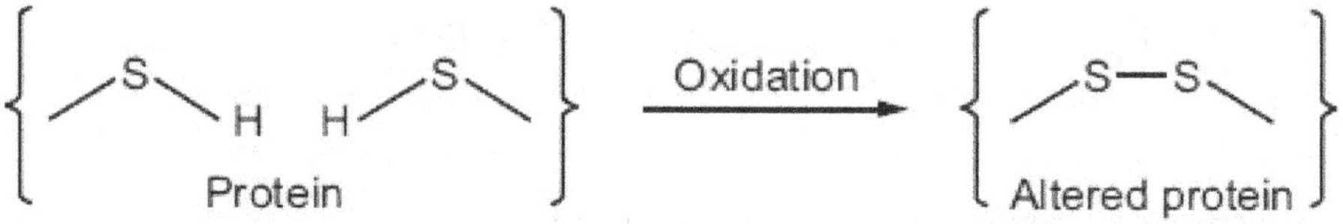

Fig. 3.3.1 Mechanism of oxidation of proteins

(2) Halogenation Mechanism: Compounds which are able to liberate chlorine or hypochlorite, or iodine act by this mechanism. These agents act on peptide linkage and alter its potential and property. The destruction of the specific function of protein causes the death of microorganisms.

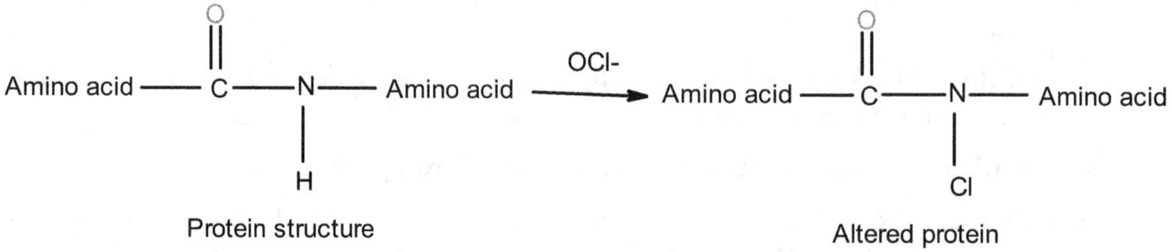

Most of the enzymes are proteinaceous in nature. A protein molecule is composed of a variety of amino acids bound by a peptide (−CONH−) linkage. As hypochlorides (OCl⁻) are found to chlorinate peptide linkage, antiseptics having hypochlorite functional group exert their antimicrobial activity by chlorination of peptide linkage in protein molecules.

The substitution of the chlorine atom in place of nitrogen of the peptide linkage causes a change in H-bonding forces responsible for the proper orientation of the protein molecule.

Hence, the functions of protein cannot be carried out.

(3) **Protein Binding or Precipitation:** Many metal ions exhibit protein binding or protein precipitation properties. The interaction of metal ions with protein is non-specific and, at sufficient concentrations, will react with host as well as microbial proteins.

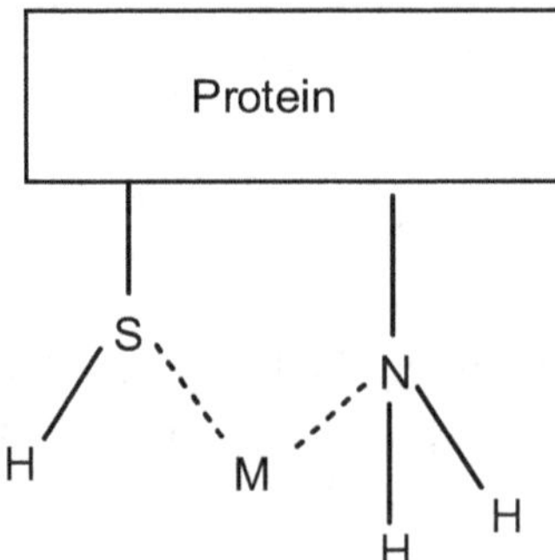

Fig. 3.3.2 Mechanism of protein precipitation

Certain metals (e.g., mercury, arsenic and antimony) show some enzyme specificity and form strong covalent bonds with particular enzyme systems.

3.3.1.3 Commonly used Antimicrobial Agents

3.3.1.3.1 Silver Nitrate: AgNO₃, M. Wt: 169.87

Silver nitrate is an inorganic compound and is a versatile precursor to many other silver compounds,

Properties: Silver nitrate appears as a colourless or white crystalline solid becoming black on exposure to light or organic material.

Preparation: silver nitrate can be prepared by dissolving silver in nitric acid followed by evaporation of the solution. The stoichiometry of the reaction depends upon the concentration of nitric acid used.

$$3\ Ag + 4\ HNO_3\ (cold\ and\ diluted) \rightarrow 3\ AgNO_3 + 2\ H_2O + NO$$

$$Ag + 2\ HNO_3\ (hot\ and\ concentrated) \rightarrow AgNO_3 + H_2O + NO_2$$

Test for Purity: Clarity and colour of the solution Acidity and alkalinity

Foreign salts: Al, Bi, Cu and Pb

Storage: Affected by light, store in tightly closed light-resistant containers.

Uses: It is a topical anti-infective, antiseptic, antibacterial, and cauterizing agent.

Antiseptic properties. Until the development of antibiotics, dilute solutions of $AgNO_3$ were used to be dropped into newborn babies' eyes at birth to prevent the contraction of gonorrhoea from the mother.

It is used to cauterize (burning or removing a part of a body) infected tissues around a skin wound. For example, to remove granulation tissue around a stoma. It creates a scab (protective tissue covering that forms after skin damage) stop bleeding from a minor skin wound. Silver nitrate is also used to help remove warts or skin tags.

Assay: Powder about 1 g of Silver Nitrate and dry in the dark over silica gel for 4 hours. Weigh accurately about 700 mg of this dried salt, dissolve in 50 mL of water, add 2 mL of nitric acid and 2 mL of ferric ammonium sulfate TS, and titrate with 0.1 N ammonium thiocyanate VS.

3.3.1.3.2 Ionic Silver: It is a chemical element with the symbol Ag (from the Latin *argentum*). At.Wt.107.87; At. No. 47.

Properties: Pure silver is nearly white, lustrous, soft, very ductile, and malleable, it is an excellent conductor of heat and electricity. It is not a chemically active metal, but it is attacked by nitric acid (forming the nitrate) and by hot concentrated sulfuric acid.

Uses:

i. In medicine, silver is incorporated into wound dressings and used as an antibiotic coating in medical devices. Wound dressings containing silver sulfadiazine or silver nanomaterials are used to treat external infections.

ii. Silver is also used in some medical applications, such as urinary catheters (where tentative evidence indicates it reduces catheter-related urinary tract infections) and in endotracheal breathing tubes (where evidence suggests it reduces ventilator associated pneumonia). The silver ion is bioactive and, in sufficient concentration, readily kills bacteria *in vitro*. Silver ions interfere with enzymes in the bacteria that transport nutrients, form structures, and synthesise cell walls; these ions also bond with the bacteria's genetic material.

iii. Silver and silver nanoparticles are used as antimicrobial in a variety of industrial, healthcare, and domestic application: for example, infusing clothing with nanosilver particles thus allows them to stay odourless for longer. Bacteria can, however, develop resistance to the antimicrobial action of silver.

iv. Silver and its alloys are used in cranial surgery to replace bone, and silver–tin–mercury amalgams are used in dentistry.

v.   Silver diammine fluoride, the fluoride salt of a coordination complex with the formula $[Ag(NH_3)_2]F$, is a topical medicament (drug) used to treat and prevent dental caries (cavities) and relieve dentinal hypersensitivity.

3.3.1.3.3 Chlorhexidine Gluconate (CHG)

Chlorhexidine is a disinfectant and antiseptic used for skin disinfection before surgery and sterilising surgical instruments. It may be used both to disinfect the skin of the patient and the hands of the healthcare providers.

Medicinal Uses:

Topical: Chlorhexidine gluconate is used as a skin cleanser for surgical scrubs, as a cleanser for skin wounds, for preoperative skin preparation, and for germicidal hand rinses. Chlorhexidine eye drops have been used to treat eyes affected by Acanthamoeba keratitis.

Dental: Chlorhexidine gluconate is usually prescribed by a dentist. Chlorhexidine gluconate is a germicidal mouthwash that reduces bacteria in the mouth. Chlorhexidine gluconate oral rinse is used to treat gingivitis (swelling, redness, bleeding gums). Use of a CHG-based mouthwash in combination with normal tooth care can help reduce plaque build-up and improve mild gingivitis. About 20 mL twice a day of concentrations of 0.1% to 0.2% is recommended for mouth-rinse solutions with a duration of at least 30 seconds.

Antiseptic: CHG is active against Gram-positive and Gram-negative organisms, facultative anaerobes, aerobes, and yeasts. It is particularly effective against Gram-positive bacteria (in concentrations $\geq$ 1 µg/l). Significantly higher concentrations (10 to more than 73 µg/ml) are required for Gram-negative bacteria and fungi. Chlorhexidine is ineffective against polioviruses and adenoviruses. The effectiveness against herpes viruses has not yet been established unequivocally.

There is strong evidence that chlorhexidine is more effective than povidone-iodine for clean surgery. Evidence shows that it is the most effective antiseptic for upper limb surgery.

Side Effects: Such mouthwash also has several, including damage to the mouth lining, tooth discolouration, tartar build-up, and impaired taste. Extrinsic tooth staining occurs when chlorhexidine rinse has been used for 4 weeks or longer.

3.3.1.3.4 Hydrogen Peroxide

Molecular Formula: H_2O_2, Mol. wt: 34.016

It has not less than 6% w/w of H_2O_2, which corresponds to about 20 times its volume of available oxygen.

Preparation

Hydrogen peroxide is obtained in two ways:

1. It is obtained by adding a thick paste of barium peroxide in ice-cold water to a calculated quantity of ice-cold dilute sulphuric acid. The insoluble barium sulphate is filtered out.

$$BaO_2 + H_2SO_4 \longrightarrow BaSO_4^- + H_2O_2$$

2. H_2O_2 is also obtained by treating sodium peroxide with dil. H_2SO_4 at low temperature. Sodium sulphate crystallises out, and H_2O_2 is distilled under 10 mm pressure.

$$Na_2O_2 + H_2SO_4 \longrightarrow Na_2SO_4 + H_2O_2$$

Properties

Hydrogen peroxide is a colourless and odourless liquid having a slightly acidic taste. The solution decomposes when it comes in contact with oxidisable matter or when made alkaline.

$$2H_2O_2 \longrightarrow 2H_2O + O_2$$

The decomposition is promoted by a catalyst such as Cu, Fe, Mn etc., whereas a small quantity of acids such as H_2SO_4, H_3PO_4 and alcohol if added, retard the decomposition of H_2O_2.

Therefore, they act as negative catalysts and are used as preservatives or stabilisers in commercial preparations.

Hydrogen peroxide is a strong oxidising agent and is miscible in water from which it can be extracted with solvent ether.

Storage

It is preserved in a light-resistant container with a stopper made of glass or plastic resistant to hydrogen peroxide. It is kept in a dark and cool place.

Assay

It is carried out by the permanganate method, in which 10 ml of the sample is diluted to 250 ml in a volumetric flask with purified water. To 25 ml of this solution, 10 ml of 5 N sulphuric acid is added. Then the contents are titrated with 0.1 N $KMnO_4$ solution until a faint pink colour is obtained.

$$2\ KMnO_4 + 3H_2SO_4 + 5H_2SO_4 \longrightarrow K_2SO_4 + 8H_2O + 5O_2 + 2MnSO_4$$

Each ml of 0.1 N $KMnO_4$ is equivalent to 0.001701 gm of H_2O_2.

Note: H_2O_2 and $KMnO_4$ are both oxidising agents. These two oxidising agents reduce one another with the evolution of gaseous oxygen. H_2O_2 reduces $KMnO_4$ solution and causes its decolouration. At the endpoint, an excess drop of $KMnO_4$ gives pink colour. $KMnO_4$ itself acts as an indicator.

Uses

1. It is a strong oxidising agent.
2. It is used for bleaching.
3. It acts as an antiseptic and a germicide and hence it is used for cleaning cuts and wounds.
4. It is an effective antidote for phosphorus and cyanide poisoning.
5. It also finds use for cleaning ears and during the removal of surgical dressings.

3.3.1.3.5 Boric Acid

Molecular Formula: H_3BO_3; M. Wt. 61.83

It has not less than 99.5% of H_3BO_3.

Preparation

In the laboratory, boric acid is obtained by adding a mixture of conc. H_2SO_4 and water to a boiling solution of borax.

$$Na_2B_4O_7 + H_2SO_4 + 5H_2O \longrightarrow Na_2SO_4 + 4H_3BO_3$$

The solution is filtered and kept aside for crystallisation. The crystals of boric acid are separated and then washed until they become free from sulphate ions. Finally, they are dried at room temperature.

Properties

Boric acid is a solid which is available in three forms:

1. Colourless, odourless pearly scales.
2. Six-sided triclinic crystals.
3. White, odourless powder.

It is odourless with a slightly acidic and bitter taste and unctuous touch.

It is stable in the air and has a density of 1.46. Boric acid is a weak acid.

Assay

It is assayed by a titrimetric method. An accurately weighed quantity of boric acid is dissolved in a mixture of 50 ml of water and 100 ml glycerine previously neutralised to phenolphthalein. Now, this solution is titrated with 1 N NaOH using phenolphthalein as an indicator.

Each ml of 1 N NaOH is equivalent to 0.06183 gm of H_3BO_3

Uses

1. It is used as a local anti-infective agent.
2. It is used in dusting powders, local antiseptic creams, ointments, lotions, etc., applied to the skin, mucous, or eye.
3. Aqueous solutions have been used as a mouthwash, eye lotions, and skin lotions.
4. It possesses weak bacteriostatic and fungistatic action.

3.3.1.3.6 Chlorinated Lime

(Synonyms: **Bleaching powder** or Chloride of lime or Calcium oxychloride).

A relatively unstable chlorine carrier in solid form; a complex chemical compound of indefinite composition, presumably consisting of varying proportions of $Ca(OCl)_2$, $CaCl_2$, $Ca(OH)_2$ and H_2O in its molecular structure. Maximum available chlorine content approaches 39%. Commercial products usually range between 24% and 37% of available chlorine.

Keywords

Antiseptic/Disinfectant; Halogens/Halogen Containing Compounds.

Properties

White or greyish-white powder; strong odour of chlorine. On exposure to air, it becomes moist and rapidly decomposes. Most of it dissolves in water or alcohol. Keep dry and tightly closed.

Caution

Strong solutions irritate the skin. Inhalation of fumes may cause laryngeal and pulmonary irritation, pulmonary oedema and death. Ingestion may produce severe oral, oesophagal, and gastric irritation.

Uses

Bleaching of wood pulp, linen, cotton, straw, oils, soaps, and in laundering; oxidiser in calico printing to obtain white designs on a coloured ground; destroying caterpillars; disinfecting drinking water, sewage, etc.; as a decontaminant for mustard gas and similar substances.

Therapeutic Category

Disinfectant. Has been used as a topical antiseptic for superficial wounds.

Antimicrobial Efficacy

The broad-spectrum effectiveness of bleach, particularly calcium hypochlorite, is owed to the nature of its chemical reactivity with microbes. Rather than acting in an inhibitory or toxic fashion in the manner of antibiotics, bleach quickly reacts with microbial cells to irreversibly denature and destroy many pathogens. Bleach has been shown to react with a microbe's heat shock proteins, stimulating their role as an intracellular chaperone and causing the bacteria to form into clumps (much like an egg that has been boiled) that will eventually die off.

In some cases, bleach's base acidity compromises a bacterium's lipid membrane, a reaction similar to popping a balloon. The range of microorganisms effectively killed by bleach is extensive, making it an extremely versatile disinfectant. The same study found that at low (micromolar), E. coli and Vibrio cholera activate a defence mechanism that helps protect the bacteria. However, the implications of this defence mechanism have not been fully investigated.

In response to infection, the human immune system will produce a strong oxidiser, hypochlorous acid, which is generated in activated neutrophils by myeloperoxidase mediated peroxidation of chloride ions and contributes to the destruction of bacteria.

3.3.1.3.7 Potassium Permanganate

Molecular Formula: $KMnO_4$; M. Wt. 158.0

It has not less than 99% w/w of $KMnO_4$.

Preparation

On a large scale, potassium permanganate is prepared by mixing a solution of KOH with powdered manganese oxide and potassium chlorate. The mixture is boiled and evaporated to yield the residue, which is heated in iron pans until it has acquired a paste-like consistency.

$$KOH + 3MnO_2 + KClO_3 \longrightarrow K_2MnO_4 + KCl + 3H_2O$$
$$\text{Potassium Manganite}$$

Potassium manganite so formed is extracted with boiling water, and a current of chlorine, CO_2 or ozonised air is passed into the liquid until it gets converted to permanganate. The MnO_2 formed is removed continuously so as to prevent its breaking down to manganate.

$$6K_2MnO_4 + 3Cl_2 \longrightarrow 6KMnO_4 + 6KCl$$

When CO_2 is passed through the solution in place of chlorine, only two-thirds of manganate gets converted into $KMnO_4$. One-third is converted into MnO_2.

$$3K_2MnO_4 + 2CO_2 \longrightarrow 2KMnO_4 + MnO_2 + 2K_2CO_3$$

The solution of $KMnO_4$ is drawn off from any precipitate of MnO_2, which is then concentrated and crystallised. The crystals are then centrifuged and dried.

Properties

Potassium permanganate occurs in the form of dark purple-coloured monoclinic prisms, almost opaque, having a blue metallic lustre. It is odourless. It is soluble in 15 parts of water and 3.5 parts of boiling water. A solution of $KMnO_4$ has a sweetish, astringent taste. It acts as a powerful oxidising agent.

Storage

It is kept in tightly closed containers. It must be handled with care because an explosion may occur when it is in contact with readily oxidisable substances.

Assay

Its assay is based upon the oxidation-reduction reaction. Potassium permanganate is a strong oxidising agent, and oxalic acid is a reducing agent. The reaction between $KMnO_4$ and oxalic acid tends to proceed slowly. Hence, warming at 70°C is required. An accurately weighed amount of about 0.8 g of sample is dissolved in water. Then it is diluted with water to 250 ml. This solution is titrated with 25 ml of 0.1 N oxalic acid, which is mixed with 25 ml of water and 5 ml of sulphuric acid. The temperature is maintained at about 70°C throughout the entire titration.

$$5H_2C_2O_4 + 2H_2O + 2KMnO_4 + 3H_2SO_4 \longrightarrow K_2SO_4 + 2MnSO_4 + 18H_2O + 10CO_2$$

Each ml of 0.1 N oxalic acid is equivalent to 0.00316 gm of $KMnO_4$.

Uses

1. It is used as an antiseptic in mouthwash.

2. As an anti-infective, it is regarded as of immense value.

3. It is used in the treatment of urethritis.

4. As it is capable of oxidising some drugs and used as an antidote in case of poisonings by barbiturates, chloral hydrate, and many alkaloids.

5. A solution of potassium permanganate can destroy poison and prevent its absorption. But it should not be kept in the stomach for a long time.

6. In veterinary practice, it has been very commonly used as an antiseptic.

QUESTION BANK

A. MULTIPLE CHOICE QUESTIONS

1. Topical agents, depending upon their action or use divided into
   A. Protectives and adsorbents
   B. Antimicrobials
   C. Astringents
   D. All of these

   Answer: D

2. Which of the following substances precipitate proteins
   A. Protectives and adsorbents
   B. Antimicrobials
   C. Astringents
   D. All of these

   Answer: C

3. The agents used to destroy or inhibit the growth of pathogenic microorganisms are
   A. Antioxidants
   B. Antimicrobial
   C. Astringent
   D. Gastrointestinal agents

   Answer: B

4. The substances applied to non-living objects to destroy microorganisms living on them are
   A. Antiseptics
   B. Antimicrobial

    C. Disinfectants

    D. Astringents

Answer: C

5. The substances applied to living tissues or skin to reduce the possibility of infection
    A. Antiseptics
    B. Antimicrobial
    C. Astringent
    D. Disinfectants

Answer: A

6. An operation of destroying all microorganisms on the surface of an object or within a fluid to prevent transmission of disease or infections due to its use is called:
    A. Filtration
    B. Suction
    C. Sterilization
    D. Irradiation

Answer: C

7. An agent only inhibiting the growth of bacteria but not killing them is called
    A. Germicidal
    B. Bactericidal
    C. Bacteriostatic
    D. All of these

Answer: C

8. An antimicrobial acts through which of the following mechanisms?
    A. Oxidation
    B. Halogenation
    C. Protein binding or precipitation
    D. Any of these

Answer: D

9. Ideal property of antimicrobials
    A. Broad spectrum of activity against microorganisms- bacteria, protozoa, fungi, viruses etc.
    B. No systemic toxicity on topical application
    C. No local cellular damage or interference with the body's defence mechanism.
    D. All of these

Answer: D

10. Chlorinated lime is the other name for
    A. Calcium hypochlorite
    B. Sodium hypochlorite
    C. Chloramine T
    D. Soda-lime

    Answer: A

11. Compounds functioning as antimicrobials through oxidation mechanism:
    A. Hydrogen peroxide
    B. Potassium permanganate
    C. Oxy-halogen ions
    D. All of these

    Answer: D

12. Astringents are therapeutically used for
    A. Promoting healing and toughening the skin
    B. Reducing inflammation through reduction of blood supply to mucosal surfaces
    C. To arrest haemorrhage through vasoconstriction and promotion of blood coagulation.
    D. All of these

    Answer: D

13. Which of the following is an astringent
    A. Alum
    B. Zinc sulfate
    C. Aluminium chloride
    D. All of these

    Answer: D

14. Correct molecular formula of boric acid is
    A. H_3BO_3
    B. H_3BO_4
    C. $Na_2B_4O_7$
    D. None of these

    Answer: A

15. Match the following
    A. Protectives   (i)   Control microbial growth
    B. Adsorbents   (ii) Cause precipitation of proteins

C. Antimicrobials

(iii) Adsorb particles, toxins, and bacteria from inflamed tissues

D. Astringents

(iv) Form protective coating over skin/mucus membrane

Answer:

(a) (iv)

(b) (iii)

(c) (i)

(d) (ii)

B. SHORT ANSWER QUESTIONS

1. Short notes (Any 3)
   - (a) Astringents;
   - (b) Protectives and adsorbents
   - (c) Astringents
   - (d) Silver nitrate
   - (e) Chlorhexidine gluconate

2. Discuss the three mechanisms of antimicrobial action with suitable examples

3. Define in one sentence (any 5)
   - (a) Antiseptics;
   - (b) Disinfectants
   - (c) Germicides;
   - (d) Bacteriostatic;
   - (e) Bactericidal;
   - (f) Sterilisers.

C. LONG ANSWER QUESTIONS

1. Discuss the mechanism of action of antimicrobial agents and discuss further one each based on these mechanisms

2. Discuss preparation, properties, uses, and an assay of
   - (a) Hydrogen peroxide,
   - (b) Potassium permanganate, and
   - (c) Boric acid.

■■■

DENTAL PRODUCTS

CONTENTS

Pharmaceutical formulations, market preparations, storage conditions and uses of Dental inorganic products:

- Calcium carbonate,
- Sodium fluoride,
- Denture cleaners,
- Denture adhesives,
- Mouthwashes

♦ LEARNING OBJECTIVES ♦

After completing this chapter, the student should be able to understand:

1. Classification of various inorganic dental products
2. Chemical Names of various inorganic dental products
3. Chemical Structures of different inorganic dental products
4. Uses, Stability and storage conditions of inorganic dental products
5. Different types of terms used in oral care and dental care

3.4.1 INTRODUCTION

The human teeth function to mechanically break down food items by cutting and crushing them in preparation for swallowing and digesting. Humans have four types of teeth: incisors, canines, premolars, and molars, which each have a specific function.

Teeth are hard, mineral-rich structures which are used to chew food. They are not made of bone like the rest of the skeleton but have their own unique structure to enable them to break down food. Tooth enamel is the most mineralised tissue in the body, consisting mainly of the rock-hard mineral hydroxyapatite.

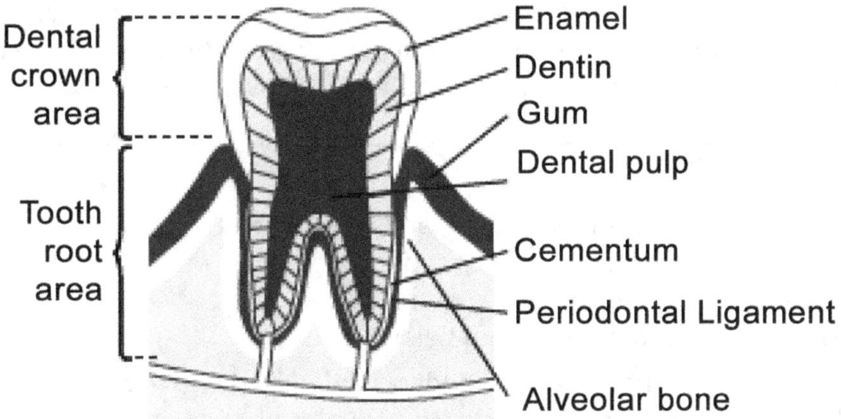

Fig. 3.4.1 Structure of the tooth

A tooth consists of enamel, dentin, cementum and pulp tissue. The portion of a tooth exposed to the oral cavity is known as the dental crown, and the portion below the dental crown is known as the tooth root. The dental pulp cavity exists in the centre of the tooth, through which the dental pulp, called the nerve, runs. In order to receive an impact on the tooth and to absorb and alleviate the force on the jaw, the surface of the tooth root area (cementum) and the alveolar bone are connected by a fibrous tissue called the periodontal ligament. The tooth is supported by the tissue consisting of the alveolar bone, gums and the periodontal ligament.

Enamel: The hardest bodily tissue covering the surface of the dental crown. It is as hard as a crystal.

Dentin: The tissue that forms the tooth from the dental crown to the tooth root, situated inside the enamel and cementum. It is softer than the enamel. A small tube filled with tissue fluid, called the dentinal tubule, runs inside the dentin.

Cementum: The tissue covering the surface of the tooth root. It connects the alveolar bone with the tooth by the periodontal ligament. Its hardness is similar to bone.

Dental pulp: The tissue is called the nerve. Blood vessels and lymph vessels, as well as nerve fibres, are located in the dental pulp, supplying nutrients to the dentin.

Periodontal ligament: Tissue consisting mainly of the fibrous tissue that connects the tooth root and the alveolar bone. It prevents force applied to the tooth from being directly imposed on the alveolar bone while chewing food.

Alveolar bone: The jaw bone supports the tooth; the tooth is planted into this bone. When a large part of the alveolar bone is destroyed by periodontal disease or other causes, the tooth becomes loose.

Gingiva: The soft tissue covering the alveolar bone. It is generally called "gum".

Gingival sulcus: The small space between the tooth and the gums. Even people with healthy teeth usually have a depth of 1 to 2 mm in this space. When this space deepens due to inflammation, it is called the periodontal pocket or gingival pocket.

3.4.2 DENTAL PRODUCTS

Dental products are mainly used for treating and cleaning teeth. Dental products are mainly used in the form of toothpaste, transparent toothpaste, mouthwash, tooth cleaning powder, gel, cleaning agents for dentures, chewing gum etc.

Some important definitions are mentioned below:

(a) **Abrasives:** Solid materials that are added to dentifrices to facilitate the mechanical removal of dental plaque, debris, and stains from tooth surfaces.

(b) **Anticaries drugs:** A drug that aids in preventing and prophylactic treatment of dental cavities (decay, caries).

(c) **Dental caries:** A disease of calcified tissues of the teeth characterised by demineralisation of the inorganic portion and destruction of the organic matrix.

(d) **Dentifrices:** Dentrifice can be defined as any substance especially prepared for cleaning the surfaces of the teeth. An abrasive-containing dosage form (gel, paste, or powder) for delivering an anticaries drug to the teeth.

(e) **Fluoride:** The inorganic form of the chemical element fluorine in combination with other elements.

(f) **Fluoride ion:** The negatively charged atom of the chemical element fluorine.

3.4.2.1 Abrasives

An abrasive is a substance that is used for abrading, grinding, or polishing. The degree of abrassion depends on the hardness of the abrasive, the morphology of the particles and the concentration of an abrasive in the paste. The abrasives found in toothpaste are often not as hard as the enamel but as hard as or harder than the dentine. Abrasives are often found as crystals. Small and smooth particles are preferred to avoid tooth wear.

Transparent toothpaste, commonly called gel toothpaste, is obtained by mixing certain abrasives. The amount and type of abrasive in toothpaste contribute to giving the toothpaste its creamy consistency. The abrasive effect is measured in the RDA (Radioactive Dentine Abrasion) scale, ranging from 40-80 in most toothpastes. Hydrated silica is a common abrasive in dentifrices; alumina and calcium carbonate may also be used.

3.4.2.2 Cleaning Agent or Dentifrices

It is a material, a powder or paste used to clean teeth, which can be applied with a brush. Cleaning property depends on the rubbing force used.

Drawback: Cannot clean the surface, inside cavities, and crevices between the teeth. Dentifrices with useful substances are known as medicated dentifrices. Flavours and colours are used to improve acceptance. A good cleaning agent removes the stains from the teeth.

Examples of cleaning agents or dentifrices are calcium phosphate dibasic and tribasic calcium phosphate, calcium carbonate and sodium metaphosphate.

3.4.2.3 Dental Caries and Anticaries Agents

Dental caries is a pathologic process of microbial etiology that results in the localised destruction of tooth tissues. The process of tooth destruction involves the dissolution of the mineral phase, consisting primarily of hydroxyapatite crystals by organic acids produced by bacterial fermentation.

3.4.2.3.1 Caries Prevention

1. The prevention of dental caries is based on attempts to –
   a. Increase the resistance of the host (fluoride therapy, occlusal sealants, immunisation).
   b. Lower the number of cariogenic microorganisms in contact with the tooth (plaque control agents and antiplaque agents).
   c. Modify the substrate by selecting noncaricogenic foods.
   d. Reduce the time that the microflora is provided with a substrate by limiting the frequency of intake of a fermentable substrate.
2. The formation of bacterial plaque also helps the decay process by forming pockets or crevices on the tooth surface where the food particles can stick and be decayed by the bacteria. If plaque is not removed, it calcifies into calculus when calcium salt precipitates from the saliva. Brushing the teeth helps remove the material from the tooth surface before it hardens into calculus.
3. Dental caries can be prevented, and oral and dental hygiene can be maintained with the help of dentifrices. These are the products that enhance the removal of stains and dental plaque by a toothbrush.
4. The most accepted approach to prevent caries includes flossing and brushing accompanied by the administration of fluoride either internally or topically to the teeth.

3.4.2.4 Fluoride

Fluorides obtained from food and water are very effective in the prevention of dental caries.

Fluoridation: Fluoridation can be carried out by the addition of fluoride to the water. However, high fluoride causes a mottling of teeth, increased density of bones, gastric disturbance, muscular weakness, convulsions and heart failure.

Role of fluoride when salt or solution of fluoride is taken internally: Fluoride is absorbed, transported, and deposited in bones or developing teeth. The deposited fluoride on the surface of the teeth does not allow the action of acids or enzymes. A very less amount of fluoride (1 ppm) is required for this purpose.

Route of administration:

The route of administration is orally and topically. Drinking water (0.5 to 1 ppm), fruit juice (1 ppm), sodium fluoride tablets or solution, 2.2 mg or topical application of 2% solution are the sources.

A wide range of therapeutic fluoride concentrations is used as topical agents to prevent caries.

Method	Fluoride concentration (ppm)
Dentifrices, adult	1000-1500
Dentifrices, children	250-500
Mouth rinse	230
Self applied gels or rinses, prescription	5000

Fluoride is considered to be the most effective caries-inhibiting agent, and almost all toothpastes today contains fluoride in one form or the other. The most common form is sodium fluoride (NaF). However, mono-fluoro-phosphate (MFP) and stannous fluoride (SnF) are also used. The fluoride amount in toothpaste is usually between 0.10 – 0.15%. Fluoride is most beneficial when the mouth is not rinsed with water after tooth brushing. In this way, a larger amount of fluoride is retained in the oral cavity. Toothpaste is the main vehicle for fluoride. The combined therapeutic and cosmetic mouthwashes usually also contain fluoride but in a non-therapeutic dose. However, there are fluoride rinses with higher fluoride concentrations.

There are three main theories considering the positive action of fluoride in preventing caries.

1. It is claimed that fluoride, incorporated into the enamel during tooth development in the form of fluorohydroxyapatite (FAP), reduces the solubility of the apatite salt. This theory implies that "caries resistance", once obtained, will always last the fluoride provided during the formation of the teeth is significantly more effective than when given later on. This theory has some draw backs since individuals who are born and raised in an area with fluoridated water, and therefore, have their teeth dehydrate under optimum fluoride conditions; quickly achieve a caries incidence characteristic of their new location if they leave the fluoridated area. Too much fluoride during tooth development can cause dental fluorosis.

2. It is also suggested that fluoride has anti-bacterial action. In an acidic environment, if fluoride is present, hydrogen fluoride (HF) is formed. HF is an undissociated, weak acid that can penetrate the bacterial cell membrane. The entry of HF into the alkaline cytoplasmic compartments results in dissociation of HF to H and F. This has two separate, major effects on the physiology of the cell. The first is that the released F interacts with cellular constituents, including various F-sensitive enzymes. The second

effect is an acidification of the cytoplasmic compartment caused by the released protons. Normally, protons are pumped out of the cell, but fluoride inhibits these processes. The decreased intracellular pH will make the environment less favorable for many of the essential enzymes required for cell growth.

3. Today, the most important anti-caries effect is claimed to be due to the formation of calcium fluoride (CaF_2) in plaque and on the enamel surface during and after rinsing or brushing with fluoride. CaF_2 serves as a fluoride reservoir. When the pH drops, fluoride and calcium are released into the plaque fluid. Fluoride diffuses with the acid from plaque into the enamel pores and forms fluoroapatite (FAP). FAP incorporated in the enamel surface is more resistant to a subsequent acid attack since the critical pH of FAP (pH = 4.5) is lower than that of hydroxyapatite (HA) (pH = 5.5). Fluoride decreases the dehydration and increases the remineralization of the enamel between pH 4.5 – 5.5, and hence, the dehydration period is shortened.

3.4.2.5 Polishing Agents

Dentifrices contain agents for cleaning tooth surfaces and providing a polishing effect on the cleaned teeth. These agents are abrasive in nature. They are responsible for physically removing plaque and debris. Pumice is too abrasive for daily use in a dentifrice.

e.g.; Dicalcium phosphate, sodium metaphosphate, calcium pyrophosphate, calcium carbonate, and calcium monohydrogen phosphate.

3.4.2.6 Desensitizing Agents

Desensitizing agents reduce the pain in sensitive teeth caused by cold, heat or touch. These products should be non-abrasive and should not be used on a regular basis unless directed by a dentist.

e.g.; Strontium chloride and zinc chloride.

3.4.2.7 Anti-calculus Agents

The crystal growth inhibitors have been the most extensively tested clinically. These agents act by delaying dental plaque calcification, thereby, promoting plaque removal with normal tooth brushing.

e.g.; Triclosan

3.4.2.8 Whitening Agents

Whitening toothpastes do not lighten the colour of the tooth structure; they simply remove surface stains with abrasives or special chemical or polishing agents, or prevent stain formation.

e.g.; abrasives and bleaching agents such as silica, pyrophosphates, hydrogen peroxide or carbamide peroxide.

3.4.2.9 Pumice

Pumice is a substance of volcanic origin, produced when lava with a very high content of water and gases is thrown out of a volcano. As the gas bubbles escape from the lava, it becomes frothy. When this lava cools and hardens, it results in a very light rock material filled with tiny bubbles of gas. It consists mainly of complex silicates of aluminum, potassium, and sodium. It is very light, hard, rough, porous, grayish mass. It is odourless and stable in air.

3.4.3 MOUTHWASH, MOUTH RINSE, ORAL RINSE OR MOUTH BATH

It is a liquid which is held in the mouth passively or swilled around the mouth by contraction of the perioral muscles and/or movement of the head, and may be gargled, where the head is tilted back and the liquid bubbled at the back of the mouth.

Usually mouthwashes are antiseptic solutions intended to reduce the microbial load in the mouth, although other mouthwashes might be given for other reasons such as for their analgesic, anti-inflammatory or anti-fungal action. Additionally, some rinses act as saliva substitutes to neutralize acid and keep the mouth moist in xerostomia (dry mouth). Cosmetic mouth rinses temporarily control or reduce bad breath and leave the mouth with a pleasant taste.

Rinsing with water or mouthwash after brushing with a fluoride toothpaste can reduce the availability of salivary fluoride. This can lower the anti-cavity re-mineralization and antibacterial effects of fluoride. Fluoridated mouthwash may mitigate this effect or in high concentrations increase available fluoride, but is not as cost effective as leaving the fluoride toothpaste on the teeth after brushing.

Ingredients of Mouthwash: Some of the common ingredients used-(Not all in one single type)

  i.   Alcohol- To destroy bacteria and also to dissolve active ingredients like thymol, eugenol etc.
  ii. NSAID's: To reduce pain and inflammation, e.g., betamethasone, benzydamine, methyl salicylate, etc.
  iii. Local Anaesthetics: To treat mucositis e.g., lidocaine, xylocaine
  iv. Nystatin- As an antifungal to oral candidiasis
  v.   Anti-infective/Antibiotics- Povidone iodine, tetracycline
  vi. Antiseptic- Triclosan, chlorhexidine digluconate, cetyl pyridinium chloride,
  vii. Anticavity: Sodium fluoride against tooth decay
  viii. Anti-plaque agents- Sanguinarine
  ix. Mucosal coating protective- Sucralfate
  x.   Desensitizers- Potassium oxalate

xi. Antibacterial essential oils-contain phenolic compounds like menthol, eugenol, thymol, xylitol etc.

xii. Flavoring agents: include sweeteners such as sorbitol, sucralose, sodium saccharin, and xylitol, which stimulate salivary function due to their sweetness and taste and helps restore the mouth to a neutral level of acidity.

xiii. Edible oils: Sesame oil, coconut oil and ghee are traditionally used,[68] but newer oils such as sunflower oil are also used.

xiv. Salts for saline taste and action- Sodium bicarbonate, sodium chloride

xv. Foaming agents- Sodium lauryl sulfate

3.4.4 PREPARATION AND ASSAY OF SOME IMPORTANT ANTI-CARIES AGENTS

3.4.4.1 Sodium Fluoride: NaF Mol. wt: 41.99

It contains not less than 98.5 percent and not more than 100.5 percent of NaF, calculated with reference to the dried substance.

Preparation

It is prepared by reacting hydrofluoric acid with sodium carbonate. Sodium fluoride being not very soluble precipitates out.

$$2HF + Na_2CO_3 \rightarrow 2NaF + H_2O + CO_2\uparrow$$

The precipitate is contaminated with fluorosilicate and the acid salt. It is made alkaline to phenolphthalein with sodium carbonate and then heated to dehydrate   the acid salt and decompose the fluorosilicate.

$$Na_2SiF_6 + 2H_2O \rightarrow 2NaF + 4HF + SiO_2$$

Assay

It is assayed by the non-aqueous titration method.

Weigh accurately about 80 mg of sodium fluoride and add a mixture of 5 ml of acetic anhydride and 20 ml of anhydrous glacial acetic acid to it. Heat to dissolve, cool, add 20 ml of dioxan and titrate with 0.1M perchloric acid using crystal violet solution as an indicator until a green colour is produced. Carry out a blank determination and make any necessary corrections.

Each ml of 0.1M perchloric acid is equivalent to 0.004199 of NaF.

Uses

1. It is used as a preventive for dental caries because of its fluoride ion content.

2. 5-3.0 ppm (equivalent to 0.7-1.3 ppm of fluoride ion) in drinking water, 2% solution as a topical application to the teeth is the common means of providing fluoride.

Sodium fluoride because of its fluoride ions is an important agent in dental practice for retarding or preventing dental caries.

3.4.4.2 Stannous Fluoride: SnF_2 Mol. wt: 156.69

It contains not less than 71.2 percent of stannous (Sn^{2+}) ions and not less than 22.3 percent and not more than 22.5 percent of fluoride, calculated with reference to the dried substance.

Preparation

Stannous fluoride is prepared by heating stannous oxide with gaseous hydrofluoric acid in the absence of oxygen.

$$SnO + 2HF \rightarrow SnF_2 + H_2O$$

Assay

Pipette 20 ml of each standard preparation and the assay preparation into separate plastic beakers. Add 20 ml of buffer solution into each beaker. Concomitantly measure the potential of the solutions from the standard preparations and assay preparation using a pH meter equipped with a fluoride specific ion indicating electrode and a calomel reference electrode. Plot the logarithms of the fluoride ion concentrations, in µg per ml of the standard preparations versus potential, in mV. Determine the concentration, C, in µg per ml, of fluoride ion in the assay preparation from the measured potential of the assay preparation. Calculate the percentage of fluoride by the formula.

$$\text{Percentage of fluoride} = 125 \, C/W$$

Where C is the concentration of fluoride determined in assay preparation and W is the weight of the stannous fluoride taken.

Uses

1. It is used as a preventive for dental caries.
2. A freshly prepared 8% solution is used at 6 to 12 month intervals.

3.4.4.3 Zinc chloride

$ZnCl_2$ Mol. wt: 136.29.

It contains not less than 95 percent and not more than 100.5 percent of $ZnCl_2$.

Properties

A white or practically white, crystalline powder; odourless; very deliquescent.

Preparation

It is prepared by heating granulated zinc with hydrochloric acid. When the evolution of hydrogen ceases, the solution is filtered and evaporated to dryness

$$Zn + 2\,HCl \rightarrow ZnCl_2 + H_2\uparrow$$

Assay

It is assayed by complexometry using strong ammonia–ammonium chloride solution as buffer, eriochrome black T solution as an indicator and titrating with 0.1M disodium edetate.

$$ZnCl_2 + C_{10}H_{14}N_2Na_2O_8 \rightarrow C_{10}H_{14}N_2Na_2Zn + 2NaCl$$

Weigh accurately about 3 g of $ZnCl_2$ dissolved in 125 ml of water, add 3 g of ammonium chloride and make up the volume to 250 ml with water. To 25 ml of the resulting solution add 100 ml of water and 10 ml of strong ammonia-ammonium chloride solution. Titrate with 0.1 M disodium edetate using eriochrome black T solution as an indicator until a deep blue colour is produced.

Each ml of 0.1 M disodium edetate is equivalent to 0.01363 g of $ZnCl_2$.

Uses

1. It is used as an antiseptic astringent to the skin and mucous membrane as a 0.5 – 2.0% solution.

2. It ranks very low among disinfectants.

3. It is used as an active ingredient to prepare magnesia cements for dental fillings and certain mouthwashes.

4. It is also used as a dental desensitizer, topically as a 10% solution to the teeth.

3.4.4.4 Calcium Carbonate (Precipitated Chalk)

$CaCO_3$ Mol. wt: 100.1

Calcium carbonate contains not less than 98.0 per cent and not more than 100.5 per cent of $CaCO_3$, calculated on the dried basis.

Properties

A fine, white, microcrystalline powder practically insoluble in water and in ethanol (95 per cent); slightly soluble in water containing carbon dioxide or any ammonium salt. It is soluble with effervescence in dilute acids.

Preparation

It is prepared by mixing and boiling a solution of calcium chloride and sodium carbonate.

$$CaCl_2 + Na_2CO_3 \rightarrow CaCO_3 + 2 NaCl$$

The precipitate is collected on a filter, washed with boiling water and dried.

Assay

Weigh accurately about 0.1 g of $CaCO_3$ and dissolve in 3 ml of dilute hydrochloric acid and 10 ml of water. Boil for 10 minutes, cool, dilute to 50 ml with water. Titrate with 0.05 M disodium edetate. Within a few ml of the expected end-point, add 8 ml of sodium hydroxide solution and 0.1 g of calcon mixture and continue the titration until the colour of the solution changes from pink to a full blue colour.

1 ml of 0.05 M disodium edetate is equivalent to 0.005004 g of $CaCO_3$.

Uses

The precipitated chalk is used externally as dentifrices because it has a mild abrasive quality. It forms common ingredients of tooth powder and tooth paste.

3.4.4.5 Dibasic Calcium Phosphate (Calcium Hydrogen Phosphate)

$CaHPO_4$; Mol. wt 136.1 (anhydrous)

$CaHPO_4$, $2H_2O$; Mol. wt 172.1 (dehydrate)

Dibasic Calcium Phosphate is anhydrous and contains two molecules of water of hydration.

Dibasic Calcium Phosphate contains not less than 98.0 per cent and not more than 105.0 per cent of $CaHPO_4$ (for anhydrous material) or of $CaHPO_4$, $2H_2O$ (for the dihydrate).

Properties

A white crystalline, odourless powder.

Preparation

It may be prepared by reacting a natural solution of calcium chloride with disodium hydrogen phosphate.

$$CaCl_2 + Na_2HPO_4 \rightarrow CaHPO_4 + 2NaCl$$

Assay

Weigh accurately about 0.3 g of dibasic calcium phosphate and dissolve in a mixture of 5 ml of water and 1 ml of 7 M hydrochloric acid, add 25.0 ml of 0.1 M disodium edetate and dilute to 200 ml with water. Neutralise with strong ammonia solution, add 10 ml of ammonia buffer pH 10.0 and 50 mg of mordant black 11 mixtures and titrate the excess of disodium edetate with 0.1 M zinc sulphate.

1 ml of 0.1 M disodium edetate is equivalent to 0.01361 g of $CaHPO_4$ or 0.01721 g of $CaHPO_4$, $2H_2O$.

Uses

1. This calcium salt has a 1:1 ratio of calcium to phosphorus and is most frequently recommended for oral administration as an electrolyte replenishes.

2. As a salt it supplies both calcium and phosphorus, which is required for the growth in children, pregnant women and lactating mothers.

3. Externally, it is used as a dentifrice having cleaning action and the moderate abrasive quality makes it suitable for tooth paste and tooth powders.

3.4.5 DENTAL ADHESIVES: ZINC EUGENOL (ZOE) DENTAL CEMENT

Zinc eugenol cements are considerably better tolerated by tissues and patients than other dental materials, as they alleviate pain and are bacteriostatic as well as, antiseptic.

These cements are good insulators and possess better sealing properties than zinc phosphate cements. Because of their poor mechanic properties, the conventional zinc oxide eugenol cements are mainly used as temporary fixing contents and filling materials, for gingival dressings and together with filling materials as impression materials. However, recently reinforced zinc oxide-eugenol cements and cements containing ethoxy benzoic acid (EBA) have been developed. These new cements have considerably better mechanic properties and are therefore used for cement bases, indirect capping, long-term temporary fillings and in selected cases as definite fixing cements.

Zinc Eugenol Dental Cement is a material created by the combination of zinc oxide and eugenol through an acid-base reaction that takes place with the formation of zinc eugenolate chelate. Therefore, they are also called as zinc oxide-eugenol cements (ZOE).

ZOE cements were introduced in the 1890s. They are classified as intermediate restorative materials and have anesthetic and antibacterial properties. ZOE can be used as a filling or cement material in dentistry. It is often used in dentistry when the decay is very deep or very close to the nerve or pulp chamber. Because the tissue inside the tooth, i.e., the pulp, reacts badly to the drilling stimulus (heat and vibration), it frequently becomes severely inflamed and precipitates a condition called acute or chronic pulpitis. This condition usually leads to severe chronic tooth sensitivity or actual toothache and can then only be treated with the removal of the nerve (pulp) called root canal therapy.

The placement of a ZOE "temporary" for a few to several days prior to the placement of the final filling can help to sedate the pulp. It is classified as an intermediate restorative material and has anesthetic and antibacterial properties. It is sometimes used in the management of dental caries as a "temporary filling".

ZOE is also used as an impression material during construction of complete dentures and is used in the mucostatic technique of taking impressions, usually in a special tray, (acrylic) produced after primary alginate impressions. However, ZOE is not used if the patient has large undercuts or tuberosities, where silicone impression materials would be better suited.

Types:

According to ANSI/ADA Specification no: 30 (ISO 3107) and depending on intended use and individual formulation designed for each specific purpose, they are classified into four types:

- Type I: Temporary ZOE Luting cement
- Type II: Long term ZOE Luting cement

- Type III: Temporary ZOE Restoration
- Type IV: Intermediate ZOE Restoration

Composition

The chemical composition of ZOE is typically:

- Zinc oxide, ~69.0%
- White rosin, ~29.3%
- Zinc acetate, ~1.0% (improves strength)
- Zinc stearate, ~0.7% (acts as accelerator)
- Liquid (eugenol, ~85%, olive oil ~15%)

ZOE impression pastes are dispensed as two separate pastes. One tube contains zinc oxide suspended in vegetable or mineral oil; the other contains eugenol and rosin. The vegetable or mineral oil acts as a plasticizer and aids in offsetting the action of the eugenol as an irritant.

Oil of cloves, which contains 70% to 85% eugenol, is sometimes used in preference to eugenol because it produces less burning sensation for patients when it contacts the soft tissues. The addition of rosin to the paste in the second tube facilitates the speed of the reaction and yields a smoother, more homogenous product.

The widely used five dental cements presently are

1. Glass Ionomer Cement.
2. Composite Resin Cement.
3. Zinc Oxide-Eugenol Cement.
4. Polycarboxylate Cement.
5. Zinc Phosphate Cement.

QUESTION BANK

A. MULTIPLE CHOICE QUESTIONS

1. ________________ are used for cleaning the teeth depending upon their action or used divided into
   A. Abrassives
   B. Anticarries agents
   C. Dentrifices
   D. Antiplaque agents

Answer: C

2. ____________ remove the stain from the teeth
    A. Dentifrices
    B. Abrassives
    C. Desensitising Agents
    D. Antiplaque agents

Answer: B

3. Calcium carbonate is an example of ________________
    A. Dentrifices
    B. Abrassives
    C. Polishing Agent
    D. All the above

Answer: D

4. Role of Fluoride in treatment of denatal carries is
    A. Promoting remineralization
    B. Slowing down demineralization
    C. Action on Plaque
    D. All of above

Answer: D

5. Molecular Weight of Calcium carbonate is ______
    A. 100
    B. 103
    C. 105
    D. 107

Answer: A

6. ___________ is not a dental desensitizing agent.
    A. strontium chloride
    B. zinc chloride
    C. silver nitrate
    D. sodium fluoride

Answer: D

7. Find the odd one which is not used as dental cement
    A. Glass Ionomer Cement
    B. Composite Resin Cement
    C. Reinforced Resin Cement
    D. Zinc Oxide-Eugenol Cement

Answer: C

8. Dental polishing substance from Volcanic origin?
    A. Rochelle's Salt
    B. Black Salt
    C. Pumice
    D. Basalt

Answer: C

9. Chemical formula of Strontium chloride is _________
    A. SrCl
    B. $SrCl_2$
    C. $SrCl_3$
    D. $SrCl_4$

Answer: B

10. Chemical formula of Stannous fluoride is _________
    A. SnF
    B. SnF_2
    C. SnF_3
    D. SnF_4

Answer: B

11. Preventive measures for Dental carries include_________
    A. Fluoride use
    B. Antiplaque agents
    C. Avoid or reduce intake of fermentable substances
    D. All of these

Answer: D

12. Dental tissues of tooth are_________
    A. Enamel
    B. Cementum
    C. Dentine
    D. All of these

Answer: D

13. Which of the following is dental desensitizing agent
    A. Sodium Fluoride
    B. Zinc Chloride
    C. Strontium chloride
    D. Both b & c.

Answer: D

14. _______ is used as an antifungal agent in oral mouthwash to combat oral candidiasis
    A. povidone iodine
    B. nystatin
    C. chlorhexidine digluconate
    D. cetyl pyridinium chloride

Answer: B

15. ZOE cement contains
    A. Zinc acetate
    B. Zinc stearate
    C. Zinc oxide
    D. All of the above

Answer: D

B. SHORT ANSWER QUESTIONS

1. Write notes on any three-
    (a) Anticaries agents.
    (b) Calcium compounds as dentifrices.
    (c) Preparation, properties, uses and assay of sodium fluoride
    (d) Role of fluoride as anticaries agent.
    (e) Zinc eugenol cements
2. Discuss mechanism of action of anti-caries agents.
3. Explain the role of fluoride and phosphate in tooth decay.
4. Explain the role of fluorides in tooth decay.
5. Give composition of ZOE
6. Discuss Dentifrices.
7. Discuss Desensitising agents
8. Discuss the composition of Oral mouthwashes

C. LONG ANSWER QUESTIONS

1. Discuss various dental products in details with their roles, composition as well as functions
2. Give an elaborate account on following terms and agents requires to take care
    (a) Plaque   (b) Calculus;
    (c) Caries   (d) Sensitization
    (e) Tartar

■■■

MEDICINAL GASES

CONTENTS

Pharmaceutical formulations, market preparations, storage conditions and uses of medicinal gases

- Carbon dioxide,
- Nitrous Oxide,
- Oxygen

♦ LEARNING OBJECTIVES ♦

After completing this chapter, the student should be able to understand:

1. The term medicinal gases
2. Names of various medicinal gases, their properties, storage
3. Chemical Structures of different medicinal gases
4. Storage conditions of medicinal gases

3.5.1 INTRODUCTION

Medical gases are those which are manufactured, packaged, and intended for administration to a patient in anaesthesia, therapy, or diagnosis. The officially listed therapeutic gases include Oxygen, Helium, Carbon dioxide, Nitrous oxide, Medical air and Nitrogen. These gases are usually given to the pre-operated, intra-operated and post-operated patients and to the patients in case of emergency. These gases are to be supplied in the airtight, colour-coded, well-labelled container as required by the respective regulatory authorities of every country or *via* a central line which runs through the entire hospital. To avoid mixing up of gases, there are special colour codes for storing these gases separately.

Color Coding & Storage of Cylinders Medical gases are stored in heavy steel cylinders made to withstand the pressure of about 200bars. The cylinders are similar in appearance. To reduce confusion, each type of gas is given a colour coding that readily identifies the gas. The symbol or name of the gas is stencilled on the cylinder shoulder.

Table 3.1.1 Storage of Medicinal Gases

Gas	Color code	Liquid or gas in cylinder
Carbon dioxide	Grey	Liquid
Cyclopropane	Orange	Liquid
Helium	Brown	Gas
Nitrogen	Grey body Black top	Gas
Nitrous oxide	Blue	Liquid
Oxygen	Black body White top	Gas
Oxygen and carbon dioxide mixture	Black body Grey & White top	Gas
Helium and Oxygen mixture 79 : 21	Black body Brown and white top	Gas

Note: *The name of the gas or its chemical symbol to be stencilled or painted on the shoulder of the cylinder*

Storage:

1. Cylinders should be stored in a cool, well-ventilated room free from inflammable materials.
2. The room should be large enough for proper grouping according to the cylinder contents, hence avoiding confusion.
3. A special storage rack is designed to ensure the use of old stock first and is lined to prevent defacement (damage) of identifying labels and colours.
4. The colour chart is put up in a visible place in the storage room.

3.5.2 CARBON DIOXIDE

Properties: Carbon Dioxide (CO_2) is a colourless & odourless gas, acidic in nature, with a density about 53% higher than that of dry air. Carbon dioxide molecules consist of a carbon atom covalently double bonded to two oxygen atoms.

$$O=C=O$$

Solubility: 1 part in 10 volumes of water at room temperature and pressure to form carbonic acid (H_2CO_3).

Occurrence, Preparation and Availability

It occurs naturally in Earth's atmosphere as a trace gas. The current concentration is about 0.04% (417 ppm) by volume. Carbon dioxide can be obtained by distillation from air, but the method is inefficient. Industrially, carbon dioxide is predominantly an unrecovered waste product, produced by several methods which may be practised at various scales. The combustion of all carbon-based fuels, such as methane (natural gas), petroleum distillates (gasoline, diesel, kerosene, propane), coal, wood and generic organic matter, produce carbon dioxide and, except in the case of pure carbon, water. An example is a chemical reaction between methane and oxygen. Iron is reduced from its oxides with coke in a blast furnace, producing pig iron and carbon dioxide.

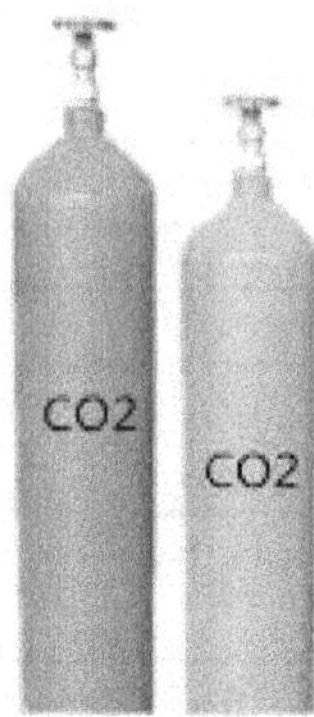

Uses: Carbon dioxide is administered to the patient *via* a face mask.

1. It is used in emergencies to induce and improve the respiration rate in newborn babies, drowning persons, and in cases of poisoning by carbon monoxide, morphine, hypnotics and other depressants.
2. Mixtures of oxygen and carbon dioxide may be administered to hasten the exhalation of anaesthetic gases after surgical operations and thus reduce the post-operative vomiting and bronchitis.
3. Frozen carbon dioxide gas (dry ice) is used to remove warts.
4. Other Medical Uses:

   - As a medical gas for noninvasive surgery, to inflate and stabilise body cavities for increased visibility and to increase blood flow to the brain;
   - to remedy bronchial spasms, to stimulate respiration, and for clinical and physiological examinations.
   - Carbon dioxide is also used as an atmosphere for organs which are artificial or awaiting transplant and as a tracer gas for pharmaceutical package testing.

3.5.3 NITROUS DIOXIDE

Nitrous Oxide (N_2O, Laughing gas), Mol. wt 44,

It is a chemical compound, an oxide of nitrogen. At room temperature, it is a colourless, non-flammable gas and has a slightly sweet scent and taste. At elevated temperatures, nitrous oxide is a powerful oxidiser similar to molecular oxygen.

$$N\equiv\overset{+}{N}-O^- \longleftrightarrow {}^-N=\overset{+}{N}=O$$

Preparation-Industrial method

Nitrous oxide is prepared on an industrial scale by careful heating of ammonium nitrate at about 250 °C, which decomposes into nitrous oxide and water vapour.

$$NH_4NO_3 \rightarrow 2\,H_2O + N_2O$$

Laboratory method

The decomposition of ammonium nitrate is also a common laboratory method for preparing the gas. Equivalently, it can be obtained by heating a mixture of sodium nitrate and ammonium sulfate

$$2\,NaNO_3 + (NH_4)_2SO_4 \rightarrow Na_2SO_4 + 2\,N_2O + 4\,H_2O$$

Another method involves the reaction of urea, nitric acid and sulfuric acid

$$2\,(NH_2)_2CO + 2\,HNO_3 + H_2SO_4 \rightarrow 2\,N_2O + 2\,CO_2 + (NH_4)_2SO_4 + 2\,H_2O$$

Uses: This is the oldest and safest gaseous weak general anaesthetic. At least 21% of the gas mixture must be oxygen Induction is quick and pleasant, and recovery is rapid, hence it is the most popular gas for producing light anaesthesia or analgesia in dentistry and obstetrics.

Disadvantages: It has low anaesthetic potency. A large dose of this gas is required to produce anaesthesia; this may cause a lack of oxygen.

N_2O also helps alleviate anxiety in patients. Nitrous oxide is commonly referred to by its nickname "laughing gas" because of its intoxicating effects, and when used as an analgesic, it is usually a precursor to other intravenous or oral painkillers. It is considered an alternative therapy for narcotic addiction.

3.5.4 OXYGEN

Oxygen is the chemical element with the symbol O and atomic number 8. It is a member of the chalcogen group in the periodic table, a highly reactive nonmetal, and an oxidising agent that readily forms oxides with most elements and other compounds. Oxygen is Earth's most abundant element, and after hydrogen and helium, it is the third-most abundant element in the universe. At standard temperature and pressure, two atoms of the element bind to form dioxygen, a colourless and odourless diatomic gas with the formula O_2. Diatomic oxygen gas currently constitutes 20.95% of the Earth's atmosphere, though this has changed considerably over long periods of time. Oxygen makes up almost half of the Earth's crust in the form of oxides.

Properties: It is in form of a colourless gas normally; while when compressed, it is a pale blue liquid.

Liquid Oxygen   **Oxygen Cylinder**

Oxygen dissolves more readily in water than in nitrogen and in freshwater more readily than in seawater. Water in equilibrium with air contains approximately 1 molecule of dissolved O_2 for every 2 molecules of N_2 (1:2), compared with an atmospheric ratio of approximately 1:4. The solubility of oxygen in water is temperature-dependent and about twice as much (14.6 mg/L) dissolves at 0 °C than at 20 °C (7.6 mg/L).

Preparation: The most common method is a fractional distillation of liquefied air, with N_2 distilling as a vapour while O_2 is left as a liquid.

The other primary method of producing O_2 is passing a stream of clean, dry air through one bed of a pair of identical zeolite molecular sieves, which absorbs the nitrogen and delivers a gas stream that is 90% to 93% O_2.

Chemical oxygen generators: They store oxygen in their chemical composition and can be used only one time.

Oxygen Candles contain a mix of sodium chlorate and iron powder. When ignited, it smoulders at about 600 °C (1,112 °F) and results in sodium chloride, iron oxide, and oxygen, about 270 litres per kg of mixture.

Some commercial airliners use emergency oxygen generators containing a mixture of sodium chlorate ($NaClO_3$), 5 per cent barium peroxide (BaO_2) and 1 per cent potassium perchlorate ($KClO_4$), which after ignition, reacts, releasing oxygen for 12 to 22 minutes while the unit reaches 500 °F (260 °C).

Tetramethylammonium ozonide (($CH_3)_4NO_3$) is proposed as a source of oxygen for generators because of its low molecular weight, being 39% oxygen.

Uses: Oxygen is used for a variety of purposes:

(A) During anaesthesia, because a person's need for oxygen continues when he/she is anaesthetised.

(B) To relieve anoxia (lack of oxygen in the tissues) that may be caused by:–

1. Inadequate oxygenation of the blood by the lungs. This may occur in pneumonia, chronic bronchitis, emphysema, pulmonary oedema, postoperative pulmonary complications, asphyxia in newborns and barbiturate poisoning.

2. Reduction of the circulating blood volume. This may result from coronary failure, the collapse of the peripheral circulation and shock.

3. Reduction of the oxygen-carrying capacity of the blood. This may be caused by severe anaemia, haemorrhage and carbon monoxide poisoning.

(C) To increase radiation efficiency in tumour therapy, the Oxygenation of tumours sensitises them to radiation. The oxygen must be used under high pressure, with the patient enclosed in a special chamber.

QUESTION BANK

A. MULTIPLE CHOICE QUESTIONS

1. _______________ is the colour of cylinder used to store Oxygen
    A. Green
    B. Grey
    C. Black
    D. Blue

Answer: C

2. _______________ is the colour of cylinder used to store Carbon dioxide
    A. Green
    B. Grey
    C. Black
    D. Blue

Answer: B

3. _______________ is the colour of cylinder used to store Nitrous oxide
    A. Green
    B. Grey
    C. Blac
    D. Blue

Answer: D

4. Nitrous oxide is called "Laughing gas" as it
    A. makes the person laugh
    B. makes the person euphoric
    C. makes the person depressed
    D. All of above

Answer: B

5. The gas not used during anaesthesia is ______
    A. Oxygen
    B. Nitrous Oxide
    C. Carbon dioxide
    D. None of the above

Answer: C

B. SHORT ANSWER QUESTIONS

1. Discuss in short the following Medicinal Gases-Any two
    (a) Carbon dioxide.   (b) Nitrous oxide
    (c) Oxygen

2. Discuss the medicinal uses of any two.
    (a) Carbon dioxide.   (b) Nitrous oxide
    (c) Oxygen

3. Explain the preparation and properties of any two.
    (a) Carbon dioxide.   (b) Nitrous oxide
    (c) Oxygen

C. LONG ANSWER QUESTIONS

1. Give a detailed account of medicinally used gases.

■■■

Chapter 4

INTRODUCTION TO NOMENCLATURE

CONTENTS

Introduction to the nomenclature of organic chemical systems with particular reference to heterocyclic compounds containing up to three rings

♦ LEARNING OBJECTIVES ♦

After completing this chapter, the student should be able to understand:

1. To know different types of Classes and Organic Compounds.
2. To understand the IUPAC Rules for Nomenclature of Organic as well as heterocyclic compounds.
3. To draw the chemical structure from the given chemical name.
4. To give the chemical name for the given chemical structure.

4.1 CLASSIFICATION OF ORGANIC COMPOUNDS

Organic chemistry is known as the chemistry of carbon compounds. Due to the unique nature of carbon atoms, it gives rise to the formation of a large number of compounds. Thus, this demands a separate branch of chemistry. The word organic is derived from the fact that in the 19th century, most of the then-known carbon compounds were considered to have originated in living organisms. Organic chemistry is important because it is the study of life and all of the chemical reactions related to life.

Broadly, organic compounds can be defined as those compounds which consist mainly of two elements- C (Carbon) & H (Hydrogen). Besides, these two main elements of organic compounds may contain in small proportion and number the elements mainly like N (nitrogen), O (oxygen), S (Sulphur) & X (Halogens), which are called as heteroatoms. The classification of organic compounds in different groups is based on the availability or absence of one or more of these heteroatoms. Many a times, these heteroatoms are also the part of the functional group of the organic compounds owing to their higher electronegativity.

A. Classification of organic compounds on the basis of functional group and elemental composition:

1. **Compounds containing carbon and hydrogen atoms only**:

   Hydrocarbons (Alkanes, Alkenes, Alkynes, Aromatic Hydrocarbons, Arylalkyl Hydrocarbons, Alicyclic Hydrocarbons).

2. **Compounds containing carbon, hydrogen and oxygen atoms only:**

   Alcohols, Phenols, Ethers, Epoxides, Carbonyl compounds, Aldehydes and Ketones, Carboxylic acids, Esters, Anhydrides.

3. **Compounds containing Carbon, Hydrogen and Nitrogen atoms only:**

   Amines and Imines, Nitriles, Hydrazines.

4. **Compounds containing Carbon, Hydrogen, and Halogens with or without Oxygen:**

   Alkyl Halides, Aryl Halides, Acyl Halides.

5. **Compounds containing Carbon, Hydrogen, Oxygen and Nitrogen atoms only:**

   Amides, Imides, Aldoximes, Ketoximes, Nitro compounds.

6. **Compounds containing Carbon, Hydrogen and Sulphur with/without Nitrogen, Oxygen and Halogen:**

   Sulphonic acids, Sulphonylhalides, Sulphonamides.

B. Classification of organic compounds on the basis of Chemical Structures: Organic compounds are also classified as per their chemical structures, mainly into two main categories; open chain or acylic or aliphatic compounds and closed chain or cyclic compounds. Further, the cyclic compounds having conjugated endo-cyclic double bond systems and following Huckel's rule for a number of π-electrons are classified as aromatic or heteroaromatic compounds, with the latter type containing one or more heteroatoms essentially. **[Fig. 4.1]**

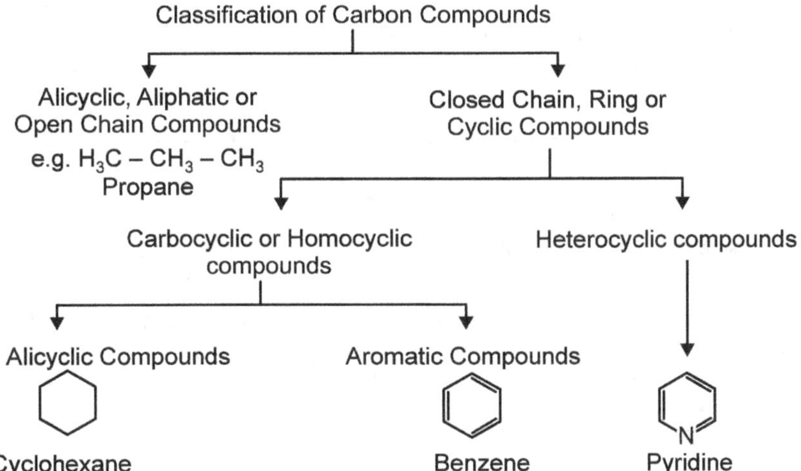

Fig. 4.1 Broad classification of Organic Compounds based on the chemical structures

Further, the classification can be made more systematically by considering the functional groups present in the compounds and accordingly dividing them into various subcategories. **[Fig. 4.2]**

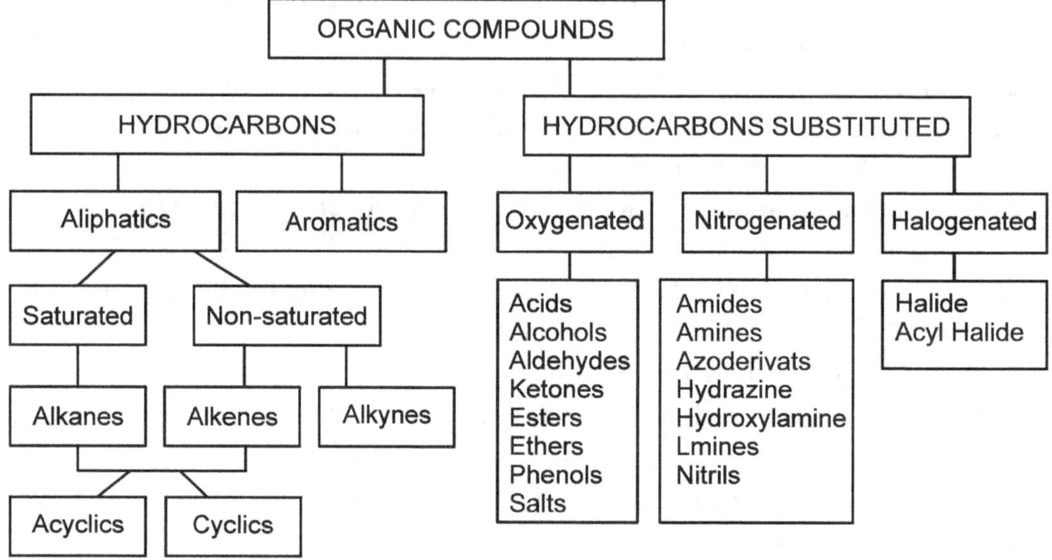

Fig. 4.2 Systematic classification of Organic Compounds based on the chemical structures & functional groups

4.2 NOMENCLATURE OF ORGANIC COMPOUNDS

4.2.1 Nomenclature based on Common Name System

4.2.1.1 Compounds Containing Carbon and Hydrogen Only

(a) Alkanes: C_nH_{2n+2}

Where n = 1, 2, 3, 4 etc., as a number of carbon atoms.

CH_4	H_3C-CH_3	$H_3C-H_2C-CH_3$	$H_3C-H_2C-H_2C-CH_3$	$H_3C-\overset{\overset{\displaystyle CH_3}{\textstyle\vert}}{\underset{\underset{\displaystyle H}{\textstyle\vert}}{C}}-CH_3$
1. Methane	2. Ethane	3. Propane	4. n-Butane	5. Isobutane

Prefix n is used for those alkanes, where all carbons are in a continuous chain. Prefix iso is used for that alkane, where a methyl group is attached to the second last carbon atom of the continuous chain.

Types of carbon atoms:

There are four types of carbon atoms as follows:

1. 1° Primary carbon: A carbon atom attached to only one (or all 4 hydrogens) carbon.
2. 2° Secondary carbon: A carbon atom attached to two other carbon atoms.
3. 3° Tertiary carbon: A carbon atom attached to three other carbon atoms.
4. 4° Quaternary carbon: A carbon atom attached to four other carbon atoms.

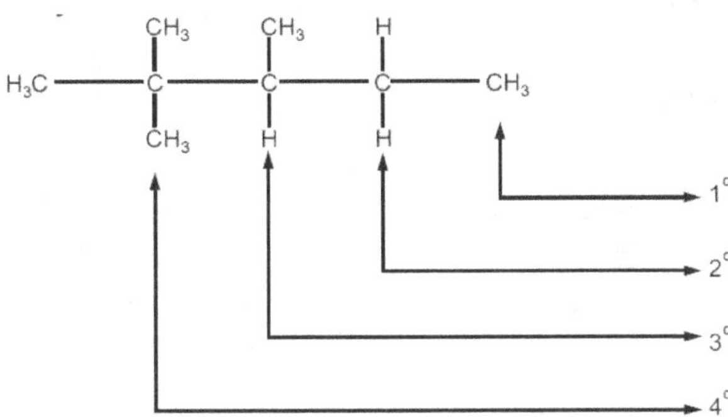

(b) Cycloalkanes: (C_nH_{2n})

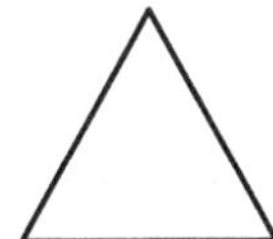

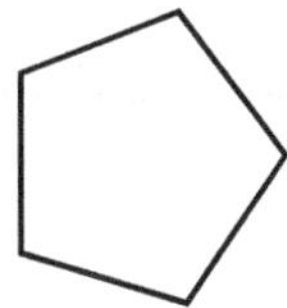
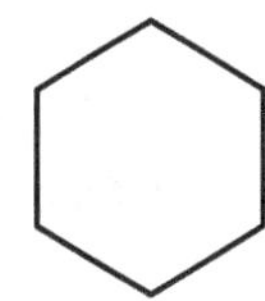

1. Cyclopropane   2. Cyclobutane   3. Cyclopentane   4. Cyclohexane

(c) Alkenes: (C_nH_{2n})

$CH_2=CH_2$ $CH_2=CH-CH_3$ $CH_2=CH-CH-CH_3$ $CH_3-CH=CH-CH_3$ $H_3C-C(CH_3)=CH_2$

1. Ethylene   2. Propylene   3. 1-Butene   4. 2-Butene   5. Isobutene

(d) Alkynes: (C_nH_{2n}-2)

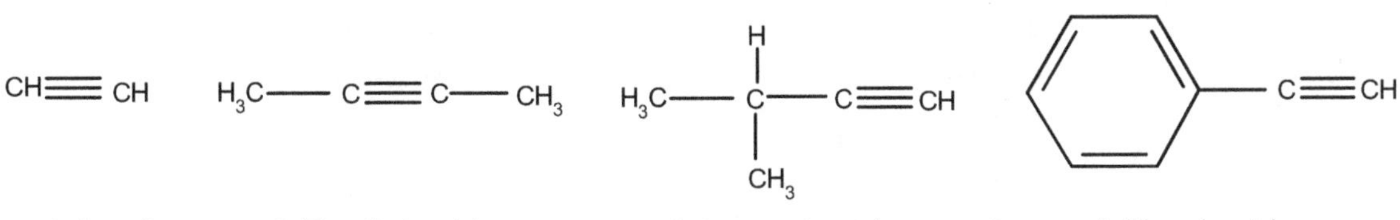

1. Acetylene   2. Dimethylacetylene   3. Isopropylacetylene   4. Phenylacetylene

(e) Aromatic Hydrocarbons

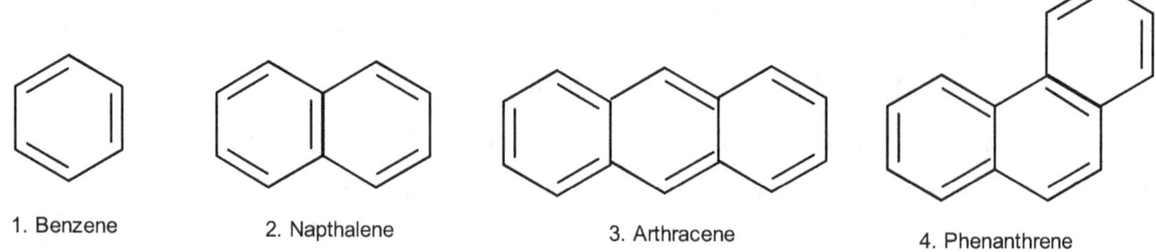

1. Benzene   2. Napthalene   3. Arthracene   4. Phenanthrene

(f) Arylalkyl Hydrocarbons

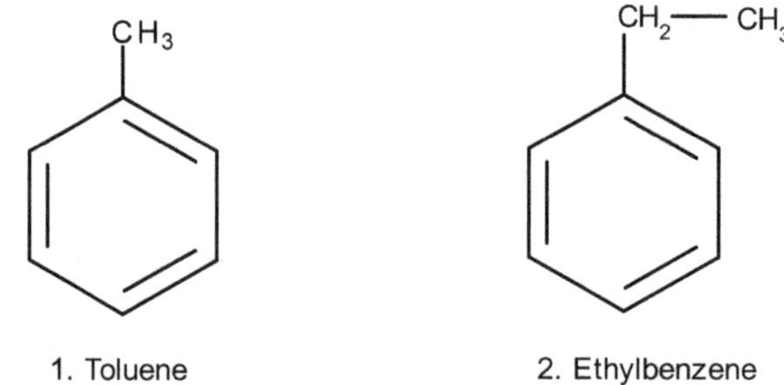

1. Toluene   2. Ethylbenzene

4.2.1.2 Compounds containing Carbon, Hydrogen and Oxygen

(a) Alcohols: R-OH; (R = Alkyl or aryl) or R-OH; where R = alkyl or aryl or arylalkyl

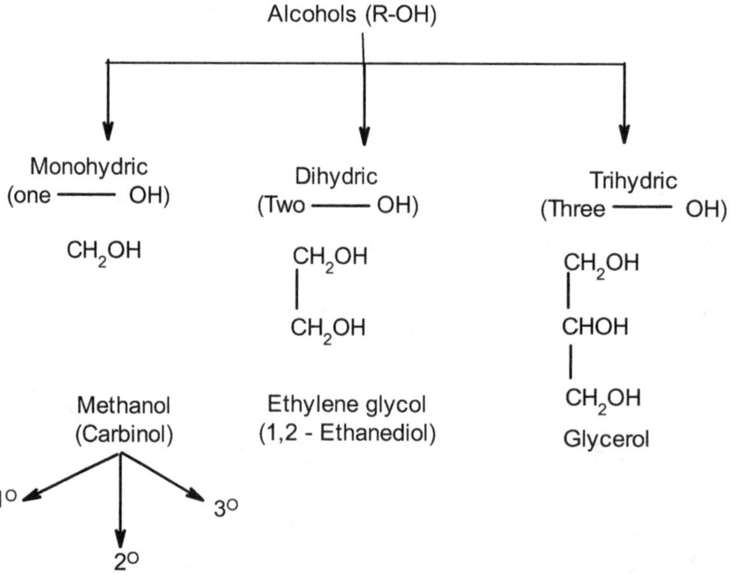

Monohydric alcohols are classified as primary (1°), secondary (2°), or tertiary (3°) depending upon –the OH group attached to a primary, secondary or tertiary carbon atom.

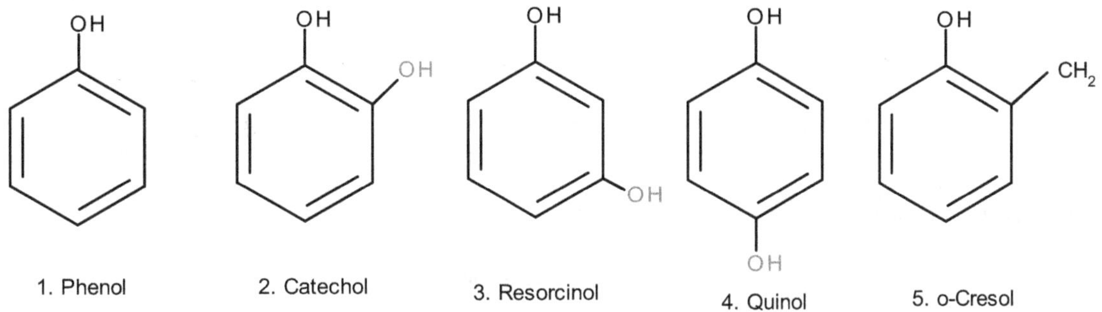

1. Methanol (Carbinol)

1. Isopropylalcohol

1. *tert*-Butylalcohol

CH_3—CH_2—OH

2. Ethanol

$CH_3CH_2CH_2OH$

3, *n*-Propylalcohol

(b) Phenols: (Ar-OH)

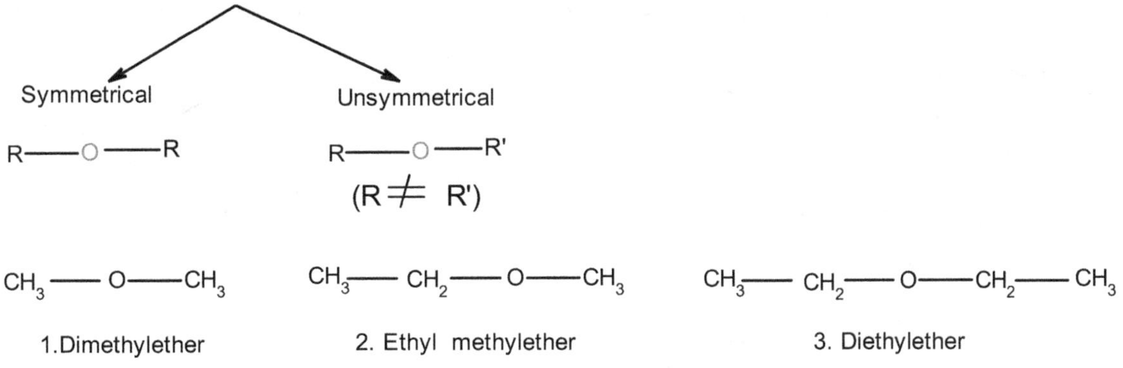

1. Phenol   2. Catechol   3. Resorcinol   4. Quinol   5. o-Cresol

(c) Ethers: R-O-R'; where R and R' = alkyl, acyl or arylalkyl

Symmetrical

Unsymmetrical

R——O——R

R——O——R'

(R ≠ R')

CH_3——O——CH_3

1. Dimethylether

CH_3——CH_2——O——CH_3

2. Ethyl methylether

CH_3——CH_2——O——CH_2——CH_3

3. Diethylether

(d) Epoxides: (cyclic ethers)

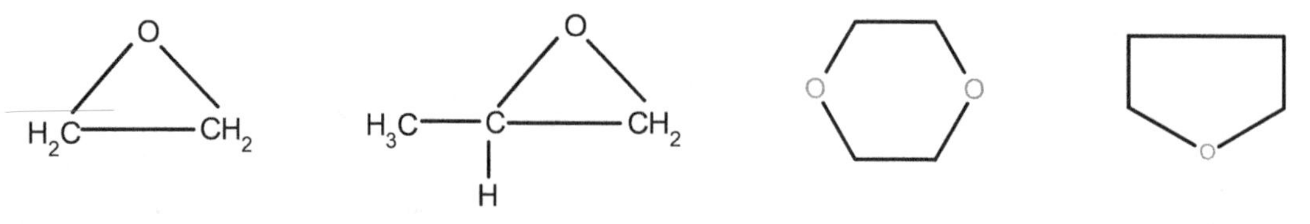

1. Ethyleneoxide (Epoxide)   2. Propyleneoxide   3. 1, 4-Dioxane   4. Tetrahydrofuran

(e) Crown Ethers: These are symmetrical cyclic polyalkyl ethers. Here 18 represents the number of atoms in the cycle while 6 refers to a number of oxygen atoms.

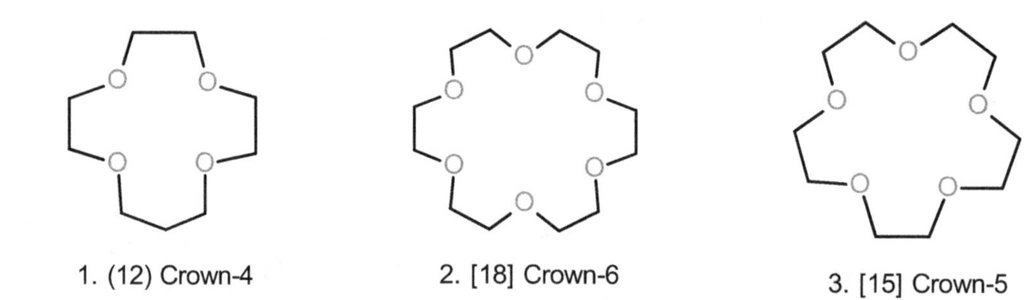

1. (12) Crown-4   2. [18] Crown-6   3. [15] Crown-5

(f) Carbonyl Compounds:

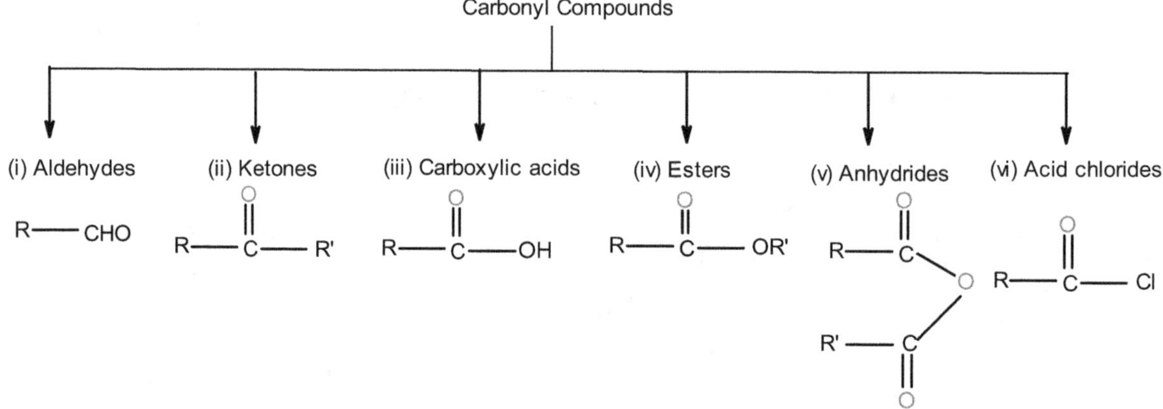

(i) Aldehydes: R-CHO; where R = Alkyl or aryl

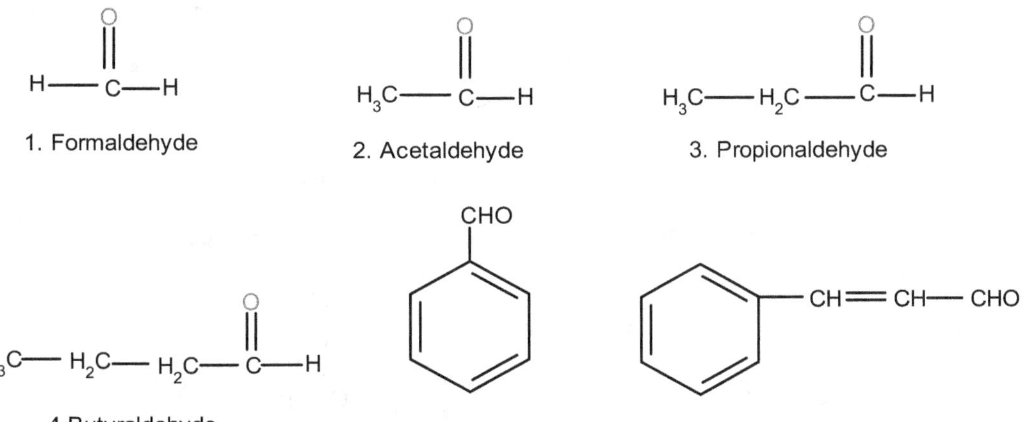

The carbon atom adjacent to the carbonyl group in an aldehyde is called α-carbon, the subsequent one is b and so on.

α-Chloropropionaldehyde

(ii) Ketones

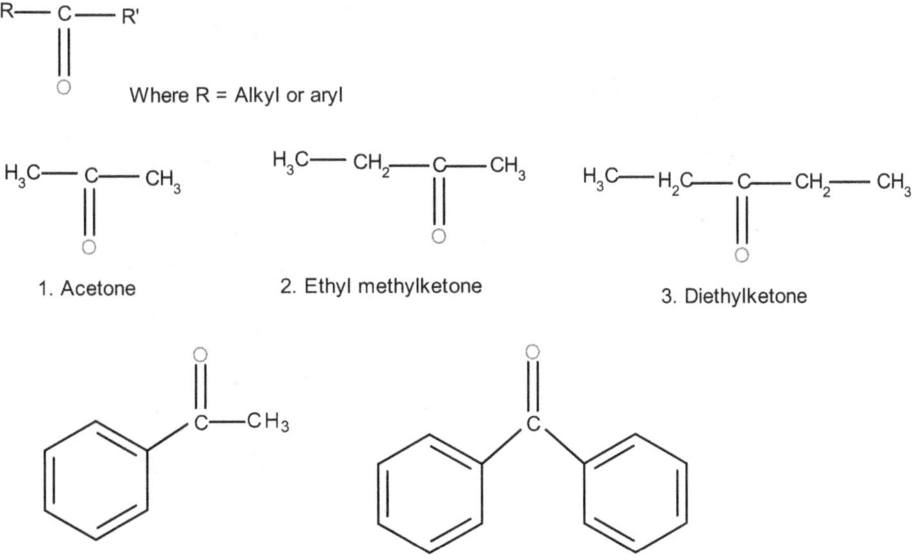

(iii) Carboxylic Acids: R-COOH; R = Alkyl or aryl

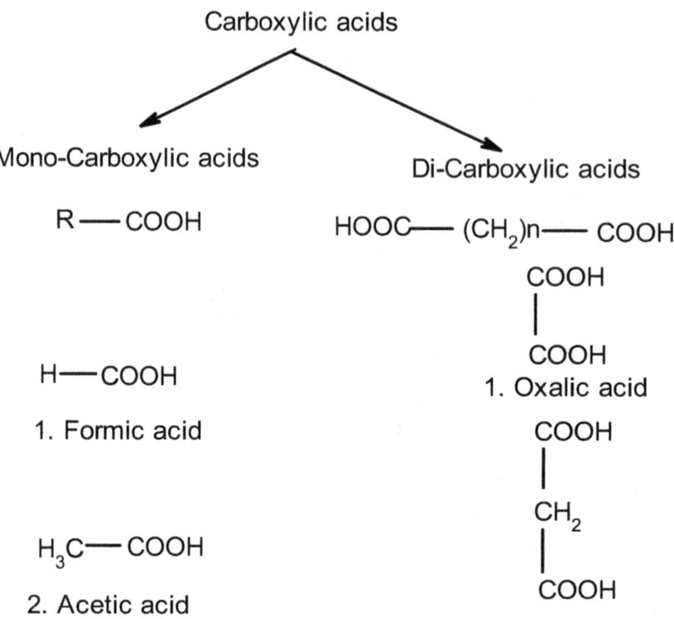

(iv) Esters

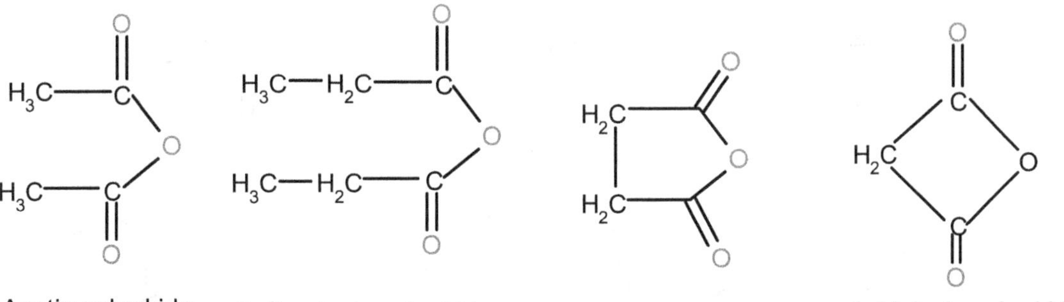

where R = Alkyl

1. Methyl formate

2. Ethyl acetate

3. Diethyl acetate

4. Methyl butyrate

5. Ethyl acetoacetate

(v) Anhydrides: $R\text{-}(CO)_2\,O\text{-}R'$; where, R, R' = alkyl, arylalkyl, aryl.

1. Acetic anhydride

2. Propionic anhydride

3. Succinic anhydride

4. Malonic anhydride

(vi) Acid chlorides

1. Acetyl chloride

2. Benzoyl chloride

4.2.1.3 Compounds containing Carbon, Hydrogen and Nitrogen only

(a) Amines: R-NH-R′; R = H, alkyl or aryl

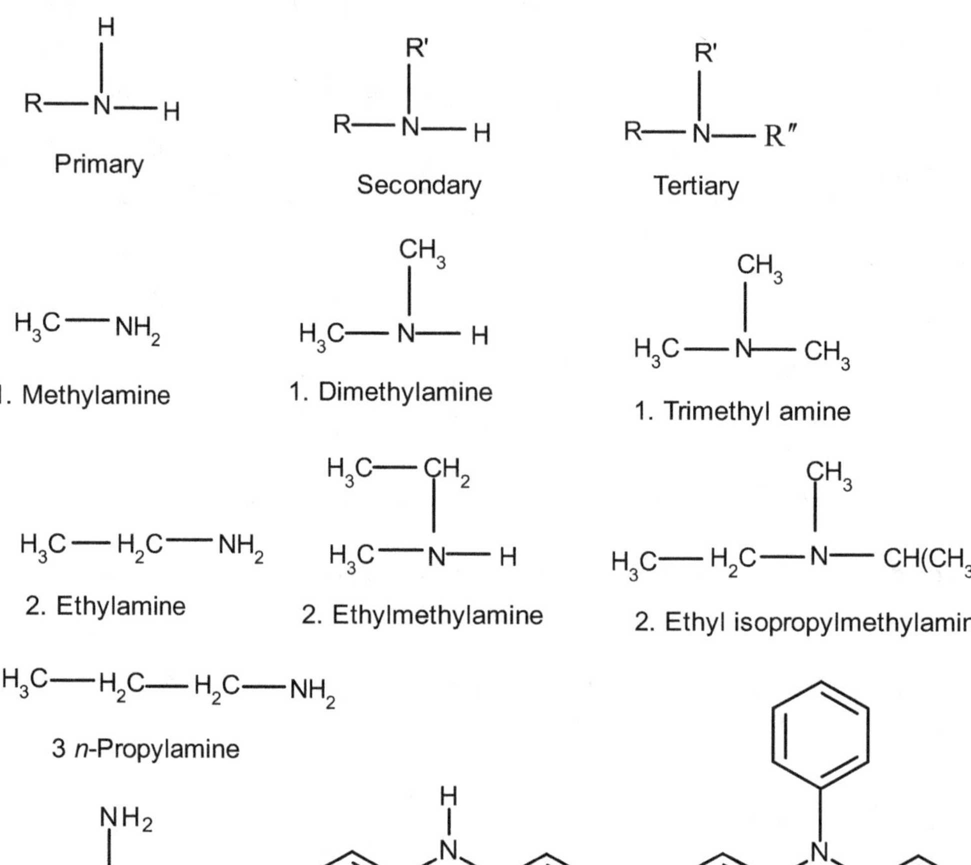

(b) Imines:

$$R-\overset{\overset{\displaystyle NH}{\|}}{C}-R'$$

(Shiff's base); where, R, R' = alkyl, arylalkyl, aryl.

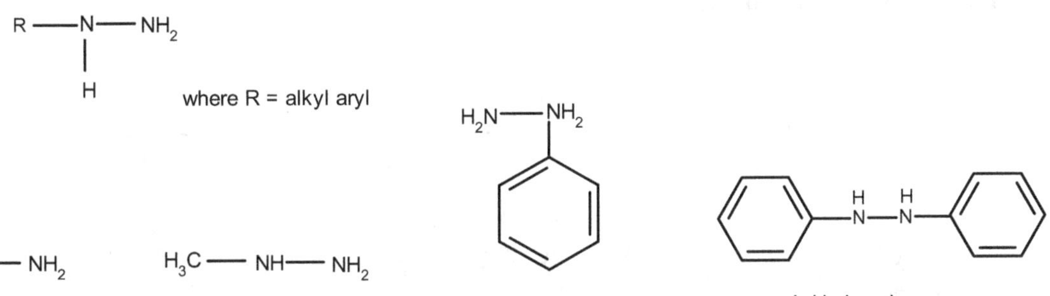

1. Diphenylamine

(c) Nitriles: R – C $^{\circ}$ N; where R = alkyl, aryl, arylalkyl.

$$H_3C\!-\!\!-\!\!-C\!\equiv\!N$$

1. Methyl cyanide or Acetonitrile

(d) Isonitriles: (Isocyanates) R – N $^{\circ}$C; where R = alkyl, aryl, arylalkyl.

$$C_6H_5\!-\!\!-\!\!NC \qquad\qquad H_3C\!-\!\!-\!\!N\!\equiv\!C$$

Phenyl isocyanide   Methyl isocyanide

(e) Hydrazines:

$$R\!-\!\!-\!\overset{\displaystyle N}{\underset{\displaystyle H}{|}}\!-\!\!-\!NH_2$$

where R = alkyl aryl

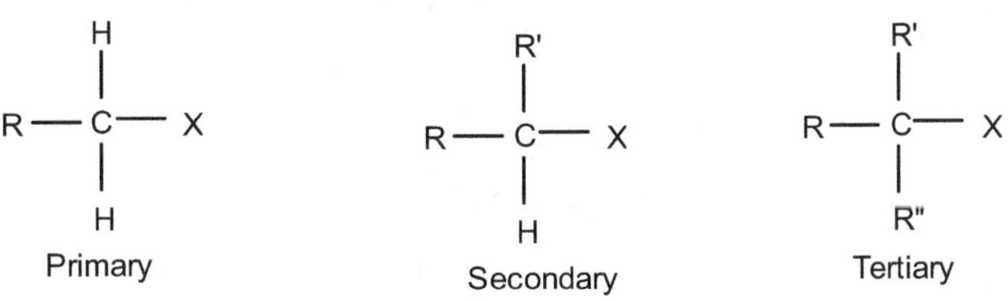

$$H_2N\!-\!\!-\!NH_2 \qquad H_3C\!-\!\!-\!NH\!-\!\!-\!NH_2$$

1. Hydrazine   2. Methyhydrazine   3. Phenythydrazine   4. Hydrazobenzene

4.2.1.4 Compounds containing Carbon, Hydrogen, Oxygen and Halogens only

(a) Alkyl-halides: R –X where, X = F/Cl/Br/I

Primary   Secondary   Tertiary

i.   Monohalogen compounds:

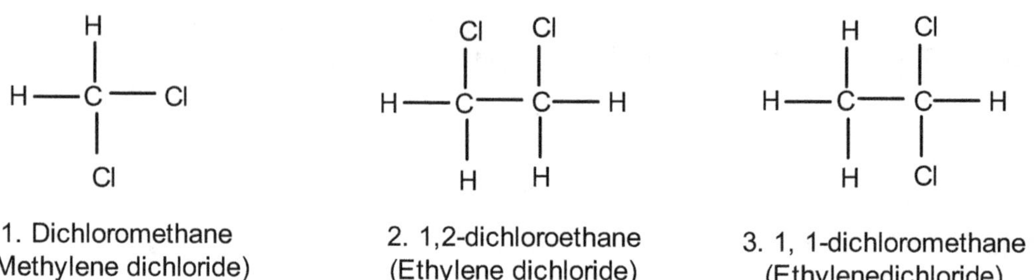

1. Methyl chloride   2. Ethyl bromide   3. *iso*-Propyl chloride   4. *tert*-Butyl chloride

ii.   Dihalogen compounds:

1. Dichloromethane
(Methylene dichloride)

2. 1,2-dichloroethane
(Ethylene dichloride)

3. 1, 1-dichloromethane
(Ethylenedichloride)

If two halogen atoms are attached to the same carbon atoms, it is called a geminal halide.

If two halogen atoms are attached to adjacent carbon atoms, it is called a vicinal halide.

iii.   Trihalogen compounds:

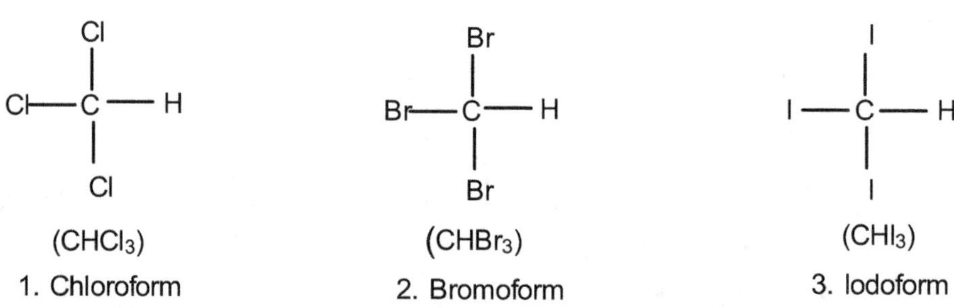

$(CHCl_3)$
1. Chloroform

$(CHBr_3)$
2. Bromoform

(CHI_3)
3. Iodoform

iv.   Tetrahalogen compounds

(CCl_4)
Carbontetrachloride

v. Unsaturated Halides:

$H_2C = CH — Cl$

1. Vinyl chloride

$H_2C = CH — Br$

2. Vinyl bromide

$H_2C = CH — I$

3. Vinyl iodide

$H_2C = CH — CH_2 — Cl$

4. Allyl chloride

$H_2C = CH — CH_2 — Br$

5. Allyl bromide

$H_2C = CH — CH_2I$

6. Allyl iodide

(b) Aryl halides: (Ar − X)

1. Chlorobenzene

2. Bromobenzene

3. Iodobenzene

4. Benzyl chloride

5. Benzyl bromide

6. Benzyl iodide

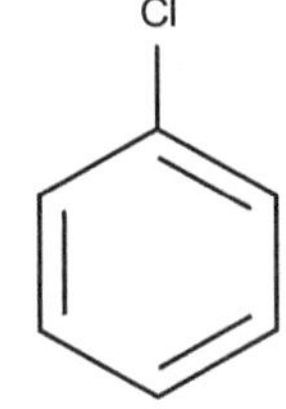

7. Halotoluene

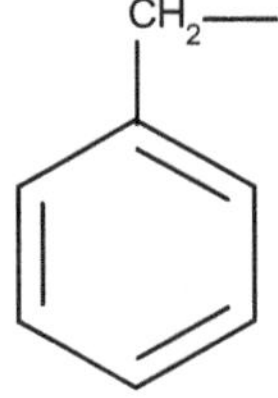

8. Haloamines

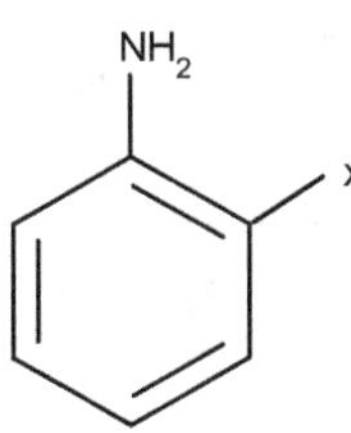

9. Halophenol

10. Benzalchloride

11. Benzotrichloride

(c) Acyl halides:

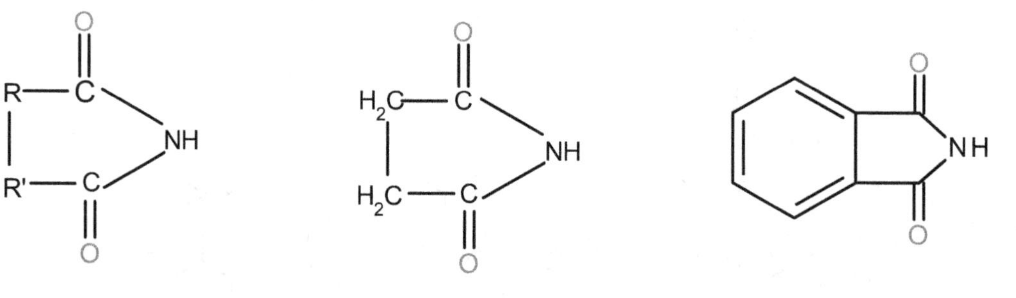

CH$_3$COCl

1. Acetyl chloride

CH$_3$CH$_2$COCl

2. Propionyl chloride

3. Benzoyl chloride

4.2.1.5 Compounds containing Carbon, Hydrogen, Oxygen, Nitrogen and Halogens only

a. Amides

1. Formamide

2. Acetamide

3. Benzamide

4. Urea

b. Imides

1. Succinimide

2. Phthalimide

c. Aldoximes

1. Acetal oxime

2. Propanal oxime

d. Ketoxime

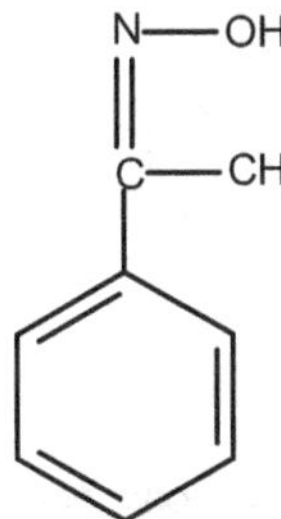

1. Acetone oxime

2. Butanone oxime
(Ethylmethyl ketoxime)

3. Acetophenone oxime

e. Nitro Compounds

1. Nitromethane

2. Nitroethane

3. Nitrobenzene

4. *m*-Dinitrobenzene
(1, 3-Dinitrobenzene

5. 1, 3, 5-Trinitrobenzene

6. *O*-Nitrotoluene

7 o-Nitrophenol

4.2.1.6 Compounds containing Carbon, Hydrogen, Oxygen, Nitrogen, Sulphur and/or Halogens

a. Sulphonic acids: R – SO₃H

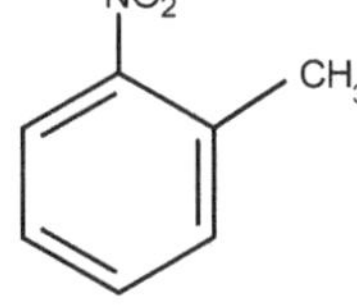

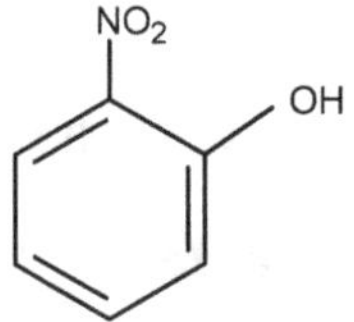

1. Methane sulphonic acid

2. Ethane sulphonic acid

3. Benzene sulphonic acid

4. *m*-Bromobenzene
sulphoric acid

5. *m-Nitro*mobenzene
sulphoric acid

6. Benzene, 1,3-disulphonic acid

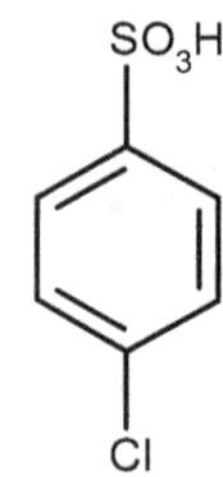

7. o-Toluene sulphonic acid
(o-methyl benzene
sulphonic acid)

8. *p*-Chlorobenzene
sulphonic acid

9. Sulphanilic acid

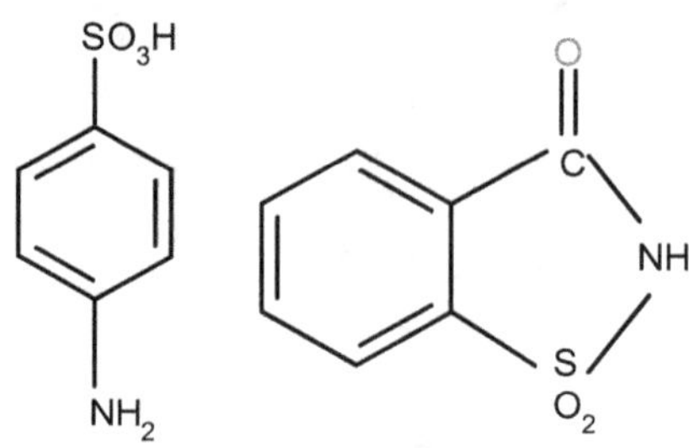

10. Saccharin

b. Sulphonyl halides

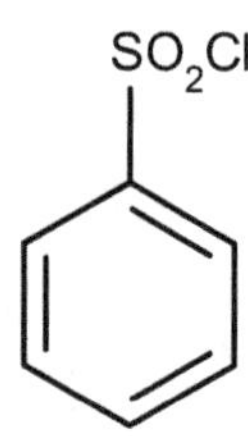

1. Benzene sulphonyl chloride

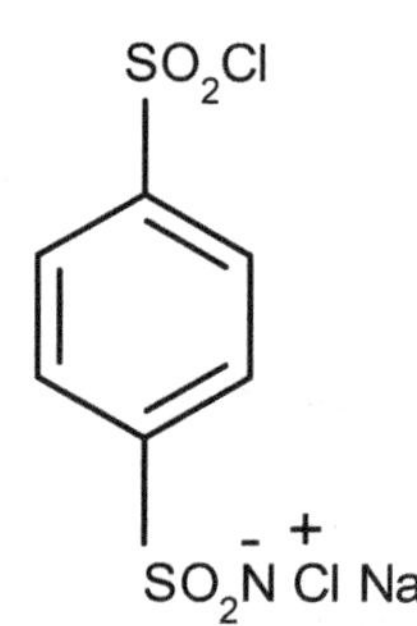

2. Chloramine-T

c. Sulphonamides

General structure

1. Sulphanilamide

2. Sulphacetamide

3. Sulphapyridine

4. Sulphadiazine

4.2.2 IUPAC Nomenclature System for Organic Compounds

GENERAL RULES

Millions of organic compounds have been discovered to date. It is very difficult to remember every compound by a particular name. Hence, as seen above in this chapter, organic compounds are classified based on their chemical structures and properties. To be more precise, the compounds are classified on the basis of their functional groups. Thus, each compound has a unique identity, and it is imperative that each should have a unique name throughout the world. Thus, IUPAC (International Union of Pune and Applied Chemistry, Geneva) came up with a harmonised system of naming the organic compounds, which is acceptable, followed and understood all across the world. This is called as the "IUPAC Nomenclature System".

General rules for writing the IUPAC name:
- Locate the longest chain containing the principal functional group and as many of the secondary functional groups and multiple bonds as possible.
- Select the root word corresponding to the length of the chain, e.g., pent for 5 carbon atom chain.
- Number the longest chain selected from the end nearer to the principal functional group or substituents or the side chain if there is no functional group.
- Depending upon the nature of the carbon-carbon bond, attach the suffix 'ane' for $-C-C-$, 'ene' for $-C=C-$ and 'yne' for $-C\equiv C-$ to the root word.
- Add suitable prefixes to indicate the number and position of each side chain, substituent or functional group.

Examples:

4-hydroxy-2-pentanone
or
4-hydroxypentan-2-one

IUPAC system of nomenclature is a highly systematic nomenclature system for chemical compounds. For a given IUPAC name, only one structure can be written. This helps in the translation of a structure to a compound and vice-versa. The IUPAC name consists of a base name (a prefix may or may not be present). The base name may comprise of a root name and prefix.

Examples:

$$H_3C\text{———}CH_3$$

Ethane

Here ethane is the base name without a prefix.

Here methyl is a prefix, while chlorohexane is the base name. The numbers indicate the positions of methyl group and chloroatom, respectively.

2-Methyl-3-Chlorohexane

Below is provided with a very novel and crisp yet very useful way of remembering the systematic IUPAC nomenclature of various organic compounds encountered in our syllabus.

This can indeed serve as a very useful "ready-reckoner" for this purpose.

General Rules for writing IUPAC names:

i.   Locate the longest chain containing the principal functional group and as many of the secondary functional groups and multiple bonds as possible.

ii. Select the root word corresponding to the length of the chain, e.g., pent, for the 5-C chain.

iii. Number the longest chain selected from the end nearer to the principal functional group or substituents or the side chain if there is no functional group.

iv. Depending on the nature of carbon-carbon bonds (C–C, C=C, C≡C); attach the suffix -ane (for-C-C-); -ene (for -C=C-); -yne (for -C≡C-) to the root word.

v.   Add suitable prefixes and suffixes with numericals to indicate the number and position of each side chain, substituent or functional group.

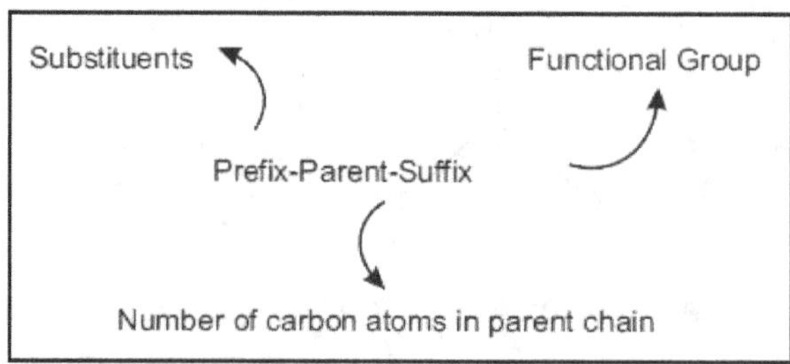

Priority of Groups

— COOH > — COOR > — SO_3H >— COX >— $CONH_2$ >— CHO >— CN > $\rangle C{=}O$ >

— OH > —SH > — NH_2 > — O >— $N{\equiv}N$ >— NO_2 > — NO > alkenyl > alkyl >— X

Example:

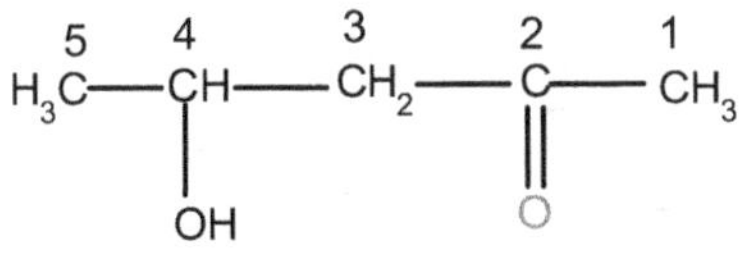

4-Hydroxy-2-pentanone

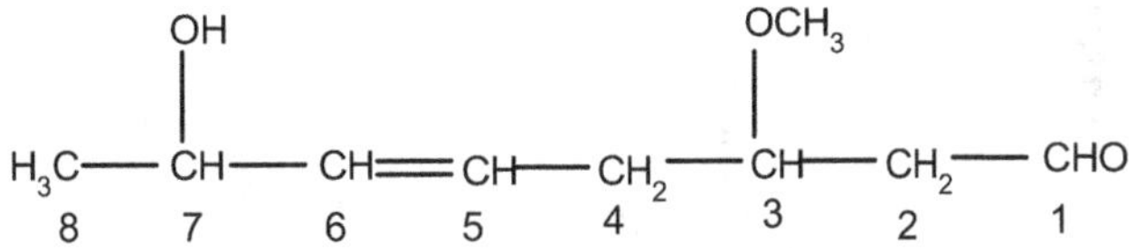

7-Hydroxy-3-methoxy-oct-5-ene-1-al

Writing the Structural Formula from the given IUPAC Name:

i.   Locate the parent alkane from the name and write the number of carbon atoms of this alkane in a straight chain and number them.

ii. Now locate the suffix, which gives information about the name and the number of the functional groups, along with their positions.

iii. Lastly, locate groups or/and substituents mentioned in the prefix and the carbon atoms to which these are attached as indicated by the locants.

iv. This gives the skeleton formula. Now add H atoms to satisfy the carbon bonds.

Example: Write the structural formula of 3-ethyl-2, 5-dimethyl-1, and 4-heptadiene.

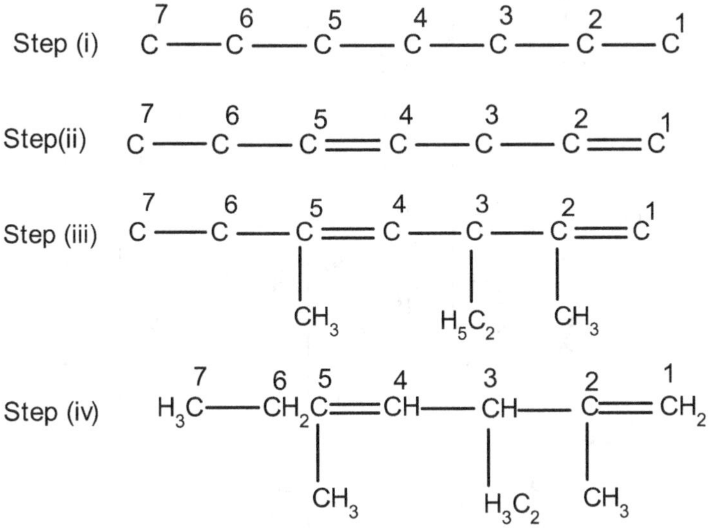

Step (i)

Step(ii)

Step (iii)

Step (iv)

The below **table 4.1** can serve as a ready reckoner for the nomenclature of all classes of simple organic compounds covering various functional groups with reference to the IUPAC rules.

Table 4.1 IUPAC nomenclature of all class compounds; Nomenclature of Mono-substituted and Poly-substituted compounds. (Recent rules of IUPAC referred).

Nomenclature as per Functional Group

Name of the functional group	General formula	Example	IUPAC name (example)	Remark suffix	General/ Common name	Remark suffix/prefix
Alkane (Carbon-Carbon Single bond)	C_nH_{2n+2}	CH_4	Alkane / Methane	– ane	Paraffin	–ane e.g. Methane
Alkene (Carbon-Carbon Double bond)	C_nH_{2n}		Alkane / Ethane	– ane to ene	Olefin	ane to –ylene e.g. Ethylene
Alkyne (Carbon-Carbon Triplebond)	C_nH_{2n-2}	$H-C\equiv C-H$	Alkane / Ethane	ane to yne	Acetylene	Prefix: alky acetylene e.g. Methyl acetylene
Halo alkanes	R—X, X= halide	H_3C-Cl	Haloalkane / Chloro methane	Prefix: Halo Suffix: ane	Alkyl halide	ane to –yl halide e.g. Methyl chloride
Hydroxyl	R—OH	H_3C-OH	Alkanol / Methanol	e to ol	Alkyl alcohol	ane to –yl alcohol e.g. Methyl alcohol
Ether	R—O—R'	$H_3C-O-CH_3$	Alkoxyalkane / Methoxy-methane	ane to oxyalkane	Alkyl ether	ane to ylether e.g. Dimethyl ether

Table 4.1 *Contd...*

Name of the functional group	General formula	Example	IUPAC name (example)	Remark suffix	General/ Common name	Remark suffix/prefix
Aldehyde	$R-CHO$	$H-CHO$	Alkanal Methanal	ane to al	Aldehyde	aldehyde e.g. Formaldehyde
Keto	$R-CO-R'$	$H_3C-CO-CH_3$	Alkanone Propanone	e to one	Ketone	Ketone e.g. Acetone
Ester	$R-CO-OR'$	$H_3C-CO-OC_2H_5$	Alkyl alkanote Ethyl ethancate	e to oate	Alkyl ester	Prefix R' = alkylic acid to ate e.g. Ethyl acetate
Amino	$R-NH_2$	H_3C-NH_2	Alkanamine Methamamine	e to amine	Alkyl amine	ane to -ylamine e.g. Methylamine
Nitro	$R-NO_2$	H_3C-NO_2	Nitro alkane Nitromethane	Prefix Nitro	Nitroalkane	Prefix nitro Suffix alkane e.g. Nitro-methane
Cyano	$R-C\equiv N$	$H_3C-C\equiv N$	Alakane nitrile Methane nitrile	nitrile	Alkyl cyanide	ane to -ylcyanide e.g. Methyl cyanide
Sulfonic acid	$R-SO_3H$	H_3C-SO_3H	Alkane sulfonic acid Methane sulfonic acid	Sulfonic acid	Alkane sulphonic acid	Suffux-sulphonic acid e.g. Methane sulphonic acid
Mercapto	$R-SH$	H_5C_2-SH	Alkane thiol Ethane thiol	Thiol	Alkyl mercaptan	ane to -ylmercaptan e.g. Ethyl mercaptan

4.3 NOMENCLATURE OF HETEROCYCLIC COMPOUNDS

There are two main systems of nomenclature for heterocyclic compounds:

1. Systematic nomenclature system – (The Hansch - Wildman system)
2. Trivial nomenclature system.

4.3.1 The Systematic Nomenclature System (Hansch-Wildman nomenclature system)

It is the most common systematic method of nomenclature, which specifies;

- Ring size
- Nature/Type
- Positions of hetero atoms.

Rules

(a) Combination of prefix with stem

(i) Prefix indicates the heteroatom. It is derived from the heteroatom(s) present in the heterocyclic compound under consideration.
- Oxygen (O) - Oxa
- Nitrogen (N) - Aza
- Sulfur (S) - Thia
- Phosphorous (P) - Phospha
- Boron (B) - Bora

  e.g., Oxazole and Thiazole.

(ii) If heteroatom is present twice, thrice, or four times etc.., then prefix becomes di-, tri, tetra-, penta- etc.

  e.g., Diazole, Diazine, Triazole, Tetrazole and Pentazole.

(iii) Preference for prefix is to be given in order O > S > N > P > Si

  e.g., Oxazole, Thiazole, Oxathiole.

(b) Stem indicates ring size and whether it is saturated or unsaturated.

(i) Suffix used for nitrogen-containing heterocyclic compounds

Ring size	Unsaturated	Saturated
3 member	-irine	-iridine
4 member	-ete	-etidine
5 member	-ole	-olidine
6 member	-ine	-inane
7 member	-epine	-epane
8 member	-ocine	-ocane
9 member	-onine	-onane
10 member	-ecine	-ecane

(ii) Suffix used for non-nitrogen containing heterocyclic compounds

Ring size	Unsaturated	Saturated
3 member	-iren	-iran
4 member	-et	-etan
5 member	-ole	-olan
6 member	-in	-ane
7 member	-epine	-epane
8 member	-ocine	-ocane
9 member	-onine	-onane
10 member	-ecine	-ecane

(c) State of Hydrogenation also to be indicated by the prefix, dihydro, tetrahydro etc., and also a number of atoms which are reduced or position that is reduced should be indicated. e.g., 2H, 3H, 4H etc.

(d) Numbering:

  (i) The numbering is always started from the heteroatom and again in the same preference:

$$O > S > N > P > Si$$

  (ii) If more than one heteroatom is present, then both heteroatoms should get minimum numbers in the same preference order.

4.3.2 Trivial Nomenclature System

(a) Monocyclic heterocyclic compounds:

(i) 3-Member with 1-heteroatom

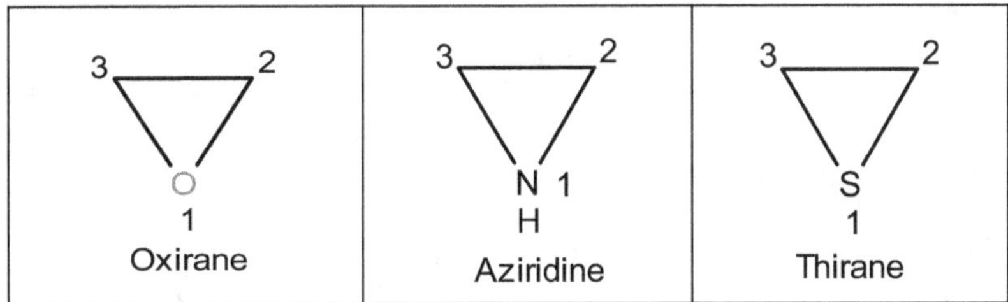

(ii) 4-Member with 1 heteroatom

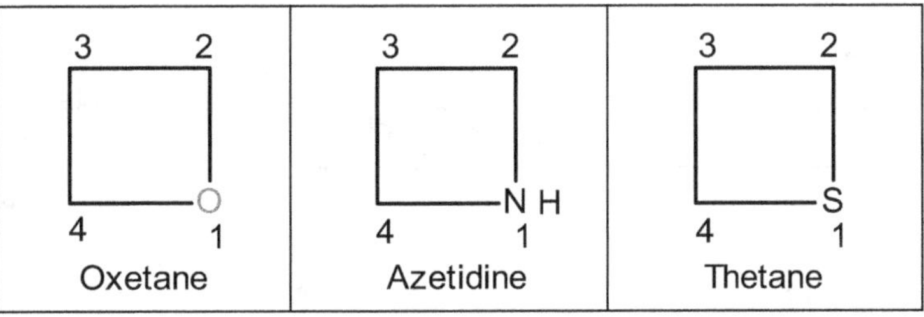

(iii) 5-Member 1 heteroatom

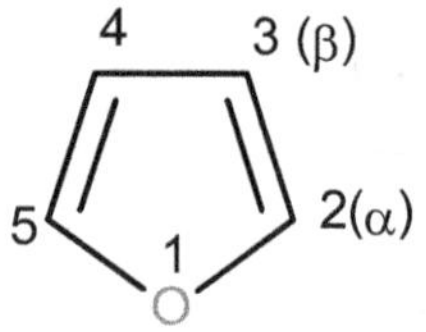 Furan	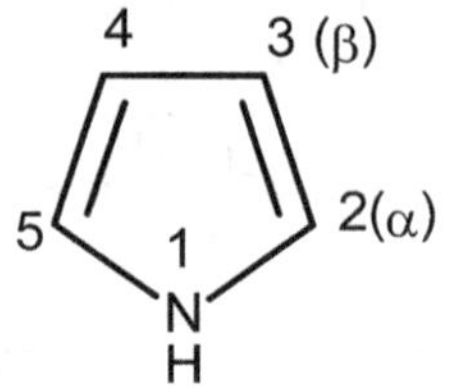 Pyrrole (1*H*-pyrrole)	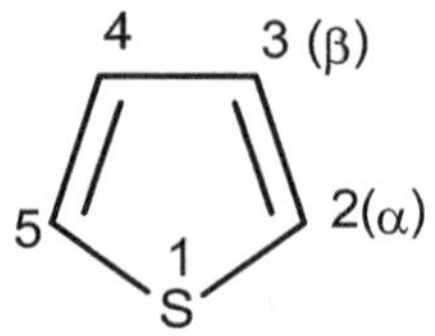 Thiophene
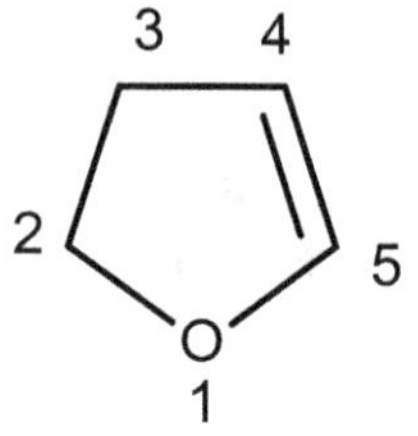 2, 3-Dihydrofuran	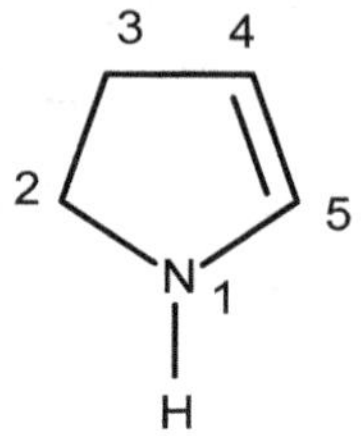 2,3-Dihydro-1*H*-pyrrole (2-Pyrroline)	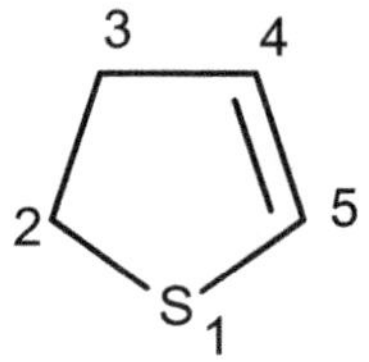 2, 3-Dihydrothiophene
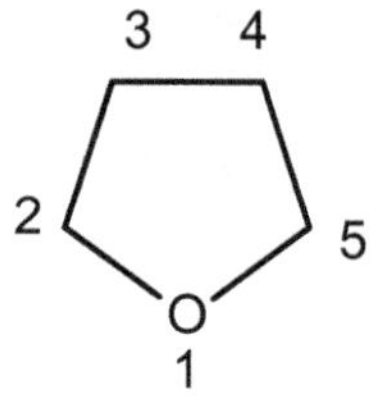 Tetrahydrofuran (THF)	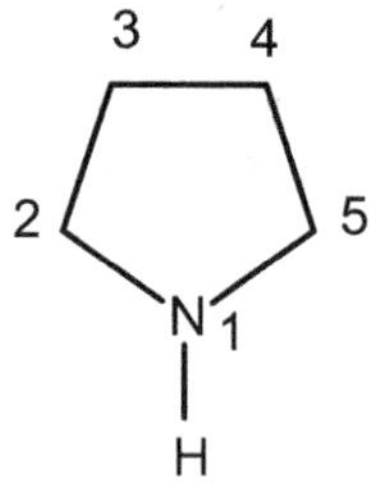 Tetrahydropyrrole (Pyrrolidine) (THP)	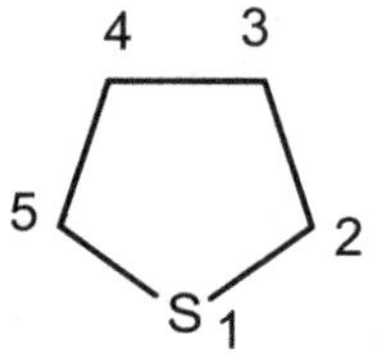 Tetrahydrothiophene (THT)

(iv) 5-Member with 2 heteroatoms

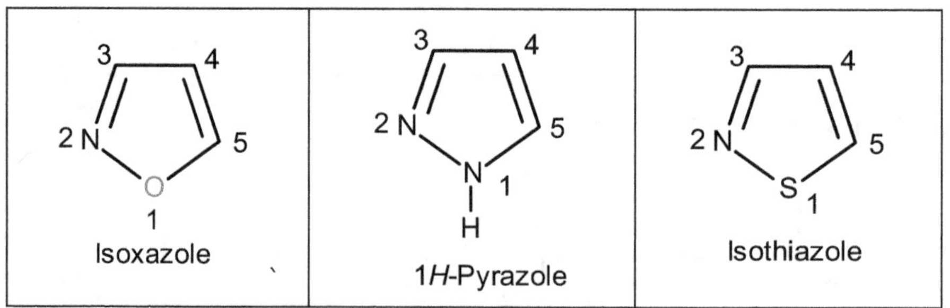

Isoxazole	1*H*-Pyrazole	Isothiazole

Contd...

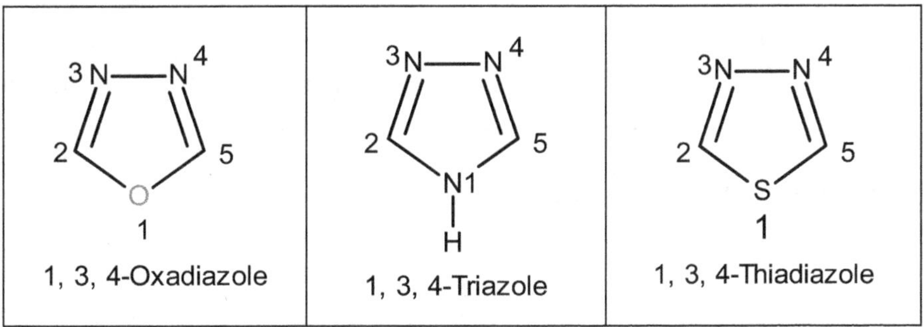

(v) 5-Member with 3 heteroatoms

(vi) 5-Member with 4 heteroatoms

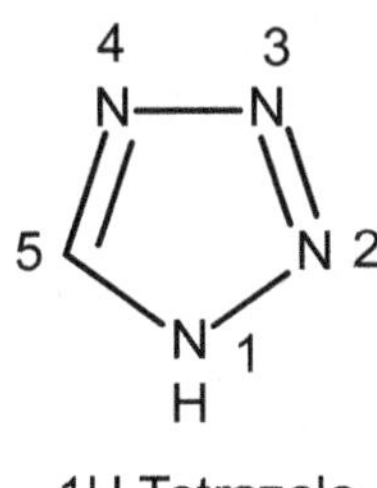

(vii) 6-Member with 1 heteroatom

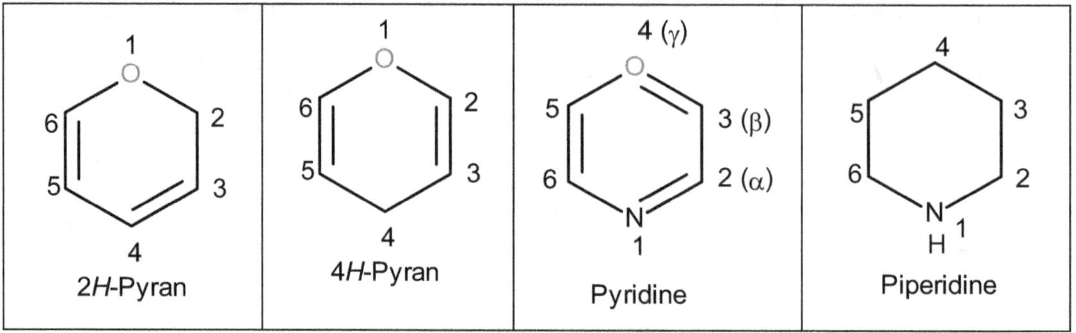

(viii) 6-Member with 2 heteroatoms

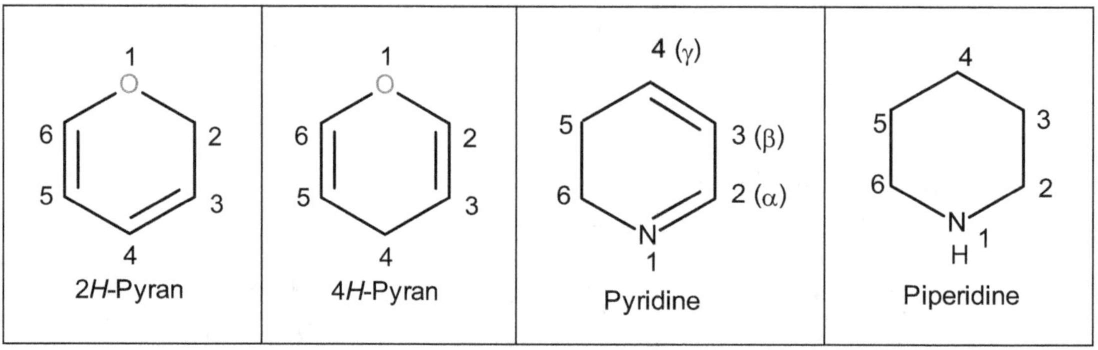

(ix) 6-Member with 3 heteroatoms

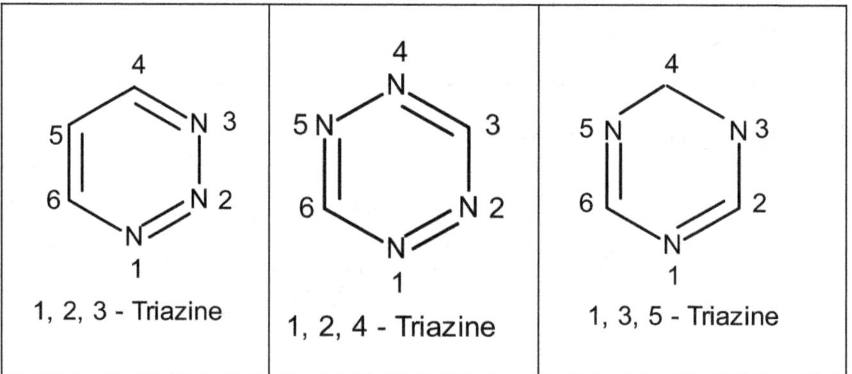

(x) 6-Member with 4 heteroatoms

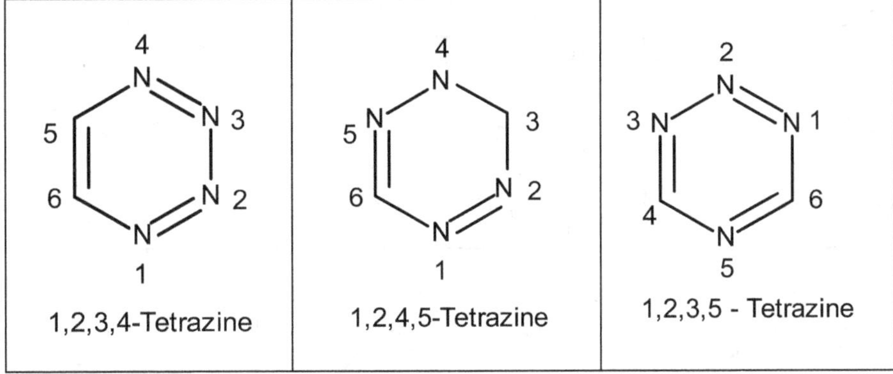

1,2,3,4-Tetrazine   1,2,4,5-Tetrazine   1,2,3,5 - Tetrazine

(xi) 7-Member with 1 or 2 heteroatoms

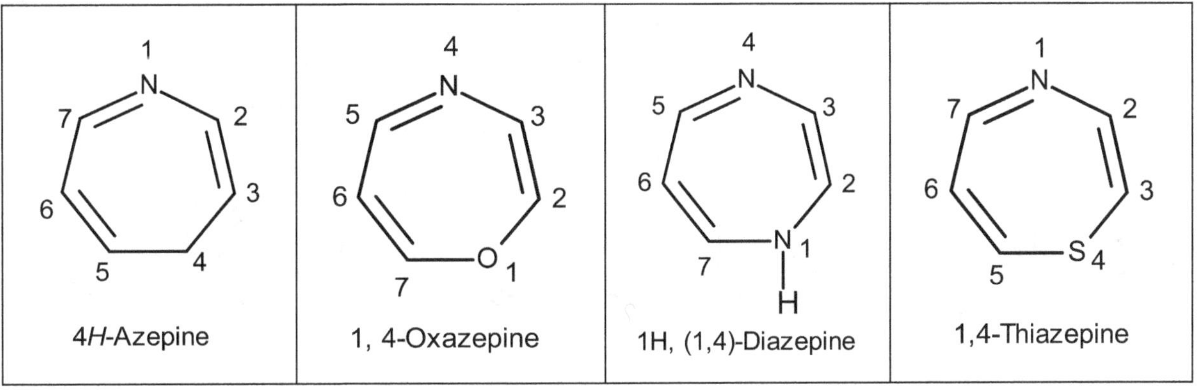

4*H*-Azepine   1, 4-Oxazepine   1H, (1,4)-Diazepine   1,4-Thiazepine

(b) Fused heterocyclic compounds

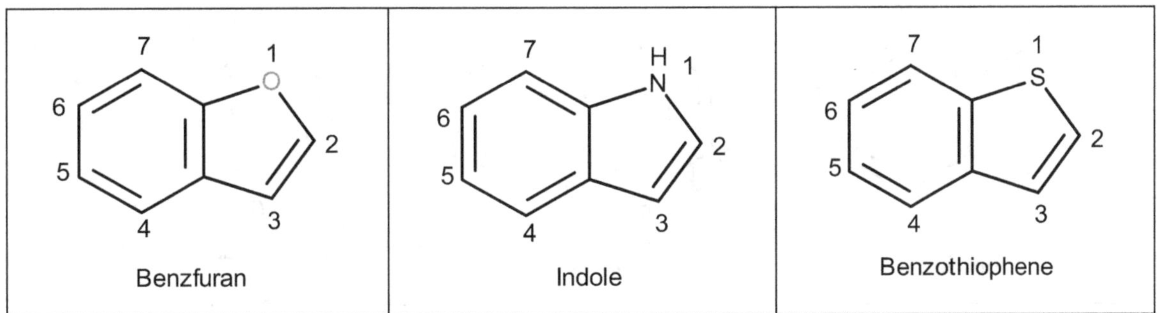

Benzfuran   Indole   Benzothiophene

Contd...

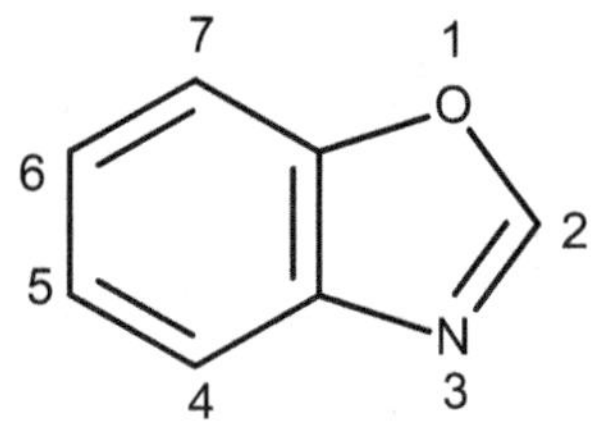 Benzoxazole	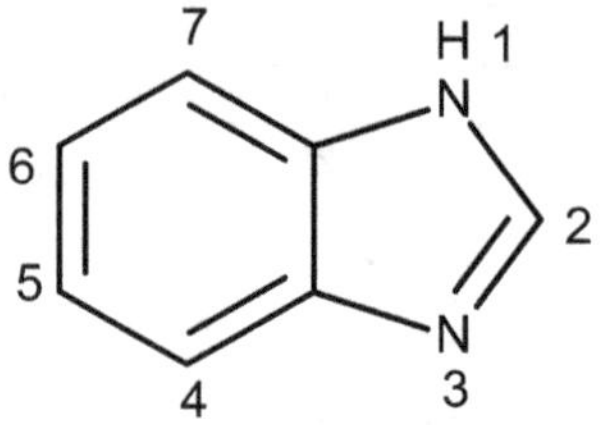 Benzimidazole	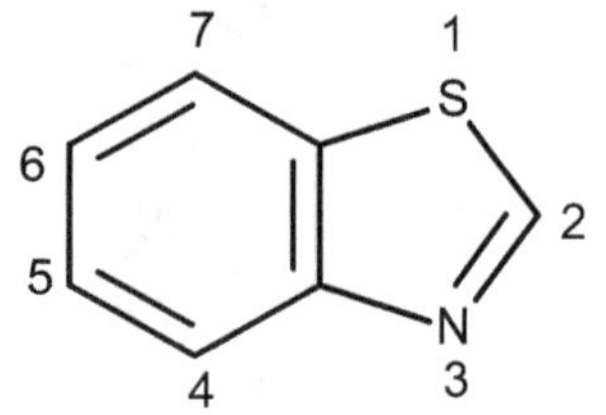 Benzthiazole
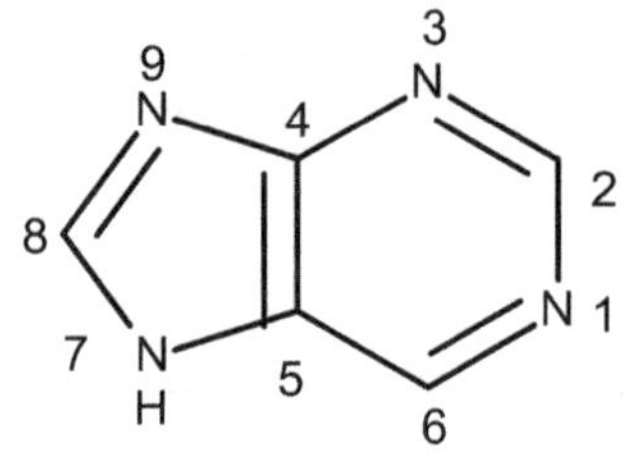 Purine	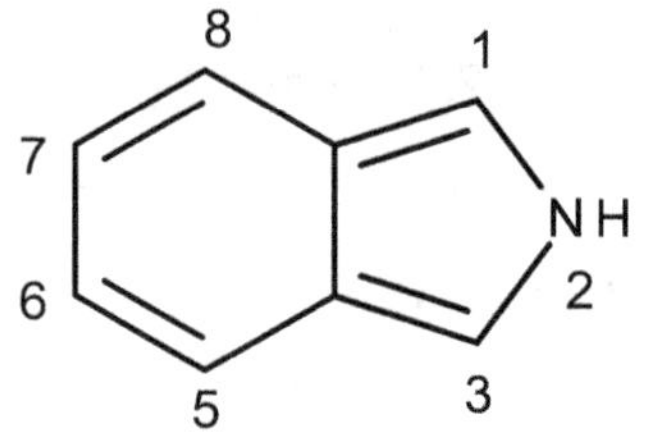 Isoindole	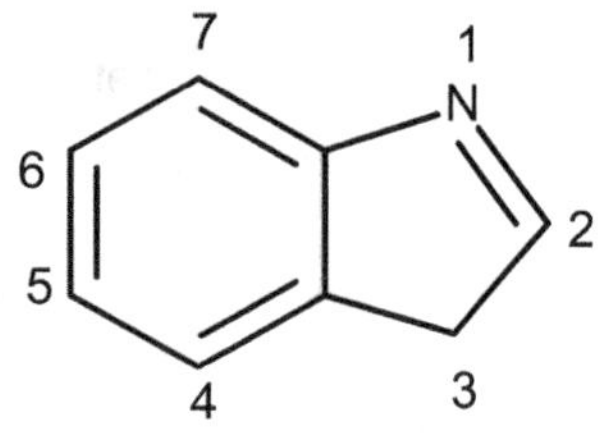 3*H*-Indole
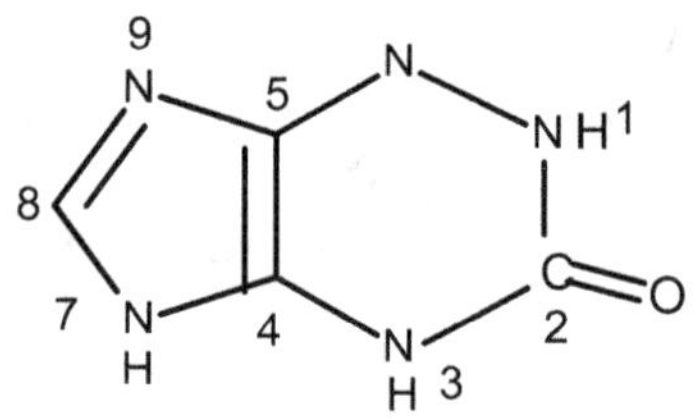 Xanthine	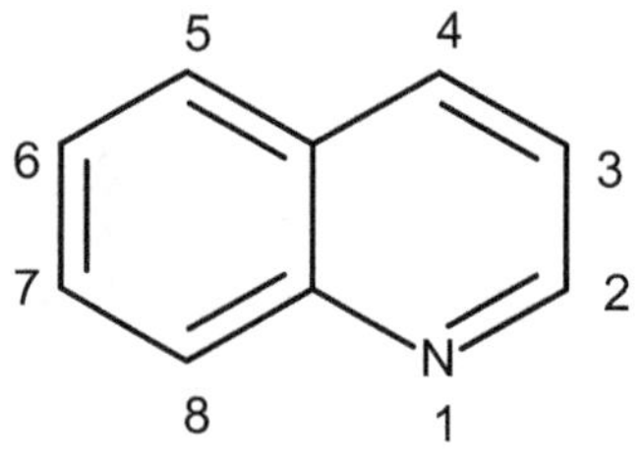 Quinoline	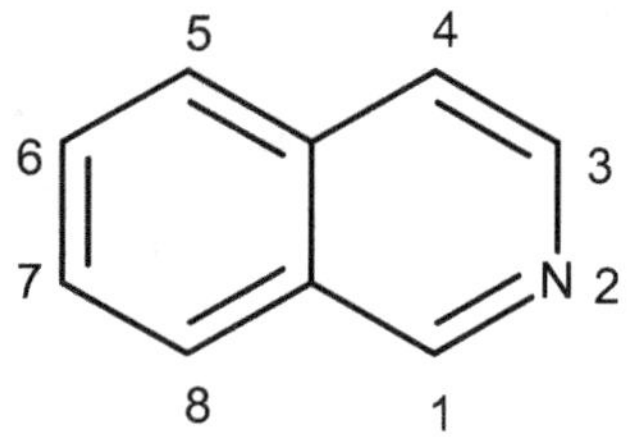 Isoquinoline
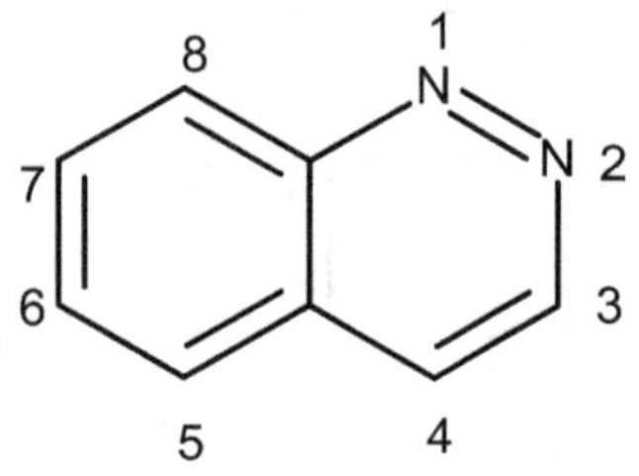 Cinnoline	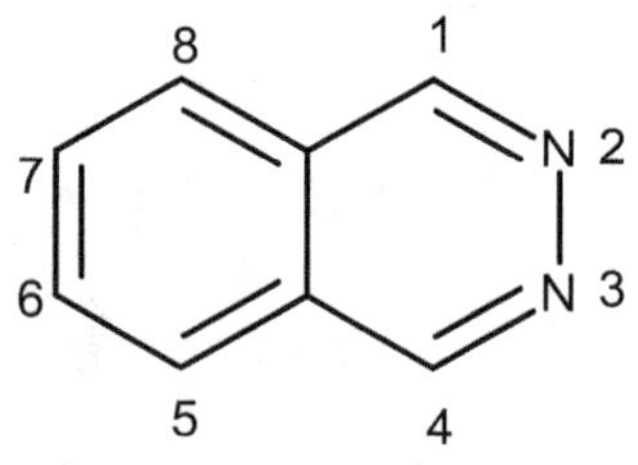 Phthalazine	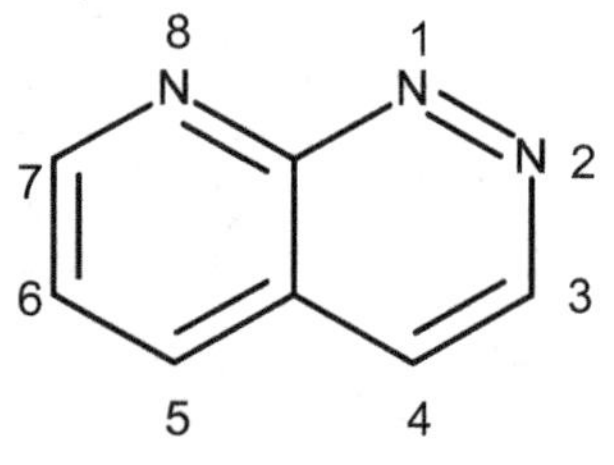 1,8-Naphthyridine

Contd...

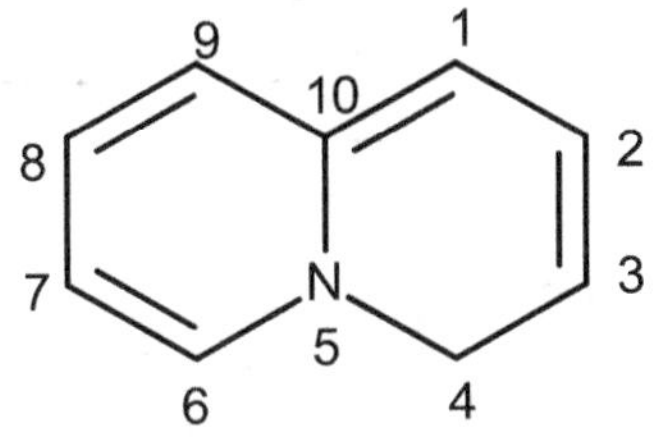

4*H*-Quinolizine

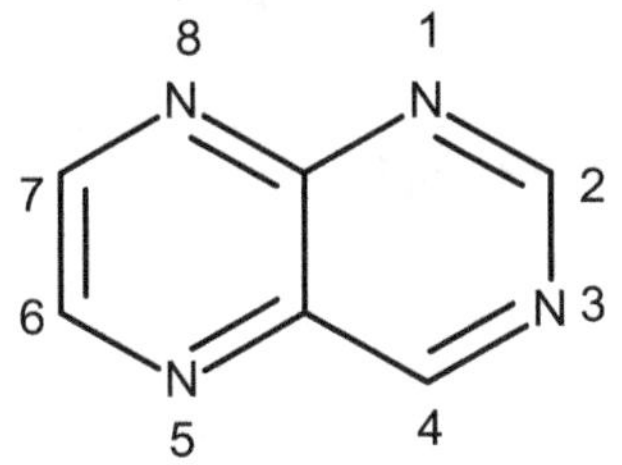

Pteridine
(Pyrimidine + Pyrazine)

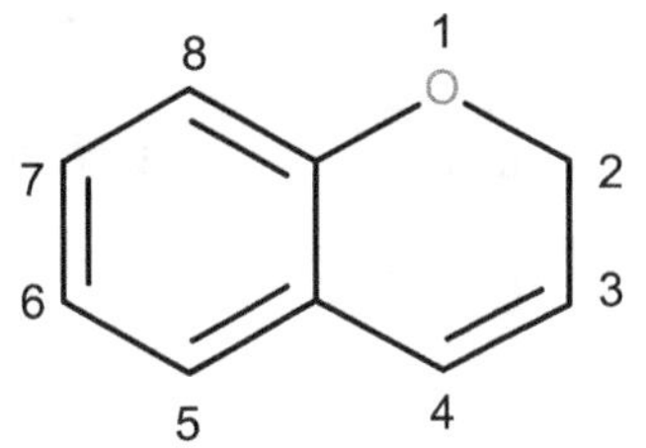

2*H*-Chromene

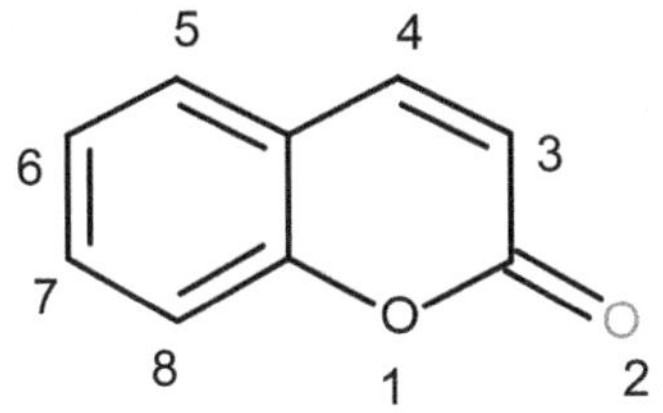

Coumarine

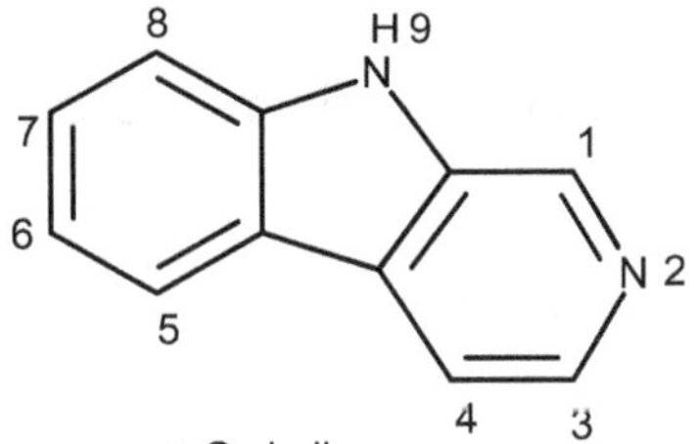

β-Carboline

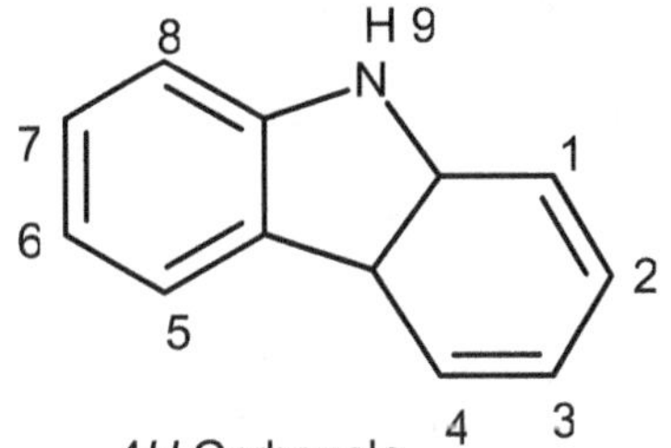

4*H*-Carbazole

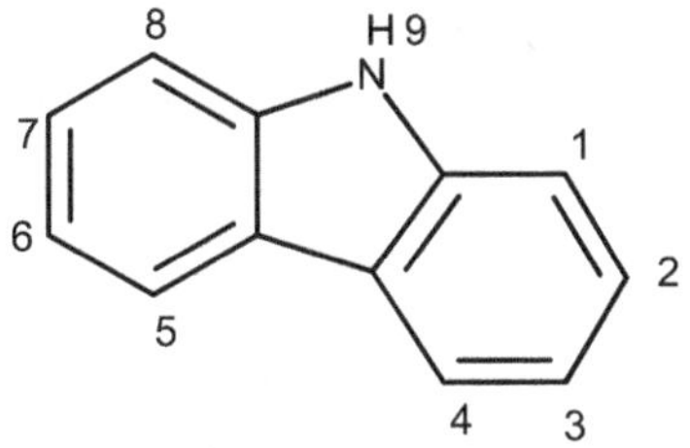

9*H*-Carbazole

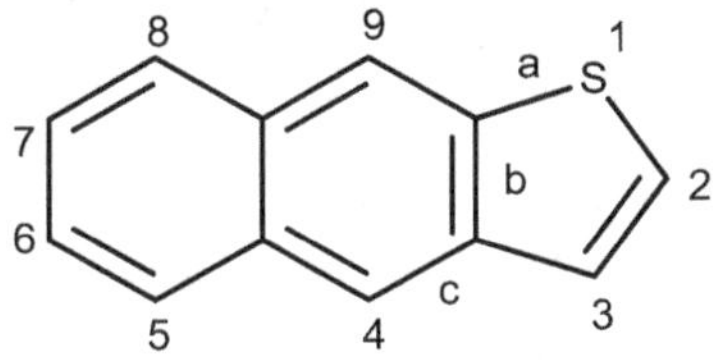

Naphtho (2, 3-b) thiophene

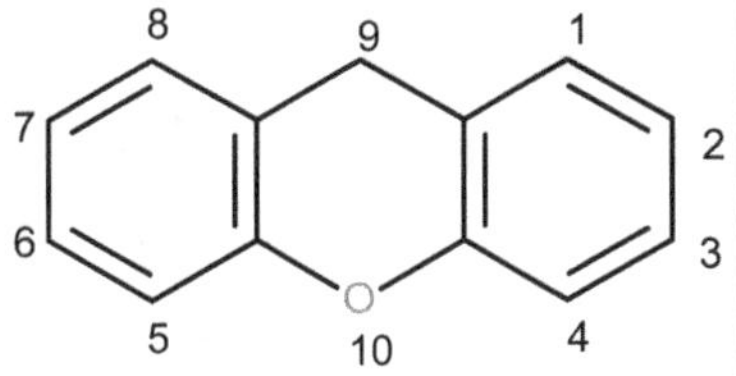

Xanthene

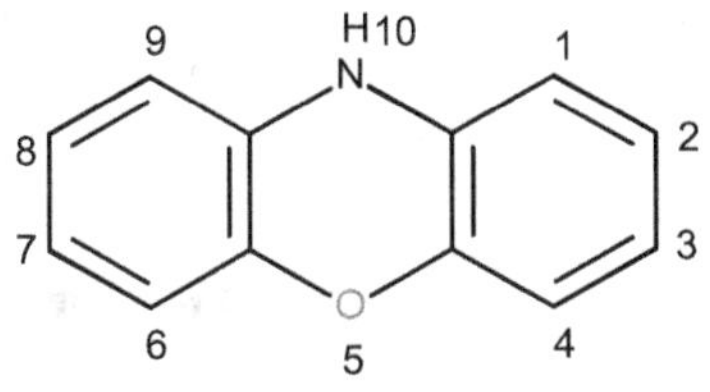

Phenoxazine

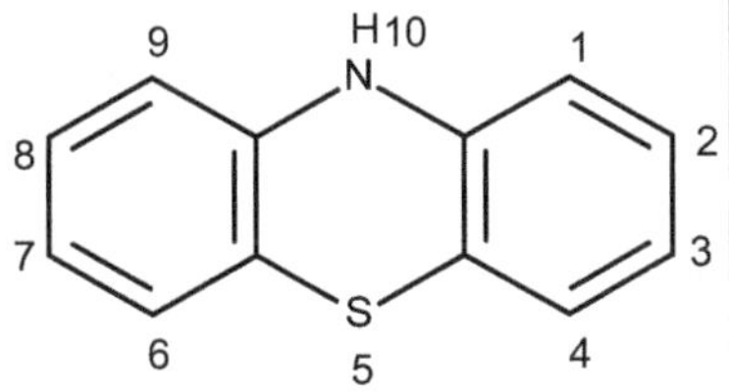

Phenothiazine

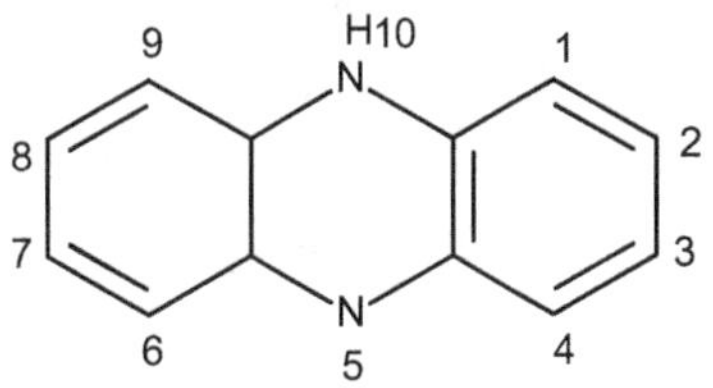

Phenazine

QUESTION BANK

A. MULTIPLE CHOICE QUESTIONS

1 What are heterocyclic compounds?
    A. Cyclic ring
    B. Aromatic
    C. Cyclic ring having one hetero atom or more
    D. All of these

Answer: C

2 Which of the following are five-membered heterocyclic compounds?
    A. Furan
    B. Pyrrol
    C. Thiophene
    D. All of the above

Answer: D

3. Which of the following are six-membered heterocyclic compounds?
    A. Pyridine
    B. Benzene
    C. Phenol
    D. None of these

Answer: A

4. Which of the following compounds is called benzopyridine?
    A. Quinoline
    B. Isoquinoline
    C. Indol
    D. None of these

Answer: A

5. Saturated hydrocarbons are otherwise referred as _____________
    A. Alkanes
    B. Alkenes
    C. Alkynes
    D. Alkaloids

Answer: A

6. Identify the smallest alkane which can form a ring structure (cycloalkane).
    A. Cyclomethane
    B. Methane
    C. Cyclopropane
    D. Propane

Answer: C

7. The first step in IUPAC nomenclature is to identify the total number of carbon atoms present in the compound.
    A. True
    B. False

Answer: A

8. The substituent in the chain is named by replacing the "ane" in the alkanes by ________
    A. ene
    B. ic
    C. one
    D. yl

Answer: D

9. The C=C bond in the chain of the compound considered is shown by ________
    A. Specifying the number of carbon atoms associated with the bond
    B. Specifying the number of carbon atoms at the beginning of the C=C bond
    C. Specifying the number of carbon atoms at the end of the C=C bond
    D. Specifying the number of carbon atoms in the entire chain

Answer: B

10. Dienes are the name given to compounds with ________
    A. Exactly a double bond
    B. Exactly a triple bond
    C. Exactly two double bond
    D. More than two double bond

Answer: C

11. Triple bond with two carbon atoms on either side is called ________
    A. Methnyl group
    B. Ethynyl group
    C. Propargyl group
    D. None of these

Answer: C

12. The substituent groups that are commonly associated with benzene ring are _______
    A. Phenyl and benzyl
    B. Propyl and phenyl
    C. Methyl and benzyl
    D. Butyl and phenyl

Answer: A

B. SHORT ANSWER QUESTIONS

1. Name the following heterocyclic compounds?

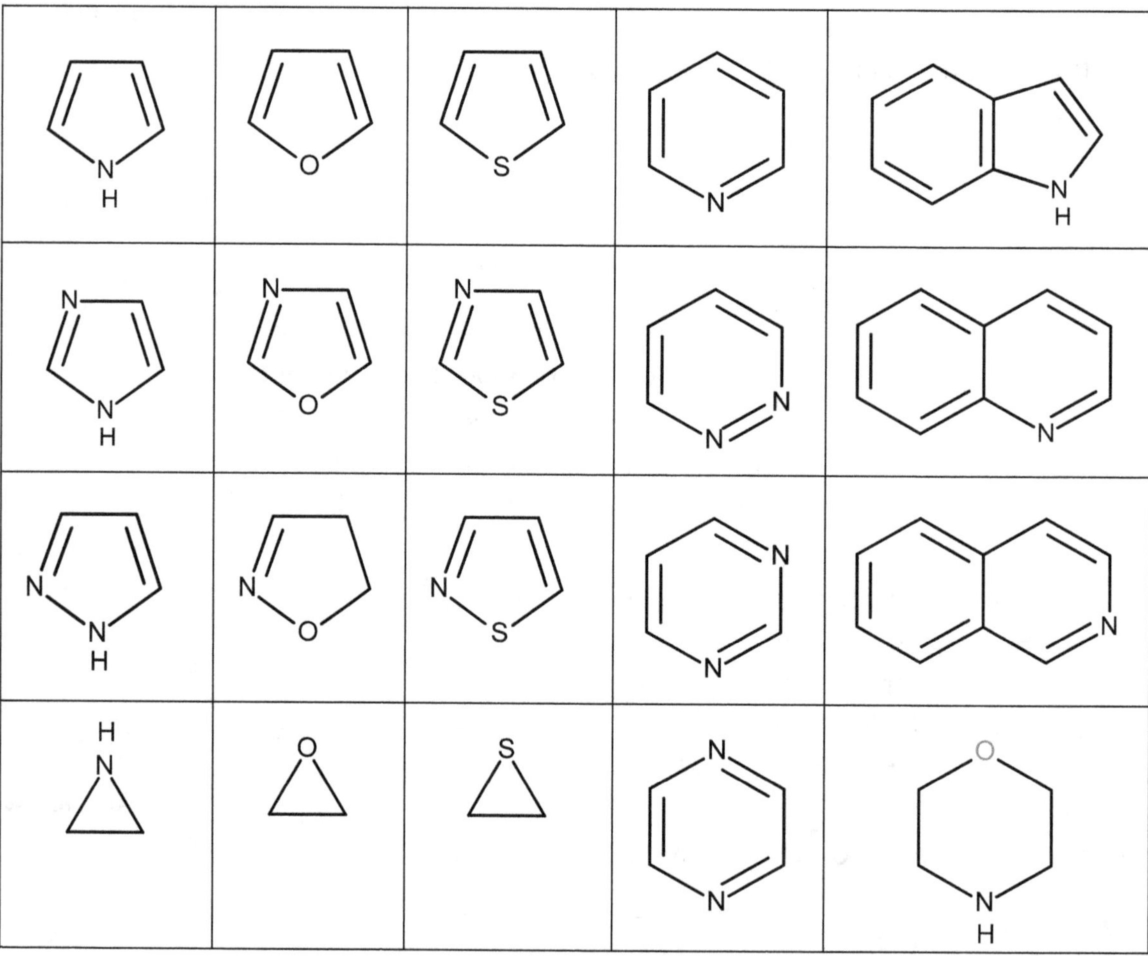

2. What is the IUPAC nomenclature system for organic compounds? What are its standard rules?

3. Give an example for each of the 1°, 2° and 3° amines. Draw the structures.

4. Classify organic compounds on the basis of functional group and elemental composition

5. Classify organic compounds based on Chemical Structures and functional groups.

C. LONG ANSWER QUESTIONS

1. Summarize the IUPAC nomenclature of organic compounds in a straight forward and easy-to-understand way.

2. Discuss the Systematic Nomenclature System (Hansch-Wildman nomenclature system)

■■■

DRUGS ACTING ON CENTRAL NERVOUS SYSTEM: ANAESTHETICS

CONTENTS

Brief introduction of Drugs acting on Central Nervous System.
Anaesthetics or General Anaesthetics:

- Thiopental Sodium*
- Ketamine Hydrochloride*
- Propofol

♦ LEARNING OBJECTIVES ♦

After completing this chapter, the student should be able to understand:

1. The term anaesthesia and its stages
2. Role of selected anaesthetic drugs
3. Chemical names and structures and synthesis of selected anaesthetic drugs
4. Mechanism of action and uses of selected anaesthetic drugs
5. Stability and storage conditions and brands of selected anaesthetic drugs.

5.1.1 INTRODUCTION

The **Nervous System** is composed of nerve tissues in the body. The functions of nerve tissues are to receive stimuli, transmit stimuli to nervous centres and initiate a response to the stimuli.

The **Central Nervous System (CNS)** consists of the brain and spinal cord and serves as the collection point of nerve impulses.

The **brain** is an extremely complex organ that can receive sensory input, can originate and coordinate motor activity and is responsible for experience, intelligence, moral and social behaviour, as well as physical activity.

The **spinal cord** is comprised of neural pathways. It also contains neural circuits that can independently control numerous reflexes and central pattern generators. The spinal cord functions primarily in the transmission of neural signals between the brain and the rest of the body. In order for a message or an impulse to be transmitted from the brain to the body, that message, or impulse, is passed from neuron to neuron through junctions called **synapses**. This process continues until the message reaches its final destination, such as a muscle, gland, or other non-neural cell.

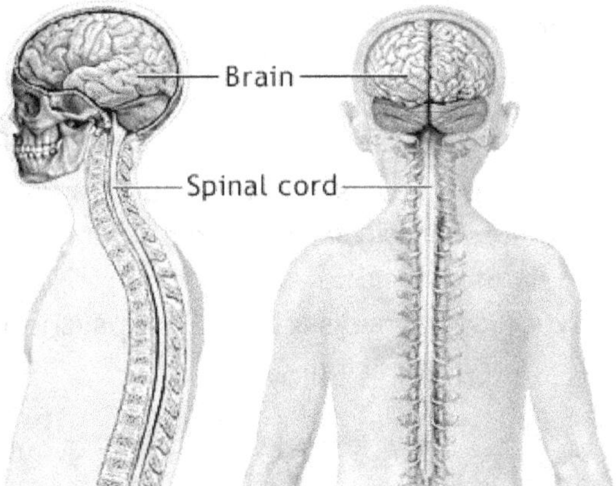

Fig. 5.1.1 The Components of Central Nervous System

Classification of Drugs Acting on CNS:

There are many drugs which act on the CNS in a general or selective manner. Their action can be positive (stimulant0 or negative (depress). Drug acting on CNS is divided into two major classes:

A. CNS Stimulants

B. CNS Depressants

A. CNS Stimulants [↑]:

CNS stimulants speedup mental and physical processes in the body. This can be useful in the treatment of certain medical conditions. These are the substances that quicken the activity of the central nervous system by increasing the rate of neuronal discharge or by blocking an inhibitory neurotransmitter. Many natural and synthetic compounds stimulate the central nervous system, but only a few are used therapeutically. CNS stimulants include drugs which are categorised as analeptics, central sympathomimetic agents, memory enhancers, hallucinogens, antidepressants, and miscellaneous CNS-acting drugs. e.g., Caffeine is a routinely used CNS stimulant in day-to-day life.

The classification of CNS stimulants is as follows;

a) Classification according to the site of action

1. Drugs acting on the cerebral cortex and subcortical regions, including the thalamus
2. Drugs acting on sensory areas of the brain
3. Drugs acting directly or reflexively on the medulla
4. Drugs acting on the spinal cord

b) Classification according to uses

1. Analeptics/ Respiratory Stimulants
2. Central Sympathomimetic Agents/Psychomotor Stimulants/Mood Elevators:
3. Nootropics/Memory Enhancers
4. Psychedelics/Hallucinogens
5. Antidepressants

B. CNS Depressants [↓]:

CNS depressants, sometimes referred to as sedatives and tranquillisers, are substances that can slow or depress brain activity. This property makes them useful for treating anxiety and sleep disorders. CNS depressants include drugs categorised as general anaesthetics, sedatives –hypnotics, anxiolytics, anticonvulsants and antipsychotics.

The classification of CNS depressants is as follows;

a) General Anesthetics

1. Inhalational anaesthetics
2. Injectable anaesthetics

b) Sedatives & Hypnotics

c) Anticonvulsants/Antiepileptic

d) Anxiolytics/Antianxiety agents/Minor Tranquillizers

e) Antipsychotics/Major Tranquillizers, which are further sub-classified as

i)   Typical Antipsychotics
ii)   Atypical Antipsychotics
iii) Rauwolfia Alkaloids

5.1.2 ANAESTHETICS OR GENERAL ANAESTHETICS

5.1.2.1 Introduction

General anesthetic is a group of drugs that belongs to CNS depressants. They produce loss of consciousness and, therefore, loss of all sensations. The absolute loss of sensation is termed anesthesia. General anesthetics bring about descending depression of the CNS, starting from the cerebral cortex, the basal ganglia, the cerebellum and finally, the spinal cord. These drugs are used in surgical operations to induce unconsciousness and therefore abolish/ stop the sensation of pain. Since, the anesthesia is felt all over the body, it is termed as general anesthesia. The purpose of general anesthesia is to undertake certain medical and surgical procedures.

General anesthesia or **anesthesia** is a medically induced coma with loss of protective reflexes resulting from the administration of either intravenous or inhalational general anesthetic medications, often in combination with an analgesic and neuromuscular blocking agent. It is generally performed in an operating theatre to allow surgical procedures that would otherwise be intolerably painful for a patient or in an intensive care unit or emergency department to facilitate endotracheal intubation and mechanical ventilation in critically ill patients.

Clinical importance of General anesthesia

The purpose of general anesthesia is to undertake certain medical and surgical procedures. Thus, it has the clinical importance of-

- Minimising the potentially harmful direct and indirect effects of surgical agents and techniques
- Sustaining physiologic homeostasis during surgical procedures
- Improving postoperative outcomes

5.1.2.2 Stages of General Anesthesia

Guedel (1920) described four stages with ether anesthesia. These clear-cut stages are not seen with the use of nowadays faster-acting general anesthetics.

Stage I. The stage of analgesia

Patient remains conscious, can hear and see and also feel a dreamlike state.

At the end of this stage, analgesia (loss of pain sensation), as well as amnesia, *i.e.,* partial or total loss of memory, develops. Reflexes and respiration remain normal. Some minor operations/ surgeries can be carried out during this stage.

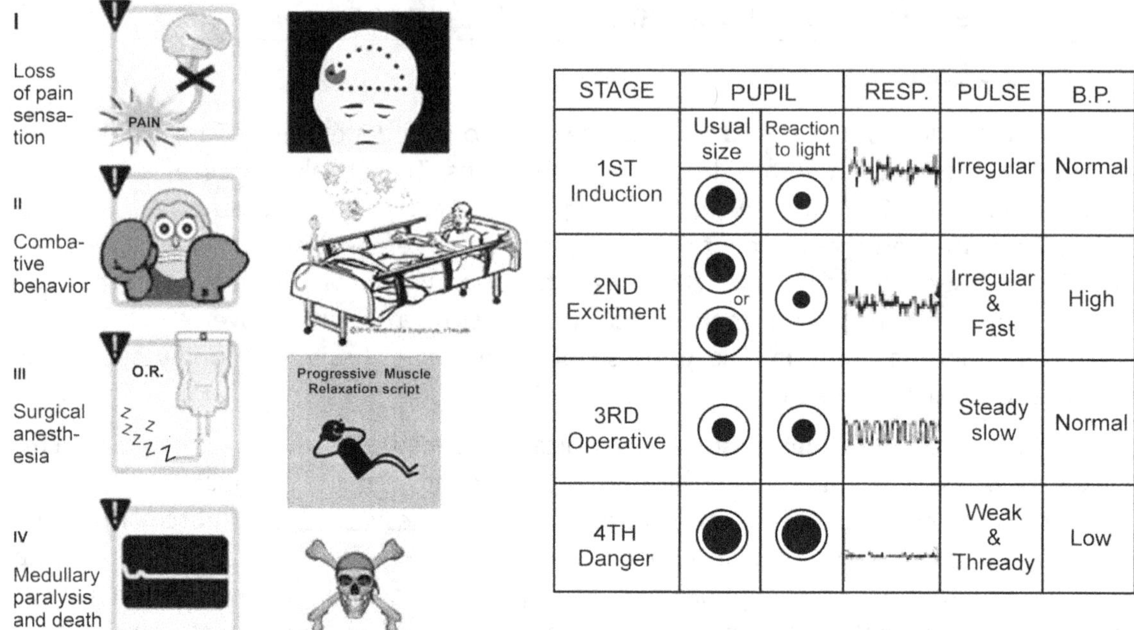

STAGE	PUPIL		RESP.	PULSE	B.P.
	Usual size	Reaction to light			
1ST Induction	●	●		Irregular	Normal
2ND Excitment	● or ●	●		Irregular & Fast	High
3RD Operative	●	●		Steady slow	Normal
4TH Danger	●	●		Weak & Thready	Low

Fig. 5.1.2 Physiological changes during stages of General Anesthesia (e.g., induced by ether)

Stage II. Stage of delirium

Apparent excitement is seen, the patient may exhibit fighting or combative behaviour and shout, struggle and hold his breath, muscletone increases, jaws are tightly closed, breathing is jerky, vomiting, involuntary micturition (urination), or defecation may also occur. Heart rate and BP may rise, and pupils dilate due to sympathetic stimulation. No stimulus should be applied, or operative procedure carried out during this stage. This stage is inconspicuous in modern anesthesia. Modern anesthetic agents help to avoid the use of large doses of a single drug that was required to produce sleep.

Stage III. Surgical anesthesia

This has been divided into 4 planes which may be distinguished as:

- Plane 1: Revolving eyeballs. This plane ends when the eyes become fixed.
- Plane 2: Loss of corneal and laryngeal reflexes
- Plane 3: Pupils start dilating, and light reflex is lost.
- Plane 4: Paralysis of intercostal muscles, minimal abdominal respiration, dilated pupils.

As anesthesia passes to deeper planes, progressively muscle tone decreases, BP falls, heart rate increases with weak pulse, and respiration decreases in depth and later in frequency also.

Stage IV. Medullary paralysis

Pupils are widely dilated, muscles are totally flabby, the pulse is imperceptible, and BP is very low.

Summary

Table 5.1.1 Signs and stages of general anesthesia

I.   Stage of analgesia
    - Analgesia
    - Unaltered conseciousness
    - Normal pupils

II. Stage of excitement
    - Disturbed consciousness (incoordinate movements, incoherent talk)
    - Irregular respiration
    - Retching and vomiting
    - Incontinence (sometimes)
    - Increased blood pressure
    - Mydriasis

III. Stage of surgical anesthesia
    - Loss of consciousness
    - Regular respiration
    - Progressive decrease of skeletal muscle tone
    - Progressive loss of somatic and autonomic reflexes
    - Progressive decrease in blood pressure
    - Miosis

IV. Stage of medullary depression
    - Loss of consciousness
    - No spontaneous respiration
    - Cardiovascular collapse
    - Mydriasis

5.1.2.3 Mechanism of Action of General Anaesthetics

Different theories have been proposed to describe the mechanism of action of general anaesthesia. These theories are briefly explained here.

I)   Meyer-Overton Theory

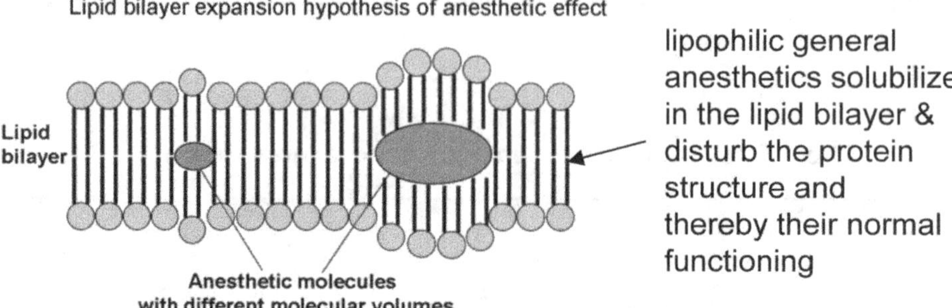

Fig. 5.1.3 Meyer-Overton Theory of Anesthetic Effect

Meyer & Overton have postulated that general anesthetics exert their action by acting on the plasma membrane. The lipophilic general anesthetics solubilise in the lipid bilayer of the neuron, disturb the lipid-protein structure of the membrane and cause disruption of normal functioning. Anesthetics can bind to hydrophobic proteins present on the membrane. Bulky and hydrophobic anesthetic molecules accumulate inside the neuronal cell membrane causing its distortion and expansion (thickening) due to volume displacement. Membrane thickening reversibly alters the function of membrane ion channels, thus, providing an anesthetic effect.

II)   Ion Channel and Protein Receptor Hypothesis

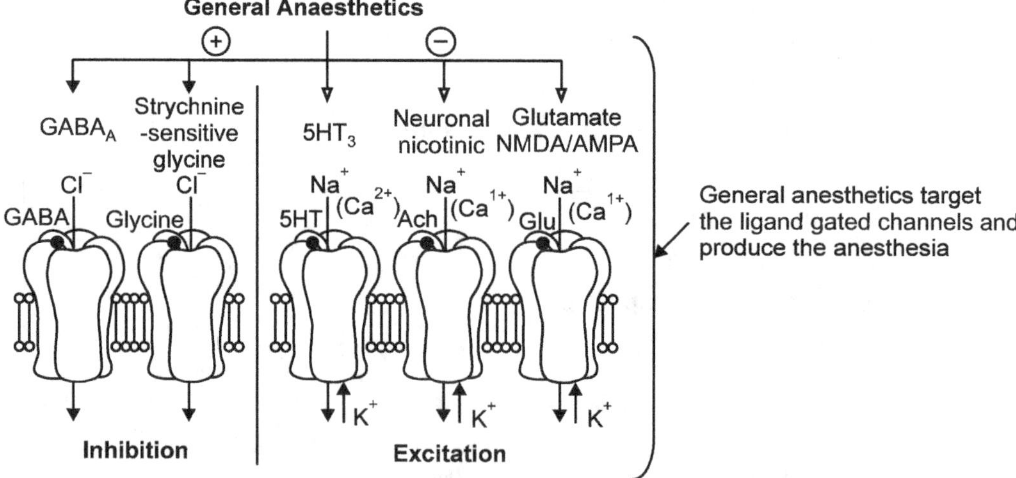

Fig. 5.1.4 Ion Channel and Protein Receptor Hypothesis

General anesthetics target the ligand-gated channels and produce the anesthesia. They exert their action by acting on either the chloride channel or sodium channel, or potassium channel.

The GABA_A receptors (the receptors that respond to the neurotransmitter gamma-amino butyric acid) and glycine receptors are the ligand-gated ion channels and are linked to **chloride channels** which normally mediate inhibitory responses within the CNS. Anesthetics (e.g., Benzodiazepine, Barbiturates) are responsible for synaptic destruction of GABA, which results in increasing neurotransmission that is inhibitory in action.

Anesthetic agents also exert action on ligand-gated **sodium channels** within NMDA (N-methyl-D-aspartate receptor) complex. This complex gets activated by glutamate and increases the conductance of sodium which promotes neuronal depolarisation. Anesthetic agents (e.g., Ketamine) show antagonistic effects on NMDA receptors.

Increased potassium conductance is required for the polarised state of the neuron as well as its repolarisation after each depolarization. Certain anesthetic agents (e.g., Dexamedetomidine) enhance **potassium channel** activity, which results in decreased neuronal excitation.

5.1.3 CLASSIFICATION OF ANESTHETICS

General anesthetics are divided into two main types depending on route administration, Inhalation anaesthesia and Intravenous anesthesia.

5.1.3.1 Inhalation anesthetics: Anesthesia induced by inhalation of certain gases or vapours of certain liquids is commonly termed as Inhalation anesthesia. These agents are classified as follows:

A) Gases:

a) Inorganic gases: *(e.g., Nitrousoxides)*

b) Hydrocarbons: *(e.g., Cyclopropane, ethylene)*

B) Volatile Liquids: *(e.g., Halothane, isoflurone, deflurane, sevaflurane, methoxyflurane.)*

5.1.3.2 Intravenous anesthesia: It is given intravenously to induce anesthesia. Several drugs are used either alone or in combination with other drugs to achieve an anesthetic effect for a minute surgery.

These include,

a) Barbiturates: *(e.g., Thiopentqal, Methohexitol)*

b) Benzo diazepines: *(e.g., Imidazole, Diazepam)*

c) Opoid analgesics: *(e.g.,Morphine, Fentanyl, Sulfenatil ,Remifentanil)*

d) Ketamine

e) Propofol

F) Miscellaneous *(e.g., Droperidol, Etomidate, Dexamedetomidine)*

5.1.4 SOME IMPORTANT DRUGS USED FOR GENERAL ANAESTHESIA

5.1.4.1 Thiopental Sodium: (*Syn:* **Sodium thiopental, Sodium Pentothal**)

Chemical Name: Sodium 5-ethyl-5-pentan-2-yl-2-sulfanylidene1,3-diazinane-4,6-dione

Structure:

Systematic (IUPAC) Name: Sodium; 5-ethyl-4,6-dioxo-5-pentan-2-yl-1*H*-pyrimidine-2-thiolate

Uses:

i. **Anesthesia:** Used commonly in the induction phase of general anesthesia. A normal dose of sodium thiopental (usually 4–6mg/kg) is given to a pregnant woman for operative delivery. Used to induce anesthesia in animals.

ii. **Medically induced coma:** Patients with brain swelling causing elevation of intracranial pressure (either secondary to trauma or following surgery) may benefit from this drug.

iii. **Status epilepticus:** Thiopental may be used to terminate a seizure.

iv. **Psychiatry:** Used to desensitise patients with phobias and to facilitate the recall of painful repressed memories

Side Effects: Respiratory depression, myocardial depression, cardiac arrhythmias, prolonged drowsiness and recovery, sneezing, coughing, bronchospasms, laryngospasms, and shivering.

Stability and storage conditions:

1. When stored in dry form, thiopental sodium is stable indefinitely.

2. Thiopental should be diluted with only sterile water for injection, sodium chloride injection.

3. Concentrations of less than 2% in sterile water should not be used as it may cause hemolysis.

4. After reconstitution, solutions are stable for 3 days at room temperature and for 7 days if refrigerated. However, as no preservative is present, it is recommended that it be used within 24 hours after reconstitution.

5. After 48 hours, the solution has been reported to attack the glass of the bottles it is stored in.

6. Thiopental may also adsorb onto plastic IV tubings and bags.

7. Do not administer any solution that has a visible precipitate.

Incompatibility: Following agents/reagents have been reported to be incompatible when mixed with thiopental;

Ringer's injection, Ringer's injection lactate, amikacinsulfate, atropine sulfate, benzquinamide, hephapirin sodium, chlorpromazine, codeine phosphate, diphenhydramine, ephedrine sulfate, glycopyrrolate, hydromorphone, insulin (regular), meperidine, metaraminol, morphine sulfate, norepinephrine, penicillin G potassium, promazineHCl, promethazine HCl, succinylcholine chloride and tetracycline HCl.

Incompatibility is dependent upon factors such as pH, concentration, temperature and diluents used.

Different types of formulations:

Thiopental sodium for Injection, USP (Powder for reconstitution)

Popular brands: It is manufactured by many companies. Some of the notable brand names in India are:

Sr. No.	Brand Name	Manufacturers	Type
1	Anesthal	Jagsonpal Pharmaceuticals Ltd	Injection
2	Intraval Sodium	Piramal Healthcare	Injection
3	Pentothal	Pharmacia & Upjohn	Injection
4	Pentone	Chandra Bhagat Pharma Pvt. Ltd.	Injection
5	Thiosal	Neon Laboratories Ltd	Injection
6	Thipen	Samarth Pharma Pvt. Ltd.	Injection

5.1.4.2 Ketamine Hydrochloride: (*Syn:* **Ketalar**)

Also categorised especially as **Dissociative Anesthetic.**

Structure:

Systematic (IUPAC) Name: 2-(2-Chlorophenyl)-2-(methylamino)cyclohexan-1-one

Dissociative anesthesia implies dissociation from the surrounding with only superficial sleep mediated by interruption of neuronal transmission from unconscious to conscious parts of the brain.

Drugs that inhibit pain by cutting off or dissociating the brain's perception of pain, Induce a state of sedation, immobility, amnesia and analgesia.

During dissociative anesthesia, the animal maintains it's pharyngeal, laryngeal, corneal, palpebral, and swallowing reflexes. The eyes also remain open.

Dissociative anesthetic agents increase muscle tone and spontaneous involuntary muscle movement (occasionally, seizures are seen in some species). Salivation and lacrimation are also increased. Somatic analgesia is good.

Examples of dissociative anesthetics are Ketamine HCl, Phencyclidine (PCP), Dextromethorphan (DXM)

Uses:

i)   **Anesthesia:** It can be used as an anesthetic for a minor procedure. It can be used as a sedative for physically painful procedures in emergency departments. It can be used to supplement spinal or epidural anesthesia/analgesia using low doses.

ii) **Pain:** Ketamine issued for postoperative pain management. Low doses of ketamine reduce morphine use as well as nausea and vomiting after surgery. Low-dose ketamine is sometimes used in the treatment of complex regional pain syndrome (CRPS).

iii) **Depression:** At doses less than those used in anesthesia can be used to treat depression.

Side Effects:

In addition to its indicated effects, some unwanted effects may be caused by ketamine. They are abnormal heart rhythms, slow heart rate or fast heart rate, high blood pressure or low blood pressure, increased intracranial pressure, transient erythema, nausea, increased salivation, vomiting, increased skeletal muscle tone (tonic-clonic movements), Double vision, increased intraocular pressure, tunnel vision, airway obstruction, increased bronchial secretions, respiratory depression, laryngospasm, anaphylaxis, dependence, emergence reaction.

Stability and storage conditions:

Stable at normal temperatures and pressures. Store at room temperature below 30°C, but do not freeze. Protect from sunlight. Store in a well-closed container.

Incompatibility:

Ketamine is chemically incompatible with barbiturates and diazepam, leading to precipitate formation when mixed with them.

Different types of formulations:

For direct intravenous administration: The 100-mg-per-mL concentration of ketamine must be diluted with an equal volume of sterile water for injection, 0.9% sodium chloride injection, or 5% dextrose injection prior to injection.

For intravenous infusion: Add 10 ml of the 50-mg-per-ml concentration or 5 mL of the 100-mg-per-mL concentration, of ketamine (base) to 500 mL of 5% dextrose injection or 0.9% sodium chloride injection and mix well. The resultant solution will contain 1 mg of ketamine (base) per ml. If fluid restriction is necessary, 250 ml of the diluent may be used to provide a solution containing 2 mg of ketamine (base) per ml)

Popular brands: Ketamine formulations are manufactured by many companies in India. Some of them are given below:

Sr. No.	Brand Name	Manufacturers	Type
1	Aneket	Neon Laboratories	Injection
2	GB -Ket	GreencoBiologicalsPvt Ltd	Injection
3	Hypnoket	Chandra BhagatPharma Pvt. Ltd.	Injection
4	Pentyl	Aci ltd.	Injection
5	Keta	V Swiss Pharma Pvt. Ltd.	Injection
6	Ketolide, Ketafast	Taj Pharmaceuticals	Injection
7	Ketalar	Parke-Davis (India)	Injection
8	Ketam	Sun Pharma Ltd	Injection
9	Ketam	Sun Pharma Ltd	Injection
10	Ketolide	Indus Pharma Pvt. Ltd.	Injection

5.1.4.3 Propofol: (*Syn:* **Diprivan**)

Structure:

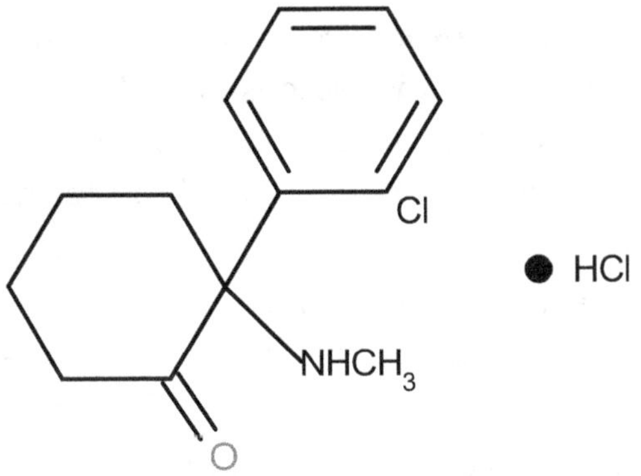

Systematic (IUPAC) Name: 2,6-Bis(propan-2-yl)phenol

Propofol was discovered in 1977 and approved for use in the United States in 1989. It is on the World Health Organization's List of Essential Medicines. It is available as a generic medication.[4] It has been referred to as milk of amnesia (a play on "milk of magnesia") because of the milk-like appearance of the intravenous preparation and because of its tendency to suppress memory recall.

Uses: It is given by injection into a vein, and the maximum effect takes about two minutes to occur and typically lasts five to ten minutes. Recovery from propofol-induced anaesthesia is generally rapid and associated with less frequent side effects (e.g. drowsiness, nausea, vomiting) than with thiopental, methohexital, and etomidate.

i.   **Anesthesia:** Its uses include the starting and maintenance of anesthesia. Propofol is also used in veterinary medicine for anesthesia

ii. **Sedation**: Propofol may also be used to sedate coronavirus (COVID-19) patients who are mechanically ventilated in ICUs.

   Propofol has also been suggested as a sleep aid in critically ill adults in the ICU; however, the effectiveness of this medicine at replicating the mental and physical aspects of sleep for people in the ICU is not apparent.

iii. **Status epilepticus:** It is also used for status epilepticus if other medications have not worked.

iv. **Pre-surgery**: Used to relax or sleep before and during surgery or other medical procedures.

v.   **Procedural sedation:** Propofol is also used for procedural sedation. Its use in these settings results in a faster recovery compared to midazolam. It can also be combined with opioids or benzodiazepines. Because of its rapid induction and recovery time, propofol is also widely used for sedation of infants and children undergoing MRIs. It is also often used in combination with ketamine with minimal side effects.

In critically ill patients, propofol is superior to lorazepam; both in effectiveness and overall cost. Propofol is relatively inexpensive compared to medications of similar use due to its shorter ICU stay length. One of the reasons propofol is thought to be more effective (although it has a longer half-life than lorazepam) is because studies have found that benzodiazepines like midazolam and lorazepam tend to accumulate in critically ill patients, prolonging sedation.

Mechanism of Action: The action of propofol involves a positive modulation of the inhibitory function of the neurotransmitter gama-aminobutyric acid (GABA) through GABA-A receptors.

Side Effects: Common side effects of propofol include:
- fast or slow heart rate,
- high or low blood pressure,

- injection site reactions (burning, stinging, or pain),
- apnea,
- rash, and

Stability and storage conditions: Propofol is compatible with commonly used inhalant anaesthetics.

Propofol has been reported to have high stability in glass and relatively high stability for up to 24 hours in polyvinyl chloride-based medical plastics.

Propofol undergoes oxidative degradation in the presence of oxygen and is therefore packaged under nitrogen to eliminate this degradation path. Store between 4 and 25°C (40-77°F).

Incompatibility: 4-Hydroxybutyric acid, midazolam hydrochloride, piritramide, and remifentanil hydrochloride are physically incompatible when mixed with propofol.

Different types of formulations:

i) **Propofol injection emulsion:** Propofol injectable emulsion is a single access parenteral product (single patient infusion vial) which contains benzyl alcohol 1.5 mg/mL and sodium benzoate 0.7 mg/mL to inhibit the growth of microorganisms.

ii) **Propofol plain injection:** Propofol with saline is administered using a closed-loop algorithm which permits to reach of a Bispectral Index target of 50. The blinded syringe contains 45 ml of propofol and 5 ml of saline.

iii) **Propofol plus lidocaine injection:** Propofol with lidocaine 1% is administered using a closed-loop algorithm which permits to reach of a Bispectral Index target of 50. The blinded syringe contains 45 ml of propofol and 5 ml of lidocaine. It is used to reduce Propofol induced pain.

Popular brands: It is manufactured by many companies. Some of the notable brand names in India are:

Sr. No.	Brand Name	Manufactures	Type
1	Propovan (10 ml)	Bharat Serums and Vaccines Limited	Injection
2	Propofol (10 ml)	Neon Laboratories Limited	Injection
3	Diprivan	ICI Healthcare Pvt. Ltd.	Injection
4	Celofil	Thesmis Pharmaceuticals Ltd.	Injection
5	Critifol (1%)	AHPL	Injection
6	Thipen	Samarth Pharma Pvt. Ltd.	Injection

QUESTION BANK

A. MULTIPLE CHOICE QUESTIONS

1. Which of the following General Anaesthetics is most likely to cause bronchodilation? :
   A. Thiopentone
   B. Ketamine
   C. Propofol
   D. Fentanyl

   Answer: B

2. After I.V., administration of which of the following anaesthetic agents, there is rapid recovery and less headedness?
   A. Propofol
   B. Diazepam
   C. Droperidol
   D. Midazolam

   Answer: A

3. In which of the following stage general anesthesia surgery is done____________?
   A. Stage I
   B. Stage II
   C. Stage III plane 2
   D. Stage III plane 3

   Answer: D

4. Ketamine is a_____________?
   A. General anaesthetic agent
   B. Local anaesthetic agent
   C. Antidepressive agent
   D. Hypnotic agent

   Answer: A

5.   Match the following

1. Inhalation Anaesthetic,	a) Midazolam
2. Intravenous Anaesthetic	b) Ketamine
3. Dissociative Anaesthetic	c) Thiopental
4. Premedication	d) Sevoflurane

Answer: 1-d; 2-c; 3-b; 4-a

B. SHORT ANSWER QUESTIONS

1.   Discuss the CNS in brief and classify the CNS-acting drugs
2.   Write a note on general anesthetics
3.   Define and classify general anesthetics.
4.   Write the structure of Sodium thiopental and give its stability and storage.
5.   Write the structure of Ketamine hydrochloride and give its stability and storage
6.   Write the structure of Propofol and give its stability and storage
7.   Write the uses of: a) Sodium thiopental b) Ketamine hydrochloride c) Propofol
8.   Enlist the brand names of: a) Sodium thiopental b) Ketamine hydrochloride c) Propofol.

C. LONG ANSWER QUESTIONS

1.   What is general anaesthesia? Discuss the induction and various stages involved. Define general anaesthetics & classify them.
2.   Write the structures, chemical names, uses, stability and storage of the following:

     a) Sodium thiopental   b) Ketamine hydrochloride   c) Propofol.

■■■

DRUGS ACTING ON CENTRAL NERVOUS SYSTEM: SEDATIVES AND HYPNOTICS

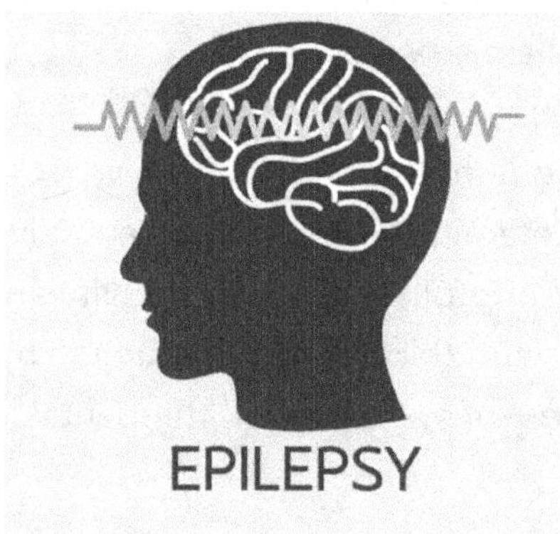

CONTENTS

Sedatives and Hypnotics:

- Diazepam*
- Alprazolam*
- Nitrazepam
- Phenobarbital*

♦ LEARNING OBJECTIVES ♦

After completing this chapter, the student should be able to understand:

1. The terms sedatives and hypnotics and their classification and difference between them.
2. Roles of sedative and hypnotic drugs and their mechanism of action.
3. Chemical names and structures of selected sedative and hypnotic drugs
4. Mechanism of action and uses of selected sedative and hypnotic drugs
5. Stability and storage conditions and brands of selected sedative and hypnotic drugs

5.2.1 INTRODUCTION

5.2.1.1 Sedatives and Hypnotics

Sedatives & hypnotics are the agents that depress the CNS in a dose-dependent manner, inducing sedation, sleep, unconsciousness, surgical anaesthesia, coma, and finally, fatal depression of respiration and circulation. Sedative hypnotics are xenobiotics that limit excitability and are responsible for sedation and/or inducing drowsiness and sleep and responsible for hypnosis.

5.2.1.2 Difference between Sedatives and Hypnotics

Sr. No.	Parameter	Sedatives	Hypnotics
1	Definition	A sedative agent decreases activity, moderates excitement and causes calmness.	A hypnotic drug produces drowsiness and facilitates the sleep. Also called as somnifacient.
2	Meaning	Sedation refers to decreased responsiveness to any level of stimulation; is associated with some decrease in motor activity and ideation or thinking process.	Induces and/or maintains sleep, similar to normal wake able sleep.
3	Dose & Action	Lower dose, more slowly acting drugs with flatter dose-response curves.	Higher dose, quicker onset, shorter duration and steeper dose-response curves
4	Time	Sedatives are usually given during daytime hours.	Hypnotics are given at night or hours of sleep (HS)

5.2.1.3 The Physiology of Sleep

Sleep is a state of altered consciousness or partial unconsciousness from which an individual can be awakened by stimuli like strong light, sound, touch, pain etc.

The duration and pattern of sleep varies considerably among individuals. Age has a significant effect on the quantity and depth of sleep.

Stages of Sleep:

There are two stages of normal sleep those are **Non-Rapid Eye Movement (NREM)** sleep and **Rapid Eye Movement (REM)** sleep. Normally 75-80%sleep is NREM, and 20-25% is REM sleep. NREM sleep consists of four stages, 1 to 4 and precedes REM sleep.

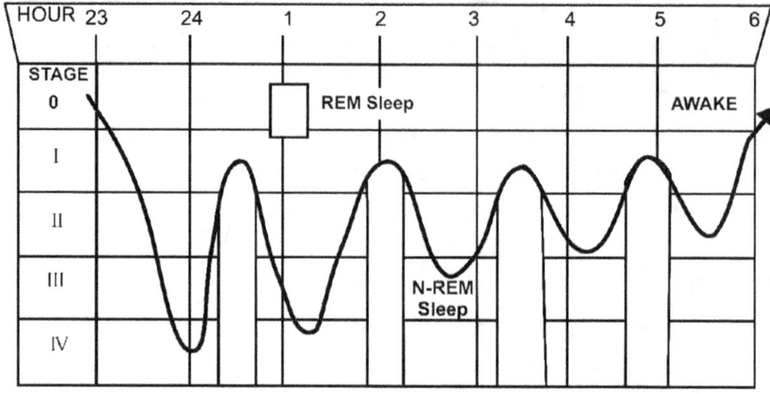

Fig. 5.2.1 Normal Sleep Cycle

Stage 0 (Awake):

It starts from lying down to falling asleep. This stage constitutes 1–2% of sleep time.

Stage 1 (Dozing):

During this stage, there is an experience of drifting in and out of sleep, and one can be easily woken up. Eye movements and body movements are slowed down. Sometimes sudden jerky movements of muscles known as myoclonic jerks may be observed. This stage occupies 3–6% of sleep time.

Stage 2 (Unequivocal sleep):

During this stage, eye movement stops, and brain waves become slower. There will also be an explosion of rapid brain activity called sleep spindles. Subjects are easily wake able. This stage comprises 40–50% of sleep time.

Stage 3 (Deep sleep transition):

It can be very difficult to wake up someone during this stage and also known as the first step of deep sleep. If someone is woken up during this stage, they may feel confused and disoriented for several minutes. This stage comprises 5–8% of sleep time.

Stage 4 (Cerebral sleep):

This stage of sleep is the second step of deep sleep. It can be very difficult to wake someone up during this stage. Both stages (3 &4) of deep sleep are important for feeling refreshed in the morning. During stages 2, 3 and 4, heart rate, BP and respiration are steady, and muscles are relaxed. Stages 3 and 4 together are called **Slow Wave Sleep (SWS).**

REM sleep (Paradoxical sleep):

REM sleep is the sleep stage in which dreaming occurs. There are marked, irregular, and darting eye movements, and breathing becomes fast, irregular and shallow.

5.2.2 CLASSIFICATION OF SEDATIVES AND HYPNOTICS

Depending on the chemical structure, time required for action, and pharmacological activity, they are classified as:

A) Barbiturates:

1. Long-acting: (e.g., Phenobarbitone)
2. Short acting: (e.g., Butobarbitone, Pentobarbitone)
3. Ultra-Short acting: (e.g., Thiopentone, Methohexitone)

B) Benzodiazepines:

1. Hypnotics: (e.g., Diazepam, Flurazepam, Nitrazepam, Alprazolam, Temazepam, Triazolam)
2. Antianxiety: (e.g., Diazepam, Chlordiazepoxide, Oxazepam, Lorazepam, Alprazolam)
3. Anticonvulsant: (e.g., Diazepam, Lorazepam, Clonazepam, Clobazam).

C) Agents having diverse chemical structures:

1. Aldehydes: (e.g., Chloral hydrate)
2. Acetylene derivatives: (e.g., Methylpentynol, Ethchlorvynol)
3. Piperidinediones: (e.g., Methylprylon, Glutethimide)
4. Cyclopyrrolones: (e.g., Zopiclone)
5. Imidazopyridines: (e.g., Zolpidem, Alpidem)
6. Antihistaminics: (e.g., Doxylamine, Diphenhydramine, Niaprazine)
7. Acyclic nitrogen containing Hypnotics:
   a) Carbamates: (e.g., Mepromate)
   b) Ureides: (e.g., Carbromal, Bromisovalum)
8. Miscellaneous: (e.g., Etomidate, Fenadiazole, Clomethiazole)

D) Agents from Plant Origin: (e.g., **Kava**, Lemon balm, Passiflora)

5.2.2.1 Mechanism of Action of Sedatives and Hypnotics

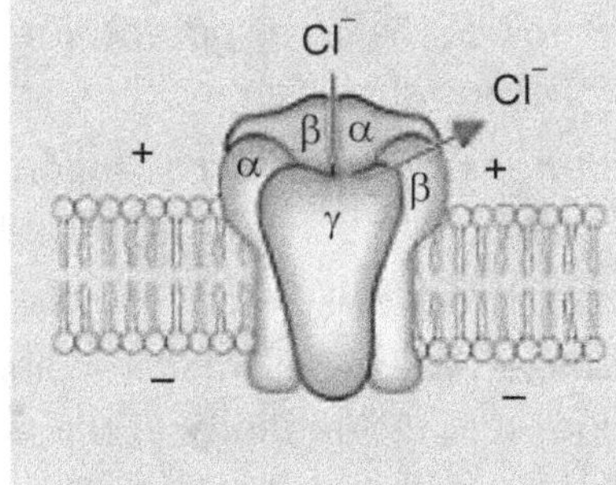

A. Receptor empty (no agonist)

Empty receptor is inactive, and the coupled chloride channel is closed.

Contd...

B. Receptor bhinding GABA

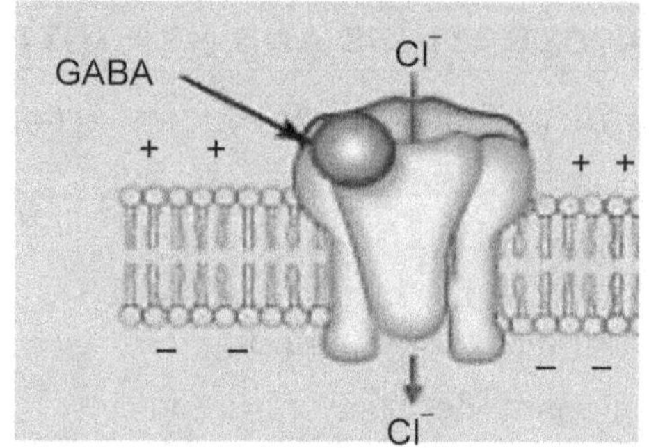

Binding of GABA cause the chloride ion channel to open, leading to hyperpolarization of the cell.

C. Receptor binding GABA and benzodiazepine

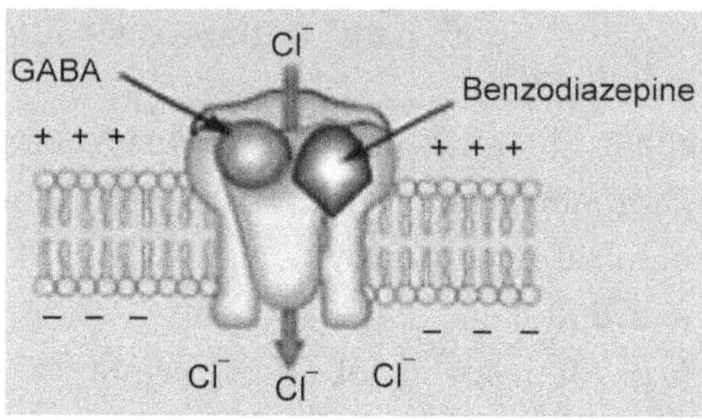

Binding of GABA is enhanced by benzodiazepine, resulting in a greater entry of chloride ion. Entry of Cl⁻ hyperpolarization the cell, making it more difficult to polarize, and reduces neural excitability

Fig. 5.2.2 Mechanism of Action of Sedatives & Hypnotics

Most of the hypnotic & sedative drugs facilitate the actions of GABA (γ-Amino Butyric Acid), a major inhibitory transmitter in the CNS. The subtype, $GABA_A$ receptor activation, leads to an increased influx of Cl⁻ ions, while $GABA_B$ receptor activation leads to increased efflux of K- ions. Both of these contribute to membrane hyperpolarisation.

Class of drugs	Mechanism of Action
Benzodiazepines	• The benzodiazepines interact with a macromolecular membrane complex which is known as site for GABA and a chloride ionophore. • Benzodiazepines have a potent interaction with the GABA inhibitory neurotransmitter system. • They bind non-specifically to benzodiazepine receptors which mediate sleep, affects muscle relaxation, anticonvulsant activity, motor coordination. • As benzodiazepine receptors are bind with $GABA_A$ receptors which enhances GABA affinity of the GABA receptor. • Binding of GABA to the site or receptor opens the chloride channel, which is responsible for hyperpolarization of cell membrane that prevents further excitation of the neuron cell.

Contd...

Class of drugs	Mechanism of Action
Barbiturates	• Barbiturates alter the mechanism of synaptic transmission or neurotransmission which is essential for the process of communication between two neurons. • In sufficient concentration, these drugs alter the permeability of the cell membrane after binding with gamma-aminobutyric acid-A (GABA$_A$) receptors and reduce the excitability of the post synaptic cell. • In general, excitatory synaptic transmission is depressed by barbiturates.

5.2.3 IMPORTANT DRUGS USED AS SEDATIVES AND HYPNOTICS

5.2.3.1 Diazepam: (*Syn.* **Valium**)

Structure:

Systematic (IUPAC) Name: 7-Chloro-1-methyl-5-phenyl-2,3-dihydro-1H-1,4-benzodia-zepin- 2-one

Uses:

i.   **Anxiety Disorders:** It is useful in the management of anxiety disorders and also for getting short-term relief from anxiety associated with depressive symptoms.

ii. **Surgery:** It is useful to provide preliminary sedation before giving general anaesthesia.

iii. **Seizure Disorders:** Diazepam is a drug of choice for the termination of status epilepticus or acute seizure episodes resulting from drug overdosage and poisons.

iv. **Alcohol withdrawal:** It is useful in the prevention or to get symptomatic relief from acute delirium and hallucinations associated with acute alcohol withdrawal.

v. **Skeletal muscle spasticity:** Adjunct to rest, physical activity, analgesics and other measures for the relief of discomfort associated with acute, painful musculoskeletal conditions.

vi. **Sedation in critical-care settings:** One of several benzodiazepines recommended for sedation of acutely disturbed patients because of its rapid onset and short duration of action when given in single doses.

vii. **Night Terrors:** Diazepam has been used effectively to prevent night terrors. (Night terrors are a form of sleep disorder in which a person partially awakens from sleep in a state of terror.)

viii. **Labour and delivery:** It is used as an adjunct to local anaesthetics and systemic analgesics during labour and delivery to reduce requirements for opiate analgesics and to produce amnesia (forgetfulness).

ix. **Myocardial Infarction:** It has been used to relieve anxiety associated with myocardial infarction

x. **Drug-induced cardiovascular emergencies:** Adjunct in the management of specific drug-induced cardiovascular emergencies (e.g., drug-induced tachycardia, hypertensive emergency, acute coronary syndrome, or acute anticholinergic syndrome). Adjunct in the initial treatment of the cocaine-induced acute coronary syndrome.

Side Effects:

Drowsiness, ataxia (it means inability to coordinate voluntary muscle movements), fatigue. With parenteral therapy, local reactions (venous thrombosis, phlebitis) at the injection site may be observed.

Stability and storage conditions:

Should be stored in tight, light-resistant containers at 15–30°C; avoid freezing. The half-life of Diazepam: is 20–50 hours.

Incompatibility:

Alcohol and caffeine have been reported to be incompatible when mixed with diazepam. Some products that may interact with this drug include certain anti-depressants (e.g., fluoxetine, fluvoxamine, nefazodone), cimetidine, clozapine, digoxin, disulfiram, kava, ketoconazole, omeprazole, phenytoin, sodium oxybate.

Different types of formulations:

1. Oral: Solution, Solution concentrates, Tablets
2. Parenteral: Injections
3. Rectal: Gels

Popular brands:

In India, diazepam formulations are manufactured by many companies. Some of the notable are as follows:

Sr. No.	Brand Name	Manufacturers	Type
1	Almpose	Ranbaxy Diagnostics	Tablet
2	Alpazepam	Alpa Laboratories Ltd	Injection
3	Alzepam	Psyco Remedies.	Tablet
4	Calmpose	Ranbaxy Diagnostics	Tablet
5	Calmtack	Indus Pharma Pvt. Ltd.	Injection
6	Diaze	Ind-Swift Limited	Injection
7	Dizep	Intas Laboratories Pvt Ltd	Tablet
8	Placidox -10	Lupin Laboratories Ltd. (Pinnacle)	Tablet
9	Sparlium	Cadila Pharmaceuticals Ltd.	Injection
10	Valium -2	Piramal Healthcare	Tablet
11	Zepose	Cipla Limited	Tablet

5.2.3.2 Alprazolam: (*Syn.* Xanax)

Structure:

Systematic (IUPAC) Name: 8-Chloro-1-methyl-6-phenyl-4*H*-[1,2,4]triazolo[4,3-a][1,4] benzodiazepine.

Uses:

Alprazolam is mostly used to treat anxiety disorders, panic disorders and nausea due to chemotherapy.

i. **Panic disorder:** Alprazolam is effective in the relief of moderate to severe anxiety and panic attacks.

ii. **Nausea due to chemotherapy:** Alprazolam may be used in combination with other medications for chemotherapy-induced nausea and vomiting.

iii. **Anxiety disorders:** Anxiety associated with depression is treated by alprazolam.

Side Effects:

More common side effects that can occur with alprazolam are drowsiness, feeling tired, slurred speech, lack of balance or coordination, memory problems, or feeling anxious early in the morning. Absent, missed or irregular menstrual periods, decreased appetite, decreased interest in sexual intercourse, decreased sexual performance or desire, abnormal ejaculation, difficulty having a bowel movement (stool), increased appetite, increased weight, loss of sexual ability, desire, drive, or performance, stopping of menstrual bleeding, watering of mouth, weight loss, etc., are reported side effects.

Stability and storage conditions:

Store immediate-release tablets & capsules below 77°F (25°C). Store extended-release tablets, orally disintegrating tablets and oral solutions at controlled room temperature (59° to 86°F or 15°C to 30°C). Protect orally disintegrating tablets from moisture.

Incompatibility & Contraindications:

Alprazolam is chemically incompatible with other benzodiazepines.

Alprazolam extended-release may be contraindicated in patients with acute narrow-angle glaucoma.

Alprazolam extended release is contraindicated with ketoconazole and itraconazole.

Different types of formulations: Oral Tablets, Capsules, and Liquid oral solutions.

Popular brands:

In India, Alprazolam combinations and formulations are manufactured by many companies. Some of the notable brands are as follows:

Sr. No.	Brand Name	Manufacturers	Type
1	Fludep Plus	Cipla Limited	Tablet
2	Alfresh	Medico Health Care	Tablet
3	Alltop	Torrent Labs (P)Ltd	Capsule
4	Alpraquil	Lupin Laboratories Ltd	Tablet
5	Alprawin F T	Mediwin Pharmaceuticals	Tablet
6	Alprax Forte	Torrent Pharmaceuticals Ltd. (Vista)	Tablet
7	ALP	Casca Remedies Pvt. Ltd.	Tablet
8	Alzot	Pfizer Ltd. (Pharmacia India Pvt. Ltd.)	Tablet
9	Alep SR (0.5mg)	D.R. Johns Labs	Tablet
10	Alzex (0.5mg)	Pharmacia India Ltd	Tablet

5.2.3.3 Phenobarbital: (*Syn.* **Phenobarbitone, Luminal**)

Structure:

Systematic (IUPAC) Name: 5-Ethyl-5-phenyl-1,3-diazinane 2,4,6-trione

Uses: Phenobarbitalis mostly used to treat anxiety disorders, panic disorders, and nausea due to chemotherapy.

1. **Seizures:** Phenobarbital is used in the treatment of all types of seizures except absence seizures. Phenobarbital is the first-line choice for the treatment of neonatal seizures.

   Phenobarbital properties can effectively reduce tremors and seizures associated with abrupt withdrawal from benzodiazepines.

2. **Detoxification:** Phenobarbital is sometimes used for alcohol detoxification and benzodiazepine detoxification due to its sedative and anticonvulsant properties.

3. **Cyclic vomiting syndrome symptoms:** Cyclic vomiting syndrome is characterised by episodes of severe vomiting that have an apparent cause. Phenobarbital can also be used to relieve cyclic vomiting syndrome symptoms.

Side Effects: Central nervous system effects, such as dizziness. In elderly patients, it may cause excitement and confusion, while in children, it may result in paradoxical hyperactivity.

Stability and storage condition:

Preserve and dispense in tight, light-resistant containers as defined in the USP. Store at a controlled room temperature of 15°C- 30°C.

Incompatibility:

Phenobarbital may interact with valproic acid to cause an increase in the risk of its own side effects. Effectiveness of beta-adrenergic blockers (e.g., propranolol), clozapine, corticosteroids (e.g., hydrocortisone), digitoxin, doxycycline, estrogens, metronidazole, oral contraceptives (e.g., ethinylestradiol), quinidine, theophyllines, voriconazole, or warfarin may be decreased by phenobarbital.

Different types of formulations:

Oral Tablets, Oral Elixir, Rectal Tablets, Parenteral solutions (intramuscular and intravenous)

Popular brand names:

The generic phenobarbital is manufactured by many companies. Some of them are as follows:

Sr. No.	Brand Name	Manufacturers	Type
1	Phenetone	Cipla Limited	Tablet
2	Barbee	Ind-Swift Limited	Tablet
3	Gardenal	Piramal Healthcare	Tablet
4	Barbinol	Human Antibiotic Pharmaceuticals Pvt Ltd	Tablet
5	Luminal	Bayer India Ltd	Tablet
6	Phenobarb	Intas Laboratories Pvt Ltd	Tablet
7	Phenobarbitone	Nicholas Piramal India Ltd.	Injection
8	Epikon	Konark Biochem	Tablet
9	Emgard	Medopharm	Tablet
10	Epitan	Reliance Formulation Pvt Ltd	Tablet

5.2.3.4 Nitrazepam: (*Syn.* Mogadon)

Structure:

Systematic (IUPAC) Name: 1,3-Dihydro-7-nitro-5-phenyl-2*H*-1,4-benzodiazepin-2-one.

Uses: Nitrazepam is used to treat short-term sleeping problems (insomnia). It is also sometimes used for refractory epilepsies.

Side Effects: Central nervous system depression, including dizziness, depressed mood, fatigue, headache, impairment of memory, impairment of motor functions, hangover feeling in the morning, slurred speech, decreased physical performance, numbed motions, reduced alertness, muscle weakness, and double vision. Unpleasant dreams and rebound insomnia have also been observed. High levels of confusion and stiffness also occur after the administration of nitrazepam. Impairment of psychomotor function may especially occur after repeated administration.

Stability and storage conditions: Protect from light. Store at 15 to 30°C.

Incompatibility: The sedative effects may be enhanced when the product is used in combination with alcohol. Subsequent use of antipsychotics (neuroleptics), tranquillisers, hypnotics, sedatives, analgesics, narcotic analgesics, anxiolytics/sedatives, antidepressants, antiepileptics, anaesthetics, antihistamines may lead to enhancement of the central depressive effects. The elderly require special supervision.

Different types of formulations: Oral Tablets, Oral Elixir, Rectal Tablet, Parenteral (intramuscular and intravenous)

Popular brands: The generic nitrazepam is manufactured by many companies. Some of the popular brands are as follows:

Sr. No.	Brand Name	Manufacturer	Type
1	Stressban	Lupin Laboratories Ltd.	Tablet
2	Nitrosum	Sun Pharmaceutical Industries Ltd.	Tablet
3	Nitraz -SR	Triton Healthcare Pvt Ltd	Tablet
4	Nitrapam	Elite Pharma Pvt Ltd	Tablet
5	Nite	Talent Healthcare	Tablet
6	Nicare	ConsernPharma Pvt Ltd.	Capsule
7	Baronite	Baroda Pharma Pvt Ltd	Tablet
8	Dreem	Esteem Pharma.	Tablet
9	Konit -10	Konark Life Sciences	Tablet
10	Nitrosh	Osho Pharma Pvt. Ltd.	Tablet

Therapeutic advantages and disadvantages of some sedative and hypnotic agents:

Drug	Advantages	Disadvantages
Benzodiazepines		
Clonazepam **Clorazepate**	Potential use in chronic therapy for seizures	Disturb intellectual functioning and motor dexterity (), produced dependence as well as withdrawal seizures may occur.
Flurazepam **Quazepam**	Less potent, slowly eliminated drugs, no rebound insomnia.	
Lorazepam **Temazepam**	Less potent, slowly eliminated drugs, no rebound insomnia.	
Barbiturates		
Phenobarbital **Pentobarbital** **Amobarbital**	Suitable for long term management, useful for young children with febrile seizures.	Induce tolerance, physical dependence, shows severe withdrawal symptoms.

Contd...

Other agents		
Buspiron	Useful in long term therapy, do not produce CNS depression, low potential for addiction	Slower onset of action, no muscle relaxation or anticonvulsant activity
Zaleplon **Zolpidem**	Minimal withdrawal symptoms, minimum rebound insomnia, no tolerance with prolong use	No muscle relaxation or anticonvulsant activity

Fig. 5.2.3 Therapeutic advantages and disadvantages of some Sedative and Hypnotic drugs

QUESTION BANK

A. MULTIPLE CHOICE QUESTIONS

1. Which of the following is most long acting barbiturate? :
    A. Phenoaritone
    B. Butobarbitone
    C. Pentoarbitone
    D. Thiopentone

Answer: A

2. Gardenal is the popular brand name for ___________
    A. Phenoaritone
    B. Butobarbitone
    C. Pentoarbitone
    D. Thiopentone

Answer: A

3. Which of the following drug contains benzodiazepine ring in structure?
    A. Phenoaritone
    B. Butobarbitone
    C. Pentoarbitone
    D. Diazepam

Answer: D

4. Chemical Name for Phenoaritone is ___________________.
    A. 5-methyl-5-phenyl-1,3-diazinane-2,4-dione
    B. 5-ethyl-5-phenyl-1,3-diazinane-2,4-dione

    C. 5-ethyl-5-phenyl-1,3-diazinane-2,4,6-trione

    D. 5-methyl-5-phenyl-1,3-diazinane-2,4,6-trione

Answer: C

5.   Chemical Name for Alprazolam is _________________________.

    A. 8-Fluro-1-methyl-6-phenyl-4*H*-[1,2,4]triazolo[4,3-a][1,4] benzodiazepine.

    B. 8-Chloro-1-methyl-6-phenyl-4*H*-[1,2,4]triazolo[4,3-a][1,4] benzodiazepine.

    C. 7-Chloro-1-methyl-6-phenyl-4*H*-[1,2,4]triazolo[4,3-a][1,4] benzodiazepine.

    D. 7-Fluro-1-methyl-6-phenyl-4*H*-[1,2,4]triazolo[4,3-a][1,4] benzodiazepine.

Answer: B

B. SHORT ANSWER QUESTIONS

1. Write a note on Sedatives and Hypnotics.
2. Give the difference between sedatives and hypnotics.
3. Define and classify Sedative and Hypnotics.
4. Write the structure of Diazepam and give its stability and storage.
5. Write the structure of Alprazolam and give its stability and storage.
6. Write the structure of Phenobarbital and give its stability and storage.
7. Write the structure of Nitrazepam and give its stability and storage.
8. Write the uses of: a) Diazepam b) Phenobarbital c) Nitrazepam d) Alprazolam (Any 2).
9. Write an brief account on incompatibilities of a) Diazepam b) Phenobarbital c) Nitrazepam d) Alprazolam (Any 2).
10. Enlist the popular formulations and brand names of a) Diazepam, b) Phenobarbitalc) Alprazolam d) Nitrazepam (Any 2).

C. LONG ANSWER QUESTIONS

1. Discuss the physiology of sleep and the terms Sedatives and Hypnotics and differentiate between them.
2. Define the terms Sedatives and Hypnotics and give detailed accounts on selected drugs; diazepam, alprazolam, phenobarbital and nitrazepam, with respect to their: chemical name, chemical structure, uses, stability and storage conditions, different types of formulations and their popular brand names.

■■■

DRUGS ACTING ON CENTRAL NERVOUS SYSTEM: ANTIPSYCHOTICS

CONTENTS

Antipsychotics:

- Chlorpromazine Hydrochloride*
- Haloperidol*
- Risperidone*
- Sulpiride*
- Olanzapine
- Quetiapine
- Lurasidone

♦ LEARNING OBJECTIVES ♦

After completing this chapter, the student should be able to understand:

1. The term Antipsychotic drugs and their classification.
2. Roles of Antipsychotic drugs and their mechanism of action.
3. Chemical names and structures of selected Antipsychotic drugs
4. Mechanism of action and uses of selected Antipsychotic drugs
5. Stability and storage conditions and brands of selected Antipsychotic drugs

5.3.1 INTRODUCTION

Antipsychotics (also known as neuroleptics or major tranquillisers) are a class of psychiatric medications primarily used to manage **psychosis**, **schizophrenia**, **bipolar disorder**, and other emotional conditions. Anti-psychotic drugs control the symptoms of psychosis and, in many cases, are effective in controlling the symptoms of other disorders that may lead to psychosis, including bipolar mood disorder (formerly termed manic-depressive).

Psychosis refers to an abnormal condition of the mind involving a "loss of contact with reality". Psychosis is an end-stage condition arising from a variety of possible causes, such as drug and alcohol addictions, reactions to severe stress, physical illness, etc.

Schizophrenia is a mental disorder often characterised by abnormal social behaviour and failure to recognise what is real. Common symptoms include false beliefs, unclear or confused thinking, auditory hallucinations, reduced social engagement and emotional expression and lack of motivation. Schizophrenia is caused by some inherent dysfunction of the brain.

Bipolar disorder or bipolar affective disorder or manic depression is a mental disorder characterised by periods of elevated mood and periods of depression.

Symptoms of psychosis:

Four main symptoms are associated with psychosis, namely, hallucinations, delusions, confused and disturbed thoughts and lack of insight & self-awareness.

1. **Hallucinations:** meaning a person perceives something that doesn't exist in reality. They can occur in all five senses, namely sight, sound, touch, smell and taste.
2. **Delusions:** A delusion is where a person has an unshakeable belief in something doubtful, unconventional or obviously untrue, like someone trying to kill /hurt him or possession of some imaginary power or authority.
3. **Confused and disturbed thoughts:** Signs of these include rapid and constant speech, spontaneous speech – switching from one topic to another mid-sentence, and sudden loss in the pattern of thoughts resulting in an abrupt pause in conversation or activity.
4. **Lack of insight:** Persons with psychotic episodes are often totally unaware of their strange behaviour, or that their delusions or hallucinations are not real.
5. **Postnatal psychosis:** a type of depression some women experience after having a baby. It most commonly occurs during the first few weeks after having a baby.

5.3.2 CLASSIFICATION OF ANTIPSYCHOTIC DRUGS

Antipsychotic drugs are classified as follows;

   1. **Typical Antipsychotics/First Generation Antipsychotics:**
       a. Phenothiazines: *e.g., chlorpromazine, thioridazine,fluphenazine*
       b. Thioxanthenes: *e.g., chlorprothixene,thiothexene*
       c. Butyrophenone: *e.g., haloperidol,droperidol*

2. **Atypical Antipsychotics/Second Generation Antipsychotics:**
   a. Diphenylbutylpiperidines: *e.g., pimozide*
   b. Dibenzazepines: *e.g., clozapine, loxapine, metiapine*
   c. Reduced indolones: *e.g., oxypertine, molindone*
   d. Benzamides: *e.g., sulpiride*
   e. Benzoqunolizines: *e.g., tetrabenazine*
   f. Benzisoxazoles: *e.g., risperidone*

3. **Rauwolfia Alkaloids:** *e.g., reserpine*

   Current antipsychotic therapy commonly employs the use of atypical agents to minimise the risk of debilitating movement disorders associated with typical drugs that act primarily at the D_2 dopamine receptors.

5.3.3 MECHANISM OF ACTION OF ANTIPSYCHOTIC DRUGS

During the psychotic effect, it has been seen that there is an excess release of dopamine. The patient suffering from schizophrenia or bipolar disorder shows less release of dopamine in the prefrontal cortex and excess release from all other pathways.

Antipsychotic drugs suppress the release of dopamine throughout their pathways and allow dopamine receptors to function normally. In addition to their antagonistic effects on dopamine, antipsychotics (*in particular atypical neuroleptics*) also antagonise $5\text{-}HT_{2A}$ receptors.

Typical antipsychotics are not particularly selective but also block dopamine receptors in other pathways, namely, the mesocortical, tuberoinfundibular and the nigrostriatal pathways or even other receptors. Blocking D_2 receptors in other pathways or blocking other receptors such as adrenergic, cholinergic and histamine binding receptors is responsible for some unwanted side effects, such as sedation, headaches, dizziness, diarrhoea, anxiety, that the typical antipsychotics can produce.

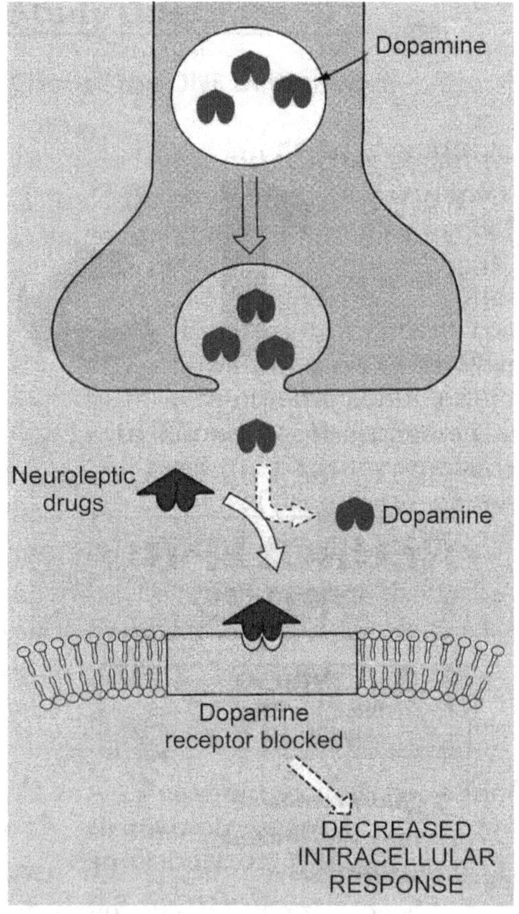

Fig. 5.3.1 Antipsychotic drugs act by blocking dopamine receptor

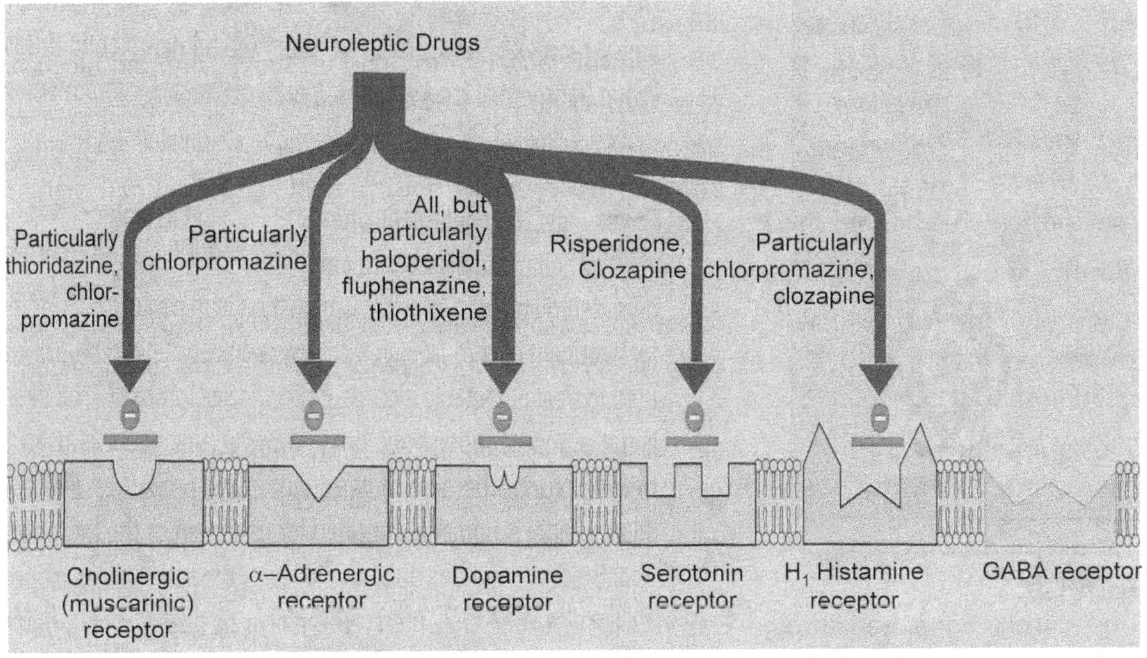

Fig. 5.3.2 Blocking of other receptors such as adrenergic, cholinergic and histamine binding receptors by antipsychotic drugs

5.3.4 SOME IMPORTANT DRUGS USED AS ANTIPSYCHOTIC AGENTS

5.3.4.1 Chlorpromazine hydrochloride: (*Syn.* Thorazine)
Structure:

Systematic (IUPAC) Name: 3-(2-Chloro-10*H*-phenothiazin-10-yl)-N,N-dimethyl-propan-1-amine.

Uses:

i) Chlorpromazine hydrochloride is used to treat certain mental/mood disorders (such as schizophrenia, psychotic disorders, manic phase of bipolar disorder, and severe behavioural problems in children).

ii) Chlorpromazine helps to think more clearly, feel less nervous, and also helps to reduce aggressive behaviour.

iii) It also helps to decrease hallucinations.

iv) Chlorpromazine is also used to control nausea/vomiting, relieve prolonged hiccups, relieve restlessness/anxiety before surgery, and help treat tetanus.

Side Effects: Extrapyramidal reactions (e.g., Parkinson-like symptoms, dystonia, akathisia, drowsiness, dizziness, skin reactions or rash, dry mouth, hypotension, amenorrhea, galactorrhea, weight gain, etc.

Stability and storage conditions:

Chlorpromazine hydrochloride oral solutions, tablets, and injections should be stored at a temperature less than 40 °C, preferably between 15-30° C; freezing of the oral solutions and Injection should be avoided.

Chlorpromazine suppositories should be stored in well-closed containers between 15-30°C. Chlorpromazine hydrochloride oral concentrate solution should be dispensed in amber glass bottles.

Incompatibility:

Food, alcohol and benztropine can reduce the absorption of chlorpromazine. Tricyclic antidepressants decrease the clearance of chlorpromazine and may lead to increase serum levels. Chlorpromazine hydrochloride is responsible for the increase in depression with benzodiazepines, anaesthetic drugs, opioids, barbiturates and lithium.

Different types of formulations: Tablet, Capsule, Injection

Popular brands: It is manufactured by many companies. Some of the notable brand names in India are;

Sr. No.	Brand Name	Manufacturers	Type
1	Chlorpromazine	Sun Pharmaceuticals	Capsule/ Tablet
2	Clozine Forte	Psychotropics India	Capsule/ Tablet
3	Megatil	Intas Pharmaceuticals	Capsule/ Tablet/Injection
4	Sun Prazin	Sun Pharmaceuticals	Capsule/ Tablet

5.3.4.2 Haloperidol: (*Syn.* **Haldol**)
Structure:

Systematic (IUPAC) Name: 4-[4-(4-chlorophenyl)-4-hydroxypiperidin-1-yl]-1-(4-fluoro phenyl)-butan-1-one

Uses:

i) Haloperidol is used to treat certain mental/mood disorders (e.g., schizophrenia, schizoaffective disorders).

ii) It can decrease negative thoughts and hallucinations.

iii) It can also be used to treat uncontrolled movements and outbursts of words/sounds related to Tourette's disorder. (Tourette's disorder is a problem with the nervous system that causes people to make sudden movements or sounds which they can't control)

iv) Haloperidol is also used for severe behaviour problems in hyperactive children when other treatments or medications have not worked.

v) Haloperidol may be used in hospitalised patients who have severe behaviour problems or confusion for short periods of time.

vi) This medication may also be used to prevent or treat nausea and vomiting due to cancer treatment.

Side Effects:

Difficulty with speaking or swallowing, inability to move the eyes, loss of balance control, muscle spasms, especially of the neck and back, restlessness, shuffling walk, stiffness of the arms and legs, trembling and shaking of the fingers and hands, twisting movements of the body, weakness of the arms and legs.

Stability and storage conditions:

Store in a cool and dry place where the temperature is below 25°C. Do not freeze or refrigerate it. Protect it from light.

Incompatibility:

Haloperidol is incompatible with anticholinergic medications, such as scopolamine and lithium; drugs used for Parkinson's disease, drugs acting on CNS; antidepressants, sleep medications, anti-anxiety medications, psychiatric medications, narcotic analgesics, amphetamine, adrenaline, etc. Also, it is incompatible with drugs acting on CVS, e.g., antiarrhythmic drugs. Avoid taking muscle relaxants and alcohol during the treatment.

Different types of formulations:

Tablets and Injections

Popular brands:

It is manufactured by many companies. Some of the notable brand names in India are:

Sr. No.	Brand Name	Manufacturers	Type
1	Agidol	ConsernPharmaPvt Ltd	Tablet
2	Brain Rest	Ind-Swift Limited	Tablet
3	Cizonil	Konark Life Sciences	Tablet
4	Depidol	Torrent Labs (P) Ltd	Tablet
5	Halidol	GlaxoSmithkline Pharmaceuticals Ltd	Tablet
6	Norma	Sunrise Remedies Pvt Ltd	Tablet
7	Zor	Pifer Pharmaceuticals Pvt Ltd	Tablet

5.3.4.3 Droperidol: (*Syn.* Droleptan, Inapsine, Dridol)

Structure:

Systematic (IUPAC) Name: 3-[1-[4-(4-Fluorophenyl)-4-oxobutyl]-3,6-dihydro-*2H*-pyridin-4-yl]-1*H*-benzimidazol-2-one

Uses:

It has a central antiemetic action and effectively prevents postoperative nausea and vomiting.

i) It has also been used as an antipsychotic.

ii) It is used for intramuscular sedation.

iii) It is useful for the treatment of vertigo in healthy elderly patients.

Side Effects:

Blurred vision, confusion, dizziness, faintness, lightheadedness when getting up suddenly from a lying or sitting position, sweating, unusual tiredness or weakness, dizziness, and tightness in the chest.

Stability and storage conditions:

It should be protected from light and stored at room temperature (15-25°C).

Incompatibility:

Droperidol injection is chemically incompatible with parenteral barbiturates. Droperidol may increase the CNS depressant activities of clobazam and clomipramine, while its own central neurotoxic activities may get enhanced due to donepezil. The risk or severity of adverse effects can be increased when fluvoxamine is combined with droperidol.

Different types of formulations: Injections, solutions.

Popular brands:

It is manufactured by a few companies. One of the notable brand names in India is Droperol (injection), manufactured by Troikaa Parenterals Pvt. Ltd., SG Pharma. Pvt. Ltd.

5.3.4.4 Risperidone: (*Syn.* **Risperidal**)

Structure:

Systematic (IUPAC) Name:3-[2-[4-(6-Fluoro-1,2-benzoxazol-3-yl) piperidin-1-yl] ethyl]-2-methyl-6,7,8,9-tetrahydropyrido[1,2-*a*] pyrimidin-4-one.

Uses:
i)   **Schizophrenia:** Risperidone is effective in treating the acute exacerbations of schizophrenia.
ii) **Bipolar disorder:** Risperidone is more effective in the treatment of manic symptoms in acute manic or mixed exacerbations of bipolar disorder.
iii) **Autism:** Risperidone treatment reduces certain problematic behaviours in autistic children, including aggression toward others, self-injury, temper tantrums, and rapid mood changes.

Side Effects:

Commonly reported side effects of risperidone are extrapyramidal reaction, akathisia, rhinitis, dystonia, dizziness, anxiety, drowsiness, agitation, constipation, nausea, and weight gain. Other side effects include abdominal pain, tachycardia, skin rash, and xeroderma.

Stability and storage conditions: Should be stored at a controlled room temperature of 15 - 25 ° C, protected from light and moisture. Avoid freezing the oral solution.

Incompatibility: Carbamazepine and other enzyme inducers may reduce plasma levels of risperidone. It is incompatible with antihypertensive drugs because it is responsible for hypotension.

Different types of formulations: Tablet, Solution, Injection

Popular brands: It is manufactured by many companies. Some of the notable brand names in India are:

Sr. No.	Brand Name	Manufacturers	Type
1	Benzix	Moxy Laboratories Pvt. Ltd.	Tablet
2	Neudon	Mankind Pharmaceuticals Pvt. Ltd (Magnet Labs Pvt Ltd)	Tablet
3	Genrest	Gentech Healthcare	Tablet
4	Psydon	Psyco Remedies.	Tablet
5	Rozidal	Ranbaxy Laboratories Ltd (Solus Pharmaceuticals)	Tablet
6	Risnia	Ranbaxy Laboratories Ltd (Solus Pharmaceuticals)	Tablet
7	Zon	Elikem Pharmaceuticals Pvt. Ltd.	Tablet

5.3.4.5 Sulpiride: (*Syn.* **Sulpyrid**).

Levosulpiride is used, and chemically, it is the (*S*)-(−)-enantiomer of sulpiride.

Structure:

Systematic (IUPAC) Name: N-[[(2S)-1-ethylpyrrolidin-2-yl]methyl]-2-methoxy-5-sulfamoyl benzamide

Uses:

i) Primarily in the management of the symptoms of schizophrenia. It has been used as both a   monotherapy and adjunctive therapy in schizophrenia.

ii) It has also been used in the treatment of dysthymia.

iii) It is also useful in treating panic disorder.

Side Effects:

Dizziness, headache, extrapyramidal side effects (tremors, dystonia, akathisia, parkinsonism), insomnia, weight gain or loss, hyperprolactinemia (elevated plasma levels of the hormone prolactin which can, in turn, lead to sexual dysfunction, galactorrhea, amenorrhea, gynecomastia, etc.), nausea, vomiting, nasal congestion, anticholinergic adverse effects (dry mouth, constipation, blurred vision, impaired concentration).

Stability and storage conditions: Do not store above 25°C

Incompatibility: With alcohol (enhances the sedative effects of neuroleptics), anaesthetics (enhances the hypotensive effect of anaesthetics), analgesics (enhanced sedative and hypotensive effect with opioid analgesics), antacids (reduce the bioavailability of sulpiride).

Different types of formulations: Tablets, injection.

Popular brands: It is manufactured by a few companies.

Sr. No.	Brand Name	Manufacturers	Type
1	Levogold (100 & 25 mg)	Lupin Laboratories Ltd.	Tablet
2	Levipride (50 mg)	Intas Pharmaceuticals Ltd	Tablet
3	Lesuride OD (75 mg)	Sun Pharmaceutical Industries Ltd.	Tablet
4	Lesuride (25 mg)	Sun Pharmaceutical Industries Ltd.	Injection

5.3.4.6 Olanzapine (*Syn.* Zyprexa)

Structure:

Systematic (IUPAC) Name: 2-Methyl-4-(4-methyl-1-piperazinyl)-10H-thieno[2,3-b][1,5]-benzodiazepine

Uses:

i) **Schizophrenia:** Olanzapine is effective in reducing symptoms of schizophrenia, treating acute exacerbations, and treating early-onset schizophrenia.

ii) **Bipolar disorder:** Olanzapine is recommended as a first-line therapy for the treatment of acute mania in bipolar disorder.

iii) **Other uses:** used in the treatment of a range of mood and anxiety disorders. It has also been used for Tourette syndrome and stuttering. Olanzapine is frequently prescribed off-label for the treatment of insomnia, including difficulty falling asleep and staying asleep.

Side Effects:

The principal side effect of olanzapine is weight gain, which may be profound in some cases and/or associated with derangement in blood-lipid and blood-sugar profiles.

Extrapyramidal side effects, although potentially serious, are infrequent to rare from olanzapine but may include tremors and muscle rigidity.

Stability and storage conditions:

Tablet formulations of olanzapine are sensitive to temperature and moisture.

Should be stored at room temperature between 68°F and 77°F (20°C and 25°C). Keep this drug away from light. Don't store this medication in moist or damp areas, such as bathrooms.

Incompatibility:

Olanzepine is involved in interaction with drugs like; apomorphine, bromocriptine, cabergoline, dopamine, fluvoxamine, levodopa, lisuride, mefloquine, methyldopa, ondansetron, pefloxacin, pramipexole, ropinirole, sodium oxybate, etc. and should not be given along with them,

Different types of formulations:

Tablet, IM Injection, Sachets

Popular brands:

It is manufactured by many companies. Some of the notable brand names in India are:

Sr. No.	Brand Name	Manufacturers	Type
1	MS-Insta	Quality Pharma Products Pvt. Ltd.	Tablet
2	Psychozap	Cadila Pharmaceuticals Ltd.	Tablet
3	Odozap	Intralife	Tablet
4	Olandus	Zydus Cadia Healthcare Ltd.	Tablet/Capsule
5	OleanZ	Sun Pharmaceutical Industries	Injection

5.3.4.7 Quetiapine (*Syn.* Seroquel)

Structure:

Systematic (IUPAC) Name: 2-(2-(4-Dibenzo[*b,f*][1,4]thiazepine-11-yl-1-piperazinyl)ethoxy) ethanol

Uses:

Schizophrenia: Quetiapine is more effective than other antipsychotics; ziprasidone, chlorpromazine, asenapine etc., and equally as effective as haloperidol and aripiprazole.

Bipolar disorder: quetiapine is used to treat depressive episodes, acute manic episodes associated with the; acute mixed episodes; and maintenance treatment of the bipolar-I disorder.

Major depressive disorders: Quetiapine is effective when used by itself and when used along with other medications in major depressive disorder (MDD). However, sedation is often an undesirable side effect.

Other uses: Quetiapine and clozapine are the most widely used medications for treating Parkinson's disease psychosis due to their very low extrapyramidal side-effect liability.

Side Effects:

Very common side effects include; Dry mouth, Dizziness, Headache, Somnolence (drowsiness; of 15 antipsychotics, quetiapine causes the 5th most sedation. Extended-release (XR) formulations tend to produce less sedation, dose-by-dose, than the immediate release formulations.).

Stability and storage conditions:

Quetiapine fumarate (QF) suspension has around 60 days of stability at room temperature as well as refrigerated storage.

Quetiapine can be stored at room temperature (between 59- and 86-degrees F) and out of sight and reach of children

Incompatibility:

Quetiapine shouldn't be taken with the following.

Anti-arrhythmic drugs such as quinidine, procainamide, amiodarone or sotalol.

Antipsychotic drugs such as ziprasidone, chlorpromazine, or thioridazine.

Antibiotics such as gatifloxacin or moxifloxacin & antiprotozoal, Pentamidine, as well as, Analgesic Methadone.

Different types of formulations:

Tablets and sustained release tablets.

Popular brands:

It is manufactured by many companies. Some of the notable brand names in India are:

Sr. No.	Brand Name	Manufacturers	Type
1	Seroquin 25 mg.	Cipla	Tablet
2	Q-Pin 25 mg.	Alkem	Tablet
3	Adequet 25 mg.	Abbott, India	Tablet
4	Qutipin 25 mg.	Sun Pharma	Tablet
5	Q-Mind 25 mg.	Torrent Pharma	Tablet

5.3.4.8 Lurasidone: (*Syn.* **Latuda**)

Structure:

Systematic (IUPAC) Name: (3aR,4S,7R,7aS)-2-{(1R,2R)-2-[4-(1,2-benzisothiazol-3-yl) pipe-razin-1-ylmethyl] cyclohexylmethyl} hexahydro-4,7-methano-2H-isoindole-1,3-dione

Uses:

i. **Schizophrenia:** for the treatment of Schizophrenia in adults and adolescents (13 to 17 years)

ii. **Bipolar disorder:** is used to treat depressive episodes; acute manic episodes associated with bipolar-I disorders.

Side Effects:

Very common side effects include; Drowsiness, dizziness, lightheadedness, nausea, shaking, weight gain, mask-like facial expression, inability to keep still, and agitation may occur.

Stability and storage conditions:

The drug product is recommended to be stored at 25 °C (77 °F) with excursions permitted to 15-30 °C (59-86 °F) (USP Controlled Room Temperature).

Incompatibility:

Lurasidone should not be used with certain medicines such as carbamazepine, clarithromycin, ketoconazole, phenytoin, rifampin, ritonavir, or voriconazole.

Different types of formulations: Tablets.

Popular brands:

It is manufactured by many companies. Some of the notable brand names in India are:

Sr. No.	Brand Name	Manufacturers	Type
1	Tablura 40 mg	Cipla	Tablet
2	Lurasid 40 mg	Intas Ltd.	Tablet
3	Lurafic 40 mg	Lupin Ltd.	Tablet
4	Luramax 40 mg	Sun Pharma	Tablet
5	Altura 40 mg	Torrent Pharma	Tablet

QUESTION BANK

A. MULTIPLE CHOICE QUESTIONS

1. Antipsychotics also known as
   A. Neuroleptics
   B. Sedative
   C. Hypnotics
   D. Anesthetics

Answer: A

2. Which of the following is Antipsychotic agent?
   A. Diazepam
   B. Alprazolam
   C. Nitrazepam
   D. Chlorpromazine hydrochloride

Answer: D

3. Which of the following heterocycle is present in Haloperidol?
   A. Pyridine
   B. Purine
   C. Piperidine
   D. None of the above

Answer: C

4. Phenothiazin ring system is present in which of the following drug?
   A. Diazepam
   B. Chlorpromazine hydrochloride
   C. Haloperidol
   D. Risperidone

Answer: B

5. Which of the following Alkaloid is used as Antipsychotic?
   A. Reserpine
   B. Scopolamine
   C. Hyoscamine
   D. Atropine

Answer: A

B. SHORT ANSWER QUESTIONS

1. What is Psychosis? Discuss the symptoms of psychosis?

2. Write a note on antipsychotics.

3. Define and classify antipsychotics

4. Write the structure of following compounds and give its stability and storage.
   a) Chlorpromazine Hydrochloride b) Haloperidol c) Droperidol d) Risperidone
   e) Sulpiride f) Olanzapine, g) Quetiapine, h) Lurasidone

5. Write the uses of: a) Chlorpromazine Hydrochloride b) Haloperidol c) Droperidol
   d) Risperidone, e) Sulpiride f) Olanzapine, g) Quetiapine, h) Lurasidone

6. Enlist the brand names of :
    a) Chlorpromazine Hydrochloride   b) Haloperidol c) Droperidol d)   Risperidone
    e) Sulpiride f) Olanzapine, g) Quetiapine, h) Lurasidone

7. Give the side effects and incompatibilities of
    a) Chlorpromazine Hydrochloride   b) Haloperidol c) Droperidol d)   Risperidone
    e) Sulpiride f) Olanzapine, g) Quetiapine, h)Lurasidone

C. LONG ANSWER QUESTIONS

1. Discuss the terms psychosis, bipolar disorders and schizophrenia, their symptoms, the mechanism of antipsychotic drugs and their classification with suitable examples.

■■■

DRUGS ACTING ON CENTRAL NERVOUS SYSTEM: ANTICONVULSANTS

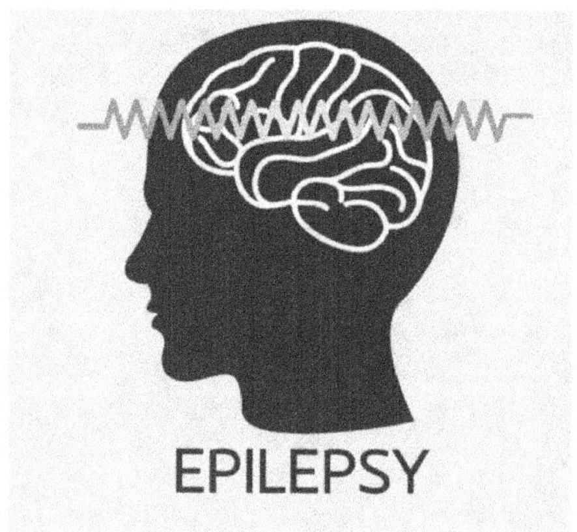

CONTENTS

Brief introduction of Drugs acting on the Central Nervous System:

Anticonvulsants

- Phenytoin*,
- Carbamazepine*,
- Clonazepam,
- Valproic Acid*,
- Gabapentin*,
- Topiramate,
- Vigabatrin,
- Lamotrigine

◆ LEARNING OBJECTIVES ◆

After completing this chapter, the student should be able to understand:

1. The term Epilepsy, Convulsions, Seizures and Anticonvulsant drugs, their classification.
2. Role of selected anticonvulsant drugs
3. Chemical names and structures and synthesis of selected anaesthetic drugs
4. Mechanism of action and uses of selected anticonvulsant drugs
5. Stability and storage conditions and brands of selected anticonvulsant drugs

5.4.1 INTRODUCTION

Anticonvulsants are drugs that prevent or reduce the severity and frequency of seizures in various types of epilepsy. They are also known as antiepileptic drugs (AEDs) or as anti-seizure drugs.

Epilepsy: A neurological disorder marked by sudden recurrent episodes of sensory disturbance, loss of consciousness, or convulsions, associated with abnormal electrical activity in the brain.

Classification of Epilepsy:

Epilepsies are classified in five ways:

1. By their first cause (or aetiology):

 a) Symptomatic epilepsy (Some discernable causes like brain injury))

 b) Idiopathic epilepsy (No cause)

 c) Cryptogenic epilepsy (Cause is not known)

2. By the location in the brain where seizures originate

 a) **Partial/Focal seizures/Localized seizures**: These are seizures which initially affect only one hemisphere of the brain.

 1. Simple partial seizure: These are seizures which affect only a tiny region of the brain, often the temporal lobes or hippocampi.

 2. Complex partial seizure: A complex partial seizure is associated with unilateral cerebral hemisphere involvement and causes impairment of awareness or responsiveness, i.e., alteration of consciousness

 3. Secondarily generalised seizures: Sometimes focal seizures spread from one side (hemisphere) to both sides of the brain

 b) **Generalised epilepsy/Primary generalised epilepsy/Idiopathic epilepsy**: Depending on the behavioural effects, generalised epilepsy is further divided into:

 1. Absence seizures: During an absence seizure, a person becomes unconscious for a short time. He may look blank and stare, or his eyelids might flutter. He will not

respond to what is happening around him. If he is walking, he may carry on walking but will not be aware of what he is doing.

2. Tonic seizures: In a tonic seizure, the person's muscles suddenly become stiff. If he is standing, he may often fall, usually backwards, and may injure the back of his head. Tonic seizures tend to be very brief and happen without warning.

3. Atonic seizures: In an atonic seizure (or 'drop attack'), the person's muscles suddenly relax, and they become floppy. If he is standing, he may often fall, usually forwards, and may injure the front of his head or face.

4. Myoclonic seizures: Myoclonic means 'muscle jerk'. Muscle jerks are not always due to epilepsy (for example, some people have them as they fall asleep). Myoclonic seizures are brief but can happen in clusters (many happening close together in time) and often happen shortly after waking.

5. Tonic-clonic (convulsive) seizures (sometimes called grand mal):

   - At the start of the seizure: The person becomes unconscious, his body goes stiff, and if he is standing up may usually fall backwards, may cry out, may bite his own tongue or cheek.

   - During the seizure: a person may jerk and shake (convulse) as his muscles relax and tighten rhythmically, breathing might be affected and become difficult or sound noisy, his skin may change colour and become very pale or bluish, and he may wet himself.

   - After the seizure (once the jerking stops): His breathing and colour return to normal, and he may feel tired, confused, have a headache or want to sleep.

6. Clonic seizures: Clonic seizures are convulsive seizures, but the person's body does not go stiff at the start.

3. By the event that triggers the seizures, as in primary reading epilepsy

Some seizures occur in response to very specific stimuli or situations. In this type, seizures occur consistently in relation to a specific trigger. For example, one type of reflex epilepsy is photosensitive epilepsy, where seizures are triggered specifically by flashing lights. Other types of reflex epilepsies may be seizures triggered by the act of reading or by noises.

4. By the observable manifestations or symptoms of the seizure

A number of causes such as illicit drug use, tumours, head injury, meningeal infection or rapid withdrawal of alcohol can produce seizures. When two or more seizures occur, the patient may be diagnosed with symptomatic epilepsy.

5.4.2 CLASSIFICATION OF ANTICONVULSANT DRUGS/AEDS

5.4.2.1 As per Chemical Structures:

I. **Barbiturates:** *e.g., Phenobarbital, mephobarbital*

  II. **Hydantoins:** *e.g., Phenytoin, ethotoin*

  III. **Oxazolidinediones:** *e.g., Trimethdione*

  IV. **Succinamides:** *e.g., Phensuximide, methsuximide, ethosuximide*

  V. **Ureas and monacylureas/diacylureas:** *e.g., Phenacmide, carbamazepine*

  VI. **Benzodiazepines:** *e.g., Clonazepam, diazepamchlorazapate*

  VII. **Miscellaneous:** *e.g., Primidone, valproic acid, gabapentine*

5.4.2.2 As per Mechanism of Action

The AEDs show their activity by different modes of action, and therefore drugs can be grouped as sodium channel blockers (e.g., *Phenytoin, carbamazepine*); calcium channel inhibitors (e.g., *Ethosuximide);* γ-aminobutyric acid (GABA) enhancers *(e.g., Clonazepam, phenobarbital);* glutamate blockers (Felbamate); carbonic anhydrase inhibitors (e.g., Acetazolamide); hormones (e.g., Progesterone) and drugs with unknown mechanisms of action. Some antiepileptic drugs work by acting on a combination of channels or through some unknown mechanisms of action.

I. **Sodium channel blockers:**

During an action potential, these channels exist in the active state and allow an influx of sodium ions. Once the activation or stimulus is terminated, sodium channels become inactive. AEDs that target the sodium channels prevent the return of these channels to the active state by stabilising them in the inactive state and preventing high-frequency neuronal firing.

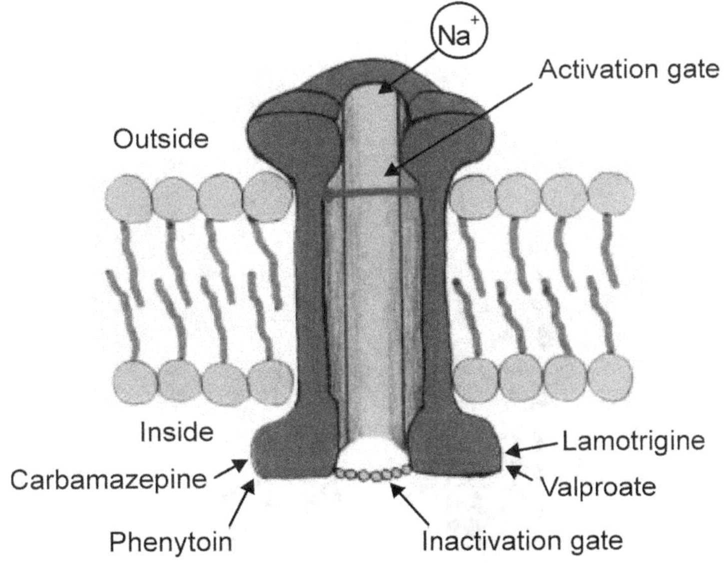

Fig. 5.4.1 Enhanced Na+ Channel Inactivation

II. Calcium channel blockers:

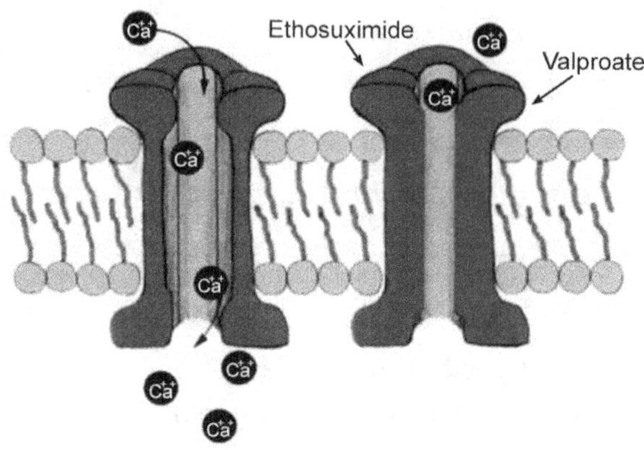

Fig. 5.4.2 Reduced action through T-type Calcium channels

Calcium channels exist in three known forms in the human brain: L, N, and T. The L-type calcium channel is part of the high-voltage activated family of voltage-dependent calcium channels. The N-type calcium channel is a type of voltage-dependent calcium channel. T-type calcium channels are low-voltage activated calcium channels that open during membrane depolarisation.

Calcium channels function as the "pacemakers" (pacemakers are nothing but specialised tissues which govern rhythmic or cyclic activities) of normal rhythmic brain activity. T-calcium channels have been known to play a role in the absence of seizures. AEDs that inhibit these T-calcium channels are particularly useful for controlling absence seizures.

III. GABA enhancers:

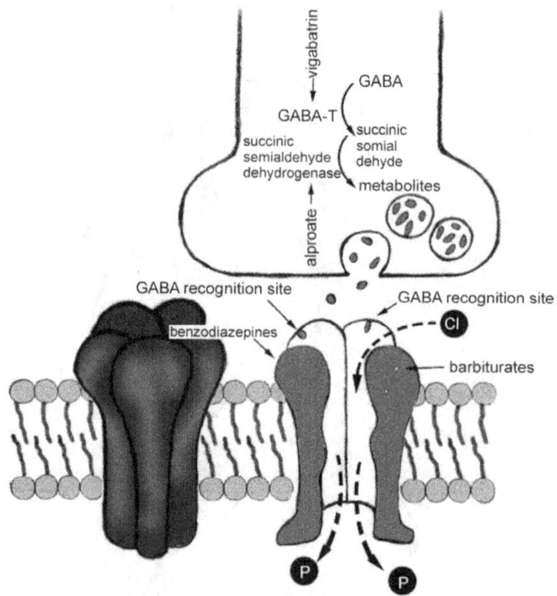

Fig. 5.4.3 Action of GABA enhancer

Gamma-aminobutyric acid (GABA) has 2 types of receptors, A and B. When GABA binds to a GABA-A receptor, the passage of chloride into the cell is facilitated through chloride channels. The GABA-B receptor is linked to a potassium channel. GABA is produced by decarboxylation of glutamate-mediated by the enzyme glutamic acid decarboxylase (GAD). Some AEDs may act as modulators of enzyme glutamic acid decarboxylase, enhancing the production of GABA. Some AEDs function as an agonist to chloride conductance, either by blocking the reuptake of GABA (e.g., tiagabine) or by inhibiting its metabolism as mediated by GABA transaminase (e.g., vigabatrin), resulting in increased accumulation of GABA at the postsynaptic receptors.

IV. Glutamate blockers:

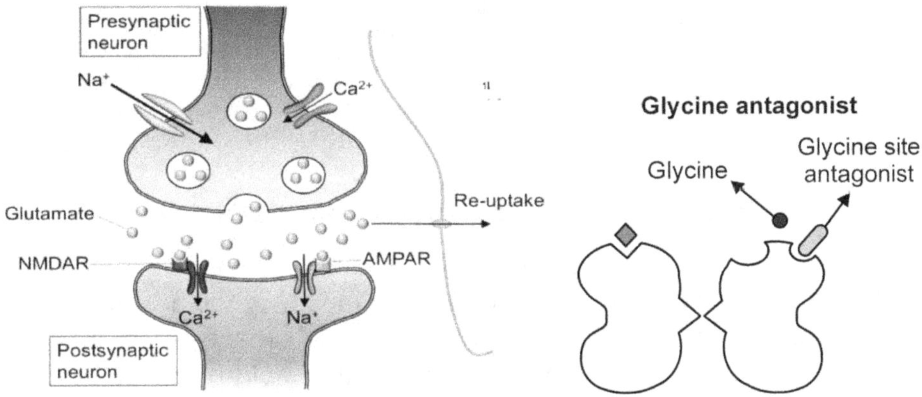

Fig. 5.4.4 Action of Glutamate blockers

Glutamate and aspartate are the two most important excitatory neurotransmitters in the brain. The glutamate system is a complex system that contains macromolecular receptors with different binding sites (i.e., alpha-amino-3-hydroxy-5-methylisoxazole-4-propionic acid [AMPA], kainate, N-methyl-D-aspartate [NMDA], glycine, and metabotropic sites). The AMPA and the kainate sites open a channel through the receptor, allowing sodium and small amounts of calcium to enter. The NMDA site opens a channel that allows large amounts of calcium to enter along with the sodium ions. This channel is blocked by magnesium in the resting state. The glycine site facilitates the opening of the NMDA receptor channel. The metabotropic site is regulated by complex reactions, and its response is mediated by second messengers.

NMDA blockers are responsible for blocking these receptors but have a limited use because they produce psychosis and hallucinations. In addition to these adverse effects, learning and memory may be impaired by blocking these receptors because NMDA receptors are associated with learning processes and long-term potentiation.

V.   Carbonic anhydrase inhibitors:

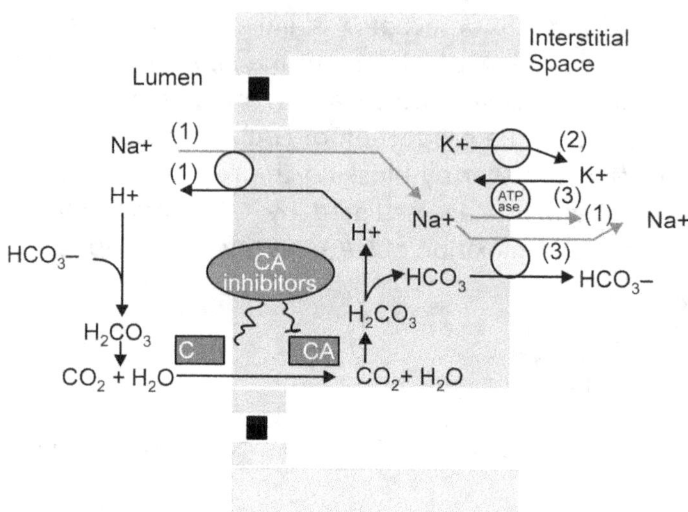

Fig. 5.4.5 Action of Carbonic anhydrase inhibitors

Inhibition of the enzyme carbonic anhydrase increases the concentration of hydrogen ions intracellularly and decreases the pH. To maintain the acid-base balance, the potassium ions shift to the extracellular compartment. This event results in hyperpolarisation and an increase in the seizure threshold of the cells. Acetazolamide has been used as adjunctive therapy in seizures which are observed around the menstrual period.

VI. Sex hormones

Progesterone is a natural anticonvulsant that acts by increasing chloride conductance at GABA-A receptors and decreasing glutamate excitatory response. It also alters messenger RNA for glutamic acid decarboxylase (GAD) and GABA-A receptor subunits.

Summary of anticonvulsant/antiepileptic drugs with mechanism of action:

Sr.	Drug	Mechanism of action
1	Carbamazepine	Block Na^+ channels
2	Ethoxsuximide	Block Ca^{++} channels
3	Felbamate	Mutiple mechanism of action
4	Gabapentin	Unkown
5	Pregabalin	Mutiple mechanisms of action
6.	Primidone	GABA receptor agonist
7.	Oxcarbazepine	Block Na^+ channels
8.	Fosphenytoin	Block Na^+ channels
9.	Topiramate	Mutiple mechanisms of action
10.	Divalproex	Mutiple mechanisms of action

5.4.3 SOME IMPORTANT DRUGS USED AS ANTICONVULSANTS

5.4.3.1 Phenytoin: (*Syn:* Dilantin)

Structure:

Systematic (IUPAC) Name: 5,5-Diphenylimidazolidine-2,4-dione

Uses:

i) **Seizures:** Phenytoin is useful in tonic-clonic seizures, focal seizures, absence seizures, seizures during surgery, *status epilepticus*

ii) **Abnormal heart rhythms-** It may be used in the treatment of ventricular tachycardia and sudden episodes of atrial tachycardia after other antiarrhythmic medications have failed.

iii) **Digoxin toxicity-** IV formulation is the drug of choice for arrhythmias caused by cardiac glycoside toxicity.

iv) **Wound healing-** Tentative evidence suggests that topical phenytoin is useful in wound healing in people with chronic skin wounds.

Side Effects:

Nausea, vomiting, constipation, tremors, slurred speech, loss of balance or coordination, rash, headache, confusion, dizziness, nervousness, or sleep problems (insomnia).

Stability and storage conditions:

Keep this medication in tightly closed containers and out of reach of children. Store it at room temperature, away from light and excess heat and moisture. Do not freeze the liquid.

Incompatibility: Many drugs can interact with phenytoins, such as antibiotics/antibacterials, cycloserine, doxycycline, isoniazid, linezolid, rifampin, and sulfa drugs. An antidepressant (such as Elavil, Vanatrip), NSAIDs; aspirin or other salicylates, birth control pills or hormone replacement therapy, Blood thinners such as warfarin (Coumadin, Jantoven), certain sedatives (Limbitrol, Valium) or antidepressants (Rapiflux, Sarafem, Selfemra, Symbyax), heart medications such as amiodarone, digoxin, furosemide, or quinidine (Quin-G), steroid medicines (prednisone and others), stomach acid reducers (Prilosec, Zegerid, Zantac, Pepcid), theophylline (Elixophyllin, Theolair, Uniphyl) etc., are known to be incompatible with phenytoin.

Different types of formulations available: Injection, solution

Popular brands: It is manufactured by many companies. Some of the notable brand names in India are:

Sr. No.	Brand Name	Manufacturers	Type
1	**Dilantin**	**Pfizer**	Capsule/Tablet
2	Epicent	Crescent	Capsule/Tablet
3	Fentoin	Sun pharmaceuticals	Capsule/Tablet
4	Eptoin	Abbott India	Capsule/Tablet
5	Gentoin	Gentech HC	Capsule/Tablet
6	Epsolin	ZydusCadila	Capsule/Tablet

5.4.3.2 Carbamazepine: (*Syn:* Tegretol)

Structure:

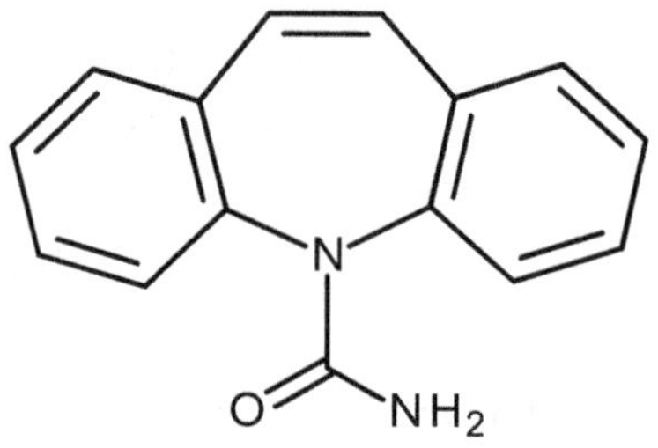

Systematic (IUPAC)Name: Benzo[b][1]benzazepine-11-carboxamide

Uses: Carbamazepine is typically used for the treatment of seizure disorders and neuropathic pain.

It is used as a second-line treatment for bipolar disorder. It is also useful in combination with an antipsychotic in the case of schizophrenia when treatment with a conventional antipsychotic alone has failed.

Side Effects: Carbamazepine increases the clearance of many drugs, decreasing their concentration in the blood to sub-therapeutic levels and reducing their desired effects. Drugs that are more rapidly metabolised due to carbamazepine include warfarin, lamotrigine, phenytoin, theophylline and valproic acid.

Stability and storage conditions: Carbamazepine tablets, extended-release tablets, and chewable tablets should be stored in tight, light-resistant containers at temperatures not exceeding 30° C.

Carbamazepine extended-release capsules should be stored in tight, light-resistant containers at 15-25°C. Keep the container closed and in a dry location, away from areas with excessive moisture.

Incompatibility: Carbamazepine may accelerate the metabolism of phenytoin, phenobarbital, primidone, valproic acid and warfarin. Triacetyloleandomycin, erythromycin, propoxyphene, isoniazid, and cimetidine may be responsible for the inhibition of the metabolism of carbamazepine.

Different types of formulations: Tablet, Capsule, Suspension

Popular brands: It is manufactured by many companies. Some of the notable brand names in India are:

Sr. No.	Brand Name	Manufacturers	Type
1	Acetol	Psychotropics India	Capsule/Tablet
2	Carbacontin	Modi Mundi Pharma	Capsule/Tablet
3	Carbatol	Torrent Pharmaceuticals	Capsule/Tablet
4	Carmaz	Sun Pharmaceuticals	Capsule/Tablet
5	Mazetol	Nicholas Piramal	Capsule/Tablet
6	Tegrital	Novartis	Liquid
7	Zen	Intas Pharmaceuticals	Capsule/Tablet

5.4.3.3 Clonazepam: (*Syn:* Rivotril, Antelepsin)
Structure:

Systematic (IUPAC) Name: 5-(2-Chlorophenyl)-7-nitro-1,3-dihydro-1,4-benzodiazepin-2-one.

Uses: Clonazepam has been found to be effective in treating epilepsy and inhibiting seizure activity in children. Clonazepam has also been found to be effective in the acute control of non-convulsive *status epilepticus*. It is also useful for the treatment of typical and atypical absences, infantile myoclonic, myoclonic and akinetic seizures.

Side Effects: Confusion, hallucinations, unusual thoughts or behaviour, unusual risk-taking behaviour, no fear of danger, weak or shallow breathing, unusual or involuntary eye movements, pounding heartbeats or fluttering in the chest, painful or difficult urination, urinating less than usual, pale skin, easy bruising or bleeding, new or worsening seizures, drowsiness, dizziness, loss of appetite, nausea, diarrhoea or constipation, headache, skin rashes etc., are the side-effects observed.

Stability and storage conditions: Clonazepam tablets and orally disintegrating tablets should be stored in air-tight, light-resistant containers at 25°C but can be exposed to temperatures ranging from 15-30°C.

Incompatibility: Clonazepam decreases the levels of carbamazepine and vice *versa*. Azole antifungals, such as ketoconazole, may inhibit the metabolism of clonazepam. Clonazepam may affect levels of phenytoin, primidone and phenobarbital.

Combined use of clonazepam with certain antidepressants or antiepileptics, such as phenobarbital, phenytoin and carbamazepine; sedative antihistamines, opiates, antipsychotics, non-benzodiazepine hypnotics like zolpidem and alcohol may result in enhanced sedative effects.

Different types of formulations available: Tablet

Popular brands: It is manufactured by many companies. Some of the notable brand names in India are:

Sr. No.	Brand Name	Manufacturers	Type
1	Adnil	Medinova Lab	Tablet
2	Clozep	Elite PharmaPvt Ltd	Tablet
3	Lonazep	Sun Pharmaceutical Industries Ltd. (Synergy)	Tablet
4	Norep	Mankind Pharmaceuticals Pvt. Ltd (Magnet Labs Pvt Ltd)	Tablet
5	Melzap	Alkem Laboratories Ltd (Pentacare)	Tablet
6	Petril	Micro Labs Ltd (Synapse)	Tablet
7	Xenotril SL	Glenmark Pharmaceuticals Ltd	Tablet
8	Zozep	Zeus Pharmaceuticals	Tablet
9	Clonaxyl	Marc Laboratories Pvt. Ltd.	Tablet
10	Cloneepam	Swede-India Healthcare	Tablet

5.4.3.4 Valproic acid: (*Syn:* Depakine, dipropylacetic acid)

Structure:

Systematic (IUPAC) Name: 2-Propylpentanoic acid

Uses: This drug is used to treat seizure disorders and mental/mood conditions (such as the manic phase of bipolar disorder) and to prevent migraine headaches.

Side Effects: Sedation and drowsiness, somnolence, asthenia, dizziness, and tremor ataxia, emotional imbalance, abnormal thinking, amnesia, and depression, increased alertness, insomnia, and nervousness, nausea, vomiting, abdominal pain, anorexia, diarrhoea, and dyspepsia may occur at the initiation of therapy.

Stability and storage conditions: Store between 15 and 30 ° C in a tight container and protect from freezing.

Incompatibility: Some products that may be incompatible with valproic acid, which include certain antidepressants (e.g., amitriptyline, nortriptyline, phenelzine), certain benzodiazepines (e.g., clonazepam, diazepam), certain antibiotics (carbapenems such as doripenem, imipenem), guanfacine, mefloquine, other medications for seizure (e.g., carbamazepine, ethosuximide, felbamate, lamotrigine, phenobarbital, phenytoin, rufinamide, topiramate), rifampin, vorinostat, warfarin, zidovudine.

Different types of formulations are available: Capsule, syrup, injection

Popular brands: It is manufactured by many companies. Some of the notable brand names in India are:

Sr. No	Brand Name	Manufacturers	Type
1	Develpin	Lifecare Innovations Pvt. Ltd.	Tablet
2	Devot-XL	Unique Chemicals (J.B. Chemicals and Pharmaceuticals Ltd)	Tablet
3	Epilex	Abbott Health Care (P) Ltd.	Tablet
4	Torvate	Torrent Pharmaceuticals Ltd.	Tablet
5	Valcot-CR	Psyco Remedies	Tablet
6	Valporate	Crescent Therapeutics Ltd.	Tablet
7	Epsoval-XR	La Pharmaceuticals	Tablet

5.4.3.5 Gabapentin: (*Syn:* Neurontin)

Structure:

$$H_2N\text{—}\overset{\displaystyle|}{\underset{\displaystyle\bigcirc}{C}}\text{—}COOH$$

Systematic (IUPAC) Name: 2-[1-(Aminomethyl)cyclohexyl]acetic acid

Uses: Gabapentin is used primarily to treat seizures and neuropathic pain.

It is also commonly prescribed for the treatment of anxiety disorders, insomnia, and bipolar disorder. It is useful as a mood stabiliser in bipolar disorder.

Side Effects: The most common side effects of gabapentin in adult patients include dizziness, fatigue, drowsiness, weight gain, and peripheral oedema (swelling of extremities). Gabapentin may also produce sexual dysfunction in some patients.

Stability and storage conditions: Store between 15 and 30 ° C in a tight container and protect from freezing.

Incompatibility: Using propoxyphene together with gabapentin may increase side effects such as dizziness, drowsiness, confusion, difficulty concentrating, and other nervous system or mental effects. Using sodium oxybate together with gabapentin may increase side effects like drowsiness, dizziness, lightheadedness, confusion, and depression.

Using buprenorphine together with gabapentin can lead to serious side effects such as respiratory distress, coma, or even death.

Different types of formulations are available: Tablet, Capsule, and Solution.

Popular brands: It is manufactured by many companies. Some of the notable brand names in India are:

Sr. No.	Brand Name	Manufacturers	Type
1	Dobin	Drukst Biotech	Tablet
2	Encentin	East West Pharma	Capsule
3	Gaba	Hanburys Health Care Ptv. Limited	Tablet
4	Gabapin	Intas Laboratories Pvt Ltd	Tablet
5	Neupent AF	Ranbaxy Laboratories Ltd	Capsule
6	Neronti	Pfizer Limited (Pharmacia India Pvt Ltd)	Capsule
7	Neropin	ConsernPharmaPvt Ltd.	Capsule
8	Nerontin	Parke-Davis (India) Ltd.	Capsule

5.4.3.6 Topiramate: (*Syn:* **Topamax, Topiramic acid**)

Structure:

Systematic (IUPAC) Name: 2,3:4,5-Bis-O-(1-methylethylidene)-β-D-fructopyranose sulfamate

Uses: Topiramate is used to treat epilepsy in children and adults. In children, it is indicated for the treatment of Lennox-Gastaut syndrome, a disorder that causes seizures and developmental delay.

It is most frequently prescribed to prevent migraines as it decreases the frequency of attacks.

Topiramate is used to treat headaches and is recommended by the European Federation of Neurological Societies as one of the few medications showing effectiveness for this indication.

Topiramate has not been shown to work as pain medicine in diabetic neuropathy, the only neuropathic condition in which it has been adequately tested.

Other Uses: One common off-label use for topiramate is in the treatment of the bipolar disorder.

Topiramate has been used as a treatment for alcoholism

Side Effects: Topiramate may impair heat regulation, especially in children. Use caution with activities leading to an increased core temperature, such as strenuous exercise, exposure to extreme heat, or dehydration.

Topiramate may cause visual field defects.

Topiramate may decrease the effectiveness of oestrogen-containing oral contraceptives.

Taking topiramate in the first trimester of pregnancy may increase the risk of cleft lip/cleft palate in the infant.

As is the case for all antiepileptic drugs, it is advisable not to discontinue topiramate suddenly as there is a theoretical risk of rebound seizures.

Stability and storage conditions: Store at controlled room temperature or in a refrigerator. Labelling: Keep out of the reach of children. Store at either room temperature or refrigerated temperature.

Incompatibility: Drinking alcohol is not allowed when under treatment for Topiramate.

Drugs not be taken simultaneously with Topiramate; include; Valproic acid, Zonisamide, Glaucoma medications, and Birth control pills.

Different types of formulations are available: Tablet, Capsule, and Solution.

Popular brands: It is manufactured by many companies. Some of the notable brand names in India are:

Sr. No.	Brand Name	Manufacturers	Type
1	Topaz	Intas	Tablet
2	Epitop	Microlabs	Tablet
3	Topirain	Unichem	Tablet
4	Topirol	Sun Pharma	Tablet
5	Nextop	Torrent	Tablet

5.4.3.7 Vigabatrin: (*Syn:* **Vigadrone, Sabril**)

Structure:

Systematic (IUPAC) Name: (*RS*)-4-aminohex-5-enoic acid

Uses: as an adjunctive treatment (with other drugs) in treatment-resistant epilepsy, complex partial seizures, secondarily generalised seizures, and for monotherapy use in infantile spasms in West syndrome.

Vigabatrin is also indicated for monotherapy use in secondarily generalised tonic-clonic seizures, partial seizures, and in fantile spasms due to West syndrome.

Other Uses: Vigabatrin is also used to treat seizures in succinic semialdehyde dehydrogenase deficiency (SSADHD), which is an inborn GABA metabolism defect that causes intellectual disability, hypotonia, seizures, speech disturbance, and ataxia through the accumulation of γ-Hydroxybutyric acid (GHB).

Side Effects: A variety of side effects, including; Sleepiness, dizziness, nervousness, memory disturbances, diplopia, aggression, ataxia, vertigo, hyperactivity, vision loss, confusion, impaired concentration, and personality issues, are reported.

Stability and storage conditions: Store at room temperature Vigabatrin far away from light and moisture. Keep out of the bathroom. Don't open the packets before preparing the dose. **Incompatibility:** Risk of side effects severity is increased when taken along with drugs like; Almotriptan, Alprazolam, Acetazolamide etc.

Different types of formulations are available: Tablet, Powder for oral solution.

Popular brands: It is manufactured by many companies. Some of the notable brand names in India are:

Sr. No.	Brand Name	Manufacturers	Type
1	Viganext	MSN Laboratories	Tablet
2	Jolvig	Jolly Healthcare	Tablet
3	Viby	Intas Pharmaceuticals	Tablet

5.4.3.8 Lamotrigine: (*Syn:* **Lamictal**)

Structure:

Systematic (IUPAC) Name: 6-(2,3-Dichlorophenyl)-1,2,4-triazine-3,5-diamine

Uses: as a first-line drug for primary generalised tonic-clonic seizures.

Lamotrigine is one of a small number of FDA-approved therapies for the form of epilepsy known as Lennox–Gastaut syndrome.

Lamotrigine is approved in the US for the maintenance treatment of bipolar I disorder and bipolar II disorder.

Other Uses: Off-label uses include the treatment of peripheral neuropathy, trigeminal neuralgia, cluster headaches, migraines, visual snow, and reducing neuropathic pain,

Side Effects: Lamotrigine prescribing information has a black box warning about life-threatening skin reactions, including Steven Johnson syndrome (SJS), DRESS syndrome, and toxic epidermal necrolysis (TEN).

Lamotrigine has been associated with a decrease in white blood cell count (leukopenia).

Stability and storage conditions: Conditions for safe storage, including any incompatibilities: Keep the container tightly closed in a dry and well-ventilated place. Recommended storage temperature: 2 - 8 deg C. Keep in a dry place.

Different types of formulations available: Tablet.

Popular brands: It is manufactured by many companies. Some of the notable brand names in India are:

Sr. No.	Brand Name	Manufacturers	Type
1	Lametec	Cipla	Tablet
2	Lamictal	Glaxo-Smith-Kline	Tablet
3	Lamez	Intas Pharmaceuticals	Tablet
4	Lamitor	Torrent	Tablet
5	Lamosyn	Sun Pharma	Tablet

QUESTION BANK

A. MULTIPLE CHOICE QUESTIONS

1. Sedative action of barbiturates is due to substituents at C-5. It is due to_____________________ :
    A. High lipophilicity of group at C-5 position
    B. Electron withdrawing effect of group at C-5 position
    C. Steric Effect
    D. Low lipophilicity of group at C-5

Answer: A

2. Which of the following is a long acting barbiturates
    A. Pentobarbital
    B. Phenobarbital
    C. Thiopental
    D. Hexabarbital

Answer: B

3. The medicine which causes drowsiness and facilitates the onset and maintenance of sleep is known as?
    A. Sedatives
    B. Anxiolytics
    C. Hypnotics
    D. Anaesthetics

Answer: C

4. Phenobarbital is an example of?
    A. Opoids
    B. Benzodiazepines
    C. Non-benzodiazepines
    D. Barbiturates

Answer: D

5. Benzodiazepine without triazolo moiety?
    A. Alprazolam
    B. Triazolam
    C. Estazolam
    D. Midazolam

Answer: D

6. '7-nitro-5-phenyl-1H-benzo[e][1,4]diazepin-2(3H)-one' is the IUPAC nomenclature of which drug?
    A. Nitrazepam
    B. Loprazolam
    C. Alprazolam
    D. Clonazepam

Answer: A

7. Which of the following doesn't contain a nitro group in its chemical structure?
    A. Nitrazepam
    B. Loprazolam
    C. Alprazolam
    D. Clonazepam

Answer: C

8. Which is most potent among   diazepam alprazolam, clonazepam and nitrazepam?
    A. Nitrazepam
    B. Loprazolam
    C. Alprazolam
    D. Clonazepam

Answer: D

9. Hypnotic benzodiazepines increase the period of time spent in following stage of sleep.
    A. Stage II
    B. Stage III
    C. Stage IV
    D. REM Stage

Answer: A

10. Which of the following barbiturates is excreted significantly unchanged?
    A. A Phenobarbital
    B. B Pentobarbital
    C. C Hexobarbital
    D. D All of the above

Answer: A

B. SHORT ANSWER QUESTIONS

1. What is Epilepsy? Discuss the different types of Epilepsy.

2. Write a note on anticonvulsants.

3. Define and classify anticonvulsants.

4. Write the structure of following compounds and give stability and storage conditions of each of them.

   a) Phenytoin   b) Topiramate   c) Carbamazepine   d) Clonazepam
   e) Vigabatrin   f) Valproic acid   g) Lamotrigine   h) Gabapentin

5. Discuss the main uses and side effects of each of the following:

   a) Phenytoin   b) Topiramate   c) Carbamazepine   d) Clonazepam
   e) Vigabatrin   f) Valproic acid   g) Lamotrigine   h) Gabapentin

C. LONG ANSWER QUESTIONS

1. What is epilepsy? Discuss the physiology involved. Define and classify various anti-convulsant drugs and discuss in detail any 4 of them.

2. Write the structures, chemical names, uses, stability and storage of the following:

   a) Phenytoin   b) Topiramate   c) Carbamazepine   d) Clonazepam
   e) Vigabatrin   f) Valproic acid   g) Lamotrigine   h) Gabapentin.

■■■

DRUGS ACTING ON CENTRAL NERVOUS SYSTEM: ANTIDEPRESSANTS

CONTENTS

Brief introduction of Drugs acting on Central Nervous System as Anti-Depressants:

- Amitriptyline Hydrochloride*,
- Imipramine Hydrochloride*,
- Fluoxetine*, Venlafaxine,
- Duloxetine,
- Sertraline,
- Citalopram,
- Escitalopram,
- Fluvoxamine,
- Paroxetine

♦ LEARNING OBJECTIVES ♦

After completing this chapter, the student should be able to understand:

1. The terms Depression, Types of depression, Mania, Antidepressants, classification of antidepressant drugs, their mechanism of action.

2. Role of selected antidepressant drugs
3. Chemical names and structures and synthesis of selected antidepressant drugs
4. Mechanism of action and uses of selected antidepressant drugs
5. Stability and storage conditions and brands of selected antidepressant drugs.

5.5.1 INTRODUCTION

Antidepressants are the drugs used to relieve or prevent depression. Antidepressants are psychiatric medications given to patients with depressive disorders to get relief from the symptoms. They correct chemical imbalances of neurotransmitters in the brain which probably cause changes in mood and behaviour.

Depression is a form of extreme sadness, because of which a person losses the ability to function normally. It is a mood disorder characterized by sadness, inactivity, difficulty with thinking and concentration, a significant increase or decrease in appetite and time spent in sleeping, feelings of dejection and hopelessness and sometimes suicidal thoughts or an attempt to commit suicide.

Signs and symptoms of depression include: Feelings of helplessness and hopelessness, loss of interest in daily activities, appetite or weight changes, sleep changes (either insomnia, especially waking in the early hours of the morning, or oversleeping also known as hypersomnia), anger or irritability, loss of energy, self-loathing, reckless behaviour, concentration problems, unexplained aches and pains.

Warning signs of suicide include: Talking about killing or harming own self, expressing strong feelings of hopelessness or being trapped, an unusual preoccupation with death or dying, acting recklessly, calling or visiting people to say goodbye, getting affairs in order (giving away prized possessions, tying up loose ends), a sudden switch from being extremely depressed to acting calm and happy.

Types of depression:

I. **Major depression:** Major depression is characterized by the inability to enjoy life and experience pleasure. The symptoms are constant, ranging from moderate to severe. Left untreated, major depression typically lasts for about six months.

II. **Dysthymia (recurrent, mild depression):** Dysthymia is a type of chronic "low-grade" depression. The symptoms of dysthymia are not as strong as the symptoms of major depression, but they last a long time (at least two years).

III. **Bipolar Disorder:** Bipolar disorder, also known as manic depression, is characterized by cycling mood changes. Episodes of depression alternate with *manic episodes*, which can include impulsive behavior, hyperactivity, rapid speech, and little to no sleep.

IV. **Seasonal Affective Disorder** (SAD): some people actually develop seasonal depression, known as seasonal affective disorder (SAD).

Mania:

Mania is characterized by the symptoms, opposite, than that of depression which include, enthusiasm, rapid thought and speech patterns, extreme self -confidence and impaired judgment.

5.5.2 CLASSIFICATION OF ANTIDEPRESSANTS

According to mode of action antidepressants are divided into following classes:

1. **Monoamine reuptake inhibitors:**
   i. Serotonin and Norepinephrine Reuptake Inhibitors (SNRI):
      a. Tricyclic Antidepressants: e.g., Imipramine, clomipramine, amitriptyline, doxepine.
      b. Nontricyclic Antidepressants: e.g., Venlafaxine, milnacipran, duloxetine
   ii. Selective Reuptake Inhibitors [SRI]:
      a. Selective Norepinephrine Reuptake Inhibitors (SNERI): (e.g., Amoxapine, maprotiline, nisoxetine)
      b. Selective Serotonin Reuptake Inhibitors [SSRI]: (e.g., Fluoxetine, paroxetine, fluvoxamine)
   iii. Serotonin receptor modulators: (e.g., Mirtazapine, trazodone, nefazodone)
   iv. Dopamine Reuptake Inhibitors (DRI): (e.g., Bupropion, amineptine)
2. **Selective Serotonin Reuptake Enhancers**: (e.g., Tianeptine)
3. **Dopamine Agonists**: (e.g., Bupropion)
4. **Monoamine Oxidase Inhibitors**:
   a. Nonselective and irreversible inhibitors: (e.g., Phenelzine, isocarboxazide, pargyline, tranylcypromine)
   b. Reversible MAOA inhibitor (RIMA): (e.g., Moclobemide, pirlindole)

5.5.3 MECHANISM OF ACTION OF ANTIDEPRESSANTS

The role of an antidepressant is to stabilize and normalize the neurotransmitters in the brain. Neurotransmitters such as serotonin, dopamine and norepinephrine play role in regulating our mood. Antidepressants inhibit the reuptake of serotonin, norepinephrine, and to a lesser extent dopamine. Serotonin receptor modulator prevents serotonin from binding to 5-HT$_{2A}$receptors. (The mammalian 5-HT$_{2A}$ receptor is a subtype of the 5-HT2 receptor that belongs to the serotonin receptor family and is a G protein-coupled receptor.)

Mechanism of action of antidepressants

<table>
<tr>
<td>

1. Selective serotonin/ Norepinephrine re-uptake inhibitors (SNRI)

</td>
<td>

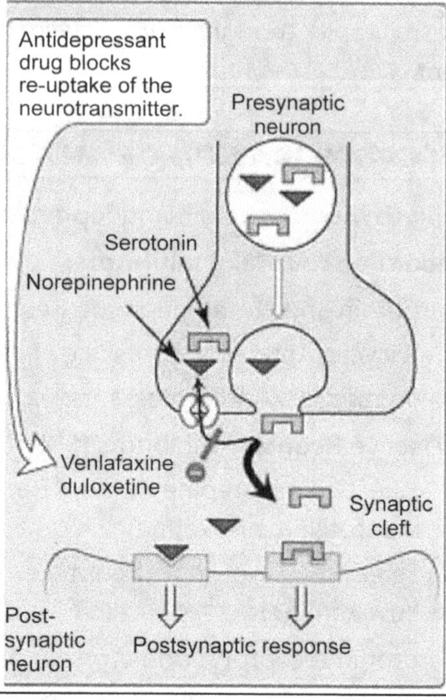

</td>
</tr>
</table>

Fig. 5.5.1 They selectively inhibit the reuptake of both serotonin and norepinephrine and are effective in treating depression cases in which, SSRIs are ineffective.

<table>
<tr>
<td>

2. Selective Serotonin reuptake inhibitors [SSRI]

</td>
<td>

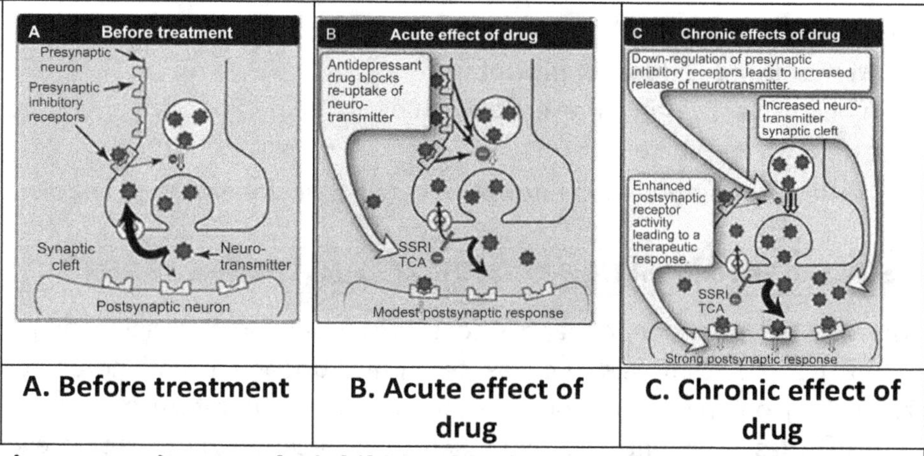

| A. Before treatment | B. Acute effect of drug | C. Chronic effect of drug |

</td>
</tr>
</table>

Fig. 5.5.2 Selective serotonin reuptake inhibitors block the reuptake of serotonin, leading to increased concentration of the neurotransmitter in synaptic cleft which ultimately responsible for greater postsynaptic neuronal activity.

3. **Monoamine oxidase inhibitors**	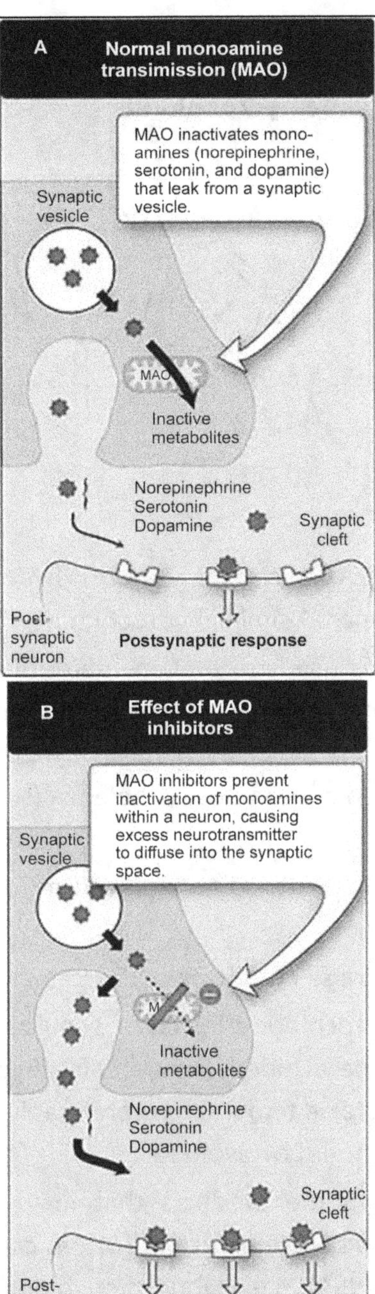	
	A. Normal monoamine (MAO) transmission	B. Effect of MAO inhibitors

Fig. 5.5.3 *Most MAO inhibitors form stable complexes with the enzyme, causing irreversible inactivation. This is responsible for increased stored levels of norepinephrine, serotonin and dopamine within the neuron and subsequent diffusion of excess nurotransmitter into the synaptic space.*

5.5.4 SOME IMPORTANT ANTIDEPRESSANT DRUGS

5.5.4.1 Amitryptyline hydrochloride (*Syn:* **Annoyltin**)

Structure:

Systematic (IUPAC) Name: 3-(5,6-Dihydrodibenzo[2,1-b:2',1'-f][7]annulen-11-ylidene)-N,N-dimethylpropan-1-amine;hydrochloride

Uses: This drug is used to treat mental/mood problems such as depression. It may help improve mood and feelings of well-being, relieve anxiety and tension. It may also be used to treat nerve pain (such as peripheral neuropathy, post therapeutic neuralgia), eating disorders, other mental/mood problems (such as anxiety, panic disorder), or migraine headaches.

Side Effects: Drowsiness, dizziness, dry mouth, blurred vision, constipation, weight gain, problem in urination.

Stability and storage conditions: Store at room temperature and protect from light.

Amitriptyline hydrochloride injection should be stored at a temperature of 30°C or less. Freezing should be avoided. Amitriptyline hydrochloride tablets should be stored in well-closed containers at a temperature preferably between 15-30°C; exposure to temperatures exceeding 30°C should be avoided.

Incompatibility: Some products that may interact with this drug include arbutamine, disulfiram, thyroid supplements, drugs that can cause bleeding (including antiplatelet drugs e.g., clopidogrel, NSAIDs e.g., ibuprofen, blood thinners e.g., warfarin), anticholinergic drugs (e.g., benztropine, belladonna alkaloids), certain drugs for high blood pressure (drugs that work in the brain such as clonidine, guanabenz, reserpine)

Different types of formulations: Tablet, Injection

Popular brands: It is manufactured by many companies. Some of the notable brand names in India are:

Sr. No.	Brand Name	Manufacturers	Type
1	Amiline	Torrent Labs (P) Ltd.	Tablet
2	Amilite	Elite PharmaPvt. Ltd.	Tablet
3	Elavin	Cyril Pharmaceuticals	Tablet
4	Eliwel	Sun Pharmaceutical Industries Ltd.	Tablet
5	Goldep SR	Glenmark Pharmaceuticals Ltd.	Tablet
6	Tadamit	Pentacare (Alkem Laboratories Ltd)	Tablet
7	Typlin	SterkemPharmaPvt. Ltd.	Tablet
8	Relidep	Reliance Formulations Pvt. Ltd.	Tablet

5.5.4.2 Impipramine hydrochloride: (*Syn:* Imidobenzyle, Melipramine)

Structure:

Systematic (IUPAC) Name: 3-(5,6-Dihydrobenzo[b][1]benzazepin-11-yl)-N,N-dimethyl-propan-1-amine

Uses:

i) This medication is used to treat depression and to improve mood, sleep, appetite and energy levels.

ii) It is also used with other therapies for the treatment of nighttime bed-wetting (enuresis) in children.

iii) It is also be used for the treatment of anxiety, panic disorders.

Side Effects: Difficulty in sleeping, weakness, confusion, dry mouth, increased appetite, nausea, vomiting, constipation, fast heartbeat, rash or itching, blurry vision.

Stability and storage conditions: Store at room temperature and protect from light.

Incompatibility: Imipramine is incompatible with potassium phosphate, drugs for bowel preparation (like potassium phosphate or potassium citrate), flumazenil, drugs for irregular heartbeat (like dronedarone), procainamide, clarithromycin, ephedra and ephedrine.

Different types of formulations: Tablet

Popular brands available: It is manufactured by many companies. Some of the notable brand names in India are:

Sr. No.	Brand Name	Manufacturers	Type
1	Depsonil	Nicholas Piramal India Ltd	Tablet
2	IRA	Elikem Pharmaceuticals Pvt. Ltd.	Tablet
3	Antidep	Torrent Labs (P) Ltd	Tablet
4	Antipres	Psyco Remedies	Tablet
5	Depsol	Intas Laboratories Pvt Ltd	Tablet
6	Depsonil PM	Abbott Healthcare Pvt Ltd (AHPL	Tablet
7	Talendep	Talent Healthcare	Tablet
8	Eldep	Psychotropics India Limited	Tablet

5.5.4.3 Fluoxetine (*Syn:* **Fluxetin**)

Structure:

Systematic (IUPAC) Name: N-Methyl-3-phenyl-3-[4-(trifluoromethyl)phenoxy]propan-1-amine

Uses:

i) **Major depressive disorders:** Depression is a mood disorder that causes a persistent feeling of sadness and loss of interest. Fluoxetine is used to treat major depressive disorder

ii) **Obsessive-compulsive disorder:** Obsessive-compulsive disorder (OCD) is characterized by unreasonable thoughts and fears (obsessions) that leads one to do repetitive behaviours (compulsions). It's also possible to have only obsessions or only compulsions and still have OCD. Fluoxetine is used to treat OCD.

iii) **Post-traumatic stress disorder (PTSD):** A condition of persistent mental and emotional stress occurring as a result of injury or severe psychological shock, typically involving disturbance of sleep and constant vivid recall of the experience, with dulled responses to others and to the outside world. Fluoxetine is effectively useful in PTSD.

iv) ***Bulimia nervosa:***

*Bulimia nervosa*is a serious, potentially life-threatening eating disorder which can be treated by Fluoxetine.

v) **Other uses:**

Fluoxetine is useful in treatment of panic disorder, premenstrual dysphoric disorder, and trichotillomania.nIt has also been used for obesity, alcohol dependence. It has also been tried as a treatment for autism spectrum disorders with moderate success in adults.

Side Effects: Commonly reported side effects of fluoxetine include diarrhoea, dyspepsia, nausea, nervousness, insomnia, weakness, anxiety, drowsiness, tremor, headache, xerostomia, decreased libido, anorexia, and decreased appetite. Other side effects include *bulimia nervosa*, dizziness, skin rash, and diaphoresis.

Stability and storage conditions: Store at room temperature and protect from light.

Incompatibility: Fluoxetine and norfluoxetine inhibit many isozymes of the cytochrome P_{450} system that are involved in drug metabolism. Its use should also be avoided in those receiving other serotonergic drugs such as monoamine oxidase inhibitors, tricyclic antidepressants, methamphetamine, 3,4-methylenedioxymethamphetamine (MDMA), triptans, buspirone, serotonin-norepinephrine reuptake inhibitors and other SSRIs due to the potential risk of serotonin syndrome.

There is also the potential for interaction with highly protein-bound drugs responsible for increasing serum concentrations of either fluoxetine or the protein-bound drugs.

Different types of formulations available: Tablet, Capsule.

Popular brands: It is manufactured by few companies.

Sr. No.	Brand Name	Manufacturers	Type
1	Nexflu-20	Next Labs. Pvt. Ltd.	Tablet
2	Flunil	Intas Laboratories Ltd	Capsule
3	Prodep	Sun Pharma Ltd	Capsule
4	Dawnex	Micro Laboratories Ltd.	Capsule
5	Fludac	Cadila Laboratories Ltd	Capsule

5.5.4.4 Venlafaxine (*Syn:* **Effexor**)

Structure:

Systematic (IUPAC) Name: (*RS*)-1-[2-Dimethylamino-1-(4-methoxyphenyl)-ethyl] cyclohe-xanol

Uses:

i) for the treatment of depression, general anxiety disorder, social phobia, panic disorder, and vasomotor symptoms, even in patients with Stroke's history.

ii) for the treatment of diabetic neuropathy and migraine prevention, albeit under observation.

iii) Due to its action on both the serotoninergic and adrenergic systems, venlafaxine is also used as a treatment to reduce episodes of cataplexy, a form of muscle weakness, in patients with the sleep disorder narcolepsy.

iv) Case reports, open trials and blinded comparisons with established medications have suggested the efficacy of venlafaxine in the treatment of obsessive–compulsive disorder.

Side Effects: Venlafaxine can increase eye pressure, so those with glaucoma may require more frequent eye checks

Stability and storage conditions: Venlafaxine hydrochloride liquid formulations (solution and suspension) are chemically stable for 30 days when stored at room temperature and protected from light.2 Store at room temperature and protect from light.

Incompatibility:

Venlafaxine is not recommended in patients hypersensitive to it, nor should it be taken by anyone who is allergic to the inactive ingredients, which include gelatin, cellulose, ethylcellulose, iron oxide, titanium dioxide and hypromellose.

It should not be used in conjunction with a monoamine oxidase inhibitor (MAOI), as it can cause potentially fatal serotonin syndrome.

Venlafaxine might interact with tramadol or other opioids, trazodone, so caution is needed while mixing multiple serotonergic agents together.

Different types of formulations available: Tablet, Capsule.

Popular brands: It is manufactured by few companies.

Sr. No.	Brand Name	Manufacturers	Type
1	Venlift-75	Torren Ltd.	Capsule
2	Ventab XL	Intas Laboratories Ltd	Tablet
3	Veniz XR-75	Sun Pharma Ltd	Capsule
4	Vnela-Xr	Sun Pharma Ltd	Capsule
5	Venlor XR 75	Cipla Ltd	Capsule

5.5.4.5 Duloxetine (*Syn:* **Cymbalta**)

Structure:

Systematic (IUPAC) Name: (+)-(*S*)-N-Methyl-3-(naphthalen-1-yloxy)-3-(thiophen-2-yl) propan-1-amine

Uses:

i) The main uses of duloxetine are in major depressive disorder, generalized anxiety disorder, neuropathic pain, chronic musculoskeletal pain, and fibromyalgia.

ii) Duloxetine is recommended as a first-line agent for the treatment of chemotherapy-induced neuropathy by the American Society of Clinical Oncology.

iii) Duloxetine is approved for the pain associated with diabetic peripheral neuropathy (DPN),

iv) The U.S. Food and Drug Administration (FDA) approved the drug for the treatment of fibromyalgia in June 2008.

Side Effects: Nausea, somnolence, insomnia, and dizziness are the main side effects, reported by about 10% to 20% of patients

There have been reports of adverse events occurring upon discontinuation of this drug, particularly when abrupt, including the following: dysphoric mood, irritability, agitation, dizziness, sensory disturbances (e.g., paresthesias such as brain zap electric shock sensations), anxiety, confusion, headache, lethargy, emotional lability, insomnia, hypomania, tinnitus, and seizures. The withdrawal syndrome from duloxetine resembles the SSRI discontinuation syndrome.

Stability and storage conditions: Keep container tightly closed in a dry and well-ventilated place. Containers which are opened must be carefully resealed and kept upright to prevent leakage. Recommended storage temperature -20 °C.

Incompatibility:

Hypersensitivity: duloxetine is contraindicated in patients with a known hypersensitivity to duloxetine or any of the inactive ingredients.

Monoamine oxidase inhibitors (MAOIs): concomitant use in patients taking MAOIs is contraindicated.

Uncontrolled narrow-angle glaucoma: in clinical trials, Cymbalta use was associated with an increased risk of mydriasis (dilation of the pupil); therefore, its use should be avoided in patients with uncontrolled narrow-angle glaucoma, in which mydriasis can cause sudden worsening.

Central nervous system (CNS) acting drugs: given the primary CNS effects of duloxetine, it should be used with caution when it is taken in combination with or substituted for other centrally acting drugs, including those with a similar mechanism of action. Duloxetine and thioridazine should not be co-administered.

In addition, the FDA has reported on life-threatening drug interactions that may be possible when co-administered with triptans and other drugs acting on serotonin pathways leading to increased risk for serotonin syndrome.

Different types of formulations available: Tablet, Capsule.

Popular brands: It is manufactured by few companies.

Sr. No.	Brand Name	Manufacturers	Type
1	Symbal	Torrent Pharmaceuticals Ltd	Tablet
2	Duvanta	Intas Laboratories Ltd	Tablet
3	Duzela	Sun Pharma Ltd	Capsule
4	Dulane	Sun Pharma Ltd	Capsule
5	Dulot	Lupin Ltd	Capsule

5.5.4.6 Sertraline (*Syn:* **Lustral Cymbalta**)

Structure:

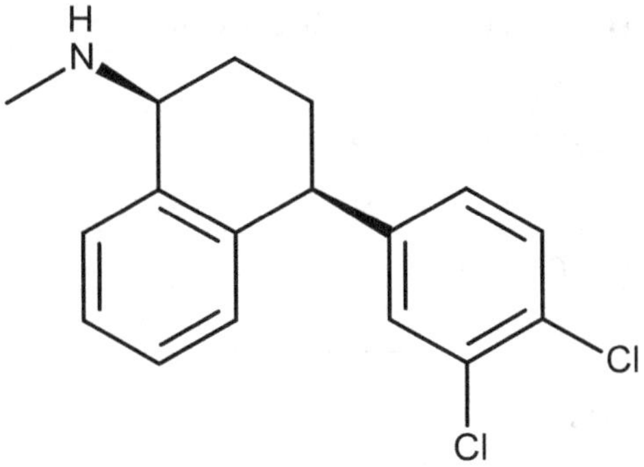

Systematic (IUPAC) Name: (1S,4S)-4-(3,4-Dichlorophenyl)-N-methyl-1,2,3,4 tetrahydro-naphthalen-1-amine

Uses: Sertraline has been approved for

- major depressive disorder (MDD),
- obsessive–compulsive disorder (OCD),
- posttraumatic stress disorder (PTSD),
- premenstrual dysphoric disorder (PMDD),
- panic disorder, and social anxiety disorder (SAD).

Side Effects: Nausea, ejaculation failure, insomnia, diarrhea, dry mouth, somnolence, dizziness, tremor, headache, excessive sweating, fatigue, and decreased libido are the common adverse effects associated with sertraline.

Those that most often resulted in interruption of the treatment are nausea, diarrhea and insomnia.

Stability and storage conditions: Keep container tightly closed in a dry and well-ventilated place. Containers which are opened must be carefully resealed and kept upright to prevent leakage. Recommended storage temperature -20 °C.

Incompatibility:

Sertraline is contraindicated in individuals taking monoamine oxidase inhibitors or the antipsychotic pimozide.

Sertraline concentrate contains alcohol and is therefore contraindicated with disulfiram.

The prescribing information recommends that treatment of the elderly and patients with liver impairment "must be approached with caution".

In addition, the FDA has reported on life-threatening drug interactions that may be possible when co-administered with triptans and other drugs acting on serotonin pathways leading to increased risk for serotonin syndrome.

Different types of formulations available: Tablet,

Popular brands: It is manufactured by few companies.

Sr. No.	Brand Name	Manufacturers	Type
1	Serdep 50	Cipla Ltd	Tablet
2	Inosert-50	IPCA Laboratories Ltd	Tablet
3	Serlift-50	Sun Pharma Ltd	Tablet
4	Sertima-50	Intas Ltd	Tablet
5	Zosert	Sun Pharma Ltd	Tablet

5.5.4.7 Citalopram (*Syn.* Celexa)

Structure: Racemic Mixture

Systematic (IUPAC) Name: (*RS*)-1-[3-(Dimethylamino)propyl]-1-(4-fluorophenyl)-1,3-dihydro isobenzofuran-5-carbonitrile.

Uses:

Celexa is an SSRI (selective serotonin reuptake inhibitor). SSRIs are mainly used to treat depression, but they are also used to treat anxiety disorders as well as OCD, bulimia, and other conditions.

Side Effects: Nausea, diarrhea, constipation, vomiting, stomach pain, heartburn, decreased appetite. weight loss are the commonly associated side effects.

Stability and storage conditions: Store citalopram tablets at room temperature, 77°F (25°C). The tablets can temporarily be stored at temperatures between 59°F and 86°F (15°C and 30°C). Keep this drug away from high temperatures.

Incompatibility:

Do not use citalopram with buspirone, fentanyl, lithium, methylene blue injection, tryptophan, St John's wort, amphetamines, or some pain or migraine medicines like rizatriptan, sumatriptan, tramadol.

Different types of formulations available: Tablet.

Popular brands: It is manufactured by few companies.

Sr. No.	Brand Name	Manufacturers	Type
1	Citadep	Cipla Ltd	Tablet
2	C-Pram	Unichem Laboratories Ltd	Tablet
3	Citara	Intas Ltd	Tablet
4	Feliz	Torrent Pharma Ltd	Tablet
5	Citopam	Sun Pharma Ltd	Tablet

5.5.4.8 Escitalopram (*Syn:* **Cipralex and Lexapro**)

Structure:

Systematic (IUPAC) Name: (*S*)-1-[3-(Dimethylamino)propyl]-1-(4-fluorophenyl)-1,3-dihydro isobenzofuran-5-carbonitrile.

Uses:

Escitalopram has FDA approval for the treatment of major depressive disorder(MDD) in adolescents and adults, and generalized anxiety disorder (GAD), social anxiety disorder (SAD), obsessive-compulsive disorder (OCD), and panic disorder with or without agoraphobia in adults.

Escitalopram is effective in reducing the symptoms of premenstrual syndrome, whether taken continuously or in the luteal phase only.

Side Effects: Nausea, headache, Ejaculation disorder (9–14%), Somnolence (4–13%), Insomnia (7–12%) are commonly observed side effects.

Escitalopram, like other SSRIs, has been shown to affect sexual functions causing side effects such as decreased libido, delayed ejaculation, and anorgasmia

Stability and storage conditions: Store the medicine in a closed container at room temperature, away from heat, moisture, and direct light. Keep from freezing. Keep out of the reach of children.

Incompatibility:

Taking MAO inhibitors with this medication may cause a serious (possibly fatal) drug interaction. Avoid taking MAO inhibitors (isocarboxazid, linezolid, metaxalone, methylene blue, moclobemide, phenelzine, procarbazine, rasagiline, safinamide, selegiline, tranylcypromine) during treatment with this medication.

Different types of formulations available: Tablet.

Popular brands: It is manufactured by few companies.

Sr. No.	Brand Name	Manufacturers	Type
1	Nexito	Sun Pharmaceutical Industries Ltd	Tablet
2	Rexipra	Intas Pharmaceuticals Ltd	Tablet
3	Stalopam	Lupin Ltd	Tablet
4	Feliz-S	Torrent Pharma Ltd	Tablet
5	Szetalo	Abbott Ltd.	Tablet

5.5.4.9 Fluvoxamine (*Syn*: **Luvox**)

Structure:

Systematic (IUPAC) Name: 2-{[(*E*)-{5-Methoxy-1-[4-(trifluoromethyl)phenyl] pentylidene} amino]- oxy} ethanamine.

Uses:

It is commonly used for major depressive disorder. Fluvoxamine is also approved in the United States for OCD and social anxiety disorder.

The drug works long-term, and retains its therapeutic efficacy for at least one year. It has also been found to possess some analgesic properties in line with other SSRIs and tricyclic antidepressants.

Side Effects: Nausea, headache, Ejaculation disorder (9–14%), Somnolence (4–13%), Insomnia (7–12%) are commonly observed side effects.

Gastrointestinal side effects are more common in those receiving fluvoxamine than with other SSRIs. Otherwise, fluvoxamine's side-effect profile is very similar to other SSRIs

Other Common Side Effects: Abdominal pain, Agitation, Anxiety, Asthenia (weakness), Constipation, Diarrhoea, Dizziness, Dyspepsia (indigestion), Headache, Hyperhidrosis (excess sweating)

Insomnia, Loss of appetite, Malaise, Nausea, Nervousness, Palpitations, Restlessness, Sexual dysfunction Somnolence (drowsiness), Tachycardia (high heart rate), Tremor, Vomiting, Weight loss, Xerostomia (dry mouth), Yawning

Stability and storage conditions: Store fluvoxamine at room temperature between 59°F and 86°F (15°C and 30°C). Keep this drug away from light. Don't store this medication in moist or damp areas, such as bathrooms.

Incompatibility:

Fluvoxamine should not be taken with or within 6 weeks of taking monoamine oxidase inhibitors (MAOIs). These include phenelzine, tranylcypromine, isocarboxazid, rasagiline and selegiline

Different types of formulations available: Tablet.

Popular brands: It is manufactured by few companies.

Sr. No.	Brand Name	Manufacturers	Type
1	Fluvoxin	Sun Pharmaceutical Industries Ltd	Tablet
2	Frext	Intas Pharmaceuticals Ltd	Tablet
3	Voxamin	Micro Labs Ltd	Tablet
4	Fluvator	Torrent Pharma Ltd	Tablet
5	Uvox	Abbott Ltd.	Tablet

5.5.4.10 Paroxetine (*Syn*: Paxil and Seroxat)
Structure:

Systematic (IUPAC) Name: (3*S*,4*R*)-3-[(2*H*-1,3-Benzodioxol-5-yloxy)methyl]-4-(4-fluoro-phenyl)piperidine

Uses: Paroxetine is used to treat depression, obsessive-compulsive disorder (OCD), panic disorder, generalized anxiety disorder (GAD), social anxiety disorder (also known as social phobia), premenstrual dysphoric disorder (PMDD), and posttraumatic stress disorder (PTSD).

Side Effects: headache, dizziness, weakness, difficulty concentrating, nervousness, forgetfulness,
Confusion, sleepiness.

Some Serious Side Effects: seeing things or hearing voices that do not exist (hallucinating), fainting, rapid, pounding, or irregular heartbeat, chest pain, difficulty breathing, seizures, etc.

Stability and storage conditions: Tablets should be kept at room temperature, 59^0F - 860F (15^0C - 30^0C). The suspension and controlled release tablets should be stored at or below 77 F (25^0C).

Incompatibility:

Contraindicated in closed angle glaucoma. liver problems. bleeding from stomach, esophagus or duodenum. severe renal impairment.

Dietary supplements of L-tryptophan (available only by prescriptions from special compounding pharmacists) taken with paroxetine caused headache, sweating, dizziness, agitation, restlessness, nausea, vomiting, and other symptoms.

Different types of formulations available: Tablet.

Popular brands: It is manufactured by few companies.

Sr. No.	Brand Name	Manufacturers	Type
1	Xet	Cadila Pharmaceuticals Ltd	Tablet
2	Pexep	Intas Pharmaceuticals Ltd	Tablet
3	Panex	Micro Labs Ltd	Tablet
4	Pari	IPCA Labs Ltd	Tablet

QUESTION BANK

A. MULTIPLE CHOICE QUESTIONS

1. Which of the following is used to treat depression?

    A. Amitriptyline Hydrochloride

    B. Diazepam

    C. Haloperidol

    D. Nitrazepam

    Answer: A

2. Which of the following is Tricyclic antidepressant?

    A. Venlafaxine

    B. Milnacipran

    C. Duloxetine

    D. Imipramine

    Answer: D

3. Which of the following is Tricyclic antidepressant?

    A. Amitriptyline

    B. Duloxetine

    C. Imipramine

    D. Clomipramine

    Answer: B

4. Amiline is popular brand name for _____________

    A. Amitriptyline Hydrochloride

    B. Diazepam

    C. Haloperidol

    D. Nitrazepam

    Answer: A

5. Duloxetine contains which of the following ring system?

    A. Anthracene

    B. Naphthalene

    C. Phenanthrene

    D. Quinidine

Answer: B

B. SHORT ANSWER QUESTIONS

1. What is mean by depression ? Discuss the symptoms and types of depression?

2. Write a note on antidepressants.

3. Define and classify antidepressants

4. Write the structure of following compounds and give its stability and storage of any 5.

    a. Amitriptyline Hydrochloride,

    b. Imipramine Hydrochloride,

    c. Fluoxetine,

    d. Venlafaxine,

    e. Duloxetine,

    f. Sertraline,

    g. Citalopram,

    h. Escitalopram,

    i. Fluvoxamine,

    j. Paroxetine

5. Write the uses of:

    a. Amitriptyline Hydrochloride,

    b. Imipramine Hydrochloride,

    c. Fluoxetine,

    d. Venlafaxine,

    e. Duloxetine,

    f. Sertraline,

    g. Citalopram,

    h. Escitalopram,

    i. Fluvoxamine,

    j. Paroxetine

C. LONG ANSWER QUESTIONS

1. What is depression? Discuss the types of depression. Define and classify various anti-depressant drugs and discuss their mechanisms of action.

2. Write the structures, chemical names, uses, stability and storage of following:
   a. Amitriptyline Hydrochloride,
   b. Imipramine Hydrochloride,
   c. Fluoxetine,
   d. Venlafaxine,
   e. Duloxetine,
   f. Sertraline,
   g. Citalopram,
   h. Escitalopram,
   i. Fluvoxamine,
   j. Paroxetine

■■■

DRUGS ACTING ON AUTONOMIC NERVOUS SYSTEM: SYMPATHOMIMETIC AGENTS

CONTENTS

Sympathomimetic Agents:

A. Direct Acting:
1. Nor- Epinephrine
2. Epinephrine
3. Phenylephrine
4. Dopamine
5. Terbutaline
6. Salbutamol (Albuterol)
7. Naphazoline
8. Tetrahydrozoline

B. Indirect Acting Agents:
1. Hydroxy Amphetamine
2. Pseudoephedrine

C. Agents with Mixed Mechanism:
1. Ephedrine
2. Metaraminol

♦ LEARNING OBJECTIVES ♦

After completing this chapter, the student should be able to understand:

1. Classification of Sympathomimetic Agents.
2. Chemical Structures of different Sympathomimetic Agents.
3. Uses, Stability and storage conditions of Sympathomimetic Agents.
4. Different types of formulations and popular brand names for Sympathomimetic Agents.

6.1.1 INTRODUCTION

Autonomic nervous system (ANS) represents the unconscious regulation and control of visceral function. It consists of afferent central connections and autonomic efferent. The afferent nerves are the first link in the reflex of ANS, which carries non-myelinated fibres from visceral parts to the cerebrospinal axis. Central connections refer to the integration of autonomic reflexes through the hypothalamus and the nucleus solitary tract. Autonomic efferent is the principal peripheral connections that are divided into sympathetic and parasympathetic outflow. Neurotransmitter in all the pre and parasympathetic postganglionic fibres is acetylcholine and in the sympathetic postganglionic release is norepinephrine.

The ANS has two divisions:

a. **Sympathetic nervous system:** It accelerates the heart rate, constricts blood vessels, and raises blood pressure. The sympathetic nervous system's primary process is to stimulate the body's fight-or-flight response (acute stress response). It is constantly active at a basic level to maintain homeostasis. The sympathetic nervous system is described as being complementary to the parasympathetic nervous system.

b. **Parasympathetic nervous system**: It slows the heart rate, increases intestinal and gland activity, and relaxes sphincter muscles. It is responsible for regulating the body's unconscious actions. The parasympathetic system is responsible for stimulation of "rest-and-digest" or "feed and breed" activities that occur when the body is at rest, especially after eating, including sexual arousal, salivation, lacrimation (tears), urination, digestion and defecation. Its action is described as being complementary to that of the sympathetic nervous system.

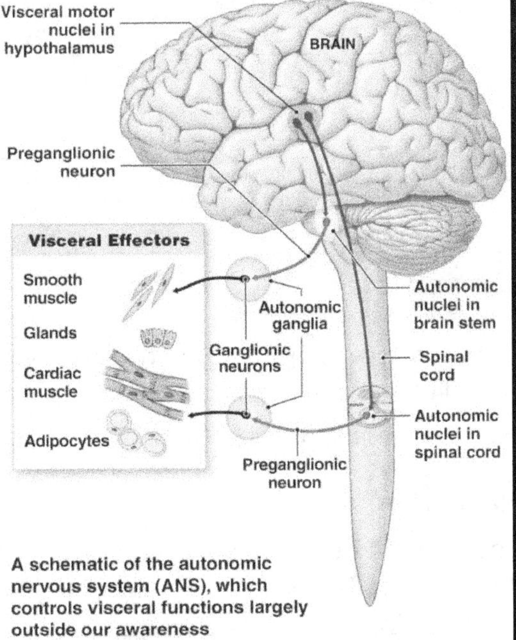

Fig. 6.1.1 Components of the autonomic nervous system

Table 6.1.1 Comparison between sympathetic and parasympathetic nervous systems

Sr. No.	Parameter	Sympathetic nervous system (SNS)	Parasympathetic nervous system (PNS)
1.	Introduction	The sympathetic nervous system (SNS) is one of two main divisions of the autonomic nervous system ANS. Its general action is to mobilize the body's fight-or-flight response.	The parasympathetic nervous system is one of the two main divisions of the autonomic nervous system (ANS). Its general function is to control homeostasis and the body's rest-and-digest response.
2.	Function	Control the body's response during perceived threat.	Control the body's response while at rest.
3.	Originates in	Spinal cord, thoracic and lumbar spinal cord	Spinal cord, medulla
4.	Activates response of	Fight-or-flight	Rest and digest
5.	Neuron Pathways	Very short neurons, faster system	Longer pathways, slower system
6.	General Body Response	Body speeds up, tenses up and becomes more alert. Functions not critical to survival shut down.	Counterbalance; restores body to state of calm.
7.	Cardiovascular System (heart rate	Decreases heart rate	Increases contraction, heart rate
8.	Musculoskeletal System	Muscles contract	Muscles relax
9.	Pupils	Dilate	Constrict
10.	Gastrointestinal System	Decreases stomach movement and secretions	Increases stomach movement and secretions
11.	Salivary Glands	Saliva production decreases	Saliva production increases
12.	Adrenal Gland	Releases adrenaline	No involvement
13.	Glycogen to Glucose Conversion	Increases; converts glycogen to glucose for muscle energy	No involvement
14.	Urinary Response	Decrease in urinary output	Increase in urinary output

Basic terminologies used in ANS:

1. **Efferent neurons:** The efferent neurons are also known as motor neurons, they bring the responses from the brain to the muscles and the glands.

2. **Afferent neurons:** The afferent neurons are also known as the sensory neurons, they bring the stimuli from the sensors (e.g., skin, eyes, and ears) to the CNS.

3. **Postsynaptic neurons** (situated after a synapse): A neuron to the cell body or dendrite of which an electrical impulse is transmitted across a synaptic cleft by the release of a chemical neurotransmitter from the axon terminal of a presynaptic neuron.

4. **Presynaptic neurons** (situated before a synapse): A neuron from the axon terminal of which an electrical impulse is transmitted across a synaptic cleft to the cell body or one or more dendrites of a postsynaptic neuron by the release of a chemical neurotransmitter.

6.1.2 ADRENERGIC RECEPTORS

Adrenergic receptors mediate the central and peripheral actions of the neurohormones norepinephrine and epinephrine. Both of these catecholamine messengers play important roles in regulating diverse physiological systems; thus, adrenergic receptors are widely distributed throughout the body. Stimulation of adrenergic receptors by catecholamines released from the sympathetic branch of the autonomic nervous system results in a variety of effects, such as increased heart rate, regulation of vascular tone, and bronchodilation. In the central nervous system, adrenergic receptors are involved in many functions, including memory, learning, alertness, and the response to stress. Adrenoceptors are membrane-bound receptors located throughout the body on neuronal and non-neuronal tissues, where they mediate a diverse range of responses to the endogenous catecholamines noradrenaline and adrenaline. They are membrane-bound G-protein coupled receptors and classified as α (alpha) and β (beta) adrenoceptors. Binding of catecholamine to the receptor is responsible **for the fight or flight response.**

Fight or flight response results in:

1. Increased BP

2. Increased blood flow to the brain, heart and skeletal muscles

3. Increased muscle glycogen for energy

4. Increased rate of coagulation

5. Pupil dilation

Adrenergic receptors are membrane-bound G-protein coupled receptors which function primarily by increasing or decreasing the intracellular production of second messengers cAMP or IP_3/DAG. In some cases, the activated G-protein itself operates K^+ or Ca^{2+} channels or increases prostaglandin production. Ahlquist (1948), based on two distinct rank orders of potencies of adrenergic agonists, classified adrenergic receptors into two types α and β. This classification was confirmed later by the discovery of selective α and β adrenergic antagonists.

6.1.2.1 α-Receptors

They are further classified into α_1 and α_2. The selective agonist for α_1 is phenylephrine, and the antagonist is prazosin. α_2 receptors are stimulated by clonidine selectively and antagonised by yohimbine. Elicitation of α_1 receptor increases phospholipase C, D, A$_2$, and inositol-triphosphate/diacylglycerol IP$_3$/DAG, which contracts smooth muscles, genitourinary tracts, and increases the secretions of glands and in the heart, it increases the force of contraction. α_2 receptor stimulation causes decrease in cyclic AMP, Ca^{2+} L type channels, and increases the IP$_3$/DAG. It decreases insulin secretion, produces platelet aggregation, decreases the release of NE as it has an auto-regulation mechanism, and causes vascular smooth muscle contraction.

6.1.2.2 β-Receptors

The β receptors are further subdivided into three subtypes and distributed in the body, that is, β_1 (heart), β_2 (bronchi, smooth muscles), and β_3 (adipose tissue). Each has a selective agonist and antagonist. For β_1 receptor, the agonist is dobutamine, and antagonists are atenolol and metaprolol. Agonists for β_2 are salbutamol and terbutaline, and the antagonist is propranolol.

The pharmacology of β receptors indicates the biochemical increase of adenyl cyclase, L-Type calcium channels increment through β_1, causing an increase in the force and rate of contraction of AV nodal conduction velocity. β_2 receptor stimulates bronchodilation through elicitation of adenyl cyclase. β_3 stimulation increases adenyl cyclase and causes lipolysis.

Table 6.1.2 Distribution and Effect of Adrenergic Receptors:

Adrenergic Receptors	Location (Organ or Tissue)	Effect of Activation	Physiological Effects
α_1	Blood vessels and skin	Vasoconstriction	Increase Blood Pressure
α_1	Mucous Membrane	Vasoconstriction	-
α_{1A}	Prostatic gland muscles	Contraction	Prostatic hyperplasia
α_2	CNS	Decrease NE Release	Decrease blood pressure
β_1 (minor β_2, β_3)	Heart Muscles	Muscles contraction	Increase heart rate and force
α_1	Bronchial smooth muscles	Smooth muscle contraction	Closes airways
β_2	Bronchial smooth muscles (Brnchodilation)	Smooth muscle relaxation	Dilate and open airways
α_1	Uterus (Pregnant)	Smooth muscle contraction	-
β_2	Uterus (Pregnant)	Smooth muscle relaxation	(-) Uterine contractions
β_1	Kidney	Increase Renin Secretion	Increase in Blood Pressure

Adrenergic drugs are the agents which act directly on the sympathetic nervous system, and these drugs are also known as sympathomimetic drugs as these agents mimic the action of the sympathetic nervous system. On the stimulation of the sympathetic nervous system following responses shows:

1. Cardiac stimulation results in an increase in heart rate and force of contraction.

2. Stimulation of CNS.

3. Relaxation of bronchial muscles.

4. Increase in rate of glycogenolysis.

5. Liberation of free fatty acids.

Table 6.1.3 Adrenergic Receptors; Drugs and their Action

Sr. No.	Receptor	Action	Stimulating drug
1	Alpha-1 (α_1), postsynaptic	Vasoconstriction, positive inotropism, antidiuresis	Adrenaline + Noradrenaline +++
2	Alpha-2 (α_2), presynaptic	Vasodilatation, inhibition of noradrenaline release.	Adrenaline + Noradrenaline +++ Dopamine +
3	Alpha-2 (α_2), postsynaptic	Constriction of coronary arteries, promotion of salt and water excretion	Adrenaline + Noradrenaline +++ Dopamine +
4	Beta-1 (β_1), postsynaptic	Positive inotropism, chronotropism, renin release	Adrenaline ++ Noradrenaline ++
5	Beta-2 (β_2), presynaptic	Noradrenaline release accelerated, positive inotropism, chronotropism	Adrenaline + Isoprenaline +++
6	Beta-2 (β_2), postsynaptic	Vasodilatation, relaxation of bronchial smooth muscle	Adrenaline + Isoprenaline +++
7	Dopamine, postsynaptic	Vasodilatation, diuresis	Dopamine ++
8	Dopamine, presynaptic	Inhibits noradrenaline release	Dopamine +

6.1.3 ADRENERGIC OR SYMPATHOMIMETIC AGENTS

Sympathomimetic or Adrenergic drug is a substance that stimulates a response from the **adrenergic receptors** or a stimulant that mimics the effects of agonists of the sympathetic nervous system such as the catecholamines. (epinephrine (adrenaline), norepinephrine (noradrenaline), dopamine, etc.). Like cholinomimetic drugs, sympathomimetics can be grouped by their mode of action and the spectrum of receptors. Some of the drugs (e.g. epinephrine, norepinephrine) act by a direct mode, that is, they interact with and activate adrenoceptors. While others act on the adrenergic beta-receptors (e.g. isoproterenol). Others have a little direct effect on the adrenergic receptors but enhance the release of natural catecholamine from the sympathetic nerve terminals (e.g. amphetamine, phenylpropanolamine).

Sympathomimetic drugs acting at alpha- or beta-adrenergic receptors within target tissues are used to treat cardiac arrest and low blood pressure or even delay premature labour.

6.1.4 CLASSIFICATION OF SYMPATHOMIMETIC AGENTS AS PER M.O.A.

Adrenergic or Sympathomimetic agents are divided into three classes based on their mechanism of action:

6.1.4.1 Direct-acting Adrenergic Agonists

They bind to and activate α_1, α_2, β_1, and β_2 receptors. Naturally occurring molecules which bind to these receptors include NE (a neurotransmitter which binds to α_1, α_2, and β_1 receptor), Epinephrine (a hormone produced in and secreted from the adrenal medulla, which binds to α_1, α_2, β_1, and β_2 receptors, it is a nonselective adrenergic agonist), and Dopamine (also a neurotransmitter, which binds to α_1, α_2, and β_1 receptor).

e.g., Nor-epinephrine, Epinephrine, Phenylephrine, Dopamine, Methyldopa, Clonidine, Dobutamine, Isoproterenol, Terbutaline, Salbutamol, Bitolterol, Naphazoline, Oxymetazoline and Xylometazoline.

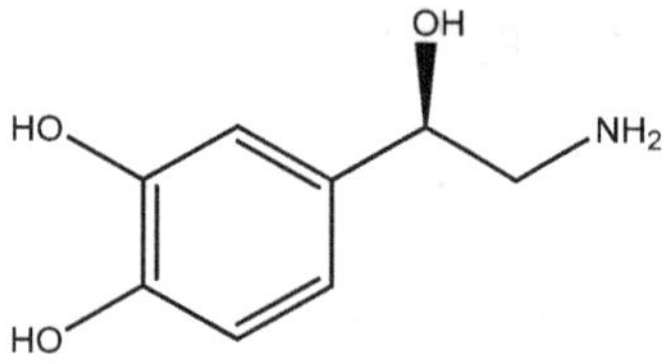

Norepinephrine (NE)/ Noradrenaline

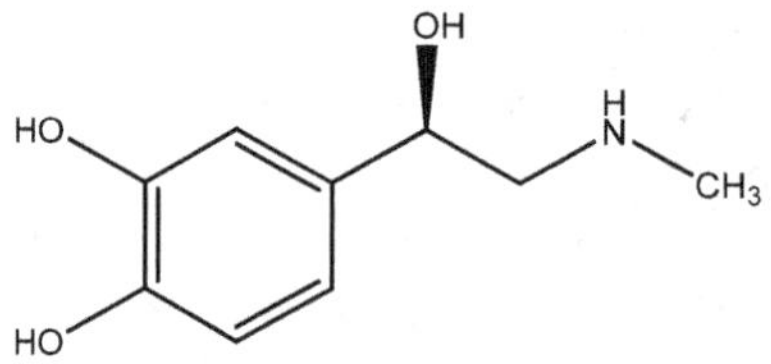

Epinephrine (E)/ Adrenaline

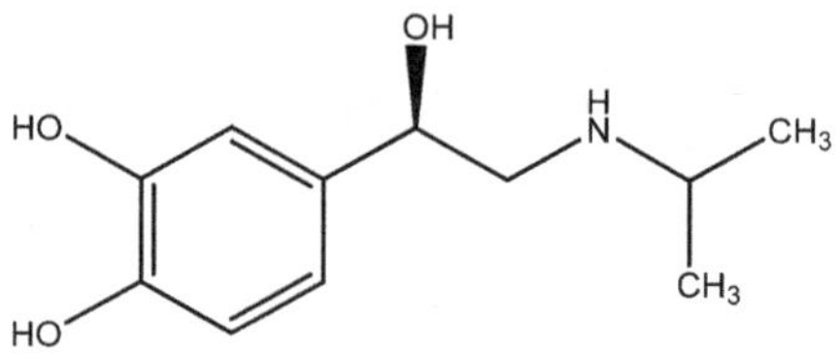

Isoproterenol (ISO)/ Isoprenaline

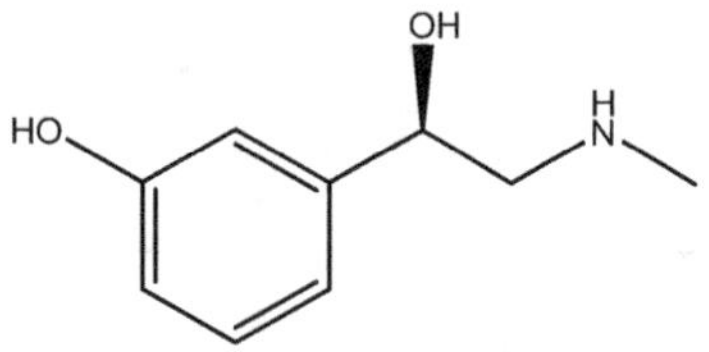

Phenylephrine

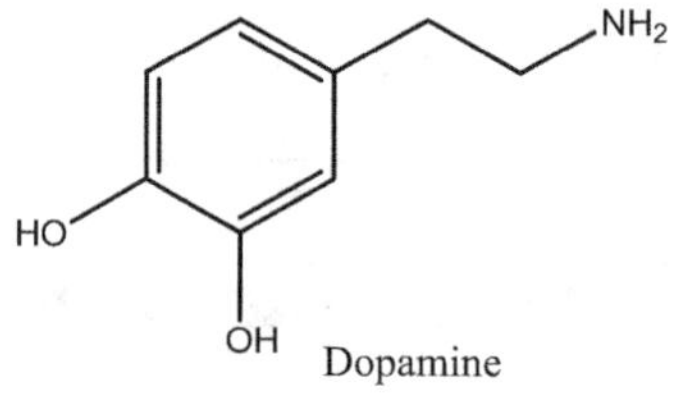

Dopamine

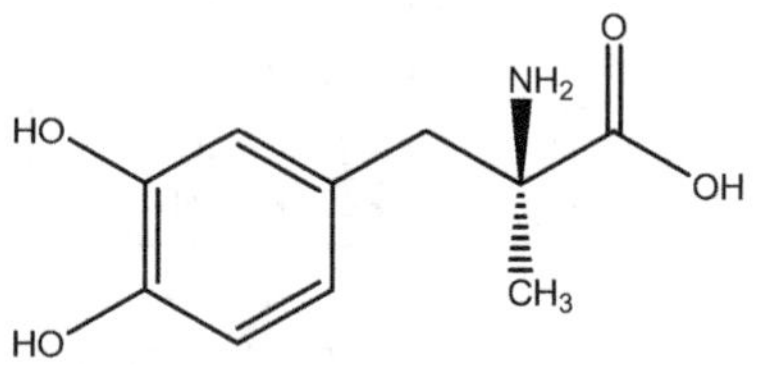

Methyl dopa

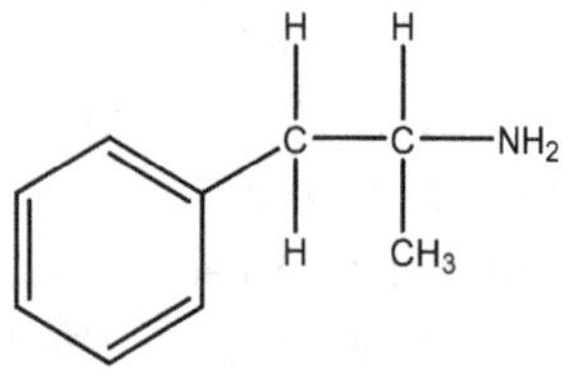

Amphetamine

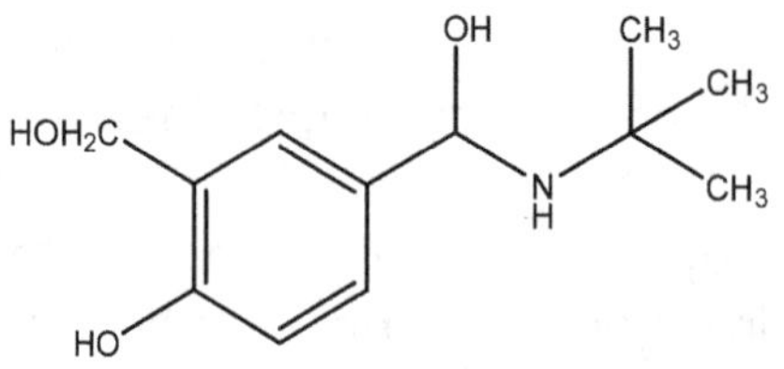

Salbutamol

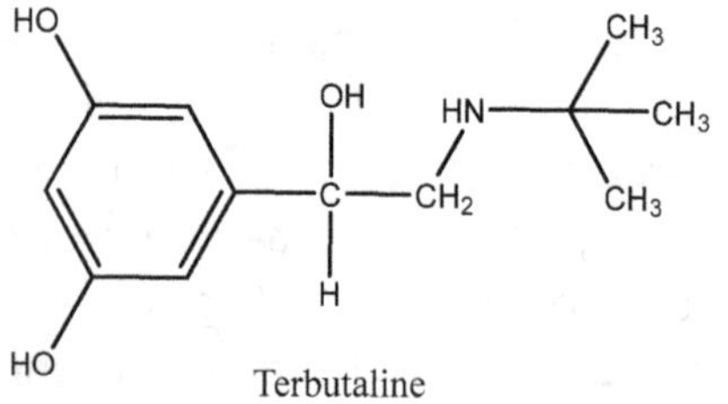

Terbutaline

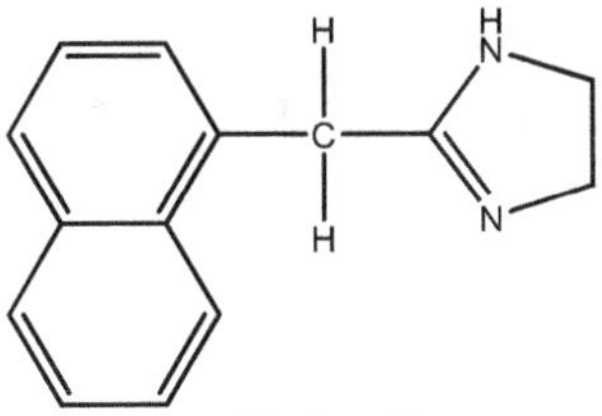

Naphazoline

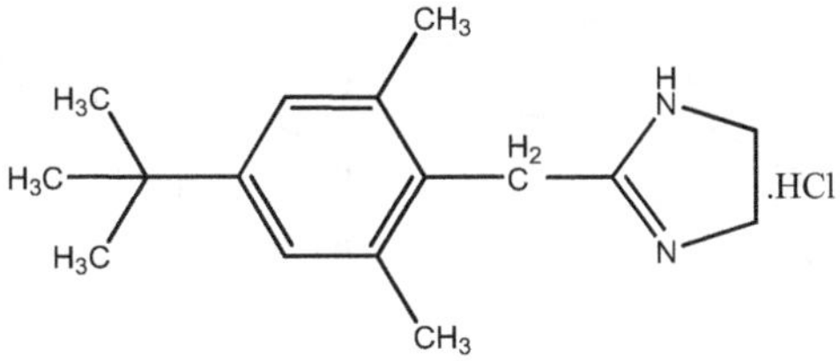

Xylometazoline

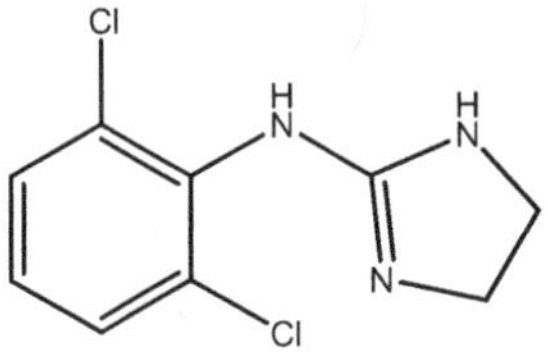

Clonidine

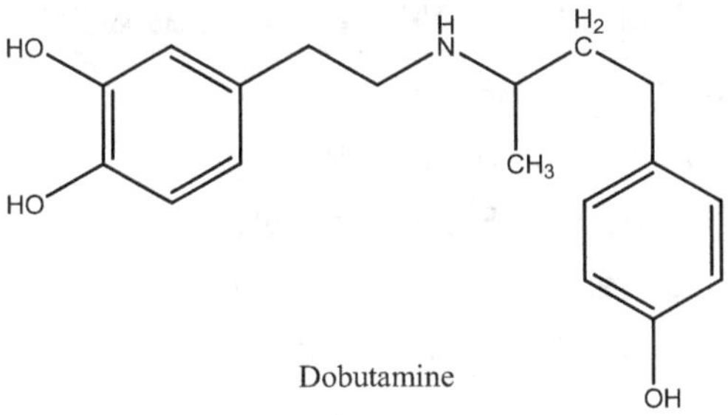

Dobutamine

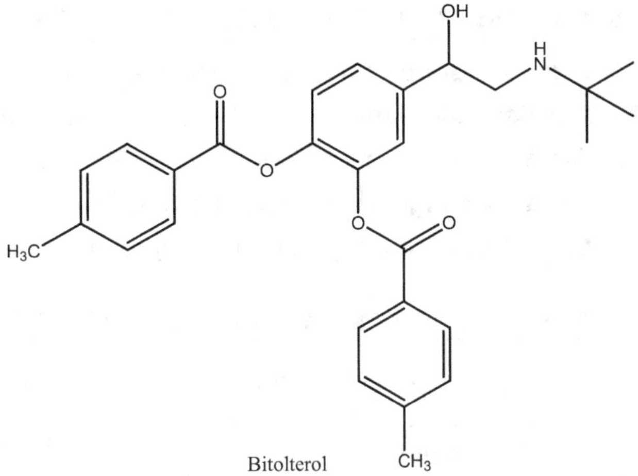

Bitolterol

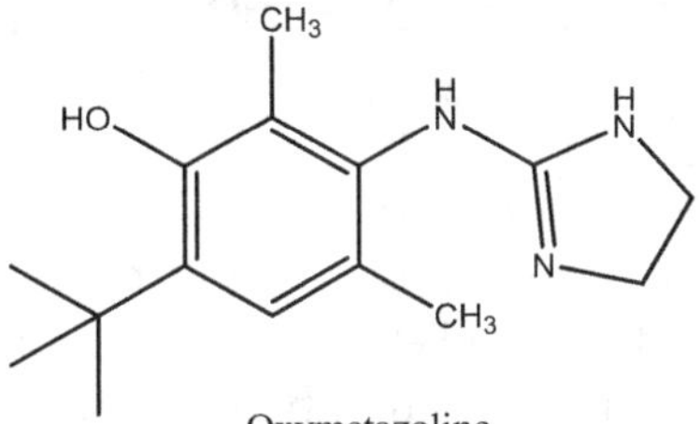

Oxymetazoline

6.1.4.2 Indirect-acting Adrenergic Agonists

They produce NE-like actions by stimulating NE release, preventing its reuptake, and producing activation.

e.g., Hydroxyamphetamine, Pseudoephedrine, Propylhexedrine.

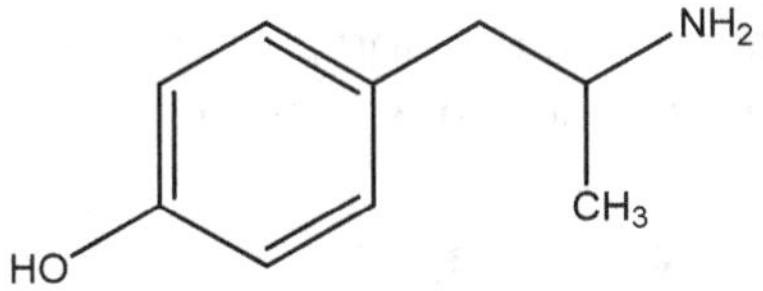

Hydroxyamphetamine

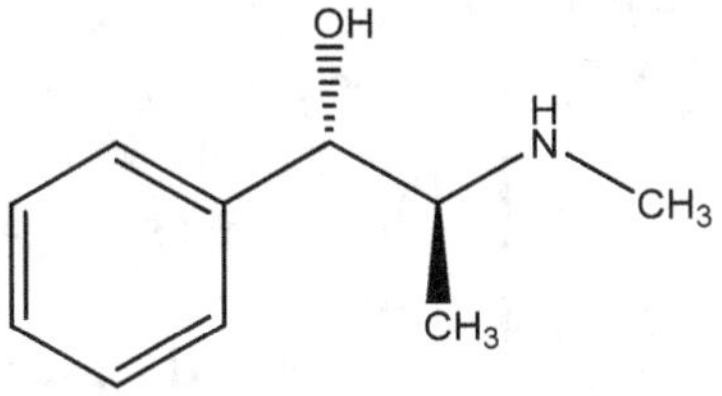

Pseudoephedrine

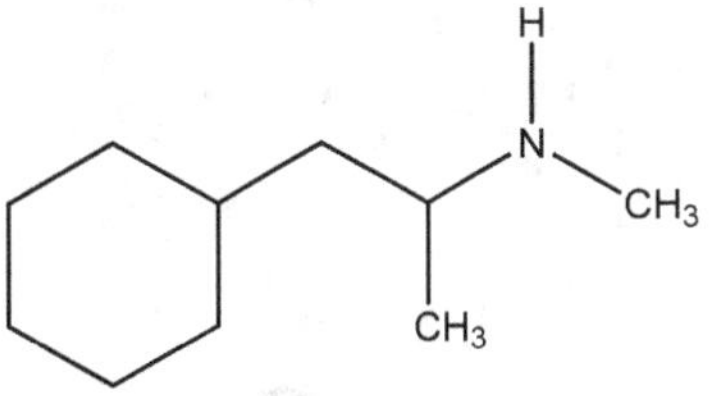

Propylhexedrine

6.1.4.3 Dual-acting Adrenergic Agonists

These agents act as dual-acting direct and indirect adrenergic agonists. They bind to adrenergic receptors and stimulate NE release.

e.g. Ephedrine, Metaraminol.

Ephedrine   Metaraminol

6.1.5 THERAPEUTIC CLASSIFICATION OF SYMPATHOMIMETIC AGENTS

1. **Pressor agents:** e.g., Noradrenaline, Ephedrine and Dopamine etc.
2. **Cardiac stimulants:** e.g., Adrenaline, Dobutamine and Isoprenaline etc.
3. **Bronchodilators:** e.g., Isoprenaline, Salbutamol, Terbutaline, Formeterol etc.
4. **Nasal decongestants:** e.g., Phenylephrine, Naphazoline, Pseudoephedrine etc.
5. **CNS stimulants:** e.g., Amphetamine, Methamphetamine, Dexamphetamine etc.

6.1.6 MECHANISM OF ACTION OF SYMPATHOMIMETIC AGENTS

Depending upon the MOA, the drug classification is discussed in the next section.

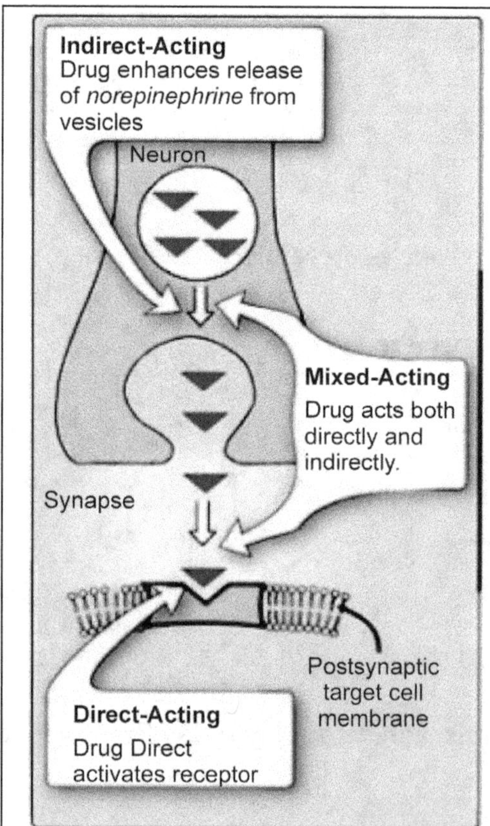

I. Direct-acting adrenergic agent:
These drugs act directly on α or β receptors and produce effects similar to those that occur after stimulation of sympathetic nerves or release of the hormone epinephrine from the adrenal medulla.

II. Indirect-acting adrenergic agent:
These agents block the uptake of norepinephrine (uptake blocker) or take it up into the presynaptic neuron and cause the release of norepinephrine from the cytoplasmic vesicles of the adrenergic neuron. As with neuronal stimulation, the norepinephrine transverses the synapse and binds to the α or β receptors.

III. Mixed-acting adrenergic agent:
Some adrenergic agents have the capacity of both to stimulate adrenergic receptors directly and to release norepinephrine from the adrenergic neuron.

Fig. 6.1.2 Mechanism of action of adrenergic agents

Based on their receptor-specific action, they are also classified as follows.

1. **β_1 Adrenergic Receptor Action:** The β adrenergic actions are mediated through cyclic AMP (cAMP). Adrenaline activates membrane-bound enzyme adenyl cyclase through regulatory protein Gs. Adenosine-triphosphate (ATP) is broken into cAMP at the inner face. In the heart, proteins, such as troponin and phospholamban, are phosphorylated and result in increased interaction with calcium at myofilaments, leading to increased force of contraction.

2. **β_2 Adrenergic Receptor Action:** β_2 receptors are predominantly found in the respiratory system. The β_2 receptor agonist produces bronchodilation by binding with the β_2 receptors.

3. **α_1 Adrenergic Receptor Action:** Stimulation of α_1 adrenergic receptor elicits a primary mode of signal transduction that involves the mobilisation of intracellular Ca^{2+} from endoplasmic stores. This increase in intracellular Ca^{2+} is thought to result from the activation of phospholipase C_β isoforms through the Gq family of G protein.

   The hydrolysis of membrane-bound polyphosphoinositides through phospholipase C results in the generation of diacylglycerol (DAG) and Inositol 1, 4, and 5-triphosphate (IP_3).

   IP_3 stimulates the release of Ca^{2+} from intracellular stores. In smooth muscles, increased intracellular Ca^{2+} cause contraction, mediated by Ca^{2+} sensitive protein kinases, that is, calmodulin dependent myosin light chain kinase.

4. **α_2 Adrenergic Receptor Action:** α_2 adrenergic receptors activate G protein gated K^+ channels resulting in hyper-polarization.

   These α_2 receptors are also capable of inhibiting voltage-gated Ca^{2+} channels mediated by G protein and also activate the mitogen-activated protein kinase (MAPK). These lead to activation of a variety of tyrosine kinase-mediated downstream events, and these inhibit the release of norepinephrine from nerve endings and suppress the sympathetic outflow to the brain.

6.1.7 PHYSIOLOGICAL EFFECTS OF SYMPATHOMIMETIC AGENTS

Stimulation of α-adrenergic receptors on smooth muscles results in:

1. Vasoconstriction of blood vessels
2. Relaxation of GI smooth muscles
3. Contraction of the uterus and
4. Male ejaculation
5. Decreased insulin release
6. Contraction of the ciliary muscles of the eye (dilated pupils)

Stimulation of β-Stimulation of β2 adrenergic receptors on the airways results in:

1. Bronchodilation (relaxation of the bronchi)
2. Uterine relaxation Uterine relaxation
3. Glycogenolysis in the liver

Stimulation of β1-adrenergic receptors on the myocardium, AV node, and SA node the results in

1. Cardiac Stimulation
2. Increased force of contraction (positive inotropic effect)
3. Increased heart rate (positive chronotropic effect)
4. Increased conduction through the AV node (positive dromotropic effect).

Side Effects:

1. Alpha-Adrenergic Effects:
   - CNS: headache, restlessness, excitement, insomnia, headache, restlessness, excitement, euphoria.
   - Cardiovascular: palpitations (dysrhythmias), tachycardia, palpitations (dysrhythmias), vasoconstriction, hypertension.
   - Other: anorexia, dry mouth, nausea, vomiting, taste changes (rare).
2. Beta-Adrenergic Effects:
   - CNS: mild tremors, headache, nervousness, dizziness mild.
   - Cardiovascular: Increased heart rate, palpitations (dysrhythmias), and fluctuations in BP.
   - Other: sweating, nausea, vomiting, muscle cramps.

6.1.8 SOME IMPORTANT SYMPATHOMIMETIC AGENTS

A. Direct Acting Drugs:

6.1.8.1 Nor-epinephrine (*Syn:* Noradrenalin or nor-adrenaline (NA))

Structure:

OH

NH₂

HO

OH

Chemical (IUPAC) name: *l*-β-[3,4-Dihydroxyphenyl]-α-methyl-aminoethanol

Uses: It is useful for blood pressure control in certain acute hypotensive states (e.g., sympathectomy means the surgical cutting of a sympathetic nerve or removal of a ganglion to relieve a condition affected by its stimulation, poliomyelitis, spinal anaesthesia, myocardial infarction and septicemia, blood transfusion, and drug reactions). As an adjunct in the treatment of cardiac arrest and profound hypotension.

Side Effects: Cardiovascular side effects from norepinephrine (NE) can be severe, it can produce profound hypertension, local vasoconstriction, and tissue hypoxia. NE-induced hypertension typically presents as headache, photophobia, stabbing chest pain, intense sweating, vomiting, weakness, dizziness, tremor, respiratory difficulty or apnea, and precordial pain.

Stability and storage conditions: Store at 20°C to 25°C, protect from light.

Incompatibility: The use of nor-adrenaline during cyclopropane and halothane anesthesia is generally considered incompatible because of the risk of producing ventricular tachycardia or fibrillation. It should be used with extreme caution in patients receiving monoamine oxidase inhibitors (MAOI) or antidepressants of the triptyline or imipramine types because severe, prolonged hypertension may result.

Different types of formulation: Injection, solution

Popular brands: It is manufactured by many companies. Some of the notable brand names in India are:

Sr. No.	Brand Name	Manufacturers	Type
1	Adrenor (4mg)	Samarth Pharma Pvt. Ltd	Injection
2	Adronis (4mg)	Neiss Labs Pvt. Ltd.	Injection
3	Epinor (4mg)	Geo Pharma Pvt Ltd	Injection
4	Nestig (4mg)	SPM Drugs Pvt. Ltd.	Injection
5	Norad (4mg)	Neon Laboratories Ltd	Injection
6	Norepirin (4mg)	Chandra Bhagat Pharma Pvt. Ltd.	Injection
7	Veraline (4mg)	Verve Formulations Pvt Ltd	Injection
8	Nodresol (4mg)	Shree Ganesh Rubber & Chemicals Co	Injection

6.1.8.2 Epinephrine (*Syn*: **L-Adrenaline, adrenaline, L-epinephrine, *adrenalin*)**

Structure:

Chemical (IUPAC) name: 1-(3,4-Dihydroxyphenyl)-2-methylaminoethanol

OR

4-[(1*R*)-1-Hydroxy-2-(methylamino)ethyl]benzene-1,2-diol

Epinephrine is prepared by Friedel Craft's acylation of catechol with chloroacetyl chloride to give α-haloacetophenone, followed by nucleophilic substitution with methylamine and catalytic reduction.

Uses: Epinephrine is used to treat a number of conditions, including cardiac arrest, anaphylaxis, and superficial bleeding. It has been used historically for bronchospasm and hypoglycemia.

It is also useful to treat very serious allergic reactions to insect bites, foods, drugs, or other substances in emergencies. Epinephrine acts quickly to improve breathing, stimulate the heart, raise dropping blood pressure, reverse hives, and reduce swelling of the face, lips, and throat.

Side Effects: Nausea or vomiting, difficulty breathing, pounding, fast or irregular heartbeat, pale skin, headache, sweating, dizziness, nervousness, anxiety, or restlessness, weakness, uncontrollable shaking.

Stability and storage conditions: Epinephrine injection should be stored at room temperature (approximately 25 °C). Preserve in tight, light-resistant containers. This chemical darkens slowly on exposure to air and light.

Incompatibility: Epinephrine should be used with caution in patients taking digoxin, quinidine, diuretics, or other alpha or beta-adrenergic agonists, as concomitant administration may lead to arrhythmias or hypertension. The effects of epinephrine may be increased when given with antihistamines, furazolidone, levothyroxine, methyldopa, reserpine, tricyclic antidepressants, or monoamine oxidase inhibitors.

Different types of formulation: Injection, solution

Popular brands: Epinephrine is available in combination with other drugs. These combinations are manufactured by many companies. Some of the notable brand names with combinations in India are:

Sr. No	Brand Name	Combination	Manufacturers	Type
1	Alergin	Caffeine, Ephedrine, Paracetamol	Cipla Limited	Tablet
2	Apricaine Adernalie	Adrenaline, Lignocaine	Indus Pharma Pvt. Ltd.	Injection
3	Cortasthma	Ephedrine Hcl, Aminophylline, Phenobarbitone	Zydus Cadila Healthcare Ltd. (Indon)	Tablet
4	Endrie	Camphor, Castor Oil, Ephedrine, Evening Primrose Oil, Menthol	Wyeth (I) Limited	Drop
5	Franklor	Ephedrine Hcl, Diphenhydramine	Franklin Laboratories (P) Ltd.	Tablet
6	Tedral	Ephedrine, Theophylline	Pfizer Limited	Tablet
7	Theocort	Ephedrine, Phenobarbitone, Theophylline	Zydus Cadila Healthcare Ltd	Table
8	Tedral E	Caffeine, Ephedrine, Paracetamol	Parke-Davis (India) Ltd.	Tablet

Other brand names: EpiPen, Anapen Bronolax, Tedral, Trox-Ex, Tusin

6.1.8.3 Dopamine: (*Syn:* **Hydroxytyramin, Oxytyramine, Dopamin,3-Hydroxytyramine)**

Structure:

$$HO-C_6H_3(OH)-CH_2CH_2-NH_2$$

Systematic (IUPAC) Name: (3,4-Dihydroxyphenylethylamine),

It differs from the other naturally occurring catecholamines, lacking the β-OH group on the ethylamine side chain. It is the metabolic precursor of noradrenaline and adrenaline and is a central neurotransmitter.

Uses: It is most commonly used as a stimulant drug in the treatment of severe low blood pressure, slow heart rate, and cardiac arrest. It is especially important in treating these in newborn infants; it is given intravenously. Dopamine HCl is indicated for the correction of hemodynamic imbalances present in the shock syndrome due to myocardial infarction, trauma, endotoxic septicemia, open-heart surgery, renal failure, and chronic cardiac decompensation, as in congestive failure.

Side Effects: Ventricular arrhythmia, atrial fibrillation, ectopic beats, tachycardia, anginal pain, palpitation, cardiac conduction abnormalities, bradycardia, hypotension, hypertension, vasoconstriction, dyspnea, nausea, vomiting, headache, anxiety.

Stability and storage conditions: Some commercially available injections of dopamine in dextrose provided in plastic containers should be stored at room temperature (25 ° C), although brief exposure to temperatures up to 40°C does not adversely affect the injections; the injections should be protected from excessive heat and should not be frozen.

Incompatibility: Do not add dopamine hydrochloride to any alkaline solution since the drug is inactivated in an alkaline solution. It is incompatible with anaesthetics and may produce ventricular arrhythmias, monoamine oxidase (MAO) inhibitors prolong and potentiates the effect of dopamine. Concurrent administration of low-dose dopamine and diuretic agents may produce an additive or potentiating effect on urine flow. Tricyclic antidepressants may potentiate the cardiovascular effects of adrenergic agents. The concomitant use of vasopressors, vasoconstricting agents (such as ergonovine) and some oxytocic drugs may result in severe hypertension. Administration of phenytoin with dopamine leads to hypotension and bradycardia.

Different types of formulation: Injection

Popular brands: It is manufactured by many companies. Some of the notable brand names in India are:

Sr. No.	Brand Name	Manufacturers	Type
1	Cupamin	Cubit (Cucard)	Injection
2	Domin	Neon Labs	Injection
3	Dopacin	Troikaa Pharmaceuticals	Injection
4	Dopacard	Nicholas Piramal	Injection
5	Dopacef	Chandra Bhagat (Nextgen)	Injection
6	Dopar	Samarth Pharma	Injection
7	Dopasol	S.G. Pharma	Injection
8	Dopamine	TTK Pharma	Injection

6.1.8.4 Terbutaline: (*Syn:* Brethaire, Brethin, Bricanyl, Terbutalin)

Structure:

Chemical (IUPAC) name: N-*tert*-Butyl-N-[2-(3,5-dihydroxyphenyl)-2-hydroxymethyl]amine

Terbutaline is a non-catecholamine, therefore, it is resistant to COMT.

Terbutaline is prepared by reduction of 2-(tert-butylamino)-3', 5'-dihydroxyacetophenone by catalytic hydrogenation:

Uses: Terbutaline is used as a fast-acting bronchodilator (often used as short-term asthma treatment) and colytic to delay premature labour. The inhaled form of terbutaline starts working within 15 minutes and can last up to 6 hours.

Side Effects: An allergic reaction (difficulty breathing, closing of the throat, swelling of the lips, tongue, or face; or hives), chest pain or irregular heartbeats, headache, dizziness, lightheadedness, or insomnia, tremor or nervousness, sweating, nausea, vomiting, diarrhoea, dry mouth.

Stability and storage conditions: Tablet stored at 15-30°C (59-86°F), Injection at 20-25°C (68-77°F). Protect from light.

Incompatibility: It is incompatible with monoamine oxidase inhibitors or tricyclic antidepressants, with β blockers producing severe bronchospasm in asthmatic patients. The ECG changes and/or hypokalemia result from the administration of nonpotassium-sparing diuretics such as terbutaline.

Different types of formulation: Tablet, liquid, syrup, injection.

Popular brands: It is manufactured by many companies. Some of the notable brand names in India are:

Sr. No.	Brand Name	Manufacturers	Type
1	Adcold-Br® [Liqd]	Adley	Liquid
2	Bricanyl	Astra Zeneca Pharma India Limited	Tablet
3	Bricanyl® [Amp]	AstraZeneca	Ampoule (inhaler)
4	Brocoter® [Syr]	Dexter Labs	Syrup
5	Etoxin-B® [Tab]	Ind-Swift	Tablet
6	Tossex-Ax® [Liqd]	Piramal HC	Liquid
7	Tussil-X® [Syr]	Stalwart	Liquid
8	Viscodril® [Syr]	Invision	Liquid

6.1.8.5 Salbutamol (Albuterol): (*Syn:* **Proventil, Sultanol, Aerolin**)

Structure:

Systematic (IUPAC) Name: (RS)-4-[2-(*tert*-Butylamino)-1-hydroxyethyl]-2-(hydroxy-methyl)- phenol

Uses: Salbuterol is used to treat or prevent bronchospasm in patients with asthma, bronchitis, emphysema, and other lung diseases. This drug is also used to prevent wheezing caused by exercise (exercise-induced bronchospasm). Salbutamol is also used in obstetrics. Intravenous salbutamol can be used as a tocolytic, also called anti-contraction medication, to relax the uterine smooth muscle to delay premature labour. It is useful in cystic fibrosis.

Side Effects: Nervousness, shaking of a part of the body, headache, nausea, vomiting, cough, irritation in the throat, muscle, bone, or back pain.

Stability and storage conditions: Albuterol sulfate tablets, extended-release tablets, and the oral solution should be stored at 2-30 deg C, unit-dose packages of the extended-release tablets should be protected from excessive moisture, Preparations containing albuterol or albuterol sulfate should be stored in well-closed, light-resistant containers. Commercially available metered-dose albuterol and albuterol sulfate inhalers should be stored at 15-30 °C and 15-25 ° C, respectively.

Incompatibility: Methyldopa (used to treat high blood pressure), theophylline, aminophylline, drugs used for depression, beta-blockers for blood pressure or heart conditions, e.g. propranolol diuretics (water tablets), steroids, digoxin, and atomoxetine (with intravenous salbutamol only) are incompatible with this drug.

Different types of formulation: Tablet, Syrup, Solution

Popular brands: It is manufactured by many companies. Some of the notable brand names in India are:

Sr. No	Brand Name	Manufacturers	Type
1	Asthalin	Cipla Limited	Tablet, Syrup, Inhaler,
2	Bemol	Bevit Pharmaceuticals Ltd.	Tablet
3	Asthacure	_ Swiss Pharma Pvt. Ltd	Tablet, Syrup
4	Asmanil	Inga Laboratories Pvt. Ltd.	Tablet
5	Bronkotab EX	Glaxo-Smithkline Pharmaceuticals Ltd.	Tablet, Inhaler
6	Bronosol -4	Alkem Laboratories Ltd	Tablet
7	Salbair	Lupin Laboratories Ltd.	Capsule, Solution
8	Salbutamol (4 mg)	Acichem Laboratories	Tablet

6.1.8.6 Naphazoline: (*Syn:* Naphthizine, Antan)

Structure:

Chemical (IUPAC) name: 2-(1-Naphthylmethyl)-2-imidazoline.

It is prepared by strong heating of 1-naphthaleneacetonitrile with ethylenediamine monochloride at 200°C.

Nephazoline

Uses: Naphazoline is an eye decongestant used to relieve redness, puffiness, and itchy/watering eyes due to colds, allergies, or eye irritations (smog, swimming, or wearing contact lenses).

Side Effects: Ongoing or worsening eye redness, eye pain, changes in your vision, chest pain, fast or uneven heart rate, severe headache, buzzing in ears, anxiety, confusion, or feeling short of breath, mydriasis, increased redness, irritation, discomfort, blurring, punctate keratitis, lacrimation, increased intraocular pressure, dizziness, headache, nausea, sweating, nervousness, drowsiness, weakness, hypertension, cardiac irregularities, and hyperglycemia.

Stability and storage conditions: Store at 20° to 25°C (68° to 77°F)

Incompatibility: Maprotiline or tricyclic antidepressants and Naphazoline are incompatible with each other and potentiate the pressor (means narrowing of an opening of a blood vessel) effect of Naphazoline. Patients under therapy with MAO inhibitors may experience a severe hypertensive crisis if given a sympathomimetic drug.

Different types of formulation: Solution/drops

Popular brands: It is manufactured by many companies. Some of the notable brand names in India are:

Sr. No	Brand Name	Manufacturers	Type
1	Naphaline	Redson Labs Pvt Ltd	Drop
2	Ocucel	FDC Limited	Drop
3	Renizol	Pfiscar India Limited	Drop

6.1.8.7 Tetrahydrozoline: (*Syn:* Tetryzoline, Tetryzolin, Tyzanol, Visine, Tetryzolinum)

Structure:

Chemical (IUPAC) name: (RS)-2-(1,2,3,4-Tetrahydronaphthalen-1-yl)-4,5-dihydro-1*H*-imidazole

Tetrahydrozoline occurs as hydrochloride.

Uses: Tetrahydrozoline is a decongestant used to relieve redness in the eyes caused by minor eye irritations (e.g., smog, swimming, dust, or smoke). Some brands of tetrahydrozoline eye drops may contain lubricants. Lubricants help protect the eyes from more irritation and dryness.

Side Effects: Stinging/redness in the eye, widened pupils, or blurred vision may occur, stinging or burning of the eye, blurred vision, increased eye redness or irritation, headache, sweating, fast or irregular heartbeat, and nervousness.

 Stability and storage conditions: Tetrahydrozoline hydrochloride nasal and ophthalmic solutions should be stored in tight containers.

Incompatibility: This drug is incompatible with tricyclic antidepressants (e.g., amitriptyline) because they may decrease tetrahydrozoline drops' effectiveness, Incompatible with cocaine, furazolidone, MAO inhibitors (e.g., phenelzine), or tricyclic antidepressants (e.g., amitriptyline) because they may increase the risk of tetrahydrozoline drops' side effects, such as headache, fever, and high blood pressure, Bromocriptine or cocaine side effects may be increased by tetrahydrozoline drops.

Different types of formulation: Drops

Popular brands: It is manufactured by a few companies. Some of the notable brand names in India are:

Sr. No.	Brand Name	Manufacturers	Type
1	Vasozine Eye	Sunways India Pvt. Ltd.	Drop
2	Visine Eye	Pfizer Limited (Pharmacia India Pvt Ltd)	Drop

B. Indirect Acting Agents:

6.1.8.8 Hydroxyamphetamine: *(Syn:* **Norpholedrine; Paredrine; Oksamfetamin; Oxampheta-mine;)**

Structure:

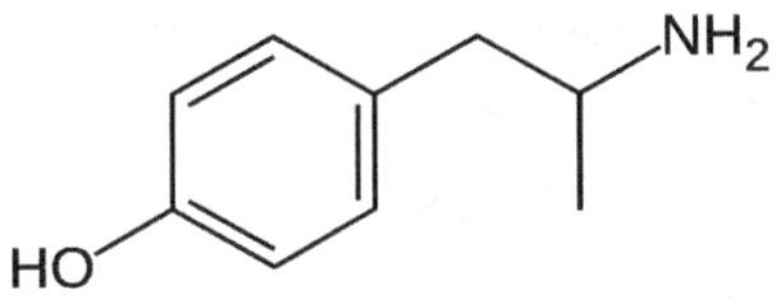

Chemical (IUPAC) name: 4-(2-Aminopropyl)phenol

Hydroxyamphetamine occurs as hydrobromide salt. Hydroxyamphetamine hydrobromide is a water-soluble, white crystalline compound.

Hydroxyamphetamine is synthesised by reducing p-methoxybenzyl methyl ketoxime followed by hydrolysis of methoxy group with HI.

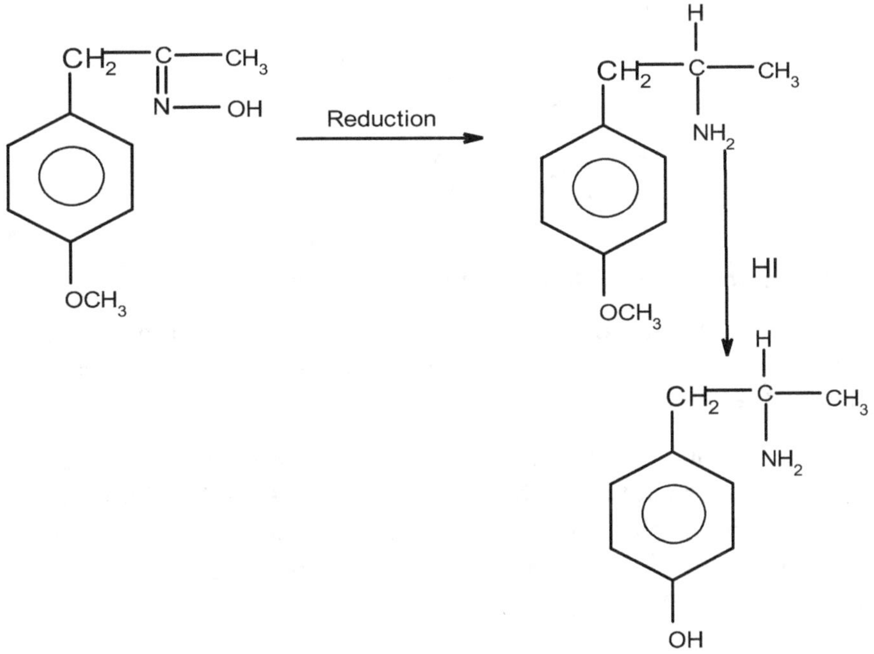

Hydroxyamphetamine possesses α-receptor stimulant activity but lacks CNS activity. It is a powerful vasoconstrictor.

Uses: Hydroxyamphetamine is used in the following conditions;

(a)   Narcolepsy (sudden attack of sleep in completely inappropriate situations)

(b)   Hyperkinetic syndrome in children

(c)   As an anorexiant in the treatment of obesity

Hydroxyamphetamine-tropicamide is a sympathomimetic and anticholinergic combination of eye drops. It works by relaxing the muscles of the eye to cause the pupil to dilate or widen (mydriasis). 4-Hydroxyamphetamine is used in eye drops to dilate the pupil (a process called mydriasis) so that the back of the eye can be examined. This is a diagnostic test for Horner's Syndrome (It is a condition marked by a contracted pupil, drooping upper eyelid, and local inability to sweat on one side of the face, caused by damage to sympathetic nerves on that side of the neck).

Storage Conditions: Store at 20° to 25°C (68° to 77°F). Protect from light.

Pharmaceutical Formulations: Eye Drops

Brand Names: Paredrine, Paremyd.

6.1.8.9 Pseudoephedrine: (*Syn:* **Norpholedrine; Paredrine; Oksamfetamin; Oxamphetamine)**

Structure:

Chemical (IUPAC) name: (S,S)-2-Methylamino- 1-phenylpropan-1-ol

It is a sympathomimetic drug of the phenethylamine and amphetamine chemical classes. It may be used as a nasal/sinus decongestant, as a stimulant, or as a wakefulness-promoting agent in higher doses.

The salts pseudoephedrine hydrochloride and pseudoephedrine sulfate are found in many over-the-counter preparations, either as a single ingredient or (more commonly) in a fixed-dose combination with one or more additional active ingredients such as antihistamines, guaifenesin, dextromethorphan, paracetamol (acetaminophen) or an NSAID (such as aspirin or ibuprofen).

Uses:

1. Pseudoephedrine is a stimulant, but it is well known for shrinking swollen nasal mucous membranes, so it is often used as a decongestant.

2. Pseudoephedrine can be used either as an oral or as a topical decongestant. Due to its stimulating qualities, however, the oral preparation is more likely to cause adverse effects, including urinary retention. According to one study, pseudoephedrine may show effectiveness as an antitussive drug (suppression of cough).

3. Pseudoephedrine is indicated for the treatment of nasal congestion, sinus congestion and Eustachian tube congestion.

4. Pseudoephedrine is also indicated for and as an adjunct to other agents in the optimum treatment of allergic rhinitis, croup, sinusitis, otitis media, and tracheobronchitis

Side Effects: Restlessness, nausea, vomiting, weakness, headache, nervousness, excitability, dizziness, insomnia, stomach pain, difficulty breathing, and changes in heart rate and activity.

Stability and storage conditions: Pseudoephedrine hydrochloride preparation should generally be stored at 15-30°C, freezing of the oral solution should be avoided. Pseudoephedrine hydrochloride tablets and extended-release capsules should be stored in tight containers, and oral solution in tight, light-resistant containers. Pseudoephedrine sulfate extended-release tablets should be protected from light and stored in well-closed containers at 2-30 ° C

Incompatibility: It is incompatible with stimulants (such as caffeine, dextroamphetamine, methamphetamine, herbal products like ephedra, and terbutaline. MAO inhibitors with this drug may cause a serious (possibly fatal) drug interaction such as isocarboxazid, linezolid, methylene blue, moclobemide, phenelzine, procarbazine, rasagiline, selegiline, tranylcypromine. Pseudoephedrine may decrease the effectiveness of blood pressure medications (such as beta-blockers, calcium channel blockers, reserpine, guanethidine, and methyldopa).

Different types of formulation: Tablet, syrup.

Popular brands: It is manufactured by a few companies. Some of the notable brand names in India are:

Sr. No.	Brand Name	Manufacturers	Type
1	Kofron-BR	Agron Remedies Pvt Ltd	Syrup
2	Pseudoephedrine	Unicure (India) Pvt.Ltd.	Tablet
3	Sucor	Ciron Drugs & Pharmaceuticals Pvt Ltd	Syrup

C. Agents with Mixed Mechanism:

6.1.8.10 Ephedrine: (*Syn:* Ephedrin, Ephedrol)

Structure:

Chemical (IUPAC) name: (R,S)-2-(Methylamino)-1-phenylpropan-1-ol

Occurs naturally in many plants, being the principal alkaloid of Ma Huang, which has been used in China for over 2000 years. It has agonist activity at both α and β-receptors. It contains two asymmetric carbon atoms, four compounds are available only racemic, and L-ephedrine is clinically in use.

Ephedrine differs from adrenaline mainly by its,

1. effectiveness after oral administration
2. longer duration of action
3. more pronounced central actions
4. much lower potency

It produces a sharp rise in systolic, diastolic and pulse pressures, with reflex bradycardia, similar to adrenaline but lasting for 10 times as long.

Uses:

1. Bronchospasm: Used orally as a bronchodilator to relieve shortness of breath, chest tightness, wheezing, and cough associated with bronchial asthma.

2. Hypotension: It has been used both for the prevention and treatment of hypotension resulting from spinal anesthesia or other types of non-topical conduction anesthesia.

3. Cardiac arrhythmias and heart block: Management of Adams- stokes syndrome with complete heart block.

4. CNS condition: Treatment of narcolepsy or depressive states.

Side Effects: Tachycardia, cardiac arrhythmias, angina pectoris, vasoconstriction with hypertension flushing, sweating, nausea, decreased urination due to vasoconstriction of renal arteries, restlessness, confusion, insomnia, mild euphoria, mania/, dyspnea, pulmonary oedema dizziness, headache, tremor, hyperglycemic reactions, dry mouth.

Stability and storage conditions: Must be stored in light-resistant containers. Ephedrine base should be stored at or below 8 °C.

Incompatibility: Ephedrine should not be compatible with certain antidepressants, namely norepinephrine-dopamine reuptake inhibitors (NDRIs), as this increases the risk of severe side effects.

Different types of formulation: Injection, tablet

Popular brands: It is manufactured by many companies. Some of the notable brand names in India are:

Sr. No.	Brand Name	Manufacturers	Type
1	Efipres	Neon Laboratories Ltd	Injection
2	Ephedrine	Unicure (India) Pvt. Ltd.	Tablet
3	Ephedrine Hydrochloride	Cyper Pharma	Tablet
4	Sulfidrin	Samarth Pharma Pvt. Ltd.	Injection
5	Tedral SA	Pfizer Limited	Tablet

6.1.8.11 Metaraminol: (*Syn:* **Aramine Metaradrine, Pressonex, L-Metaraminol, Hydroxy-norephedrine, M-Hydroxypropadrin)**

Structure:

1R, 2S-Metaraminol

Chemical (IUPAC) name: 3-[(1R,2S)-2-Amino-1-hydroxypropyl]phenol.

Metaraminol is an isomer of phenylephrine.

Metaraminol is synthesised from *m*-hydroxy benzaldehyde by a selective condensation with nitroethane using tetrabutylammonium fluoride in tetrahydrofuran as a catalyst, followed by a reduction with Raney nickel in formic acid.

Metaraminol

Uses: Metaraminol is used for its pressor action for maintaining blood pressure for the prevention and treatment of hypotension, particularly as a complication of anaesthesia. Metaraminol is also used in the treatment of priapism (Priapism means the persistent and painful erection of the penis).

Side Effects: Serious side effects include fast or slow heartbeat (tachycardia or bradycardia), heart throbbing or pounding, heart ventricle rhythm problems, high blood pressure (hypertension), or trouble breathing.

Stability and storage conditions: Metaraminol injection should be stored at room temperature and protected from light. Stable between 2-8°C for 24-48 hours in an intravenous infusion of Sodium Chloride 0.9% Solution or Glucose 5% Solution. Do not store above 25°C. After dilution, chemical and physical in-use stability has been demonstrated for 48 hours when the diluted product is stored between 2 to 8°C. Freezing and temperatures above 40 °C should be avoided.

Incompatibility: The combination of digitalis and sympathomimetic amines may cause ectopic arrhythmias. Monoamine oxidase inhibitors or tricyclic antidepressants may potentiate the action of sympathomimetic amines. Therefore, when initiating pressor therapy in patients receiving these drugs, the initial dose should be small and given with caution.

Different types of formulation: Injection

Popular brands: Popular brand name is Aramine

QUESTION BANK

A. MULTIPLE CHOICE QUESTIONS

1. β_1 receptor is present in _____.
    A. Adipose tissue
    B. Bronchi
    C. Heart Muscles
    D. All of the above

    Answer: C

2. Which of the following sympathomimetics acts indirectly?
    A. Epinephrine
    B. Norepinephrine
    C. Ephedrine
    D. Hydroxy Amphetamine

    Answer: D

3. Which of the following is a direct-acting sympathomimetic drug?
   A. Salbutamol
   B. Pseudoephedrine
   C. Ephedrine
   D. Metaraminol

   Answer: A

4. Ephedrine is having __________.
   A. Direct action
   B. Indirect action
   C. Mixed action
   D. None of the above.

   Answer: C

5. Albuterol is another name for
   A. Salbutamol
   B. Ephedrine
   C. Adrenaline
   D. Nor-adrenaline

   Answer: A

6. IUPAC name for hydro amphetamine is__________.
   A. 2-(2-aminopropyl)phenol
   B. 4-(2-aminopropyl)phenol
   C. 2-(2-aminoethyl)phenol
   D. 4-(2-aminoethyl)phenol

   Answer: B

7. Ephedrine differs from adrenaline mainly by its,
   A. effectiveness after oral administration
   B. longer duration of action & lower potency
   C. more pronounced central actions
   D. All of the above

   Answer: D

8. Which of the following heterocyclic ring system present in Tetrahydrozoline?
   A. Thiophene
   B. Pyrrole
   C. Imidazole
   D. Pyridine

   Answer: C

9. Which of the following main Active Pharmaceutical Ingredient (API) is present in Privine eye drop?
   A. Naphazoline
   B. Tetrahydrozoline
   C. Hydroxy Amphetamine
   D. Adrenaline

Answer: A

10. Chemically Dopamine is
    A. 3,4-dihydroxyphenylmethylamine
    B. 3,4-dihydroxyphenylethylamine
    C. 3,4-dihydroxyphenylproylamine
    D. 3,4-dihydroxyphenylamine

Answer: B

B. SHORT ANSWER QUESTIONS

1. Classify sympathomimetic agents.
2. Enlist names of direct-acting sympathomimetic agents.
3. Enlist names of indirect-acting sympathomimetic agents.
4. Enlist names of sympathomimetic agents with Mixed Mechanisms.
5. Enlist pharmaceutical formulations available for Epinephrine.
6. Enlist different brand names for Terbutaline.
7. Enlist different brand names for Salbutamol.
8. Write a note on Ephedrine.
9. Write a note on Nor-Epinephrine.
10. Write a note on Pseudoephedrine.

C. LONG ANSWER QUESTIONS

1. Define sympathomimetic agents. Classify it. Write a note on Adrenaline.
2. Give physical properties, pharmaceutical formulation and brand names of Salbutamol.
3. Give physical properties, pharmaceutical formulation and brand names of Terbutaline.

■■■

DRUGS ACTING ON AUTONOMIC NERVOUS SYSTEM: ADRENERGIC ANTAGONISTS

CONTENTS

Adrenergic Antagonists:

A. Alpha Adrenergic Blockers:

Tolazoline

Phentolamine

Phenoxybenzamine

Prazosin

B. Beta-Adrenergic Blockers:

Propranolol

Atenolol

Carvedilol

♦ LEARNING OBJECTIVES ♦

After completing this chapter, the student should be able to understand:

1. Classification of Adrenergic Antagonists.
2. Chemical Structures of different Adrenergic Antagonists.
3. Uses, stability and storage conditions of Adrenergic Antagonists.
4. Different types of formulations and popular brand names for Adrenergic Antagonists.

6.2.1 INTRODUCTION

An **Adrenergic antagonist** is a pharmaceutical substance that acts to inhibit the action of catecholamines at the adrenergic receptors. It is a type of sympatholytic.

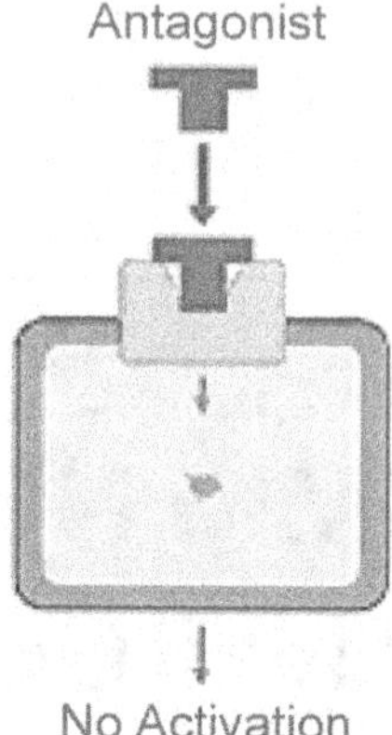

Adrenergic blockers are also called antiadrenergic drugs or sympatholytics. Adrenergic blocking agents prevent the response of effector organs to endogenous as well as exogenous adrenaline and noradrenaline. These drugs block the actions of adrenergic drugs at alpha (α) or beta (β) adrenergic receptors. Many types of adrenergic antagonists are used, and several of these are clinically useful in medicine, particularly in the treatment of cardiovascular diseases. Drugs that decrease the amount of norepinephrine released as a consequence of sympathetic nerve stimulation, as well as drugs that inhibit sympathetic nervous activity by suppressing sympathetic outflow, is also widely used in medications. Almost all of these agents are competitive antagonists in their interactions with either α or β adrenergic receptors, or one exception is phenoxybenzamine, an irreversible antagonist that binds covalently to α-adrenergic receptors. These are due to important structural differences among the various types of adrenergic receptors. Selective β1 antagonist drugs are used to act on the heart, and selective β2 antagonist drugs are used to act on the respiratory system.

These agents competitively antagonise the effects of the catecholamines at α and/or β-adrenergic receptors. Since the sympathetic nervous system is intimately involved in the modulation of a large number of homeostatic mechanisms, interference with its function impairs the capacity of the organism to generate appropriate physiological responses to provocative or adverse environmental conditions. Thus, many of the side effects of these agents are postural hypotension, sedation or depression, increased GIT motility, diarrhoea, impaired ability to ejaculate, increased blood volume and sodium retention.

6.2.2 CLASSIFICATION

Adrenergic blockers are classified on the basis of their chemical structure, as follows:

A. Alpha adrenergic blockers:
   a) Beta-haloalkylamine e.g. Phenoxybenzamine
   b) Imidazole derivatives e.g. Tolazoline, Phentolamine
   c) Quinazolines e.g. Prazosin
   d) Hydrogenated ergot alkaloids e.g. Dihydroergotamine, Methylsergide

B. Beta adrenergic blockers:

a) Nonselective β-blocker e.g. Propranolol, Metipranolol

b) Selective β1-blockers e.g. Atenolol, Metaprolol, Esmolol, Bisoprolol, Betaxolol.

c) β-Blockers with α-blocking property e.g. Labetalol, Carvedilol

Classification of Drugs	
β-Adrenergic blockers	Propranolol, metaprolol, atenolol.
α+β Adrenergic blockers	Labetalol, carvedilol, sodium
α - Adrenergic blockers	Prazosin., terazosin, phentolamine
Central sympatholytic	Clonidine, methyldopa
Vasodilators	Hydralazine, minoxidil
Renin inhibitors	Aliskiren

Fig. 6.2.1 Adrenergic blockers: Classification

6.2.3 MECHANISM OF ACTION

A. Alpha-adrenoceptors antagonists (Alpha-blockers):

These drugs block the effect of sympathetic nerves on blood vessels by binding to alpha-adrenoceptors located on the vascular smooth muscle. Most of these drugs act as competitive antagonists to the binding of norepinephrine that is released by sympathetic nerves synapsing on smooth muscle. Sometimes these drugs are referred to as sympatholytics because they antagonise sympathetic activity. Some alpha-blockers are non-competitive (e.g., phenoxybenzamine), which greatly prolongs their action, whereas others are relatively selective for one type of alpha-adrenoceptor.

The α_1-adrenoceptors are the predominant α-receptor located on vascular smooth muscle. These receptors are linked to Gq-proteins that activate smooth muscle contraction through the IP3 signal transduction pathway. α_1 adrenoceptor antagonists cause vasodilation by blocking the binding of norepinephrine to the smooth muscle receptors.

The α_1-adrenoceptors are the predominant α-receptor located on vascular smooth muscle. These receptors are linked to Gq-proteins that activate smooth muscle contraction through the IP3 signal transduction pathway. α_1 adrenoceptor antagonists cause vasodilation by blocking the binding of norepinephrine to the smooth muscle receptors.

An α_2-adrenergic receptor antagonist prevents the activation of the α2 adrenergic receptor. The α2 receptor is coupled to inhibitory G-proteins, which dissociate from the receptor following agonist binding and inhibit both secondary messenger signalling mechanisms and cell depolarization. Antagonist binding to the α_2-adrenergic receptor prevents secondary messenger inhibition and allows cell depolarization to occur.

Non-selective α_1 and α_2-adrenoceptor antagonists block postjunctional α_1 and α_2-adrenoceptors, which causes vasodilation.

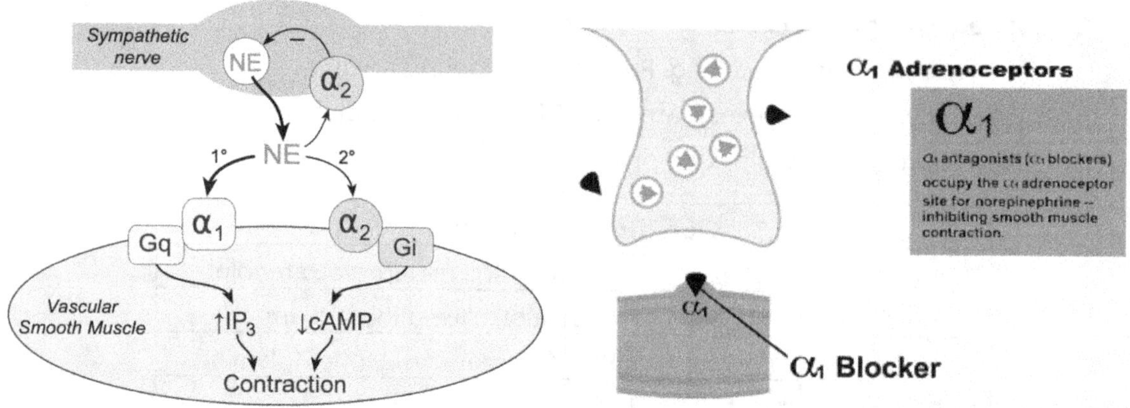

α₁-blockers

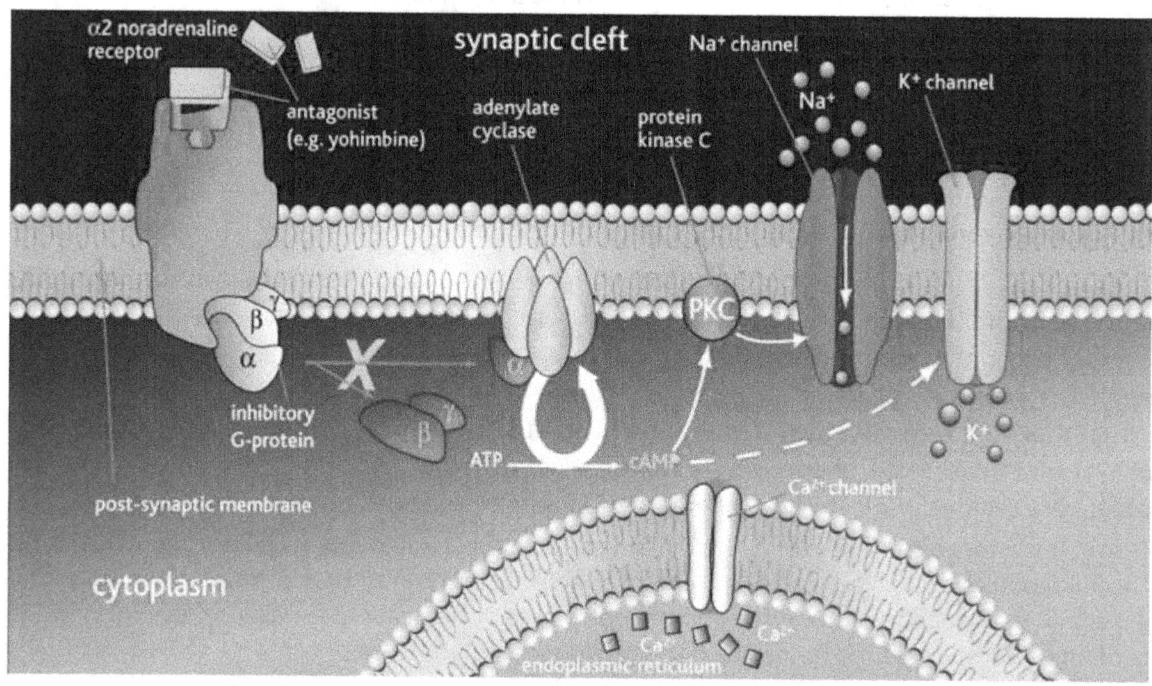

α₂- blockers

Fig. 6.2.2 Mechanism of action of alpha-blockers

B. Beta-adrenoceptors antagonists (Beta-blockers):

Beta-blockers are drugs that bind to beta-adrenoceptors and thereby block the binding of norepinephrine and epinephrine to these receptors. This inhibits normal sympathetic effects that act through these receptors. Therefore, beta-blockers are sympatholytic drugs. Beta-blockers bind to beta-adrenoceptors located in cardiac nodal tissue, the conducting system, and contracting myocytes. The heart has both β1 and β2 adrenoceptors, although the predominant receptor type in number and function is β1. These receptors primarily bind norepinephrine that is released from sympathetic adrenergic nerves. Additionally, they bind norepinephrine and epinephrine that circulate in the blood. Beta-blockers prevent the normal ligand (norepinephrine or epinephrine) from binding to the beta-adrenoceptor by competing for the binding site.

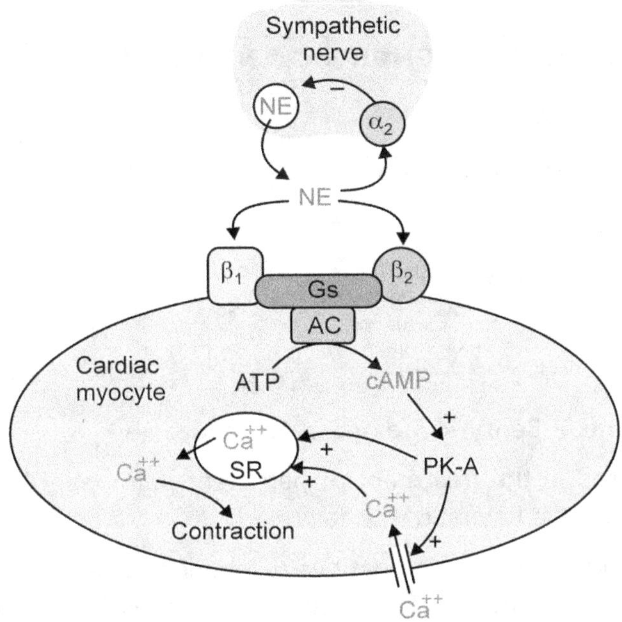

Abbreviations: NE, norepinephrine; Gs, G-stimulatory protein; AC, adenylyl cyclase; PK-A, cAMP-dependent protein kinase; SR, sarcoplasmic reticulum

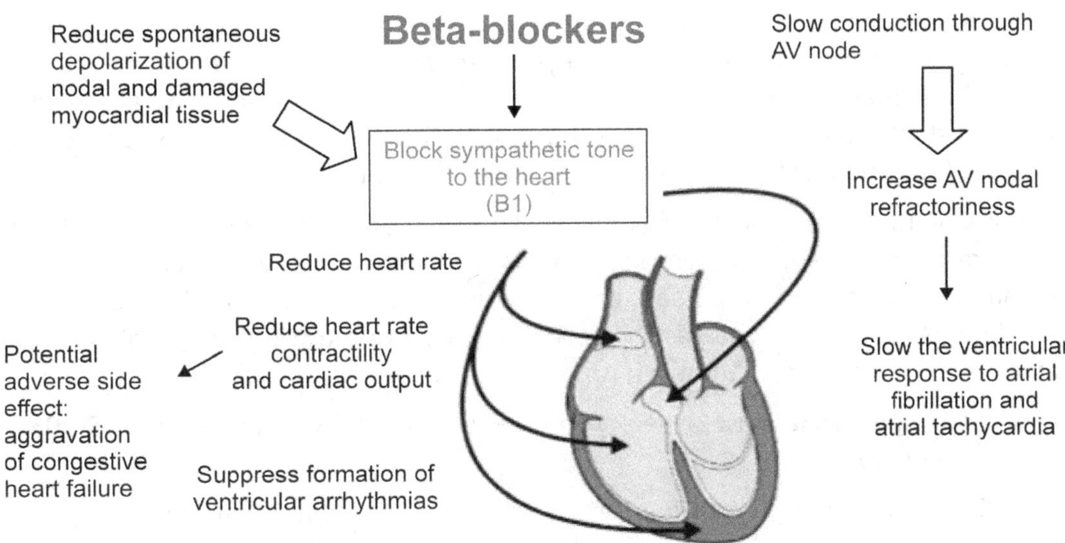

Fig. 6.2.3 Mechanism of action of beta-blockers

The first generation of beta-blockers was non-selective, meaning that they blocked both beta-1 (β_1) and beta-1 (β_2) adrenoceptors. Second-generation beta-blockers are more cardio-selective in that they are relatively selective for β_1-adrenoceptors. Note that this relative selectivity can be lost at higher drug doses. Finally, the third-generation beta-blockers are drugs that also possess vasodilator actions through the blockade of vascular alpha-adrenoceptors.

6.2.4 SOME IMPORTANT ADRENERGIC ANTAGONISTS

A. Alpha Adrenergic Blockers:

6.2.4.1 Tolazoline: (*Syn:*** Priscoline, Benzidazol)**

Structure:

Chemical (IUPAC) name: 2-Benzyl-4,5-dihydro-1H-imidazole

Tolazoline has been used in the treatment of persistent pulmonary hypertension in the newborns. It is mainly used in Raynaud's disease.

Uses: Tolazoline is a non-selective competitive α-adrenergic receptor antagonist. It is a vasodilator used to treat spasms of peripheral blood vessels (as in acrocyanosis means a condition marked by the bluish or purple colouring of the hands and feet, caused by slow circulation). It has also been used (in conjunction with Sodium nitroprusside) successfully as an antidote to reverse the severe peripheral vasoconstriction which can occur as a result of overdose with certain 5-HT$_{2A}$ agonist test drugs such as 25I-NBOMe, DOB and Bromodragonfly (prolonged severe vasoconstriction can lead to gangrene if untreated). It is also useful in the treatment of arteriosclerosis, obliterans, causalgia, peripheral vascular diseases, persistent fetal circulation syndrome, Raynaud disease, scleroderma, systemicspasm, thromboangiitis obliterans, and thrombophlebitis.

Side Effects: Common side effects are abnormally low blood pressure, bleeding of the stomach or intestines, decreased blood platelets, kidney failures, kidney problems causing a decreased amount of urine to be passed, and a low amount of chloride in the blood. Less common side effects are diarrhoea, fast heartbeat, feeling like throwing up, problems with muscles that cause goosebumps, throwing up, and widening blood vessels.

Stability and storage conditions: Store at room temperature between 59-86 ° F (15-30°C) away from light and moisture.

Incompatibility: Dopamine, ephedrine, epinephrine or norepinephrine, metaraminol, methoxamine, or phenylephrine are incompatible with tolazoline.

Different types of formulations available: Injectable Solution, Oral Tablet, Liquid

Popular brands: Tolazoline is found under the Priscoline Hydrochloride as a brand name.

Dose: 25 mg/1mL

Brand Names: Priscoline hydrochloride

6.2.4.2 Phentolamine: (*Syn:* **Regitine, Regitin, Fentolamin, Dibasin, Rogitine)**

Structure:

Chemical (IUPAC) name: 3-[(4,5-Dihydro-1*H*-imidazol-2-ylmethyl)(4-methylphenyl) amino] phenol

Phentolamine is an imidazoline derivative.

Phentolamine has similar affinities for α_1 and α_2 receptors. It produces vasodilation and a fall in blood pressure. However, very large doses of phentolamine can block other actions of 5-HT.

Uses: Phentolamine is used for the diagnosis and treatment of pheochromacytoma. It is also used to treat Raynaud's syndrome. The primary application of phentolamine is for the control of hypertensive emergencies, most notably due to pheochromocytoma. (A phaeochromocytoma is a rare tumour that secretes catecholamines.)

It also has usefulness in the treatment of cocaine-induced cardiovascular complications; it should only be given to patients who do not fully respond to benzodiazepines, nitroglycerin, and calcium channel blockers. When given by injection, it causes blood vessels to expand, thereby increasing blood flow. When injected into the penis (intracavernosal), it increases blood flow to the penis, which results in an erection.

Phentolamine also has diagnostic and therapeutic roles in complex regional pain syndrome (reflex sympathetic dystrophy). Phentolamine has recently been introduced in the dental field as a local anaesthetic reversal agent.

Side Effects: Side effects includes nasal congestion, post-treatment pain, injection site pain, diarrhoea, cardiac dysrhythmia, chest pain, hypotension, myocardial infarction, and CVA - cerebrovascular accident due to cerebral artery occlusion.

Stability and storage conditions: Store in a tight container as defined in the USP-NF. Phentolamine mesylate powder for injection should be stored at 15 - 30 deg C

Incompatibility: Phentolamine is incompatible with tadalafil oral, and there is an increase in the chance of low blood pressure.

Different types of formulations available: Injection

Popular brands: It is manufactured by many companies. Some of the notable brand names in India are:

Sr. No.	Brand Name	Manufacturers	Type
1	Fentanor	Samarth Pharma Pvt. Ltd.	Injection
2	Fentosol	Shree Ganesh Pharmaceuticals	Injection
3	Phentosol	Shree Ganesh Rubber & Chemicals Co	Injection

Other Brand Names: Rogitine, Phentolamine Mesylate

6.2.4.3 Phenoxybenzamine: (*Syn:* **Dibenyline, Dibenzyline, Dibenzyran, Bensylyt)**

Structure:

Chemical (IUPAC) name: (RS)-N-Benzyl-N-(2-chloroethyl)-1-phenoxypropan-2-amine. Phenoxybenzamine is moderately selective for α_1-receptors. It selectively blocks α_1-receptors and has no agonist activity their effects are due to a direct action on α-receptors.

Uses: It is used in the treatment of hypertension, and specifically that caused by pheochromocytoma.

It was also the first alpha-blocker to be used for the treatment of benign prostatic hyperplasia. However, it is currently seldom used for that indication due to unfavourable side effects.

It has been used in the treatment of hypoplastic left heart syndrome.

It is also used in complex regional pain syndrome type 1(CRPS) due to its anti-adrenergic effects.

Phenoxybenzamine is also used to control bladder problems such as urgency, frequency, and inability to control urination with neurogenic bladder, functional outlet obstruction, and partial prostatic obstruction.

Side Effects: Common side effects include stomach upset, nausea, stuffy nose, drowsiness, dizziness or decreased pupil size, nasal congestion, upset stomach, and sexual dysfunction (difficulty ejaculating). Serious side effects include fainting, fast heartbeat, and vomiting.

Stability and storage conditions: Store at room temperature between 59-86 ° F (15-30°C) away from light and moisture.

Incompatibility: It is incompatible and may interact with compounds that stimulate both *alpha*-and *beta*-adrenergic receptors (i.e. epinephrine) to produce an exaggerated hypotensive response and tachycardia.

Dose: 50mg

Different types of formulations are available: Capsule, Injection

Popular brands: It is manufactured by many companies. Some of the notable brand names in India are:

Sr. No.	Brand Name	Manufacturers	Type
1	Fenoben	Celon Labs	Injection
2	Fenoxene	Samarth Pharma Pvt. Ltd.	Capsule, Injection

6.2.4.4 Prazosin (*Syn:* Furazosin, Prazosine)

Structure:

Chemical (IUPAC) name: [4-(4-Amino-6,7-dimethoxy-2-quinazolinyl)-1-piperazinyl](2-furyl)methanone

Prazosin, the first known selective α_1-blocker, was discovered in the late 1960s. It is a far more potent α_1-receptor blocker.

Uses: Prazosin blocks the postsynaptic α_1-adrenoreceptors causing vasodilation and reduction in blood pressure. Prazosin is used to treat all grades of hypertension.

It is also useful in treating urinary hesitancy associated with prostatic hyperplasia (abnormal increase in number of cells), blocking alpha-1 receptors, which control constriction of both the prostate and urethra.

Side Effects: Dizziness, headache, drowsiness, lack of energy, weakness, palpitations, nausea.

Stability and storage conditions: Store at room temperature between 59-86 ° F (15-30°C) away from light and moisture.

Incompatibility: It is incompatible with Avanafil (Stendra), sildenafil (Viagra, Revatio), tadalafil (Cialis, Adcirca), vardenafil (Levitra, Staxyn), verapamil (Calan, Covera, Isoptin, Verelan).

Dose: 2.5 mg, 5 mg

Different types of formulations available: Capsule, tablet

Popular brands: It is manufactured by many companies. Some of the notable brand names in India are:

Sr. No.	Brand Name	Manufacturers	Type
1	Cyber - Cr	Raptakos Brett & Co.	Modified Release capsule, tablet
2	Minipress Xl	Pfizer	Capsule/ Tablet
3	Prazopress	Sun Pharmaceuticals	Capsule/ Tablet
4	Prazocip -1	Cipla Limited	Tablet
5	Prazovin-xl 5	Bison Biotec Pvt Ltd	Tablet

Other Brand Names: Czoprexx-XL, Prazocip-XL, Gits.

B. Beta Adrenergic Blockers:

6.2.4.5 Propranolol: (*Syn:* ***Propanolol, Dociton, Beta-Propranolol, Euprovasin, Propanalol***)

Structure:

Chemical (IUPAC) name: 1-Naphthalen-1-yloxy-3-(propan-2-ylamino)propan-2-ol.

Black and co-workers discovered propranolol, which is a structural relative to pronethalol. The levo isomer is 100 times more active than the dextro isomer. The first pass effect is substantial, and the 4-OH metabolite is active. Propranolol is a mixed β-blocker. S-isomer is the most active.

Propranolol is a prototype and non-selective β-blocker. It blocks β_1 and β_2 receptors with equal affinity.

Uses: Propranolol is used to treat tremors, angina (chest pain), hypertension (high blood pressure), heart rhythm disorders, and other heart or circulatory conditions. Hemangeol (propranolol oral liquid 4.28 milligrams) is given to infants who are at least 5 weeks old to treat a genetic condition called infantile hemangiomas. Hemangiomas are caused by blood vessels grouping together in an abnormal way.

Side Effects: More common - in children, cough producing mucus, difficulty with breathing, tightness in the chest. Others include chronic trouble sleeping, feeling weak, low energy, bronchospasm, chronic heart failure, depression, dizziness.

Stability and storage conditions: Store at room temperature between 59-86°F (15-30°C) away from light and moisture.

Incompatibility: It is incompatible with reserpine and results in hypotension, marked bradycardia, vertigo, syncopal attacks, or orthostatic hypotension, with calcium-channel blocking drug, depress myocardial contractility or atrioventricular conduction, with intravenous use of a beta blocker and verapamil has resulted in severe adverse reactions, with digitalis glycosides and beta-blockers decrease heart rate and increase the risk of bradycardia. Hypotension and cardiac arrest were observed with the concomitant use of propranolol and haloperidol. Aluminium hydroxide gel greatly reduces intestinal absorption of propranolol.

Dose: The dose is 10 mg, 20 mg, 40 mg, 80 mg, and 2 mg/mL, 5 mg/mL 20mg/0.25mg

Different types of formulations are available: Tablet, capsule, injection

Popular brands: It is manufactured by many companies. Some of the notable brand names in India are:

Sr. No.	Brand Name	Manufacturers	Type
1	Betacap	Sun Pharmaceuticals	Capsule/ Tablet
2	Ciplar	Cipla	Capsule/ Tablet
3	Corbeta	Sarabhai Chemicals	Capsule/ Tablet
4	Pranosol	SG Pharma	Injection
5	Propranolol (Samarth)	Samarth Pharma	Injection

Other Brand Names: Arminol, Syziral, Inderal, Dizepax, Alpron Plus, Besprol, Arminol Inderal, Dizepax-M, Inderal, Migrates, Syziral M-20

6.2.4.6 Atenolol: (*Syn:* **Tenormin, Tenormine, Normiten, Blokium**)

Structure:

Systematic (IUPAC) Name: 2-[4-[2-hydroxy-3-(propan-2-ylamino)propoxy]phenyl] acetamide

It has significantly fewer CNS side effects compared to propranolol. Atenolol is the hydrophilic β_1-selective antagonist. The drug belongs to the group of beta-blocker used in the treatment of cardiovascular disease and hypertension.

Uses: Atenolol is used for a number of conditions, including hypertension, angina, long QT syndrome, acute myocardial infarction, supraventricular tachycardia, ventricular tachycardia, and the symptoms of alcohol withdrawal

Side Effects: Side effects include feeling short of breath, even with mild exertion; swelling of your ankles or feet; nausea, stomach pain, low fever, loss of appetite, dark urine, clay-coloured stools, jaundice (yellowing of the skin or eyes), depression.

Stability and storage conditions: Atenolol tablets should be protected from heat, light, and moisture and stored in well-closed, light-resistant containers at 20-30°C. Atenolol injection should be stored at room temperature between 20-25°C and protected from light.

Incompatibility: It is incompatible with calcium channel blockers and produces an additive effect, with disopyramide responsible for severe bradycardia, asystole and heart failure,

Beta-blockers may exacerbate the rebound hypertension, which can follow the withdrawal of clonidine. Concomitant use of prostaglandin synthase inhibiting drugs, e.g., indomethacin, may decrease the hypotensive effects of beta blockers. Concomitant use of digitalis glycosides can increase the risk of bradycardia.

Dose: 25 mg, 50 mg

Different types of formulations are available: Tablet, Injection.

Popular brands: It is manufactured by many companies. Some of the notable brand names in India are:

Sr. No	Brand Name	Manufacturers	Type
1	Aten	(Zydus Cadila)Zydus Cadila Healthcare Ltd	Tablet
2	Atenolol	Abbott Healthcare Pvt Ltd (AHPL)	Tablet
3	Atenova	Lupin Laboratories Ltd	Tablet
4	Atepose (50 mg)	Pharma Synth Formulations Ltd	Tablet
5	Atpark (100 mg)	Pfizer Limited (Pharmacia India Pvt Ltd)	Tablet
6	BP ACT (25 mg)	Active Healthcare	Tablet
7	Dilcare (25 mg)	Ind-Swift Limited	Tablet
8	O Beta (25 mg)	Intas Laboratories Pvt Ltd	Tablet
9	Tenormin(50mg) (AHPL)Tablet	Abbott Healthcare Pvt Ltd	Tablet

Other Brand Names: Malodip-AT, Ziblok, Atecard, Amlogla-Plus, Beta-Bloc, Espil-AT, Amchek-AT, BPC-AT, Cheerten-H, Cadpres, Ampine

6.2.4.7 Carvedilol: (*Syn:* **Coreg, Dilatrend, Carvedilolum, Eucardic**)

Structure:

Chemical (IUPAC) name: (±)-1-(9*H*-Carbazol-4-yloxy)-3-[2-(methoxyphenoxy)ethylamino]propa-2-ol.

It acts on both β and α-adrenergic receptors. Only δ isomer is β–blocking, and both enantiomers have α-blocking activity.

Uses: Carvedilol is indicated in the management of congestive heart failure (CHF), commonly as an adjunct to angiotensin-converting enzyme inhibitors (ACE inhibitors) and diuretics. Carvedilol is also indicated in the treatment of hypertension. It can be used alone or with other antihypertensive agents.

Side Effects: Dizziness, fatigue, low blood pressure, diarrhoea, weakness, slowed heart rate, weight gain and shortness of breath.

Severe side effects may include bronchospasm. Safety during pregnancy or breastfeeding is unclear.

Stability and storage conditions: Store at 20° to 25°C (68° to 77° F) in a tight container as defined in the USP-NF in a dry place.

Dose: 3.125 mg, 6.25 mg, 6.5 mg, 12.5 mg, 50 mg,

Incompatibility: Conduction disturbance has been observed when COREG is coadministered with diltiazem. Agents with β -blocking properties may enhance the blood-sugar-reducing effect of insulin and oral hypoglycemics.

Different types of formulations available: Tablet, Capsule

Popular brands: It is manufactured by many companies. Some of the notable brand names in India are:

Sr. No.	Brand Name	Manufacturers	Type
1	Caditone	Kopran Pharma Ltd	Tablet
2	Carca	Intas Pharmaceuticals Ltd.	Tablet
3	Cardivas	Sun Pharmaceutical Industries Ltd.	Tablet

Table *Contd...*

Sr. No.	Brand Name	Manufacturers	Type
4	Carloc	Cipla Limited	Tablet
5	Carvas	Medley Pharmaceuticals Ltd.	Tablet
6	Carvedil	Alkem Laboratories Ltd	Tablet
7	Carvetrend	Nicholas Piramal India Ltd.	Tablet
8	Carzec	Glaxo Smith kline Pharmaceuticals Ltd.	Tablet
9	Cevas	Cadila Pharmaceuticals Ltd.	Tablet
10	Coslot	Ranbaxy Diagnostics	Tablet
11	Carvil	Zydus Cadila Healthcare Ltd	Tablet

QUESTION BANK

A. MULTIPLE CHOICE QUESTIONS

1. Which of the following heterocyclic ring system is present in Tolazoline?
   A. Imidazole
   B. Furan
   C. Pyrrol
   D. Pyridine

   Answer: A

2. Which of the following alpha-adrenergic blocker is Quinazoline derivative?
   A. Tolazoline
   B. Prazosin
   C. Phentolamine
   D. Propranolol

   Answer: B

3. Which of the following is not an alpha-adrenergic blocker?
   A. Tolazoline,
   B. Phentolamine
   C. Propranolol
   D. Prazosin

   Answer: C

4. Which of the following is not a beta-adrenergic blocker?
   A. Propranolol
   B. Atenolol

    C. Carvedilol

    D. Tolazoline

Answer: D

5. Propranolol is ______________.

    A. Nonselective β-blocker

    B. Nonselective α-blocker

    C. Selective β-blocker

    D. Selective α-blocker

Answer: A

6. Which one of the following is Selective β_1-blockers?

    A. Propranolol

    B. Atenolol

    C. Labetalol

    D. Carvedilol

Answer: B

7. Which of the following adrenergic blocker is an imidazoline derivative?

    A. Tolazoline

    B. Prazosin

    C. Phentolamine

    D. Propranolol

Answer: C

8. Phenoxybenzamine is ____________________.

    A. Beta halo alkyl amine

    B. Imidazole derivatives

    C. Quinazolines

    D. Hydrogenated ergot alkaloids

Answer: A

9. Which of the following is β–blocker with α-blocking property?

    A. Propranolol,

    B. Atenolol,

    C. Metaprolol,

    D. Labetolol

Answer: D

10. Selective β_1 antagonist drugs used to act on
    A. Respiratory system
    B. Digestive system
    C. Heart
    D. Nervous system

Answer: C

B. SHORT ANSWER QUESTIONS

1. Classify Adrenergic Blockers.
2. Enlist names of Alpha-Adrenergic Blockers.
3. Enlist names of Beta-Adrenergic Blockers.
4. Enlist pharmaceutical formulations available for Propranolol.
5. Enlist different brand names for propranolol.
6. Write a note on Prazosin.
7. Write a note on Atenolol.
8. Enlist different brand names for Carvedilol
9. Enlist pharmaceutical formulations available for Phentolamine
10. Write a note on Phentolamine.

C. LONG ANSWER QUESTIONS

1. Explain in detail the adrenergic receptor antagonist
2. Classify Adrenergic Blockers. Add a note on "Propranolol as Beta Adrenergic Blockers."
3. Classify Adrenergic Blockers. Give Storage condition, pharmaceutical formulation and popular dosage form of Atenolol.

■ ■ ■

DRUGS ACTING ON AUTONOMIC NERVOUS SYSTEM: CHOLINERGIC DRUGS AND RELATED AGENTS

CONTENTS

Brief introduction of Drugs acting on Autonomic Nervous System-the Cholinergic Drugs and Related Agents or parasympathomimetic agents

Direct Acting Agents:

- Acetylcholine*
- Carbachol *
- Pilocarpine

Cholinesterase Inhibitors:

- Neostigmine*
- Edrophonium Chloride
- Tacrine Hydrochloride
- Pralidoxime Chloride
- Echothiopate Iodide

♦ LEARNING OBJECTIVES ♦

After completing this chapter, the student should be able to understand:

1. The Cholinergic or parasympathomimetic nervous system and its classification
2. The drugs acting on the Cholinergic nervous system or the parasympathomimetics and their classification
3. Structure-activity relationship for cholinergic drugs
4. Mechanism of action of cholinergic drugs/parasympathomimetics
5. Important drugs used cholinergic drugs, their chemical structures, names, uses, incompatibilities, stability and storage conditions and brands.

6.3.1 INTRODUCTION

Cholinergic drugs are medications that produce the same effects as the parasympathetic nervous system.

Choline refers to the various quaternary ammonium movement of fats into the cells. The richest sources of choline are liver, kidney, brain, wheat germ, brewer's yeast, and egg yolk. Neurologically **cholinergic** word is derived from **acetylcholine.** Since it uses acetylcholine to send its messages, and hence the system is also known as the cholinergic system. Neuromuscular junctions, preganglionic neurons of the sympathetic nervous system, the basal forebrain and brain stem complexes are involved in the cholinergic system. In addition, the receptors for the merocrine sweat glands are also part of the cholinergic system since acetylcholine is released from postganglionic sympathetic neurons. In neuroscience and related fields, the term cholinergic is used for a substance (or ligand) if it is capable of producing, altering, or releasing acetylcholine ("indirect-acting") or mimicking its behaviour at one or more of the body's acetylcholine receptor types ("direct-acting"). Such mimics are called parasympathomimetic drugs or cholinomimetic drugs. A receptor is cholinergic if it uses acetylcholine as its neurotransmitter. A synapse is cholinergic if it uses acetylcholine as its neurotransmitter.

The cholinergic system is further classified as assaults containing the *N, N, N*-trimethylethanol ammonium cation. Choline is a primary component of the neurotransmitter acetylcholine and functions with inositol as a basic constituent of lecithin. It prevents fat deposits in the liver and facilitates the:

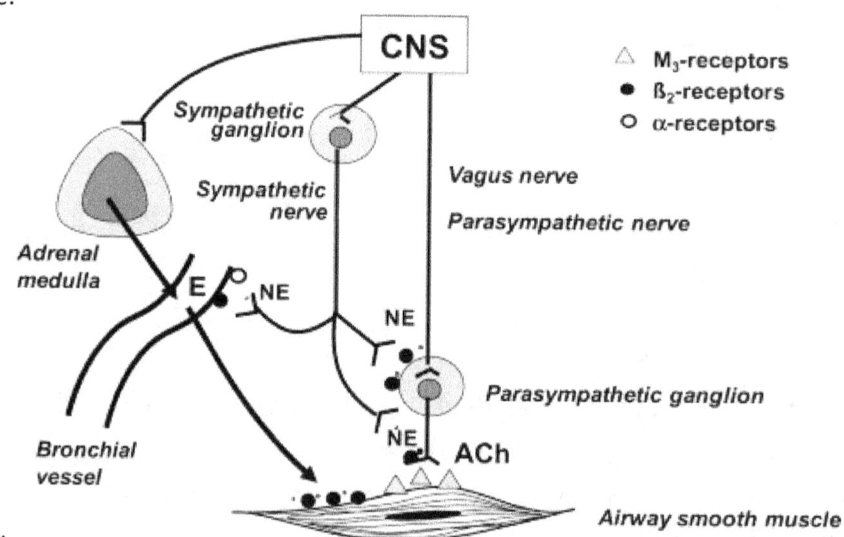

Fig. 6.3.1 Classification of Cholinergic nervous system

Cholinergic receptors are divided into two types muscarinic and nicotinic receptors. They are distinguished from each other on the basis of their different affinities for agents that mimic/increase the action of acetylcholine.

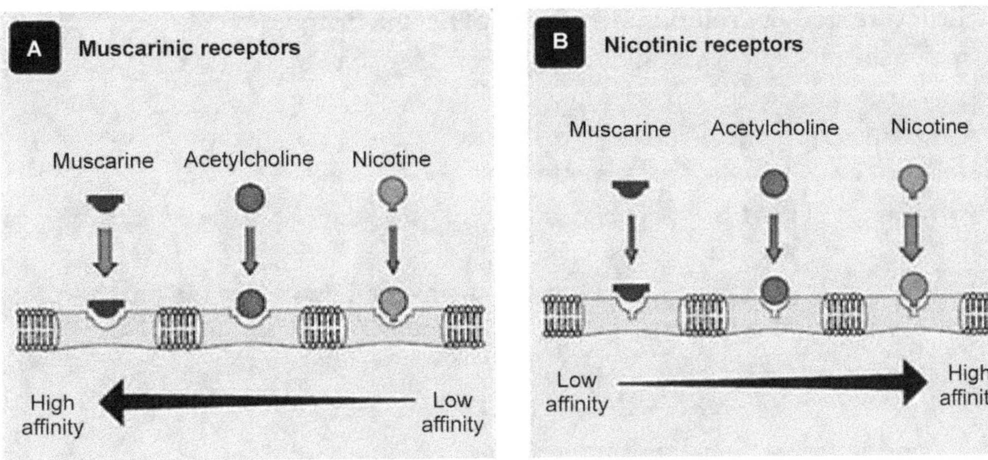

Fig. 6.3.2 Types of cholinergic receptors

6.3.2 CLASSIFICATION OF CHOLINERGIC/PARASYMPATHOMIMETIC DRUGS

Cholinergic drugs/parasympathomimetics are divided into two types. These include,

A) Direct-Acting Cholinergic Agonists:

    a)   Choline esters: *e.g.,* acetylcholine, *methacholine, carbachol, bethanechol*

    b)   Alkaloid: *e.g., Pilocarpine*

    c)   Others: *e.g. Aceclidine*

B) Indirect-Acting Cholinergic Agonists:

    a) Reversible choline esterase inhibitors: *e.g., Physostigmine, neostigmine, edrophonium*

    b) Irreversible choline esterase inhibitors: *e.g., Ecothiophate*

Cholinergic agonist- Classification		
Direct Acting Cholinergic Drug		
• Acetylcholine		
• Betanechol		
• Pilocarpine		
• Methacholine		
Indirect Acting Cholinergic Drugs		
• **Reversible:** *water soluble-*	Neostigmine, Edrophonium Pyridostigmine,	
Lipid soluble-	*Physostigmine, Donepezil, Tacrine, Gallantamine*	
• **Irreversible-**	Organophosphorous Compounds, *Echothiophate, malathion, parathion, tabum*	
Reactivation of acetylcholinesterase- Pralidoxime		

Fig. 6.3.3 Classification of cholinergic drugs

6.3.2.1 Structure-activity relationship for cholinergic drugs:

1. A molecule must possess a nitrogen atom capable of bearing a positive charge, preferably a quaternary ammonium salt.
2. For maximum potency, the size of the alkyl groups substituted on the nitrogen should not exceed the size of a methyl group.
3. The molecule should have an oxygen atom, preferably an ester-like oxygen capable of participating in a hydrogen bond.
4. A two-carbon unit should occur between the oxygen atom and the nitrogen atom.

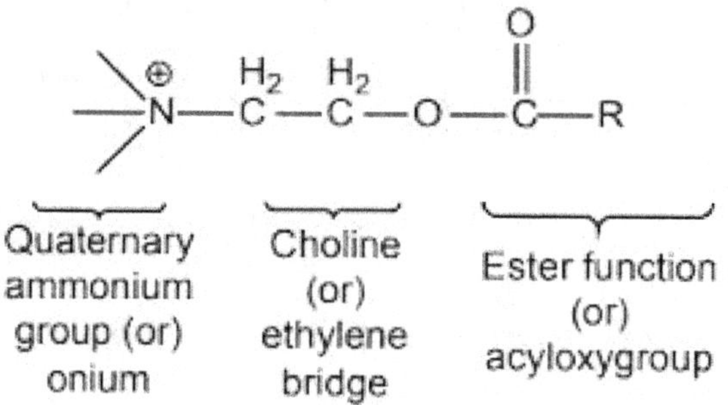

Fig. 6.3.4 Structure-activity relationship for cholinergic drugs

Cholinergic drugs act at different locations in autonomic and somatic nervous systems, as depicted below.

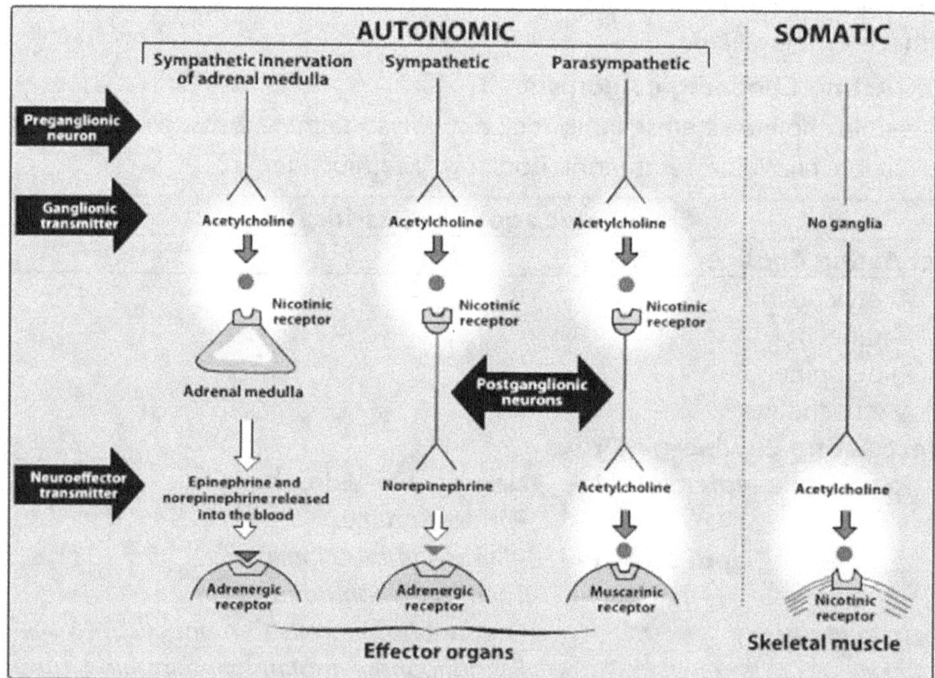

Fig. 6.3.5 Sites of actions of cholinergic agonists in the autonomic & somatic nervous systems

6.3.2.2 Mechanism of action of cholinergic drugs/parasympathomimetics:

Cholinergic agonists (also known as parasympathomimetics) mimic the effects of acetylcholine by binding directly to cholinoreceptors. All of the direct-acting cholinergic drugs have longer durations of action than acetylcholine. Some of them preferentially bind to muscarinic receptors and are sometimes referred to as muscarinic agents.

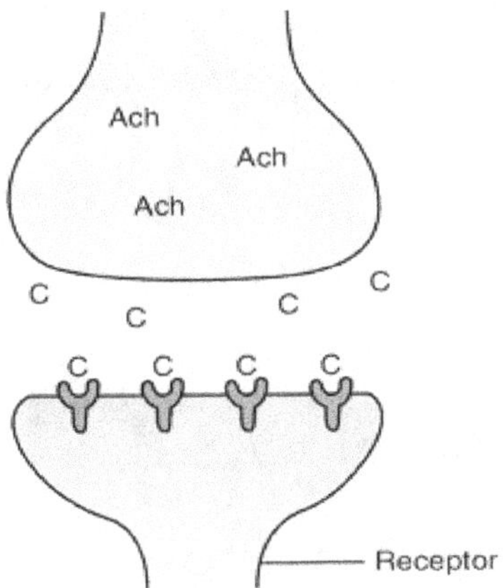

Fig. 6.3.6(a) Direct cholinergic agonists

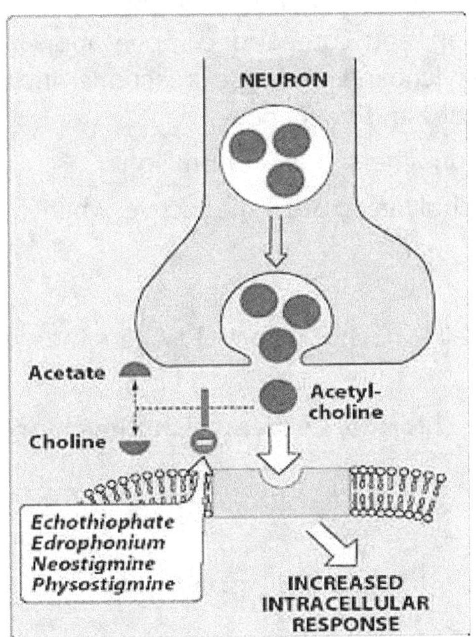

Fig. 6.3.6(b) Indirect cholinergic agonists

Acetylcholinesterase is an enzyme that specifically cleaves acetylcholine to acetate and choline and terminates its actions. Inhibitors of acetylcholinesterase indirectly provide a cholinergic action by prolonging the lifetime of acetylcholine produced endogenously at the cholinergic nerve endings. This results in the accumulation of acetylcholine in the synaptic space. These drugs can thus provoke a response at all cholinoceptors in the body, including both muscarinic and nicotinic receptors of the ANS, as well as at neuromuscular junctions and in the brain. A number of synthetic organophosphate compounds have the capacity to bind covalently to acetylcholinesterase. The result is a long-lasting increase in acetylcholine at all sites where it is released.

6.3.3   SOME IMPORTANT CHOLINERGIC/PARASYMPATHOMIMETIC DRUGS

6.3.3.1 Acetylcholine (*Syn:* Choline acetate, O-Acetylcholine, Acetylcholine ion, Choline acetate (ester), Acetylcholinum)

Structure:

Systematic (IUPAC) Name: 2-Acetyloxyethyl(trimethyl)azanium

Uses: It is useful to obtain miosis of the iris in seconds after delivery of the lens in cataract surgery, penetrating keratoplasty, iridectomy and other anterior segment surgery where rapid miosis may be required.

Side Effects: Side effects include corneal oedema, corneal clouding, corneal decompensation, bradycardia, flushing, hypotension, breathing difficulty, sweating, infrequent cases of corneal oedema, corneal clouding, and corneal decompensation have been reported with the use of intraocular acetylcholine. Adverse reactions include bradycardia, hypotension, flushing, breathing difficulties and sweating.

Stability and storage conditions: Store in airtight containers. Protect from light.

Incompatibility: Acetylcholine chloride and carbachol have been ineffective when used with topical nonsteroidal anti-inflammatory agents.

Different types of formulation: Solution, drop.

Popular brands: Popular brand names are Miochol-E, Miochol, Miochol-E/Steri-Tags, and Miochol-E System Pak.

6.3.3.2 Carbachol (*Syn:* **Miostat, Carbamylcholine chloride, Carbamoylcholinechloride, Carbacholine; Carbocholine**)

Structure:

Systematic (IUPAC) Name: 2-Carbamoyloxyethyl(trimethyl)azanium;chloride

Uses: It is useful in open-angle glaucoma and reduction of elevated intraocular pressure (iop) in patients with primary, open-angle (chronic simple, non-congestive) glaucoma. Used mainly in patients who are hypersensitive to pilocarpine. Use concurrently with sympathomimetic agents, β-adrenergic blocking agents, or carbonic anhydrase inhibitors.

Useful in ocular surgery production of miosis during surgery on the anterior chamber of the eye (e.g., cataract extraction, keratoplasty, peripheral iridectomy, cyclodialysis).

Side Effects: Side effect includes veil or curtain appearing across part of the vision, diarrhoea, stomach cramps or pain, or vomiting, fainting, flushing or redness of the face, frequent urge to urinate, increased sweating, irregular heartbeat, shortness of breath, wheezing, or tightness in chest, unusual tiredness or weakness, watering of mouth.

Stability and storage conditions: Store in airtight containers. Protect from light.

Incompatibility: Incompatible with miotics, anticholinesterase.

Different types of formulation: Injection

Popular brands: Some popular brand names are Isopto Carbachol, Miostat, Mioticol

6.3.3.3 Pilocarpine (*Syn:* **Isoptocarpine, Pilokarpin, Pilocarpin, Pilocarpol, Pilokarpol**)

Structure:

Systematic (IUPAC) Name: (3S, 4R)-3-Ethyl-4-[(3-methylimidazol-4-yl)methyl]oxolan-2-one

Uses: Pilocarpine is a drug used to treat dry mouth (xerostomia) and glaucoma. Pilocarpine is also used to reduce the possibility of glowering at night from lights when the patient has undergone implantation of phakic intraocular lenses (a lens which is placed over the existing natural lens). Pilocarpine is used to stimulate sweat glands in a sweat test to measure the concentration of chloride and sodium that is excreted in sweat. It is used to diagnose cystic fibrosis.

Side Effects: Common side effects of pilocarpine include urinary frequency, nausea, dizziness, weakness, rhinitis, headache, flushing, chills, and diaphoresis. Symptoms of overdose include chest pain, confusion, diarrhoea, fainting, fast, slow, or irregular heartbeat, headache, nausea or vomiting, shortness of breath, troubled breathing, stomach cramps or pain, tiredness or weakness, trembling or shaking, and trouble seeing. More common side effects include chills, cough, diarrhoea, feeling of warmth or heat, fever, flushing or redness of the skin, especially on the face and neck, increased need to urinate, indigestion, joint pain, muscle aches and pains, nausea, passing urine more often, runny nose, unusual tiredness or weakness, sweating.

Stability and storage conditions: Keep in the well-closed container & protected from light.

Incompatibility: Pilocarpine should be administered with caution with beta-adrenergic antagonists because of the possibility of showing incompatibility and being responsible for disturbances in conduction. Drugs with parasympathomimetic effects administered concurrently with pilocarpine would be expected to result in additive pharmacologic effects. Pilocarpine antagonises the anticholinergic effects of drugs used concurrently.

Different types of formulation: Solution, drop.

Popular brands: It is manufactured by many companies. Some of the notable brand names in India are:

Sr.No.	Brand Name	Manufacturers	Type
1	Andre Carpine	Intas Laboratories Pvt Ltd	Drop
2	Carpinol Sol.	Sunways India Pvt. Ltd.	Solution
3	CarpoMiotic	Bell Pharma Pvt. Ltd.	Drop
4	Locarp	Cadila Pharmaceuticals Ltd. (LeSante)	Drop
5	Pilagan	Allergan India Ltd	Drop
6	Pilodrops	Micro Vision	Drop
7	Pilorin Eye	Klar Sehen Pvt. Limited	Drop

6.3.3.4 Neostigmine (*Syn:* **Eustigmin, Eustigmin, Prostigmin, Prostigmin, Vagostigmine)**

Structure:

Systematic (IUPAC) Name: [3-(Dimethylcarbamoyloxy)phenyl]-trimethylazanium

Uses: Neostigmine improves muscle strength in patients with a certain muscle disease (myasthenia gravis). Also useful in anaesthesia to reverse the effects of non-depolarizing muscle relaxants such as rocuronium and vecuronium at the end of an operation, usually in a dose of 25 to 50 µg per kilogram. It can also be used for urinary retention resulting from general anesthesia and to treat drug toxicity. Sometimes administer a solution containing neostigmine intravenously to delay the effects of envenomation (the process by which venom is injected) through snakebite.

Side Effects: Side effects include nausea, vomiting, diarrhoea, abdominal cramps, increased saliva/mucus, decreased pupil size, increased urination, or increased sweating, difficulty speaking, dizziness, drowsiness, gas, headache, joint pain, muscle twitching, and weakness may occur.

Stability and storage conditions: Keep in a well-closed container & protected from light.

Incompatibility: It is incompatible with aminoglycoside antibiotics (e.g., gentamicin), antiarrhythmic medications (e.g., quinidine), and drugs that decrease the movements of the stomach (e.g., antihistamines such as diphenhydramine, antispasmodics such as dicyclomine, narcotic pain medications such as morphine).

Different types of formulation: Tablet, injection.

Popular brands: It is manufactured by many companies. Some of the notable brand names in India are:

Sr.No.	Brand Name	Manufacturers	Type
1	Beemine	Biomiicron Pharmaceuticals	Injection
2	Myostigmin	Neon Laboratories Ltd	Injection
3	Neostigmine	Piramal Healthcare	Injection
4	Prostigmin	Nicholas Piramal India Ltd.	Tablet
5	Stimin	Celon Labs	Injection
6	Tilstigmin	Tablets (India) Limited	Tablet
7	Ostig	Oboi Laboratories	Injection

6.3.3.5 Edrophonium chloride (*Syn:* Tensilon chloride, Edrofoniocloruro (DCIT), Edrophoniichloridum)

Structure:

Systematic (IUPAC) Name: Ethyl-(3-hydroxyphenyl)-dimethylazanium;chloride

Uses: Edrophonium is used to differentiate myasthenia gravis from cholinergic crisis and Lambert-Eaton myasthenic syndrome. Edrophonium, an effective acetylcholinesterase inhibitor, is useful for reducing muscle weakness. The drug may also be used for postoperative decurarization.

Side Effects: Side effects include arrhythmias, fall in cardiac output leading to hypotension, increased trachea bronchial secretions, laryngospasm, bronchiolar constriction and respiratory muscle paralysis, convulsions, dysarthria, dysphonia and dysphagia, nausea, vomiting, increased peristalsis, increased gastric and intestinal secretions, diarrhoea, abdominal cramps, increased urinary frequency, diaphoresis, increased lacrimation, pupillary constriction, diplopia, and conjunctival hyperemia.

Stability and storage conditions: Store below 40 °C (104 °F), preferably between 15 and 30 °C (59 and 86 °F).

Incompatibility: ENLON-PLUS Injection is incompatible with nondepolarising muscle relaxants. Narcotic analgesics, except when combined with potent inhaled anaesthetics, appear to potentiate the effect of edrophonium on the sinus node and conduction system, increasing both the frequency and duration of bradycardia. With beta-adrenergic blocking agents, there is an increased risk for excessive bradycardia from the unopposed parasympathetic vagal tone. Digitalis glycoside (when used concurrently with edrophonium, the additive vagomimetic effects may cause excessive bradycardia). Neuromuscular blocking agents (phase I block of depolarising neuromuscular blocking agents such as succinylcholine may be prolonged when used concurrently with edrophonium. Effects of many nondepolarising neuromuscular blocking agents are antagonised by edrophonium, especially moderate degrees of residual neuromuscular blockade.

Different types of formulation: Injection

Popular brands: Popular brand names include Antirex, Edrophonium Alliance, Enlon, and Anticude.

6.3.3.6 Tacrine hydrochloride (*Syn:* **Hydroaminacrine**)

Structure:

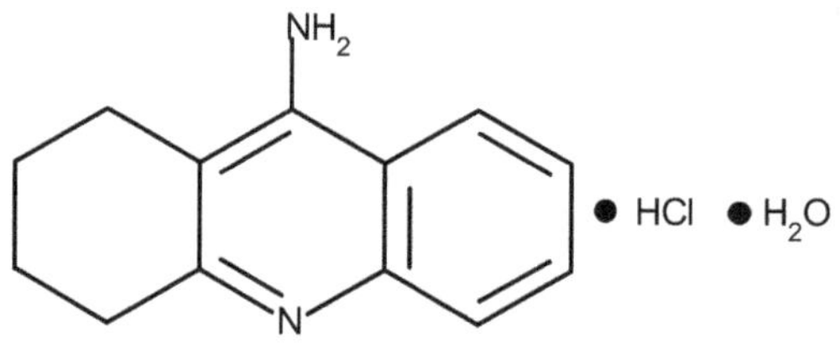

Systematic (IUPAC) Name: 1, 2, 3, 4-Tetrahydroacridin-9-amine; hydrochloride

Uses: Useful in Alzheimer's Disease Palliative treatment of mild to moderate dementia of the Alzheimer's type (Alzheimer's disease, presenile or senile dementia)

Side Effects: Common side effects include elevated aminotransferases, nausea, vomiting, diarrhoea, anorexia, dyspepsia, myalgia and ataxia.

Stability and storage conditions: Store at 15–30°C. Protect from moisture.

Incompatibility: It is metabolised primarily by CYP1A2, dosage adjustments may be necessary with concomitant administration of drugs that affect or are metabolised by this system. Antacids (magnesium- or aluminium-containing) show incompatibility and change in tacrine bioavailability is unlikely. With anticholinergic agents, it shows interference and affects the activity of anticholinergic agents. Warfarin's effect on anticoagulant activity is unlikely.

Different types of formulation: Capsule

Popular brands: Popular brand name is Cognex, manufactured by First Horizon

6.3.3.7 Pralidoxime chloride (Syn: **2-PAM chloride, Pralidoxine chloride, ComboPen, Protopam chloride, Pralidoxim**)

Structure:

Systematic (IUPAC) Name: [(E)-(1-Methylpyridin-2-ylidene)methyl]-oxoazanium; chloride

Uses: Pralidoxime Chloride is specifically indicated for intramuscular use as an adjunct to atropine in the treatment of poisoning by nerve agents having anticholinesterase activity.

Side Effects: Side effects include drowsiness, dizziness, visual disturbances, nausea, tachycardia, headache, hyperventilation and muscle weakness, potentially fatal side effects include rapid admin causes tachycardia, laryngospasm and rigidity, rapid breathing, increased muscle stiffness, a choking feeling. Large doses cause neuromuscular blockade.

Stability and storage conditions: Store at 20-25°C

Incompatibility: It is incompatible with atropine, and when atropine and pralidoxime chloride are used together, the signs of atropinisation (flushing, mydriasis, tachycardia, dryness of the mouth and nose) may occur.

Different types of formulation: Injection

Popular brands: It is manufactured by many companies. Some of the notable brand names in India are:

Sr.No.	Brand Name	Manufacturers	Type
1	2-Pam-1	VHB Life Sciences Limited	Injection
2	Adipam	Adison Laboratories	Injection
3	Aldopam	Samarth Pharma Pvt. Ltd	Injection
4	CBC-Pam	Chandra Bhagat Pharma Pvt. Ltd	Injection
5	Lyphe	VHB Life Sciences Limited	Injection
6	Neopam	Troikaa Pharmaceuticals Ltd.	Injection
7	Nispam	Neiss Labs Pvt. Ltd.	Injection
8	Pam-A Korea	SG Pharma Pvt. Ltd	Injection
9	Pam-up	Redson Pharmaceuticals	Injection
10	Unipam	VHB Life Sciences Limited	Injection

6.3.3.8 Echothiophate iodide (*Syn:* Echothiophate iodide; Phospholine iodide; Ecothiophate iodide; Echodide; Ecostigminijodidum)

Structure:

$$\left[(CH_3)_3\,N^+CH_2CH_2 - S - P\underset{OC_2H_5}{\overset{\displaystyle\uparrow O}{\overset{OC_2H_5}{\Big|}}} \right] I^-$$

Systematic (IUPAC) Name: 2-Diethoxyphosphorylsulfanylethyl(trimethyl)azanium;iodide

Uses: This drug is used alone or with other drugs to treat certain types of glaucoma and other eye conditions (e.g., accommodative esotropia, synechial formation). It is also used to test for certain eye conditions (e.g., accommodative esotropia). It works by causing the pupil to shrink, decreasing the amount of fluid within the eye, and affecting certain eye muscles.

Side Effects: Side effects include burning, stinging, redness, or tearing eyes, eyelid muscle twitches, headache or brow ache, or decreased vision in poor light, blurred vision, stinging or burning after using the eye drops, watery eyes, twitching eyelids; pain above your eyes, or red or puffy eyelids, an allergic reaction (shortness of breath, swelling of the lips, face, or tongue, or hives), abdominal cramps or diarrhoea, watering mouth, excessive sweating, urinary incontinence, muscle weakness, difficulty breathing, irregular heartbeat, severe eye redness, small white or yellow patches on the surface of your eye, vision problems, seeing flashes of light or "floaters" in vision, fast, slow, or uneven heartbeats, muscle weakness, trouble breathing, increased salivation, heavy sweating, diarrhoea, or loss of bladder control.

Stability and storage conditions: Store echothiophate eye drops at room temperature, around 77°F (25°C). Throw away any unused medicine after 4 weeks. Store away from heat, moisture, and light. Keep echothiophate eye drops out of the reach of children and away from pets.

Incompatibility: Echothiophate iodide for ophthalmic solution potentiates other cholinesterase inhibitors such as succinylcholine or organophosphate and carbamate insecticides. Patients undergoing systemic anticholinesterase treatment should be warned of the possible additive effects of echothiophate iodide for an ophthalmic solution.

Different types of formulation: Solution, drop.

Popular brands: Popular brand name is phospholine Iodide.

QUESTION BANK

A. MULTIPLE CHOICE QUESTIONS

1. Which of the following heterocyclic ring system is present in Pilocarpine?
    A. Imidazole
    B. Furan
    C. Pyrrol
    D. Pyridine

Answer: A

2. Which is not an essential structural feature for cholinergic drugs?
    A. Quaternary Ammonium group
    B. Ethylene Bridge
    C. Acyloxy group
    D. Ethylenediamine group

Answer: D

3. Which is not a Cholinesterase inhibitor?
    A. Neostigmine
    B. Edrophonium Chloride
    C. Tacrine Hydrochloride
    D. Carbachol

Answer: D

4. Which of the following is not a Direct Acting Cholinergic drug?
    A. Tacrine HCl
    B. Acetylcholine
    C. Carbachol
    D. Pilocarpine

Answer: A

5. A Cholinergic drug containing Sulphur and Phosphorous atoms _____________.
    A. Edrophonium Chloride
    B. Tacrine Hydrochloride
    C. Pralidoxime Chloride
    D. Echothiopate Iodide

Answer: D

6. Which one of the following is used to improve muscle strength in patients with a certain muscle disease (myasthenia gravis)?
    A. Pilocarpine
    B. Neostigmine
    C. Pralidoxime
    D. Endorphonium

Answer: B

7. ________________ is used as an adjunct to atropine in the treatment of poisoning by nerve agents having anticholinesterase activity.
    A. Neostigmine
    B. Pilocarpine
    C. Pralidoxime
    D. Endorphonium

Answer: C

8. ____________________ is used in treatment of Alzheimer's disease
    A. Edrophonium Chloride
    B. Tacrine Hydrochloride
    C. Pralidoxime Chloride
    D. Echothiopate Iodide

Answer: B

9. Which of the following is useful in treating open-angle glaucoma?
    A. Acetylcholine
    B. Carbachol
    C. Pilocarpine
    D. Neostigmine

Answer: B

10. ____________ has a quaternary pyridinium moiety in its chemical structure.
    A. Neostigmine
    B. Pilocarpine
    C. Pralidoxime
    D. Endorphonium

Answer: C

B. SHORT ANSWER QUESTIONS

1. Write the structures, chemical names, and uses stability and storage of the following (any 2).
   a) Edrophonium Chloride;   b) Tacrine Hydrochloride;
   c) Pralidoxime Chloride;   d) Echothiopate Iodide

2. Mechanism of action of Direct acting cholinergic agonists.

3. Mechanism of action of Indirect acting cholinergic agonists.

4. Define cholinergic drugs and classify them with examples

5. Write the structure of the following compounds with chemical names.
   a) Acetylcholine   b) Carbachol   c) Pilocarpine
   d) Tacrine Hydrochloride   e) Pralidoxime Chloride

C. LONG ANSWER QUESTIONS

1. Write classification and SAR of Cholinergic drugs.

2. Classify Cholinergic Drugs and elaborate on their mechanism of action.

3. Discuss in detail various Cholinesterase inhibitors.

■■■

DRUGS ACTING ON AUTONOMIC NERVOUS SYSTEM: CHOLINERGIC BLOCKING AGENTS: NATURAL & SYNTHETIC

CONTENTS

Brief introduction of Drugs acting on Autonomic Nervous System-the Cholinergic blocking drugs or Anti-cholinergic agents.

Naturally occurring Agents:

- Atropine Sulphate*,
- Ipratropium Bromide

Synthetic Cholinergic Blocking Agents:

- Tropicamide,
- Cyclopentolate Hydrochloride,
- Clidinium Bromide,
- Dicyclomine Hydrochloride*

♦ LEARNING OBJECTIVES ♦

After completing this chapter, the student should be able to understand:

1. The Anticholinergic or parasympatholytic agents block the neurotransmitter acetylcholine in the central and peripheral nervous systems.
2. Their classification and mechanism of action.
3. Important drugs used as Anticholinergic or parasympatholytic agents have their chemical structures, names, uses, incompatibilities, stability and storage conditions and brands.

6.4.1 INTRODUCTION

An **anticholinergic agent** is a substance that blocks the neurotransmitter acetylcholine in the central and peripheral nervous systems. They inhibit parasympathetic nerve impulses by selectively blocking the binding of the neurotransmitter acetylcholine to its receptor in nerve cells and are hence also known **as parasympatholytic agents.** The nerve fibres of the parasympathetic system are responsible for the involuntary movement of smooth muscles present in the gastrointestinal tract, urinary tract, lungs, etc.

Anticholinergic drugs are used to treat a variety of conditions:

1. Gastrointestinal disorders
   (e.g., gastritis, diarrhoea, pylorospasm, diverticulitis, ulcerative, colitis, nausea, and vomiting)

2. Genitourinary disorders (e.g., cystitis, urethritis, and prostatitis)

3. Respiratory disorders (e.g., asthma, chronic bronchitis, and chronic obstructive pulmonary disease [COPD])

4. Sinus bradycardia due to a hypersensitive vagus nerve.

5. Insomnia, although usually only on a short-term basis

6. Dizziness (including vertigo and motion sickness-related symptoms)

Anticholinergic drugs generally have *antisialagogue* effects (decreasing saliva production), and most produce some level of sedation, both being advantageous in surgical procedures. "Parasympatholytic" and sympathomimetic agents have similar but not identical effects. For example, both cause mydriasis, but parasympatholytics reduce accommodation (cycloplegia), whereas sympathomimetics do not.

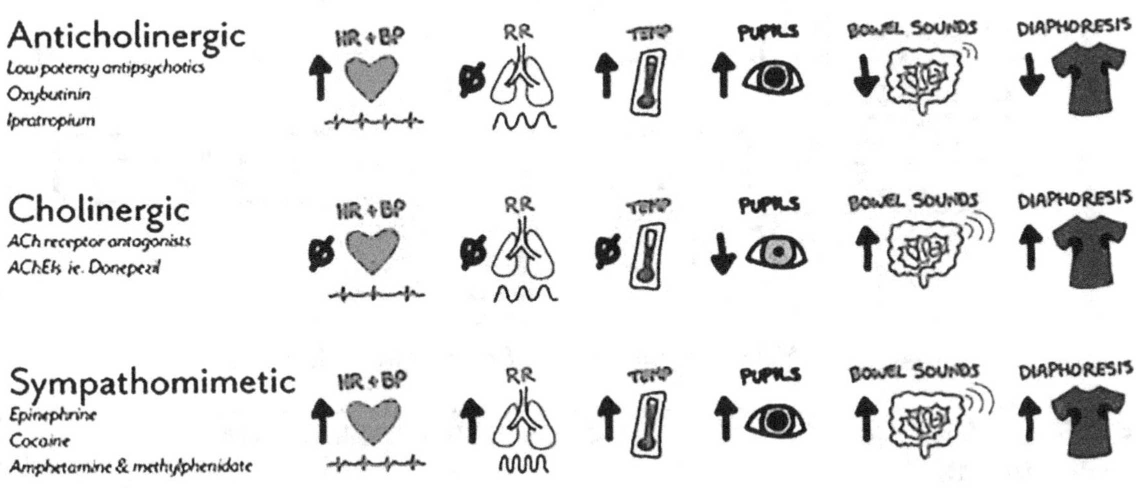

Fig. 6.4.1 Summary of effects of anticholinergics, cholinergic, sympathomimetics

6.4.2 MECHANISM OF ACTION

A) Antimuscarinic Agents

These agents block muscarinic receptors, causing inhibition of all muscarinic functions. In addition, these drugs block the few exceptional sympathetic neurons that are cholinergic, such as those innervating salivary and sweat glands.

Below is summarised mechanism of action of all types of anticholinergic agents.

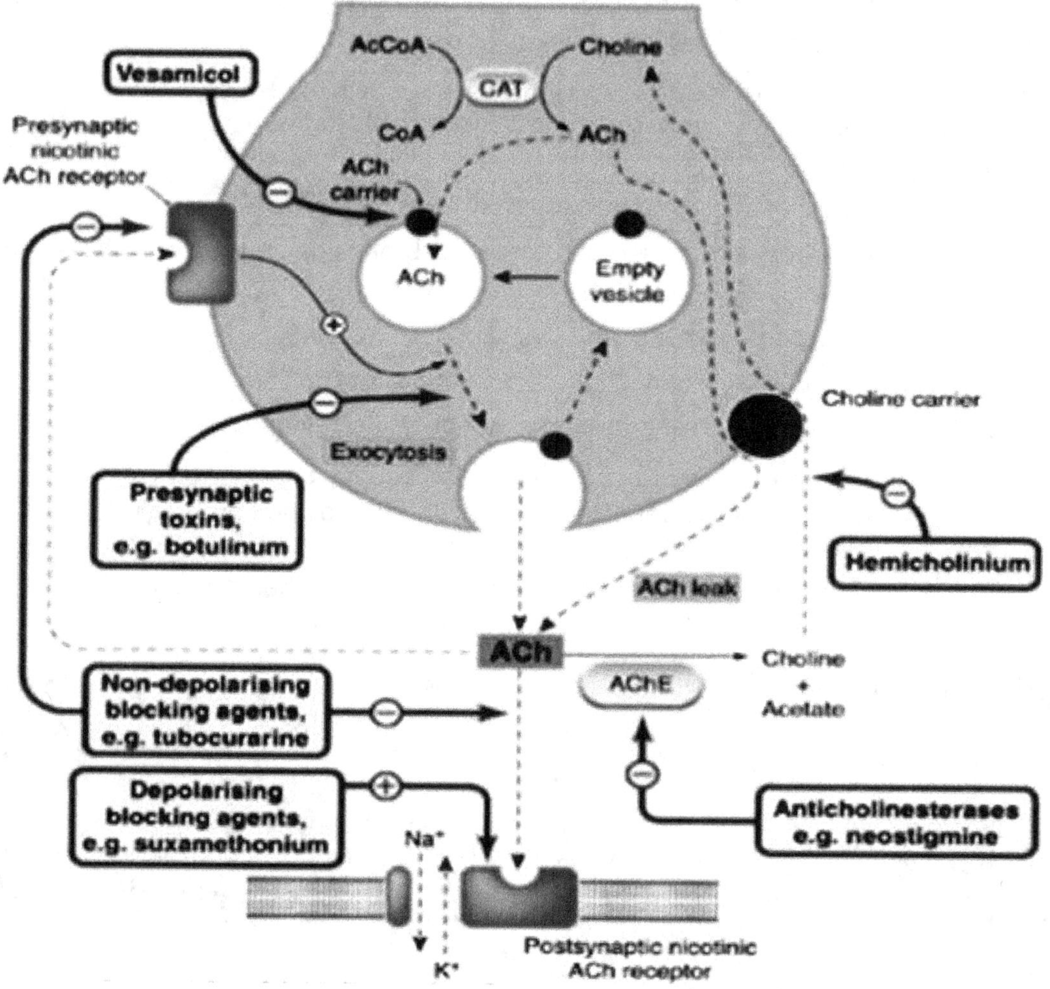

Fig. 6.4.2 Mechanism of action of anticholinergic drugs

6.4.3 CLASSIFICATION OF ANTI-CHOLINERGIC/PARASYMPATHOLYTIC DRUGS

Anticholinergics are divided into three categories in accordance with their specific targets in the central and peripheral nervous system: antimuscarinic agents, ganglionic blockers, and neuromuscular blockers.

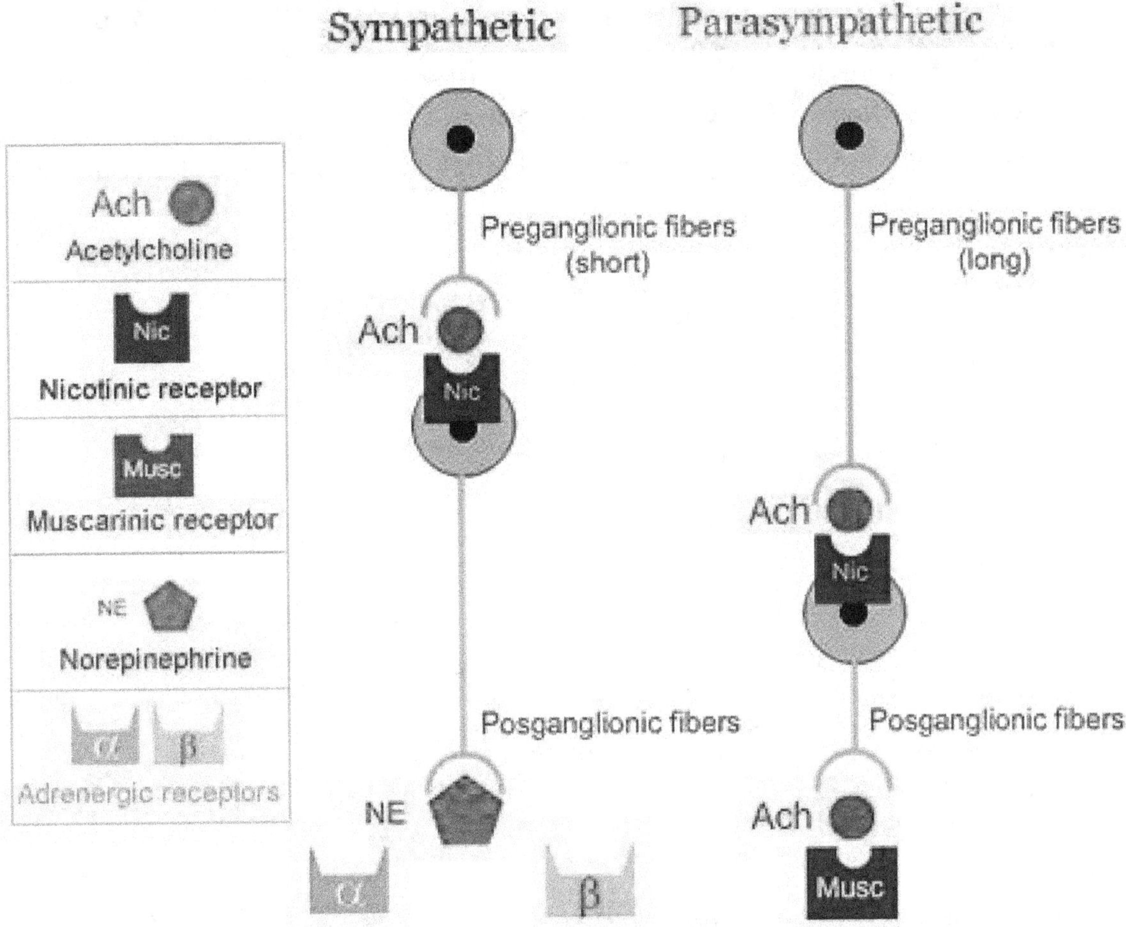

Fig. 6.4.3 Classification of Anticholinergic drugs

A) **Antimuscarinic agents**: *e.g., Atropine, cyclopentolate, ipratropium, scopolamine, tropicamide*

B) **Ganglionic blockers:** *e.g., Mecamylamine, nicotine*

C) **Neuromuscular blockers:** *e.g., Atracurium, cisatracurium, doxacurium, metocurine, mivacurium, pancuronium, tubocurarine*

They can also be classified as

I. **Natural alkaloids:** *e.g., Atropine, Hyoscine, Scopoderm*

II. **Semisynthetic derivatives:** *e.g., Homatropine, Hyoscine butyl bromide*

III. **Synthetic compounds:** *e.g., Oxyphenonium, Pirenzepine (M_1-blockers), Ipratropium, Tiotropium, Flavoxate, Oxybutynyne, Trospium, Tropicamide, Benztropine, Biperiden, Trihexyphenidyl*

6.4.4 SOME IMPORTANT ANTICHOLINERGIC/PARASYMPATHOLYTIC DRUGS

A. *Naturally occurring Anti-Cholinergic Agents:*

6.4.4.1 Atropine sulphate: (*Syn:* **Atropine sulphate; Sulfatropinol; Tropintran; Atropette; Corbella)**

Structure:

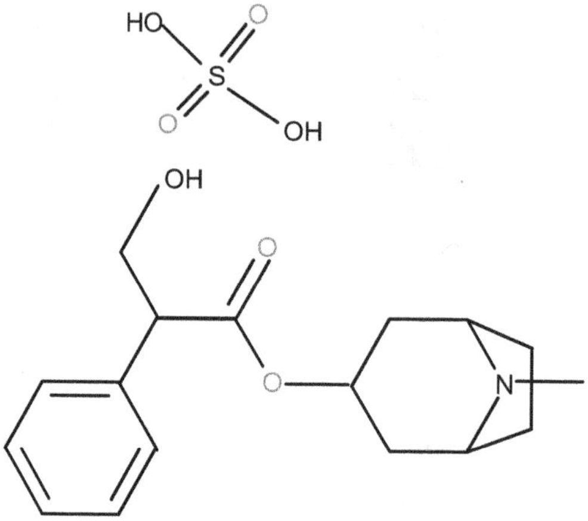

Systematic (IUPAC) Name: (8-Methyl-8-azabicyclo[3.2.1]octan-3-yl)3-hydroxy-2-phenyl-propanoate; sulfuric acid

Uses:

1. As an anti-sialogogue (a drug that reduces, slows, or prevents the flow of saliva) when reduction of secretions of the respiratory tract is thought to be needed, its routine use as a pre-anaesthetic agent is discouraged.

2. To blunt the increased vagal tone (decreased pulse and blood pressure) produced by intra-abdominal tract or ocular muscle traction, its routine use to prevent such events is discouraged

3. To temporarily increase heart rate or decrease AV-block until definitive intervention can take place when bradycardia or AV-block are judged to be hemodynamically significant and thought to be due to excess vagal tone.

4. As an antidote for inadvertent overdose of cholinergic drugs or for cholinesterase poisoning such as from organophosphorus insecticides.

5. As an antidote for the "rapid type of mushroom poisoning due to the presence of the alkaloid muscarine, in certain species of fungus such as Amanita muscaria.

6. To alleviate the muscarinic side effects of anticholinesterase drugs used for the reversal of neuromuscular blockade.

7. Atropine Sulfate is used as a preoperative medication for the reduction of salivary and bronchial secretions during cardiopulmonary resuscitation to treat sinus bradycardia or asystole.

Side Effects: Side effects of atropine are dryness of the mouth, blurred vision, photophobia and tachycardia, anhidrosis (producing heat intolerance or impairing temperature regulation), constipation and difficulty in micturition, and occasional hypersensitivity reactions have been observed, especially skin rashes. Adverse effects include palpitation, dilated pupils, difficulty in swallowing, hot, dry skin, thirst, dizziness, restlessness, tremor, fatigue and ataxia. Toxic doses lead to marked palpitation, restlessness and excitement, hallucinations, delirium and coma. Depression and circulatory collapse occur only with severe intoxication. In such cases, blood pressure declines and death due to respiratory failure following paralysis and coma.

Stability and storage conditions: Protect from light. Store at less than 25 °C.

Incompatibility: It is incompatible with many drugs, which include other antispasmodics (e.g. hyoscine, propantheline or dicycloverine), antihistamines, such as those used to treat allergies, travel sickness or cough and cold remedies (e.g. promethazine, chlorphenamine, or diphenhydramine) anti-sickness medicines (e.g. hyoscine hydrobromide, cyclizine, domperidone or metoclopramide), antidepressants (e.g. tricyclic antidepressants [tcas] such as amitriptyline or clomipramine, and any monoamine oxidase inhibitors [maois] such as moclobemide), antiarrhythmics, to control heartbeat rhythm problems (e.g. disopyramide), antifungals (e.g. ketoconazole), antipsychotics, used in conditions such as schizophrenia (e.g. haloperidol, chlorpromazine or clozapine), others: amantadine (used in Parkinson's disease), tiotropium and ipratropium (both used to treat lung conditions such as chronic obstructive pulmonary disease [COPD])

Different types of formulation: Tablet, injection drops

Popular brands: It is available in combination with other drugs. It is manufactured by many companies. Some of the notable brand names in India are:

Sr. No	Brand Name	Manufactures	Type
1	Atrisolon	Intas Laboratories Pvt Ltd	Drops
2	Atroren -P	Indoco Remedies Ltd	Injection
3	Phenotil	Allied Chemicals & Pharmaceuticals Pvt. Ltd	Tablet
4	Opthopin	Optho Remedies Pvt.Ltd	Drop
5	Dexatropine Eye	Sunways India Pvt. Ltd	Drop
6	C -Pin	Calibre Pharmaceutical Pvt. Ltd.	Drop
7	Indipa	Biomedica International	Drop
8	Lomofen	RPG Life Sciences Ltd	Tablet

6.4.4.2 *Ipratropium bromide*: (*Syn:* **Atrovent**)

Structure:

Systematic (IUPAC) Name: (8-Methyl-8-propan-2-yl-8-azoniabicyclo[3.2.1]octan-3-yl) 3-hydroxy-2-phenylpropanoate; bromide

Uses: It is an anticholinergic bronchodilator.

1. Ipratropium Bromide Inhalation Solution administered either alone or with other bronchodilators, especially beta adrenergics, is indicated as a bronchodilator for maintenance treatment of bronchospasm associated with chronic obstructive pulmonary disease, including chronic bronchitis and emphysema.

2. Ipratropium bromide is used to treat breathing difficulties due to lung conditions such as asthma or chronic obstructive airway disease (COPD).

3. It is also used to treat rhinitis (inflammation of the nose).

4. It is used to widen the airways in the lung, making it easier to breathe. Ipratropium bromide is taken using an inhaler device to treat breathing difficulties so that the drug is delivered directly to the lungs where it is needed.

5. Ipratropium bromide is taken using a nasal spray to relieve the symptoms of allergic and non-allergic rhinitis as it reduces secretions from the nose. Allergic rhinitis is a condition with inflammation of the nose arising from an allergy, such as pollen in hayfever, house dust mites, or animal fur, leading to cold-like symptoms, such as sneezing and itchiness and blocked or runny nose. Non-allergic rhinitis has similar symptoms and arises from oversensitive blood vessels in the nose.

6. In general, this drug is used to treat asthma, COPD and rhinitis.

Side Effects: It shows the side effects like headache, dizziness, vomiting, dry mouth, constipation (problem with gastrointestinal motility), diarrhoea (problem with gastrointestinal motility), throat irritation, cough, unexpected chest tightness, skin rash, itchiness, fast heart rate, itchy skin rash, palpitations (fast and irregular heartbeats), abnormal heart rhythm such as atrial fibrillation, eye problems such as eye pain, raised eye pressure and wide pupils, throat tightness, nausea, difficulty urinating, bad taste in the mouth. Severe allergic reactions include tongue, face and lips swelling or anaphylaxis (very severe allergic reaction).

Stability and storage conditions: Store between 59°F (15°C) and 86°F (30°C). Protect from light.

Incompatibility: Ipratropium bromide has been shown to be a safe and effective bronchodilator when used in conjunction with beta-adrenergic bronchodilators. Ipratropium bromide has also been used with other pulmonary medications, including methylxanthines and corticosteroids, without adverse drug interactions

Different types of formulation: Solution, inhaler, respules, rotacaps

Popular brands: It is manufactured by many companies. Some of the notable brand names in India are:

Sr.No.	Brand Name	Manufactures	Type
1	Ipramist (250mcg)	German Remedies (Zydus Cadila Healthcare Ltd)	Solution
2	Ipranase AQ (42mcg)	Cipla Limited	Inhaler
3	Ipratop (20mcg)	Astra Zeneca Pharma India Limited	Inhaler
4	Ipravent (15ml)	Cipla Limited	Solution, inhaler Respules, rotacaps

B. *Synthetic Anti-Cholinergic Agents:*

6.4.4.3 Tropicamide (*Syn:* Mydriacyl; Mydriaticum; Tropicacyl; Minims tropicamide)

Structure:

Systematic (IUPAC) Name: N-Ethyl-3-hydroxy-2-phenyl-N-(pyridin-4-ylmethyl) propane-mide

Uses: It is used to widen (dilate) the pupil of the eye in preparation for certain eye examinations.

Side Effects: Ocular: Transient stinging, blurred vision, photophobia and superficial punctate keratitis have been reported with the use of Tropicamide. Increased intraocular pressure has been reported following the use of mydriatics.

Non-Ocular: Dryness of the mouth, tachycardia, headache, allergic reactions, nausea, vomiting, pallor, central nervous system disturbances and muscle rigidity have been reported with the use of Tropicamide. Psychotic reactions, behavioural disturbances, and vasomotor or cardiorespiratory collapse in children have been reported using anticholinergic drugs.

Stability and storage conditions: Store between 15°-25°C (59°-77°F). Do not refrigerate. Do not store at high temperatures. Keep the container tightly closed.

Incompatibility: Tropicamide may interfere with the antihypertensive action of carbachol, pilocarpine, or ophthalmic cholinesterase inhibitors.

Different types of formulation: Liquid, solution, drop

Popular brands: It is available in combination with other drugs. It is manufactured by many companies. Some of the notable brand names in India are:

Sr. No.	Brand Name	Combination Generics	Manufactures	Type
1	Dilatine Eye	Tropicamide, Chlorobutol	Jawa Pharmaceuticals (India) Pvt. Ltd.	Drop
2	Drosyn –T	Tropicamide, Phenylephrine	FDC Limited	Drop
3	Itrop Plus	Tropicamide, Benzalkonium Chloride,Phenylephrine	Cipla Limited	Drop
4	Meptrop P	Tropicamide, Phenylephrine	Alna Biotech	Drop
5	Optimide Plus	Tropicamide, Phenylephrine	Micro Vision	Drop
6	Picamid Plus	Tropicamide, Chlorbutol,Phenylephrine	Indoco Remedies Ltd (Excel)	Drop
7	Tropac P Eye	Tropicamide, Phenylephrine	Optho Remedies Pvt.Ltd	Drop
8	Tropicacyl	Tropicamide, Chlorbutol	Sunways India Pvt. Ltd	Drop
9	Triocyl Plus	Tropicamide, Phenylephrine	Intas Laboratories Pvt Ltd	Drop
10	Tmde Plus	Tropicamide, Benzalkonium Chloride,Phenylephrine	Cadila Pharmaceuticals Ltd. (Le Sante)	Drop

6.4.4.4 Cyclopentolate hydrochloride (*Syn:* **Cyclopentolate HCL, AK-Pentolate, Akpentolate, Cyplegin**)

Structure:

$$\text{(1-hydroxycyclopentyl)(phenyl)CH—COOCH}_2\text{CH}_2\text{N(CH}_3)_2 \cdot \text{HCl}$$

Systematic (IUPAC) Name: 2-(Dimethylamino)ethyl-2-(1-hydroxycyclopentyl)-2-phenyl-acetate; hydrochloride

Uses: Cyclopentolate Hydrochloride Ophthalmic Solution is used to produce mydriasis and cycloplegia.

Side Effects: It shows side effects such as increased intraocular pressure, burning, photophobia, blurred vision, irritation, hyperemia, conjunctivitis, blepharon conjunctivitis, punctuate keratitis, and synechiae have been reported. The use of cyclopentolate has been associated with psychotic reactions and behavioural disturbances, usually in children, especially with a 2% concentration. These disturbances include ataxia, incoherent speech, restlessness, hallucinations, hyperactivity, seizures, disorientation as to time and place, and failure to recognise people. This drug produces reactions similar to those of other anticholinergic drugs, but the central nervous system manifestations, as noted above, are more common. Other toxic manifestations of anticholinergic drugs are skin rash, abdominal distention in infants, unusual drowsiness, tachycardia, hyperpyrexia, vasodilation, urinary retention, diminished gastrointestinal motility and decreased secretion in salivary and sweat glands, pharynx, bronchi and nasal passages. Severe manifestations of toxicity include coma, medullary paralysis and death.

Stability and storage conditions: Store between 15°-25°C (59°-77°F).

Incompatibility: Cyclopentolate shows incompatibility and interaction with the ocular anti-hypertensive action of carbachol, pilocarpine, or ophthalmic cholinesterase inhibitors.

Different types of formulation: Drop

Popular brands: Cyclopentolate is manufactured by many companies. Some of the notable brand names in India are:

Sr. No.	Brand Name	Manufactures	Type
1	Bell Pentolate	Bell Pharma Pvt. Ltd	Drop
2	C -Pen Eye	Opthal Remedies (I) Pvt. Ltd	Drop
3	Cyclate	Zydus Cadila Healthcare Ltd	Drop
4	yclofez	Entod Pharmaceuticals Ltd	Drop
5	Cyclogik	FDC Limited	Drop
6	Cyclopan	Biomedica International	Drop
7	Cyclopen Eye	Optho Remedies Pvt.Ltd	Drop
8	Cyclopent	Sun Pharmaceuticals Industries Ltd (Milmet)	Drop
9	Cycloriv Eye	East African (I) Remedies Pvt Ltd	Drop
10	Dilate	Micro Vision	Drop
11	Open Eye	Optho Remedies Pvt.Ltd.	Drop
12	Pentol	Klar Sehen Pharmaceuticals Pvt Ltd	Drop

6.4.4.5 Clindinium bromide (*Syn:* **Clidinii bromidum**)

Structure:

Systematic (IUPAC) Name: 3-[(2-Hydroxy-2,2-diphenylacetyl)oxy]-1-methyl-1-azabicyclo [2.2.2]octan-1-ium bromide

Uses: Used in fixed combination with chlordiazepoxide as adjunctive therapy in the treatment of peptic ulcer disease, in the treatment of acute enterocolitis, and in the treatment of functional GI motility disturbances (e.g., irritable bowel syndrome).

Side Effects: Side effects include dizziness, drowsiness, weakness, blurred vision, dry eyes, dry mouth, nausea, constipation, and abdominal bloating.

Stability and storage conditions: Stored at room temperature in a tightly closed container.

Incompatibility: It is incompatible with potassium tablets/capsules, sodium oxybate, and drugs that are affected by slowed gut movement (such as pramlintide). This drug affects the absorption of other products such as certain azole anti-fungal drugs (ketoconazole, itraconazole), and slowly-dissolving forms of digoxin, among others. Incompatible with certain drugs and responsible for the increase in side effects which includes amantadine, other anticholinergic drugs (such as atropine, glycopyrrolate, scopolamine), other antispasmodic drugs (such as dicyclomine, propantheline), belladonna alkaloids, certain drugs used to treat Parkinson's disease (such as benztropine, trihexyphenidyl), certain drugs used to treat irregular heart rhythms (such as disopyramide, quinidine), MAO inhibitors (isocarboxazid, linezolid, methylene blue, moclobemide, phenelzine, procarbazine, rasagiline, selegiline, tranylcypromine), phenothiazines (such as chlorpromazine), tricyclic antidepressants (such as amitriptyline).

Different types of formulation: Capsule

Popular brands: It is available in combination with other drugs. It is manufactured by many companies. Some of the notable brand names in India are:

Sr. No.	Brand Name	Manufactures	Combination	Type
1	Abdcool	Solitaire (Prominent)	[Chlordiazepoxide + Clidinium Bromide]	capsule
2	Avidip	Avalanche	[Chlordiazepoxide + Clidinium Bromide]	capsule
3	Alspas Forte	Allenge	Chlordiazepoxide + Clidinium Bromide	capsule
4	Cebrex	Psycorem	[Chlordiazepoxide + Clidinium Bromide]	capsule
5	Checkspas	Quest	[Chlordiazepoxide + Clidinium Bromide	capsule
6	Colinorm	Lancer	Pharma [Chlordiazepoxide + Clidinium Bromide]	capsule
7	Dorpep	Candor (Aileron)	[Chlordiazepoxide + Clidinium Bromide]	capsule
8	IBS	Daffohils Pharma	[Chlordiazepoxide + Clidinium Bromide]	capsule
9	Equirex	Jagsonpal Pharma	[Chlordiazepoxide + Clidinium Bromide]	capsule
10	Librax	Abbott Health-care Pvt. Ltd.	[Chlordiazepoxide + Clidinium Bromide]	capsule
11	Normaxin	Systopic	[Chlordiazepoxide + Clidinium Bromide]	capsule
12	Neurospas-CD	Ind- Swift	[Chlordiazepoxide + Clidinium Bromide]	capsule
13	Normaxin-RT	Systopic	[Chlordiazepoxide + Clidinium Bromide]	capsule
14	Serecon-D	Unimarck	[Chlordiazepoxide + Clidinium Bromide]	capsule

6.4.4.6 Dicyclomine hydrochloride (*Syn:* **Bentyl; Dicyclomine HCl**)

Structure:

Systematic (IUPAC) Name: 2-(Diethylamino)ethyl-1-cyclohexylcyclohexane-1-carboxylate; hydrochloride

Uses: Dicyclomine is used to treat a certain type of intestinal problem called irritable bowel syndrome. It helps to reduce the symptoms of stomach and intestinal cramping. This medication works by slowing the natural movements of the gut and by relaxing the muscles in the stomach and intestines.

Side Effects: Side effects include dizziness, drowsiness, lightheadedness, weakness, blurred vision, dry eyes, dry mouth, nausea, constipation, and abdominal bloating. Serious side effects include decreased sweating, dry/hot/flushed skin, fast/irregular heartbeat, loss of coordination, slurred speech, mental/mood changes (such as confusion, hallucinations, agitation, nervousness, and unusual excitement), difficulty urinating, and decreased sexual ability.

Stability and storage conditions: Store in containers protecting them from excessive heat, light, and moisture. Capsules and tablets should be stored in well-closed containers at room temperature, preferably below 30°C. Solution should be at room temperature, preferably below 30°C; avoid exposure to excessive heat. Injection stored at room temperature, preferably less than 30°C; do not freeze.

Incompatibility: Some products that may interact with this drug include potassium tablets/capsules and drugs that are affected by slowed gut movement (such as pramlintide).

Dicyclomine may affect the absorption of other products such as levodopa, certain azole anti-fungal drugs (ketoconazole, itraconazole), and slowly-dissolving forms of digoxin, among others.

It is incompatible with other drugs and responsible for increased side effects those drugs are amantadine, other anticholinergic drugs (such as atropine, glycopyrrolate, and scopolamine), other antispasmodic drugs (such as clidinium, propantheline), belladonna alkaloids, certain drugs used to treat Parkinson's disease (such as benztropine, trihexyphenidyl), certain drugs used to treat irregular heart rhythms (such as disopyramide, quinidine), MAO inhibitors (isocarboxazid, linezolid, methylene blue, moclobemide, phenelzine, procarbazine, rasagiline, selegiline, tranylcypromine), phenothiazines (such as chlorpromazine), tricyclic antidepressants (such as amitriptyline). It is incompatible with other products that cause drowsiness, including alcohol, antihistamines (such as cetirizine and diphenhydramine), drugs for sleep or anxiety (such as alprazolam, diazepam, zolpidem), muscle relaxants, and narcotic pain relievers (such as codeine).

Different types of formulation: Drop, Injection, Capsule/Tablet

Popular brands: It is manufactured by many companies. Some of the notable brand names in India are:

Sr. No.	Brand Name	Manufactures	Type
1	Clomin	Core Healthcare	Drops,injection
2	Coligon	Fourrts (India) Laboratories	Capsule/Tablet
3	Cyclozobid	Nicholas Piramal	Capsule/Tablet
4	Decolic Inj.	KonTest Chemicals	Capsule/Tablet
5	Dicyclomine	Core Healthcare	Injection
6	Duospan	Sigma Laboratories	Capsule/Table
7	Meftal Spas	Blue Cross Laboratories	Capsule/Tablet
8	Pipcol	Geno Pharmaceuticals	Injection
9	Spasmo – Proxyvon	Wockhardt	Injection,Capsule/Tablet
10	Swispas	Ind – Swift	Capsule/Tablet
11	Trigan – D	Cadila Pharmaceuticals	Injection

QUESTION BANK

A. MULTIPLE CHOICE QUESTIONS

1. Which of the following ring system is present in Atropine?
   A. Azabicyclooctane
   B. Azoniabicyclooctane
   C. Cyclopentane
   D. Pyridine

Answer: A

2. Which of the following ring system is present in Ipratropium?
   A. Azabicyclooctane
   B. Azoniabicyclooctane

    C. Cyclopentane

    D. Pyridine

Answer: B

3. Which of the following ring system is present in Tropicamide?
   A. Azabicyclooctane
   B. Azoniabicyclooctane
   C. Cyclopentane
   D. Pyridine

Answer: D

4. Which of the following ring system is present in Cyclopentanoate?
   A. Azabicyclooctane
   B. Azoniabicyclooctane
   C. Cyclopentane
   D. Pyridine

Answer: C

5. Which of the following ring system is present in Clinidium bromide?
   A. Azabicyclooctane
   B. Azoniabicyclooctane
   C. Cyclopentane
   D. Pyridine

Answer: A

6. Which of the following ring system is present in Dicyclomine?
   A. Azabicyclooctane
   B. Cyclohexane
   C. Cyclopentane
   D. Pyridine

Answer: B

7. Atropine is used as an _____________________agent
   A. Bronchodilator
   B. Antiulcer
   C. Anti-IBS
   D. Antisialogouge

Answer: D

8. Clinidium chloride is used as ___________________
   A. Bronchodilator
   B. Antiulcer adjunct

   C. Anti-IBS

   D. Antisialogouge

Answer: B

9. Ipratropium is used as a _________ _____________.
   A. Bronchodilator
   B. Antiulcer
   C. Anti-IBS
   D. Antisilagouge

Answer: A

10. Tropicamide as well as Cyclopentanoate are used to produce _____________________.
   A. Bronchodilator effect
   B. Mydriasis
   C. Antiulcer action
   D. Antisilagouge effect

Answer: B

B. SHORT ANSWER QUESTIONS

1. Write the structures, chemical names, and uses stability and storage of the following
   a) Atropine Sulphate*   b) Ipratropium Bromide
   c) Tropicamide   d) Cyclopentolate Hydrochloride
   e) Clidinium Bromide   f) Dicyclomine Hydrochloride*

2. Discuss the Mechanism of Action of Anti-muscarinic agents.

3. Discuss the Mechanism of Action Ganglionic Blocking Agents.

4. Discuss the Mechanism of Action of Neuromuscular Blocking Agents.

5. Write the structure of the following compounds with chemical names.
   a) Atropine Sulphate*   b) Ipratropium Bromide
   c) Tropicamide   d) Cyclopentolate Hydrochloride
   e) Clidinium Bromide   f) Dicyclomine Hydrochloride*

C. LONG ANSWER QUESTIONS

1. Define and classify Anticholinergic agents with suitable examples. Elaborate on their mechanism of action.

2. Discuss in detail the naturally occurring Anti-cholinergic agents.

3. Discuss in detail any 3 synthetic Anti-cholinergic agents.

DRUGS ACTING ON CARDIOVASCULAR SYSTEM: ANTI-ARRHYTHMIC DRUGS

CONTENTS

Anti-Arrhythmic Drugs:

- Quinidine Sulphate
- Procainamide Hydrochloride
- Verapamil
- Phenytoin Sodium*
- Lidocaine Hydrochloride
- Lorcainide Hydrochloride
- Amiodarone
- Sotalol

♦ LEARNING OBJECTIVES ♦

After completing this chapter, the student should be able to understand:

1. Classification of Anti-Arrhythmic Drugs
2. Chemical Names of Anti-Arrhythmic Drugs
3. Chemical Structures of different Anti-Arrhythmic Drugs
4. Uses, Stability and storage conditions of Anti-Arrhythmic Drugs
5. Different types of formulations and popular brand names Anti-Arrhythmic Drugs

7.1.1 INTRODUCTION

Cardiac arrhythmias are caused by a disturbance in the conduction of the impulse through the myocardial tissue, by disorders of impulse formation, or by a combination of these factors. The antiarrhythmic agents used most commonly affect impulse conduction by

They also depress spontaneous diastolic depolarisation, causing a reduction of automaticity by ectopic foci. Many pharmacological agents are available for the treatment of cardiac arrhythmias. Agents such as oxygen, potassium, and sodium bicarbonate relieve the underlying cause of some arrhythmias. Other agents, such as digitalis, propranolol, phenylephrine, edrophonium, and neostigmine, act on the cardiovascular system by affecting the heart muscle or on the autonomic nerves to the heart. Finally, there are drugs that alter the electrophysiological mechanisms causing arrhythmias.

Within the past 5 decades, research on normal cardiac tissues and, in the clinical setting, on patients with disturbances of rhythm and conduction has brought to light information on the genesis of cardiac arrhythmias and the mode of action of antiarrhythmic agents. In addition, laboratory tests have been developed to measure blood levels of antiarrhythmic drugs such as phenytoin, disopyramide, lidocaine, procainamide, and quinidine, to help evaluate the pharmacokinetics of these agents. As a result, it is possible to maintain steady-state plasma levels of these drugs, which allows the clinician to use these and other agents more effectively and with greater safety. No other clinical intervention has been more effective at reducing mortality and morbidity in coronary care units.

7.1.2 CLASSIFICATION

I	sodium-channel blockade	Reduce phase 0 slope and peak of action potential.
IA	- moderate	Moderate reduction in phase 0 slope; increase APD; increase ERP.
IB	- weak	Small reduction in phase 0 slope; reduce APD; decrease ERP.
IC	- strong	Pronounced reduction in phase 0 slope; no effect on APD or ERP.
II	beta-blockade	Block sympathetic activity; reduce rate and conduction.
III	potassium-channel blockade	Delay repolarization (phase 3) and thereby increase action potential duration and effective refractory period.
IV	calcium-channel blockade	Block L-type calcium-channels; most effective at SA and AV nodes; reduce rate and conduction.

Class I agents

The class I antiarrhythmic agents interfere with the sodium channel. Class I agents are grouped by what effect they have on the Na^+ channel and what effect they have on cardiac action potentials. Class I agents are called membrane-stabilizing agents, "stabilising" referring to the decrease of excitogenicity of the plasma membrane, which is brought about by these agents. (Also noteworthy is that a few class II agents like propranolol also have a membrane stabilising effect.)

Class I agents are divided into three groups (Ia, Ib, and Ic) based upon their effect on the length of the action potential.

- Ia lengthens the action potential (right shift)
- Ib shortens the action potential (left shift)
- Ic does not significantly affect the action potential (no shift)

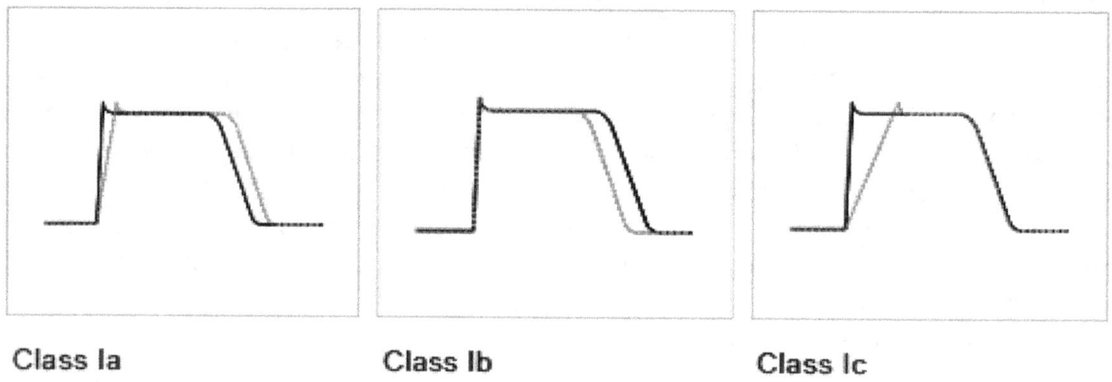

Class Ia   **Class Ib**   **Class Ic**

Class II agents

Class II agents are conventional beta blockers. They act by blocking the effects of catecholamines at the β_1-adrenergic receptors, thereby decreasing sympathetic activity on the heart, which reduces intracellular cAMP levels and hence reduces Ca^{2+} influx. These agents are particularly useful in the treatment of supraventricular tachycardias. They decrease conduction through the AV node.

Class II agents include atenolol, esmolol, propranolol, and metoprolol.

Class III agents

Class III agents predominantly block the potassium channels, thereby prolonging repolarisation. Since these agents do not affect the sodium channel, conduction velocity is not decreased. The prolongation of the action potential duration and refractory period, combined with the maintenance of normal conduction velocity, prevent re-entrant arrhythmias. (The re-entrant rhythm is less likely to interact with tissue that has become refractory).

The class III agents exhibit reverse-use dependence (their potency increases with slower heart rates and therefore improves maintenance of sinus rhythm). Inhibiting potassium channels and slowing repolarisation results in slowed atrial-ventricular myocyte repolarization. Class III agents have the potential to prolong the QT interval of the EKG and maybe proarrhythmic (more associated with the development of polymorphic VT). Class III agents includes: bretylium, amiodarone, ibutilide, sotalol, dofetilide, vernakalant and dronedarone.

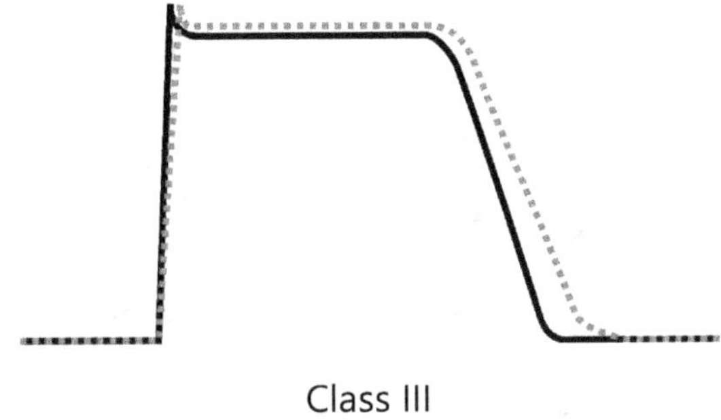

Class III

Class IV agents

Class IV agents are slow non-dihydropyridine calcium channel blockers. They decrease conduction through the AV node and shorten phase two (the plateau) of the cardiac action potential. They thus reduce the contractility of the heart, so it may be inappropriate in heart failure. However, in contrast to beta blockers, they allow the body to retain adrenergic control of heart rate and contractility.

Class IV agents include verapamil and diltiazem.

Class V and others

Since the development of the original Vaughan Williams classification system, additional agents have been used that do not fit cleanly into categories I through IV.

Such agents include:

- Digoxin, which decreases the conduction of electrical impulses through the AV node and increases vagal activity via its central action on the central nervous system, via indirect action, leads to an increase in acetylcholine production, stimulating M_2 receptors on the AV node leading to an overall decrease in speed of conduction.
- Adenosine is used intravenously for terminating supraventricular tachycardias.
- Magnesium sulfate, an antiarrhythmic drug, but only against very specific arrhythmias which, has been used for torsades de pointes.
- Trimagnesium dicitrate (anhydrous) as a powder or powder caps in pristine condition, better bioavailability than ordinary MgO.

7.1.3 DIFFERENT ANTI-ARRHYTHMIC DRUGS

7.1.3.1 Quinidine Sulfate

Structure:

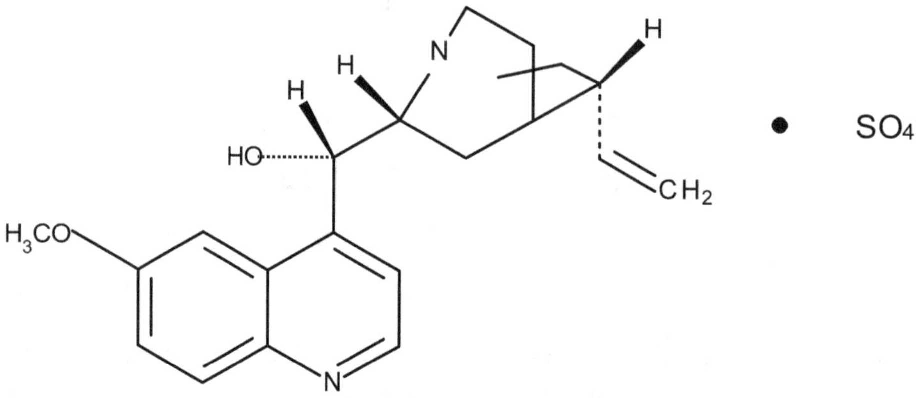

Chemical (IUPAC) name: (5-ethenyl-1-azabicyclo[2.2.2]octan-2-yl)-(6-methoxyquinolin-4-yl)methanol;sulfuric acid

Quinidine sulfate is the sulphate of an alkaloid obtained from various species of Cinchona and their hybrids. It is a dextrorotatory diastereoisomer of quinine. The salt crystallises from water as the dihydrate in the form of fine, needle-like white crystals. Quinidine sulfate contains a hydroxymethyl group that serves as a link between a quinoline ring and a quinuclidine moiety.

Quinidine sulfate is bitter and light-sensitive. Aqueous solutions are nearly neutral or slightly alkaline. It is soluble to the extent of 1% in water and more highly soluble in alcohol or chloroform.

Uses:

1. It is used in the treatment and prophylaxis of
   - Paroxysmal ventricular tachycardia
   - Cardiac arrhythmia includes atrial fibrillation, atrial fibrillation, atrial flutter and premature systoles.
   - Extra systolic atrial tachycardia.

2. In thyrotoxicosis, in which fibrillation persists after thyroidectomy

3. To maintain sinus rhythm

4. As an alternative to quinine to treat malaria.

Common side effects of quinidine sulfate include diarrhoea, nausea, vomiting, loss of appetite, stomach pain/cramp, slight headedness, fatigue, weakness, angina-like chest pain, changes in sleep habits, a burning feeling in the throat or chest (e.g., heartburn), tremor etc.

Stability and storage:

It darkens on exposure to light, and hence it is stored in a tightly-closed light resistance containers.

Dose: The dose is 100- 200 mg

Pharmaceutical Formulations: Tablet

Brand Names: Phenipam, Natacardine

7.1.3.2 Procainamide Hydrochloride

Structure:

Chemical (IUPAC) name: *p*-amino-N-[2-(diethylamino)ethyl] benzamide monohydrochloride, procainamidium chloride.

It has emerged as a major antiarrhythmic drug. It was developed in the course of research for compounds structurally similar to procaine, which had limited effect as an antiarrhythmic agent because of its central nervous system (CNS) side effects and short-lived action caused by rapid hydrolysis of its ester linkage by plasma esterases. Because of its amide structure, procainamide hydrochloride is also more stable in water than is procaine. Aqueous solutions of procainamide hydrochloride have a pH of about 5.5.

Uses:

1. It is used in the treatment and prophylaxis of
   - Paroxysmal ventricular tachycardia
   - Cardiac arrhythmia includes atrial fibrillation, atrial fibrillation, atrial flutter and premature systoles.
   - Extra systolic atrial tachycardia.
2. In thyrotoxicosis, in which fibrillation persists after thyroidectomy
3. To maintain sinus rhythm

Dose: The dose is 0.5 to 1gm orally, IM 0.25 to 0.5 gm every 2 hours.

Pharmaceutical Formulations: Tablet, Injection

Brand Names: Pronestyl

7.1.3.3 Verapamil

Structure:

CH_3, CH_3, CH_3, CH_3O, N, OCH_3, CN, CH_3O, OCH_3, $\bullet$ HCl

Chemical (IUPAC) name: 5-[(3,4-dimethoxyphenethyl) methylamino]-2-(3,4-dimethoxy-phenyl)-2-isopropylvaleronitrile

It was introduced in 1962 as a coronary vasodilator and is the prototype of the Ca^{2+} antagonists used in cardiovascular diseases. It is used in treating angina pectoris, arrhythmias from ischemic myocardial syndromes, and supraventricular arrhythmias.

Verapamil's major effect is on the slow Ca^{2+} channel. The result is a slowing of AV conduction and the sinus rate. This inhibition of the action potential inhibits one limb of the re-entry circuit believed to underlie most paroxysmal supraventricular tachycardias that use the AV node as a re-entry point. It is categorised as a class IV antiarrhythmic drug. Hemodynamically, verapamil causes a change in the preload, afterload, contractility, heart rate, and coronary blood flow. The drug reduces systemic vascular resistance and means blood pressure, with minor effects on cardiac output.

Uses:

Verapamil is used to treat high blood pressure and to control angina (chest pain). The immediate-release tablets are also used alone or with other medications to prevent and treat irregular heartbeats.

Dose: The dose is 40 mg, 80mg, 120 mg, 240mg

Pharmaceutical Formulations: Tablet

Brand Names: Calaptin 40 Tablet, Calaptin 120 SR Tablet, Calaptin 240 SR, Veramil

7.1.3.4 Phenytoin Sodium

Structure:

Chemical (IUPAC) name: 5,5-diphenyl-2,4-imidazolidinedione, 5,5-diphenylhydantoin, diphenyl hydantoin sodium

It has been used for decades in the control of grand mal types of epileptic seizures. It is structurally analogous to barbiturates but does not possess their extensive sedative properties. The compound is available as the sodium salt. Solutions for parenteral administration contain 40% propylene glycol and 10% alcohol to dissolve the sodium salt.

Phenytoin sodium's cardiovascular effects were uncovered during observation of toxic manifestations of the drug in patients being treated for seizure disorders. Phenytoin sodium was found to cause bradycardia, prolong the PR interval, and produce T-wave abnormalities on electrocardiograms.

Storage:

It is stored in a tightly-closed container protected from light.

Uses:

Phenytoin is used to control certain types of seizures and to treat and prevent seizures that may begin during or after surgery in the brain or nervous system.

Dose: The dose is 100 mg, 500 mg

Pharmaceutical Formulations: Injection, Tablet, Capsule, oral Suspension

Brand Names: Phenytoin Sodium 100mg Injection, Eptoin, Garoin.

7.1.3.5 Lidocaine Hydrochloride

Structure:

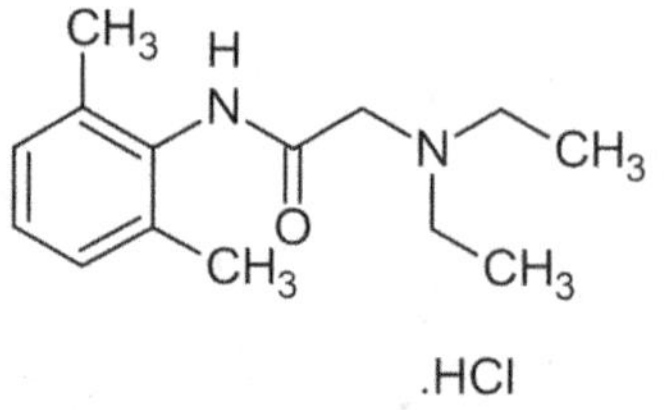

.HCl

Chemical (IUPAC) name: 2-(diethylamino)-N-(2,6-dimethylphenyl)acetamide; hydrochloride (Xylocaine).

It was conceived as a derivative of gramine (3-dimethylaminomethylindole) and introduced as a local anaesthetic. It is now being used intravenously as a standard parenteral agent for the suppression of arrhythmias associated with acute myocardial infarction and cardiac surgery. It is the drug of choice for the parenteral treatment of premature ventricular contractions.

Lidocaine hydrochloride is a class IB antiarrhythmic agent with a different effect on the electrophysiological properties of myocardial cells from that of procainamide and quinidine. It binds with equal affinity to the active (A) and inactive (I) Na^+ ion channels. It depresses diastolic depolarisation and automaticity in the Purkinje fibre network and increases the functional refractory period relative to action potential duration, as do procainamide and quinidine.

Lidocaine hydrochloride administration is limited to the parenteral route and is usually given intravenously, although adequate plasma levels are achieved after intramuscular injections. Lidocaine hydrochloride is not bound to any extent to plasma proteins and is concentrated in the tissues. It is metabolised rapidly by the liver.

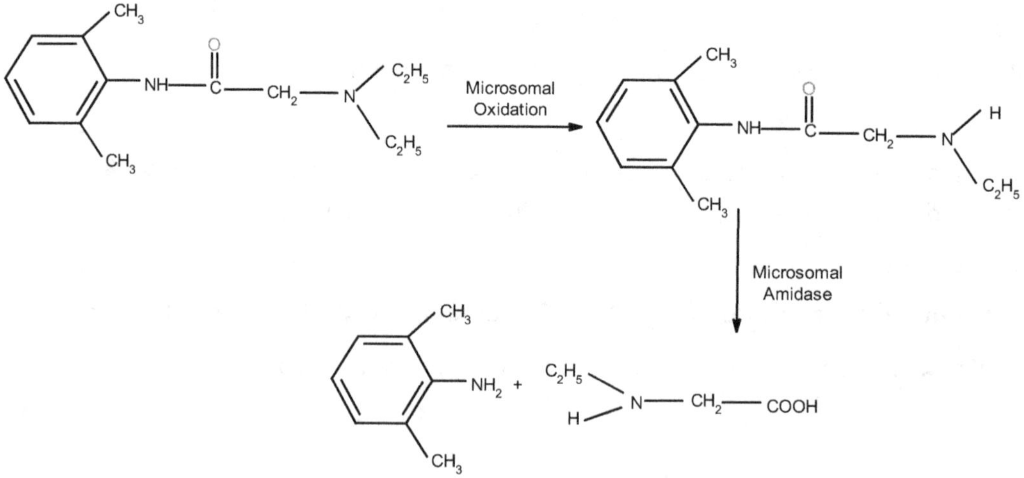

Metabolism of Lidocaine

Precautions must be taken so that lidocaine hydrochloride solutions containing epinephrine salts are not used as cardiac depressants. Such solutions are intended only for local anaesthesia and are not used intravenously. The aqueous solutions without epinephrine may be autoclaved several times, if necessary.

Uses:

This medication is used to prevent and relieve pain during certain medical procedures (such as inserting a tube into the urinary tract). It is also used to numb the mouth, throat, or nose lining before certain medical procedures (such as intubation).

Dose: The dose is 20 mg

Pharmaceutical Formulations: Solution

Brand Names: Xylocaine Viscous Solution

7.1.3.6 Lorcainide (Lorcainide hydrochloride)

Structure:

Chemical (IUPAC) name: *N*-(4-chlorophenyl)-2-phenyl-*N*-(1-propan-2-ylpiperidin-4-yl) acetamide; hydrochloride

It is a Class 1c antiarrhythmic agent that is used to help restore normal heart rhythm and conduction in patients with premature ventricular contractions, ventricular tachycardiac and Wolff-Parkinson-White syndrome. Lorcainide was developed by Janssen Pharmaceutica (Belgium) in 1968 under the commercial name Remivox and is designated by code numbers R-15889 or Ro 13-1042/001. It has a half-life of 8.9 +- 2.3 hrs, which may be prolonged to 66 hrs in people with cardiac disease.

Uses:

It is used to help restore normal heart rhythm and conduction in patients with premature ventricular contractions, ventricular tachycardiac and Wolff-Parkinson-White syndrome.

Dose: The dose is 10 mg, 20 mg

Pharmaceutical Formulations: Tablet

Brand Names: Lotensyl 10 Tablet Lotensyl 20 Tablet

7.1.3.7 Amiodarone

Structure:

Chemical (IUPAC) name: 2-butyl-3-benzofuranyl-4-[2-(diethylamino) ethoxy]-3,5-diiodo-phenyl ketone

It was introduced as an antianginal agent. It has very pronounced class III action and is especially effective in maintaining sinus rhythm in patients who have been treated by direct current shock for atrial fibrillation. Like class III antiarrhythmic drugs, amiodarone lengthens the effective refractory period by prolonging the action potential duration in all myocardial tissues. Amiodarone is eliminated very slowly from the body, with a half-life of about 25 to 30 days after oral doses.

Uses:

Amiodarone is used to treat and prevent certain types of severe, life-threatening ventricular arrhythmias (a certain type of abnormal heart rhythm when other medications did not help or could not be tolerated.

It is used in the treatment of Supraventricular Tachycardia.

It is used in the treatment of Atrial Fibrillation.

Storage:

It should be protected from light and stored at a temperature not exceeding 30^{0}C.

Dose: The dose is 100 mg, 200 mg

Pharmaceutical Formulations: Tablet, Solution, Injection

Brand Names: Cordarone, Cordarone X, Tachyra 100, Nexterone, Pancerone.

7.1.3.8 Sotalol

Structure:

Chemical (IUPAC) name: 4-[1-hydroxy-2-(isopropylamino) ethyl]methylsulfonanilide.

It is a relatively new antiarrhythmic drug, characterised most often as a class III agent. Although it has effects that are related to the class II agents, it is not therapeutically considered a class II antiarrhythmic.

The sotalol enantiomers produce different effects on the heart. Class III action of *d*-sotalol in the sinus node is associated with slowing sinus heart rate. In contrast, β - adrenergic blockade contributes to the decrease in heart rate observed with *l*- or *d, l*-sotalol. Sotalol is not metabolised, nor is it bound significantly to proteins. Elimination occurs by renal excretion, with more than 80% of the drug eliminated unchanged.

Sotalol is characteristic of class III antiarrhythmic drugs in that it prolongs the duration of the action potential and, thus, increases the effective refractory period of myocardial tissue. It is distinguished from the other class III drugs (amiodarone and bretylium) because of its β -adrenergic receptor–blocking action.

Uses:

It's used to treat atrial fibrillation and other conditions that cause an irregular heartbeat (arrhythmia).

Dose: The dose of 80 mg

Pharmaceutical Formulations: Tablet, Injection

Brand Names: Sotagard, Betapace, Sorine, Sotylize

QUESTION BANK

A. MULTIPLE CHOICE QUESTIONS

1. Which one of the following Anti-Arrhythmic Drugs are used as antimalarial?
    A. Quinidine Sulphate
    B. Verapamil

    C. Phenytoin

    D. Sotalol

Answer: A

2. Which one of the following is used to treat seizure disorders?
   A. Verapamil
   B. Lidocaine
   C. Amiodarone
   D. Phenytoin

Answer: D

3. Lidocaine Hydrochloride is also known as __________.
   A. Lorcainide
   B. Xylocaine
   C. Ketamine
   D. Thiopental sodium

Answer: B

4. Which one of the anti-arrhythmic drugs was introduced as a local anaesthetic?
   A. Verapamil
   B. Lidocaine
   C. Quinidine Sulphate
   D. Sotalol

Answer: B

5. Cordarone is brand name for ________.
   A. Amiodarone
   B. Lorcainide
   C. Sotalol
   D. Quinidine Sulphate

Answer: A

6. Quinidine sulfate is the __________.
   A. Alkaloid
   B. Tannin
   C. Glycosides
   D. Volatile oil

Answer: A

7. Betapace is brand name for _______.
    A. Amiodarone
    B. Lorcainide
    C. Sotalol
    D. Quinidine Sulphate

    Answer: C

8. Which one of the following is of class III antiarrhythmic drugs?
    A. class I
    B. class II
    C. class III
    D. class IV

    Answer: C

9. Eptoin is brand name for _______.
    A. Amiodarone
    B. Phenytoin
    C. Sotalol
    D. Quinidine Sulphate

    Answer: B

10. Lidocaine hydrochloride is a _________ antiarrhythmic agent.
    A. class I
    B. class II
    C. class III
    D. class IV

    Answer: A

B. SHORT ANSWER QUESTIONS

1. Write a note on Anti-Arrhythmic Drugs.
2. Explain the Quinidine Sulphate as Anti-Arrhythmic.
3. Draw the structure for Procainamide Hydrochloride
4. Explain the uses of Verapamil
5. Give the storage condition and marketed formulation for Phenytoin Sodium.
6. Explain the Lidocaine Hydrochloride.
7. Enlist different kinds of formulations available for Lorcainide Hydrochloride along with brand names.

8. Write a note on Sotalol.

9. Enlist different kinds of formulations available for Amiodarone along with brand names.

10. Give Storage conditions, different kinds of the formulation are available for Lidocaine Hydrochloride, along with brand names.

C. LONG ANSWER QUESTIONS

1. Give the classification of Anti-Arrhythmic Drugs with an example. Discuss Quinidine Sulphate in detail.

2. Give the classification of Anti-Arrhythmic Drugs with an example. Discuss Lidocaine Hydrochloride in detail.

■■■

DRUGS ACTING ON CARDIOVASCULAR SYSTEM: ANTI-HYPERTENSIVE AGENTS

CONTENTS

Introduction to hypertension

Classification of Anti-hypertensive agents

Selected Anti-hypertensive Agents:

- Propranolol
- Captopril
- Ramipril
- Methyldopate Hydrochloride
- Clonidine Hydrochloride
- Hydralazine Hydrochloride
- Nifedipine

♦ LEARNING OBJECTIVES ♦

After completing this chapter, the student should be able to understand:

1. Concept of hypertension and its types.
2. Classification of Anti-hypertensive Agents.
3. Chemical Structures of different Anti-hypertensive Agents.
4. Uses, Stability and storage conditions of Anti-hypertensive Agents.
5. Different types of formulations and popular brand names for Anti-hypertensive Agents.

7.2.1 INTRODUCTION

High blood pressure (hypertension) is a common condition in which the long-term force of the blood against your artery walls is high enough that it may eventually cause health problems, such as heart disease.

Blood pressure is determined both by the amount of blood your heart pumps and the amount of resistance to blood flow in your arteries. The more blood your heart pumps and the narrower your arteries, the higher your blood pressure. A blood pressure reading is given in millimetres of mercury (mm Hg). It has two numbers.

- Top number (systolic pressure). The first, or upper, number measures the pressure in your arteries when your heart beats.
- Bottom number (diastolic pressure). The second, or lower, number measures the pressure in your arteries between beats.

A person can have high blood pressure for years without any symptoms. Uncontrolled high blood pressure increases your risk of severe health problems, including heart attack and stroke. Fortunately, high blood pressure can be easily detected. And once you know you have high blood pressure, you can work with your doctor to control it.

Symptoms:

Most people with high blood pressure have no signs or symptoms, even if blood pressure readings reach dangerously high levels.

A few people with high blood pressure may have headaches, shortness of breath or nosebleeds, but these signs and symptoms aren't specific and usually don't occur until high blood pressure has reached a severe or life-threatening stage.

Causes:

There are two types of high blood pressure:

Primary (essential) hypertension:

For most adults, there's no identifiable cause of high blood pressure. This type of high blood pressure, called primary (essential) hypertension, tends to develop gradually over many years.

Secondary hypertension:

Some people have high blood pressure caused by an underlying condition. This type of high blood pressure, called secondary hypertension, tends to appear suddenly and cause higher blood pressure than does primary hypertension. Various conditions and medications can lead to secondary hypertension, including:

- Obstructive sleep apnea
- Kidney disease
- Adrenal gland tumours

- Thyroid problems
- Certain defects you're born with (congenital) in blood vessels
- Certain medications, such as birth control pills, cold remedies, decongestants, over-the-counter pain relievers and some prescription drugs
- Illegal drugs, such as cocaine and amphetamines

Risk factors:

High blood pressure has many risk factors, including:

- **Age:** The risk of high blood pressure increases as you age. Until about age 64, high blood pressure is more common in men. Women are more likely to develop high blood pressure after age 65.
- **Race:** High blood pressure is particularly common among people of African heritage, often developing at an earlier age than it does in whites. Serious complications, such as stroke, heart attack and kidney failure, also are more common in people of African heritage.
- **Family history:** High blood pressure tends to run in families.
- **Being overweight or obese:** The more you weigh, the more blood you need to supply oxygen and nutrients to your tissues. As the amount of blood flow through your blood vessels increases, so does the pressure on your artery walls.
- **Not being physically active:** People who are inactive tend to have higher heart rates. The higher your heart rate, the harder your heart must work with each contraction and the stronger the force on your arteries. Lack of physical activity also increases the risk of being overweight.
- **Using tobacco:** Not only does smoking or chewing tobacco immediately raise your blood pressure temporarily, but the chemicals in tobacco can damage the lining of your artery walls. This can cause your arteries to narrow and increase your risk of heart disease. Secondhand smoke also can increase your heart disease risk.
- **Too much salt (sodium) in your diet:** Too much sodium in your diet can cause your body to retain fluid, which increases blood pressure.
- **Too little potassium in your diet:** Potassium helps balance the amount of sodium in your cells. A proper balance of potassium is critical for good heart health. If you don't get enough potassium in your diet, or you lose too much potassium due to dehydration or other health conditions, sodium can build up in your blood.
- **Drinking too much alcohol:** Over time, heavy drinking can damage your heart. Having more than one drink a day for women and more than two drinks a day for men may affect your blood pressure.
- If you drink alcohol, do so in moderation. For healthy adults, that means up to one drink a day for women and two drinks a day for men. One drink equals 12 ounces of beer, 5 ounces of wine or 1.5 ounces of 80-proof liquor.

- **Stress:** High levels of stress can lead to a temporary increase in blood pressure. Stress-related habits such as eating more, using tobacco or drinking alcohol can lead to further increases in blood pressure.

- **Certain chronic conditions:** Certain chronic conditions also may increase your risk of high blood pressure, including kidney disease, diabetes and sleep apnea.

Sometimes pregnancy contributes to high blood pressure as well.

Although high blood pressure is most common in adults, children may be at risk, too. For some children, high blood pressure is caused by problems with the kidneys or heart. But for a growing number of kids, poor lifestyle habits — such as an unhealthy diet and lack of exercise — contribute to high blood pressure.

Complications:

The excessive pressure on your artery walls caused by high blood pressure can damage your blood vessels as well as your organs. The higher your blood pressure and the longer it goes uncontrolled, the greater the damage.

Uncontrolled high blood pressure can lead to complications, including:

- **Heart attack or stroke:** High blood pressure can cause hardening and thickening of the arteries (atherosclerosis), leading to a heart attack, stroke or other complications.

- **Aneurysm:** Increased blood pressure can cause your blood vessels to weaken and bulge, forming an aneurysm. If an aneurysm ruptures, it can be life-threatening.

- **Heart failure:** To pump blood against the higher pressure in your vessels, the heart has to work harder. This causes the walls of the heart's pumping chamber to thicken (left ventricular hypertrophy). Eventually, the thickened muscle may have difficulty pumping enough blood to meet your body's needs, which can lead to heart failure.

- **Weakened and narrowed blood vessels in your kidneys:** This can prevent these organs from functioning normally.

- **Thickened, narrowed or torn blood vessels in the eyes:** This can result in vision loss.

- **Metabolic syndrome:** This syndrome is a group of disorders of your body's metabolism, including increased waist size, high triglycerides, decreased high-density lipoprotein (HDL) cholesterol (the "good" cholesterol), high blood pressure and high insulin levels. These conditions make you more likely to develop diabetes, heart disease and stroke.

- **Trouble with memory or understanding:** Uncontrolled high blood pressure may also affect your ability to think, remember and learn. Trouble with memory or understanding concepts is more common in people with high blood pressure.

- **Dementia:** Narrowed or blocked arteries can limit blood flow to the brain, leading to a certain type of dementia (vascular dementia). A stroke that interrupts blood flow to the brain also can cause vascular dementia.

7.2.2 Classification of Antihypertensive Agents

1. **ACE Inhibitor:** e.g. Captopril, Lisnopril, Esnopril
2. **Angiotensin receptor Blocker:** e.g. Valsartan, Lomisartan
3. **Alpha Blocker:** e.g. Prazocine, Tetrazocine
4. **Alpha plus beta Blocker:** e.g. Carbediolol, Labetolol
5. **Adrenergic Blocker:** e.g. Reserpine,
6. **Beta-blocker:** e.g. Propranolol, Sotalol, Esmolol
7. **Central Sympatholytic:** e.g. Methyldopa, Clonidine
8. **Direct Renin Inhibitor:** e.g. Aliskerin
9. **Diuretics**
    a) **Thiazide derivative** e.g. Hydrochlorthiazide, Chlorthiazide
    b) **High ceiling diuretics**, e.g. Furosemide
    c) **K- sparing diuretics** e.g. Spiranolactone
10. **Ganglionic Blockers** e.g. Pentaonium
11. **Vasodilator**
    a) **Arteriodilator:** e.g. Minoxidil, Hydralazine
    b) **Artery and vein dilator:** e.g. Sodium Nitroprusside, Pentaonium
12. **Calcium Channel blockers:** e.g. Nifidipin

7.2.3 SOME IMPORTANT ANTIHYPERTENSIVE AGENTS

7.2.3.1 Propranolol

Structure:

Chemical (IUPAC) name: (RS)-1-(1-methylethylamIno)-3-(1-naphthyloxy)propan-2 ol

Uses: It is a nonselective β-adrenergic blocker and is the prototype for class II antiarrhythmics.

Its use as an antiarrhythmic is typically for the treatment of supraventricular arrhythmias, including atrial flutter, paroxysmal supraventricular tachycardia, and atrial fibrillation. Propranolol is also reported to be effective in the treatment of digitalis-induced ventricular arrhythmias.

Moreover, beneficial results can be obtained when propranolol is used in combination with other agents. For example, in some instances, quinidine and propranolol together have proved to be more successful in alleviating atrial fibrillation than either agent alone. Few serious adverse effects are associated with propranolol therapy.

Side Effects: The main side effects of propranolol are feeling dizzy or tired, cold hands or feet, difficulties sleeping and nightmares.

Storage Condition: Propranolol suspensions 2 mg/mL and 5 mg/mL stored at 25°C maintained at least 94.7% of their initial concentration for 120 days, and suspensions stored at 4°C maintained at least 93.9% of their initial concentration for 120 days.

Dose: The dose is 10 mg, 20 mg, 40 mg, 80 mg, and 2 mg/mL, 5 mg/mL 20mg/0.25mg

Pharmaceutical Formulations: Tablet, Suspension

Brand Names: Arminol, Ciplar-LA, Syziral, Inderal, Dizepax, Alpron Plus, Besprol, Arminol Inderal, Ciplar, Dizepax-M, Inderal, Migrales, Syziral M-20

7.2.3.2 Captopril

Structure:

Chemical (IUPAC) name: 1-[(2S)-3-mercapto-2-methyl-1-oxopropionyl] praline.

Uses: It blocks the conversion of angiotensin I to angiotensin II by inhibiting the converting enzyme.

The rational development of captopril as an inhibitor of ACE was based on the hypothesis that ACE and carboxy peptidase-A functioned by similar mechanisms. It was noted that *d*-2-benzylsuccinic acid was a potent inhibitor of carboxy peptidase-A, but not ACE.

By use of this small molecule as a prototype, captopril was designed with a carboxyl group on proline, and a thiol group was introduced to enhance the binding to the zinc ion of ACE.

The important binding points at the active site of ACE are thought to be an arginine residue, which provides a cationic site that attracts a carboxylate ion, and a zinc ion, which can polarise a carbonyl group of an amide function to make it more susceptible to hydrolysis.

Hydrophobic pockets lie between these groups in the active site, as does a functional group that forms a hydrogen bond with an amide carbonyl.

Side Effects: The main side effects of Captopril are dizziness or lightheadedness, salty or metallic taste, or decreased ability to taste, cough, fast heartbeat, excessive tiredness etc.

Storage Condition: Store captopril at room temperature between 68°F and 77°F (20°C and 25°C). Don't freeze captopril. Keep this drug away from light.

Dose: The dose is 12.5 mg, 25 mg, 25 mg/15mg

Pharmaceutical Formulations: Tablet

Brand Names: Captopril, Aceten, Acezide, Captopril-H, Angiopril, Angiopril-DU

7.2.3.3 Ramipril

Structure:

Chemical (IUPAC) name: (2S, 3aS, 6aS)-1-[(S)-N-[(S)-1-carboxy- 3-phenylpropyl]alanyl] octahydrocyclopenta[b]- pyrrole-2-carboxylic acid 1-ethyl ester

It is hydrolysed to ramiprilat, its active diacid form, faster than enalapril is hydrolysed to its active diacid form. Peak serum concentrations from a single oral dose are achieved between 1.5 and 3 hours. The ramiprilate formed completely suppresses ACE activity for up to 12 hours, with 80% inhibition of the enzyme still observed after 24 hours.

Uses: Ramipril is used to treat high blood pressure (hypertension). Lowering high blood pressure helps prevent strokes, heart attacks, and kidney problems.

Ramipril is also used to improve survival after a heart attack. It may also be used in high-risk patients (such as patients with heart disease/diabetes) to help prevent heart attacks and strokes.

Side Effects: The main side effects of Ramipril are Feeling dizzy or lightheaded, especially when you stand up or sit up quickly, Headaches, Diarrhoea, Being sick (vomiting), A mild skin rash, and Blurred vision.

Storage Condition: Store at controlled room temperature (59°–86°F).

Dose: The dose is 1.25 mg, 2.5 mg, 5 mg, 2.5 mg/2.5 mg

Pharmaceutical Formulations: Tablet

Brand Names: Cardace, Ramipro, Bampril-H, Ril-AH, Teram, Ramveda, Diakit-4, Loram-H, Sartace-H, Dexace, Ramihart-H, Topril, Telipril, Ramic-M. Ril.

7.2.3.4 Methyldopa

Structure:

Chemical (IUPAC) name: (S)-2-amino-3-(3,4-dihydroxyphenyl)-2-methyl-propanoic acid

Mechanism of action:

The mechanism of action of methyldopa is not fully clear. Although it is a centrally acting sympathomimetic, it does not block reuptake or transporters. It may reduce the dopaminergic and serotonergic transmission in the, and it indirectly affects norepinephrine (noradrenaline) synthesis. Methyldopa acts on alpha two receptors which are found on the pre-synaptic nerve terminal. The activation of alpha two on pre-synaptic nerve terminals inhibits the synthesis of norepinephrine by inhibiting tyrosine hydroxylase.

The S-enantiomer of methyldopa is a competitive inhibitor of the enzyme aromatic L-amino acid decarboxylase (LAAD), which converts L-DOPA into dopamine. L-DOPA can cross the blood-brain, and thus, methyldopa may have similar effects. LAAD converts it into alpha-methyldopamine, a false precursor to norepinephrine, which in turn reduces the synthesis of norepinephrine in the vesicles. Dopamine beta-hydroxylase (DBH) converts α-methyldopamine into α-methyl norepinephrine, which is an agonist of the presynaptic α_2-adrenergic receptor causing inhibition of neurotransmitter release.

Uses: Methyldopa is used in the clinical treatment of the following disorders:

- Hypertension (or high blood pressure).
- Gestational hypertension (or pregnancy-induced hypertension) and pre-eclampsia.

Side effects: Common side effect includes sleepiness. More severe side effects include red blood cell breakdown, liver problems, and allergic reactions. Methyldopa is in the α_2-adrenergic receptor agonist family of medications. It works by stimulating the brain to decrease the activity of the sympathetic nervous system.

Storage Condition: Methyldopa oral suspension should be stored in tight, light-resistant containers at a temperature less than 26 °C and protected from freezing.

Dose: The dose is 250 mg, 500 mg

Pharmaceutical Formulations: Tablet, Suspension

Brand Names: Gynapres, Alsopam, Metozide, Alphadopa, Dopagyt, Emdopa, Sembrina.

7.2.3.5 Clonidine Hydrochloride

Structure:

Chemical (IUPAC) name: 2-[(2,6-dichlorophenyl)imino]imidazolidine monohydrochloride.

It was synthesised in 1962 as a derivative of the known α-sympathomimetic drugs naphazoline and tolazoline, potential nasal vasoconstrictors, but instead, it proved to be effective in the treatment of mild-to-severe hypertension.

Uses: Clonidine is also used in the treatment of-

- dysmenorrhea (severely painful cramps during a menstrual period),
- hypertensive crisis (a condition in which your blood pressure is very high),
- Tourette's syndrome (a condition characterised by the need to perform repetitive motions or to repeat sounds or words),
- menopausal hot flashes, and
- alcohol and opiate (narcotic) withdrawal.

Side Effects: More common side effect observed with the administration of Clonidine is Constipation. Whereas other less common side effects include darkening of the skin, decreased sexual ability, dry, itching, or burning eyes, loss of appetite, nausea or vomiting.

Storage Condition: Store in a cool location.

Dose: The dose is 150 mg, 100mcg

Pharmaceutical Formulations: Injection, Tablet

Brand Names: Catapres, Arkamin, Catapres, Clodict.

7.2.3.6 Hydralazine Hydrochloride

Structure:

Chemical (IUPAC) name: 1-hydrazinophthalazine monohydrochloride

It is originated from the work of a chemist attempting to produce some unusual chemical compounds and from the observation that this compound had antihypertensive properties.

It occurs as yellow crystals and is soluble in water to the extent of about 3%. A 2% aqueous solution has a pH of 3.5 to 4.5. Hydralazine hydrochloride is useful in the treatment of moderate-to-severe hypertension.

Uses: Hydralazine is used to treat high blood pressure. Hydralazine is in a class of medications called vasodilators. It works by relaxing the blood vessels so that blood can flow more easily through the body.

Hydralazine is also used after heart valve replacement and in the treatment of heart failure.

Side Effects: Common side effects observed with the administration of Hydralazine are flushing, headache, upset stomach, vomiting, loss of appetite, diarrhoea, constipation, eye tearing etc.

Storage Condition: Oral tablets: are stored in lightproof and air-tight containers, between 15°C and 40°C (59° to 104°F). Ampules should be kept between the same temperatures described above and should not be frozen.

Dose: The dose is 25 mg

Pharmaceutical Formulations: Tablet, injection

Brand Names: Carbetazine, Corbetazine.

7.2.3.7 Nifedipine

Structure:

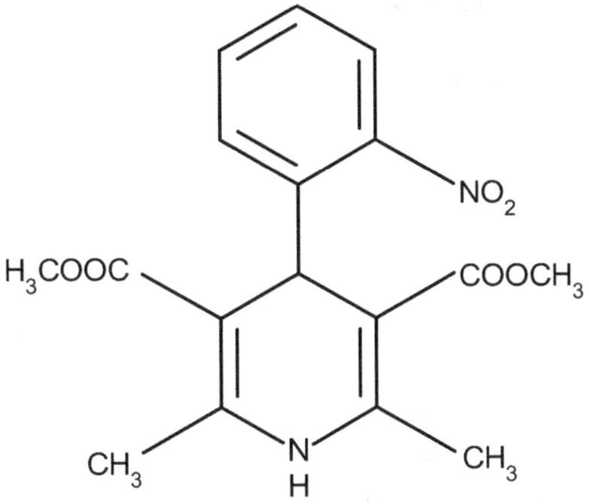

Chemical (IUPAC) name: 1,4-dihydro-2, 6-dimethyl-4-(2-nitrophenyl)-3,5-pyridinedicarboxylate dimethyl ester

It is a dihydropyridine derivative that bears no structural resemblance to the other calcium antagonists.

Uses: Nifedipine is used alone or together with other medicines to treat severe chest pain (angina) or high blood pressure (hypertension).

Nifedipine is more effective in patients whose anginal episodes are caused by coronary vasospasm and is used in the treatment of vasospastic angina as well as classic angina pectoris.

Side Effects: More common side effects of Nifedipineare Bloating or swelling of the face, arms, hands, lower legs, or feet, cough, difficult or laboured breathing, dizziness or lightheadedness, fast, irregular, pounding, or racing heartbeat or pulse, feeling of warmth, headache, muscle cramps, rapid weight gain etc.

Storage Condition: Capsules should be protected from light and moisture and stored at controlled room temperature, 59° to 77°F (15° to 25°C)

Dose: The dose is 5 mg, 10 mg, 20 mg, 30 mg, 50 mg /20 mg

Pharmaceutical Formulations: Tablet, Capsule

Brand Names: Nicardia Retard, Angifine-SR, Nilol HD, Nifcard, Myogard, Systonorm-NF, Depicor, Nicardia. Calcigard.

QUESTION BANK

A. MULTIPLE CHOICE QUESTIONS

1. Ciplar is popular brand name of __________.
   A. Propranolol
   B. Ramipril
   C. Methyldopate Hydrochloride
   D. Clonidine Hydrochloride

   Answer: A

2. Proparanolol is __________________ .
   A. Nonselective α-adrenergic blocker
   B. Nonselective β-adrenergic blocker
   C. Selective α-adrenergic blocker
   D. Selective β-adrenergic blocker

   Answer: B

3. Which of the following is an ACE Inhibitor?
   A. Clonidine Hydrochloride
   B. Hydralazine Hydrochloride
   C. Captopril
   D. Nifedipine

   Answer: C

4. Which of the following is a Calcium channel blocker?
   A. Propranolol
   B. Captopril
   C. Clonidine Hydrochloride
   D. Nifedipine

   Answer: D

5. Which of the following heterocyclic ring system is present in Nifedipine?
   A. Pyrrol
   B. Pyridine
   C. Benfuran
   D. Thiophene

   Answer: B

B. SHORT ANSWER QUESTIONS

1. Explain Propranolol as Anti-Hypertensive Agent.
2. Explain Nifedipine as Anti-Hypertensive Agent.
3. Give physical properties and structure of Captopril.
4. Enlist names of drugs from ACE Inhibitor.
5. Enlist pharmaceutical formulations available for Propranolol
6. Nifedipine belongs to which class of Antihypertensive agents?
7. Enlist different brand names for Nifedipine
8. Enlist pharmaceutical formulations available for Methyldopa.
9. Write a note on Hydralazine Hydrochloride.
10. Write a note on Clonidine Hydrochloride.

C. LONG ANSWER QUESTIONS

1. Classify Antihypertensive Agents with examples.
2. Classify Antihypertensive Agents. Add note on Propranolol.
3. Classify Antihypertensive Agents. Explain in detail Nifedipine. Explain the properties, Uses, Pharmaceutical Formulation, and Popular Brand names of Ramipril.

■■■

DRUGS ACTING ON CARDIOVASCULAR SYSTEM: ANTIANGINAL AGENTS

CONTENTS

Antianginal Agents:

- Isosorbide Dinitrate

♦ LEARNING OBJECTIVES ♦

After completing this chapter, the student should be able to understand:

1. Ischemia
2. Ischemic heart disease (IHD)
3. Angina pectoris
4. Classification of Angina pectoris
5. Signs and symptoms of Angina pectoris
6. Classification of Angina pectoris
7. A drug used for the treatment of Angina pectoris

7.3.1 INTRODUCTION

Definitions Ischemia:

It is the inadequate blood supply to a part of the body, even to the point of complete deprivation.

Ischemic heart disease (IHD):

It is defined as an acute or chronic form of cardiac disability arising from an imbalance between the myocardial supply and demand for oxygenated blood, e.g. Angina pectoris and myocardial infarction.

Angina pectoris:

It is a clinical syndrome of IHD resulting from transient myocardial ischemia. It is characterised by pain in the substernal or pre-cordial region of the chest which is aggravated by an increase in the demand of the heart and relieved by a decrease in the work of the heart. Often pain radiated to the left arm, neck, jaw or right arm.

Angina, also known as angina pectoris, is chest pain or pressure, usually due to insufficient blood flow to the heart muscle (myocardium).

Angina is usually due to obstruction or spasms of the arteries that supply blood to the heart muscle. Other causes include anaemia, abnormal heart rhythms, and heart failure. The main mechanism of coronary artery obstruction is atherosclerosis as part of coronary artery disease. The term derives from the Latin angere ("to strangle") and pectus ("chest") and can therefore be translated as "a strangling feeling in the chest".

There is a weak relationship between the severity, and the degree of oxygen deprivation in the heart muscle, where there can be severe pain with little or no risk of a myocardial infarction (heart attack) and a heart attack can occur without pain. In some cases, angina can be quite severe. In the early 20th century, this was a known sign of impending death. However, given current medical therapies, the outlook has improved substantially. People with an average age of 62 years who have moderate to severe degrees of angina (graded II, III, and IV) have a five-year survival rate of approximately 92%.

Worsening angina attacks, sudden-onset angina at rest, and angina lasting more than 15 minutes are symptoms of unstable angina (usually grouped with similar conditions as an acute coronary syndrome). As these may precede a heart attack, they require urgent medical attention and are generally treated similarly to myocardial infarction.

Classification

Stable angina

Also known as 'effort angina', this refers to the classic type of angina related to myocardial ischemia. A typical presentation of stable angina is chest discomfort and associated symptoms precipitated by some activity (running, walking, etc.) with minimal or non-existent symptoms at rest or after administration of sublingual nitroglycerin. Symptoms typically abate several minutes after activity and recur when activity resumes. In this way, stable angina may be thought of as being similar to intermittent claudication symptoms. Other recognised precipitants of stable angina include cold weather, heavy meals, and emotional stress.

Unstable angina

Unstable angina (UA) (also "crescendo angina"; this is a form of the acute coronary syndrome) is defined as angina pectoris that changes or worsens.

It has at least one of these three features:

1. It occurs at rest (or with minimal exertion), usually lasting more than 10 minutes

2. It is severe and of new-onset (i.e., within the prior 4–6 weeks)

3. It occurs with a crescendo pattern (i.e., distinctly more severe, prolonged, or frequent than before).

UA may occur unpredictably at rest, which may be a serious indicator of an impending heart attack. What differentiates stable angina from unstable angina (other than symptoms) is the pathophysiology of atherosclerosis. The pathophysiology of unstable angina is the reduction of coronary flow due to transient platelet aggregation on apparently normal endothelium, coronary artery spasms, or coronary thrombosis. The process starts with atherosclerosis and progresses through inflammation to yield an active unstable plaque, which undergoes thrombosis and results in acute myocardial ischemia, which, if not reversed, results in cell necrosis (infarction). Studies show that 64% of all unstable anginas occur between 22:00 and 08:00 when patients are at rest.

Instable angina, the developing atheroma, is protected with a fibrous cap. This cap may rupture in unstable angina, allowing blood clots to precipitate and further decrease the area of the coronary vessel's lumen. This explains why, in many cases, unstable angina develops independently of activity.

Cardiac syndrome X

Cardiac syndrome X, sometimes known as microvascular angina, is characterised by angina-like chest pain in the context of normal epicardial coronary arteries (the largest vessels on the surface of the heart, prior to significant branching) on angiography. The original definition of cardiac syndrome X also mandated that ischemic changes on exercise (despite normal coronary arteries) were displayed, as shown on cardiac stress tests. The primary cause of cardiac syndrome X is unknown, but the factors apparently involved are endothelial dysfunction and reduced flow (perhaps due to spasm) in the tiny "resistance" blood vessels of the heart. Since microvascular angina is not characterised by major arterial blockages, it is harder to recognise and diagnose. Microvascular angina was previously considered a relatively benign condition, but more recent data has changed this attitude. Studies, including the Women's Ischemia Syndrome Evaluation (WISE), suggest that microvascular angina is part of the pathophysiology of ischemic heart disease, perhaps explaining the higher rates of angina in women than in men, as well as their predilection towards ischemia and acute coronary syndromes in the absence of obstructive coronary artery disease.

Signs and symptoms

Angina pectoris can be quite painful, but many patients with angina complain of chest discomfort rather than actual pain: the discomfort is usually described as a pressure, heaviness, tightness, squeezing, burning, or choking sensation. Apart from chest discomfort, anginal pains may also be experienced in the epigastrium (upper central abdomen), back, neck area, jaw, or shoulders. This is explained by the concept of referred pain and is because the spinal level that receives visceral sensation from the heart simultaneously receives cutaneous sensation from parts of the skin specified by that spinal nerve's dermatome without an ability to discriminate the two. Typical locations for referred pain are arms (often inner left arm), shoulders, and neck into the jaw. Angina is typically precipitated by exertion or emotional stress. It is exacerbated by having a full stomach and by cold temperatures. Pain may be accompanied by breathlessness, sweating, and nausea in some cases. In this case, the pulse rate and blood pressure increase. Chest pain lasting only a few seconds is normally not angina (such as precordial catch syndrome).

Myocardial ischemia comes about when the myocardium (the heart muscle) receives insufficient blood and oxygen to function normally either because of increased oxygen demand by the myocardium or because of decreased supply to the myocardium. This inadequate perfusion of blood and the resulting reduced delivery of oxygen and nutrients are directly correlated to blocked or narrowed blood vessels.

Some experience "autonomic symptoms" (related to increased activity of the autonomic nervous system), such as nausea, vomiting, and pallor.

Major risk factors for angina include cigarette smoking, diabetes, high cholesterol, high blood pressure, a sedentary lifestyle, and a family history of premature heart disease.

A variant form of angina—Prinzmetal's angina—occurs in patients with normal coronary arteries or insignificant atherosclerosis. It is believed to be caused by spasms of the artery. It occurs more in younger women.

Coital angina, also known as angina d'amour, is angina subsequent to sexual intercourse. It is generally rare, except in patients with severe coronary artery disease.

Cause

Major risk factors

- Age (≥ 45 years for men, ≥ 55 for women)
- Smoking
- Diabetes mellitus
- Dyslipidemia
- Family history of premature cardiovascular disease (men <55 years, females <65 years old)
- Hypertension
- Kidney disease (microalbuminuria or GFR<60 mL/min)
- Obesity (BMI ≥ 30 kg/m2)
- Physical inactivity
- Prolonged psychosocial stress

Routine counselling of adults to advise them to improve their diet and increase their physical activity has not been found to alter behaviour significantly and thus is not recommended.

Conditions that exacerbate or provoke angina

- Medications
    - Vasodilators
    - Excessive thyroid hormone replacement
- Vasoconstrictors
- Polycythemia, which thickens the blood, slowing its flow through the heart muscle.
- Hypothermia
- Hypervolemia
- Hypovolemia

One study found that smokers with coronary artery disease had a significantly increased level of sympathetic nerve activity when compared to those without. This is in addition to increased blood pressure, heart rate, and peripheral vascular resistance associated with nicotine, which may lead to recurrent angina attacks. In addition, the Centers for Disease Control and Prevention (CDC) reports that the risk of CHD (Coronary heart disease), stroke, and PVD (Peripheral vascular disease) is reduced within 1–2 years of smoking cessation. In another study, it was found that, after one year, the prevalence of angina in smoking men under 60 after an initial attack was 40% less in those having quit smoking compared to those that continued. Studies have found that there are short-term and long-term benefits to smoking cessation.

Other medical problems

- Oesophagal disorders
- Gastroesophageal reflux disease (GERD)
- Hyperthyroidism
- Hypoxemia
- Profound anaemia
- Uncontrolled hypertension

Other cardiac problems

- Bradyarrhythmia
- Hypertrophic cardiomyopathy
- Tachyarrhythmia
- Valvular heart disease

Myocardial ischemia can result from:

1. A reduction of blood flow to the heart that can be caused by stenosis, spasm, or acute occlusion (by an embolus) of the heart's arteries.

2. Resistance of the blood vessels. This can be caused by narrowing of the blood vessels; a decrease in radius. Blood flow is proportional to the radius of the artery to the fourth power.

3. Reduced oxygen-carrying capacity of the blood due to several factors, such as a decrease in oxygen tension and haemoglobin concentration. This decreases the ability of haemoglobin to carry oxygen to myocardial tissue.

Atherosclerosis is the most common cause of stenosis (narrowing of the blood vessels) of the heart's arteries and, hence, angina pectoris. Some people with chest pain have normal or minimal narrowing of heart arteries; in these patients, vasospasm is a more likely cause for the pain, sometimes in the context of Prinzmetal's angina and syndrome X.

Myocardial ischemia also can be the result of factors affecting blood composition, such as the reduced oxygen-carrying capacity of the blood, as seen with severe anaemia (low number of red blood cells) or long-term smoking.

7.3.2 CLASSIFICATION ANTI-ANGINAL AGENTS

Nitrates

a) **Short-acting:** e.g. Glyceryl trinitrate (GNT, Nitogycerine)

b) **Long-acting:** e.g. Isosorbide dinitrate, Isosorbide mononitrate, Erythrityl tetranitrate

7.3.3 ISOSORBIDE DINITRATE

Isosorbide Dinitrate

Structure:

Chemical (IUPAC) name: 1,4:3,6-Dianhydro-D-glucitol dinitrate

Properties:

It occurs as a white, crystalline powder. Its water solubility is about 1 mg/mL.

Isosorbide dinitrate, as a sublingual or chewable tablet, is effective in the treatment or prophylaxis of acute anginal attacks. When it is given sublingually, the effect begins in about 2 minutes, with a shorter duration of action than when it is given orally. Oral tablets are not effective in acute anginal episodes; the onset of action ranges from 15 to 30 minutes. The major route of metabolism involves denitration to isosorbide 5-mononitrate. This metabolite has a much longer half-life than the parent isosorbide dinitrate. As such, this particular metabolite is marketed in a tablet form that has excellent bioavailability with much less first-pass metabolism than isosorbide dinitrate. Isosorbide is marketed as a combination therapy with hydralazine under the tradename BiDil, specifically for African Americans with CHF.

Mechanism of Action:

Isosorbide dinitrate is converted to the active nitric oxide to activate guanylate cyclase. This activation increases levels of cyclic guanosine 3',5'-monophosphate (cGMP). cGMP activates protein kinases and causes a series of phosphorylation reactions which leads to dephosphorylation of myosin light chains of smooth muscle fibres. Finally, there is a release of calcium ions which causes smooth muscle relaxation and vasodilation.

Uses:

Isosorbide dinitrate is used to prevent chest pain (angina) in patients with a certain heart condition (coronary artery disease). This medication belongs to a class of drugs known as nitrates. It works by relaxing and widening blood vessels so blood can flow more easily to the heart.

Side effects:

After long-term use for treating chronic conditions, tolerance may develop in patients, reducing its effectiveness. The mechanisms of nitrate tolerance have been thoroughly investigated in the last 30 years, and several hypotheses have been proposed. These include:

1. Severe allergic reactions (rash; hives; itching; difficulty breathing; tightness in the chest; swelling of the mouth, face, lips, or tongue); fainting; fast or slow heartbeat; nausea; new or worsening chest pain; vomiting.

2. Impaired biotransformation of ISDN to its active principle NO (or a NO-related species)

3. Neurohormonal activation causes sympathetic activation and release of vasoconstrictors such as endothelin and angiotensin II, which counteract the vasodilation induced by ISDN.

4. Plasma volume expansion

5. The oxidative stress hypothesis

Stability and storage conditions:

Isosorbide dinitrate should be stored at room temperature, 15^0C - 30^0C (59^0F - 86^0F).

Dose: The dose is 5 mg, 10 mg.

Pharmaceutical Formulations: Tablet

Brand Names: Sorbitrate 5 Tablet, Sorbitrate 10 Tablet, Isordil 5 Sublingual tablet, Isordil 10 Tablet

QUESTION BANK

A. MULTIPLE CHOICE QUESTIONS

1. Glyceryl trinitrate is ______________ anti-anginal agent.
   A. Short-acting
   B. Long Acting
   C. Intermediate-acting
   D. None of the above

   Answer: A

2. Isosorbide dinitrate is ______________ anti- anginal agent.
   A. Short-acting
   B. Long Acting
   C. Intermediate-acting
   D. None of the above

   Answer: B

3. Which one of the following is used to prevent chest pain?
   A. Quinidine Sulphate
   B. Phenytoin
   C. Isosorbide dinitrate
   D. Verapamil

   Answer: C

4. The chemical name for the Isosorbide dinitrate is______________________.
   A. 1,4:3,6-dehydro-D-glucitol
   B. 1,4:3,6-dehydro-L-glucitol
   C. 1,4:3,6-dianhydro-L-glucitol
   D. 1,4:3,6-dianhydro-D-glucitol

   Answer: D

5. Isosorbide denigrate is available in which of the following pharmaceutical dosage forms?
   A. Intramuscular Injection
   B. Intravenous Injection
   C. Solution
   D. Sublingual or chewable tablet

Answer: D

B. SHORT ANSWER QUESTIONS

1. Classify anti-anginal drugs.
2. Give the structure for Isosorbide dinitrate.
3. Enlist different pharmaceutical formulations for Isosorbide dinitrate.
4. Enlist different brand names for Isosorbide dinitrate.
5. Enlist different Side effects for Isosorbide dinitrate.

C. LONG ANSWER QUESTIONS

1. What are anti-anginal drugs? Give examples. Write a short account of Isosorbide dinitrate.

■■■

Chapter 8

DIURETICS

CONTENTS

Diuretics:

- Acetazolamide
- Frusemide*
- Bumetanide
- Chlorthalidone
- Benzthiazide
- Metolazone
- Xipamide

♦ LEARNING OBJECTIVES ♦

After completing this chapter, the student should be able to understand:

1. Classification of Diuretics
2. Chemical Structures of different Diuretics.
3. Uses, Stability and storage conditions of Diuretics.
4. Different types of formulations and popular brand names for Diuretics.

8.1 INTRODUCTION

Diuretics are chemicals that increase the rate of urine formation. By increasing the urine flow rate, diuretic usage leads to increased excretion of electrolytes (especially sodium and chloride ions) and water from the body without affecting protein, vitamin, glucose, or amino acid reabsorption. These pharmacologic properties have led to the use of diuretics in the treatment of edematous conditions resulting from a variety of causes (e.g., congestive heart failure, nephrotic syndrome, and chronic liver disease) and in the management of hypertension. Diuretic drugs also are useful as the sole agent or as an adjunct therapy in the treatment of a wide range of clinical conditions, including hypercalcemia, diabetes insipidus, acute mountain sickness, primary hyperaldosteronism, and glaucoma.

The primary target organ for diuretics is the kidney, where these drugs interfere with the reabsorption of sodium and other ions from the lumina of the nephrons, which are the functional units of the kidney. The amount of ions and accompanying water that are excreted as urine following administration of a diuretic, however, is determined by many factors, including the chemical structure of the diuretic, the site or sites of action of the agent, the salt intake of the patient, and the amount of extracellular fluid present. In addition to the direct effect of diuretics on impairing solute and water reabsorption from the nephron, diuretics also can trigger compensatory physiologic events that have an impact on either the magnitude or the duration of the diuretic response. Thus, it is important to be aware of the normal mechanisms of urine formation and renal control mechanisms to understand clearly the ability of chemicals to induce diuresis.

8.2 CLASSIFICATION OF DIURETICS

A. Drugs action on Proximal Convoluted Tubule

    a) Osmotic Diuretics: e.g. Isosorbide, Mannitol.

    b) Carbonic anhydrase inhibitors: e.g. Acetazolamide, Methazolamide, Dichlorphenamide.

    c) Acidifying Agents: e.g. Ammonium Chloride

B. Loop Diuretics: High ceiling diuretics

    a) Mercurials

    b) Phenoxy acetic acid derivatives: e.g. Ethacrynic acid

    c) Sulfamyl-anthranilic acid derivatives: e.g. Furosemide.

C. Drugs acting on Distal Convoluted Tubule

    a) Benzothiazdizines

      i) Thiazides: e.g. Chlorthiazide, Cyclothiazide.

      ii) Hydrothiazides: e.g. Hydrochlorothiazide Hydroflumethiazide.

D. Potassium Sparing Diuretics

    a) Pyrazine analogue: e.g. Amiloride

    b) Pteridine analogue: e.g. Triamterene.

    c) Aldosterone antagonist: e.g. Spironolactone

8.3 DIFFERENT DIURETIC DRUGS

8.3.1 Acetazolamide

Structure:

Chemical (IUPAC) name: N-(5-sulfamoyl-1,3,4-thiadiazol-2-yl)acetamide

Acetazolamide, a thiadiazole derivative, was the first of the carbonic anhydrase inhibitors to be introduced as an orally effective diuretic, with a therapeutic effect that lasts approximately 8 to 12 hours.

Molecular Weight: 222.24 g/mol

Physical Properties: Acetazolamide appears as white to yellowish-white fine crystalline powder. No odourless and tasteless powder.

Storage: It should be stored in the light-sensitive container.

Uses:

It is used in the treatment of glaucoma, drug-induced oedema, heart failure-induced oedema, epilepsy and in reducing intraocular pressure after surgery. It has also been used in the treatment of altitude sickness, Ménière's disease, increased intracranial pressure and neuromuscular disorders.

Adverse Effects:

The carbonic anhydrase inhibitors generally do not produce serious side effects. However, monitoring for electrolyte disturbances may be necessary. Gastrointestinal (e.g., nausea, vomiting), nervous system (e.g., sedation, headache), and renal effects (kidney stones) have been reported.

Dose: The dose is 250 mg, 500 mg

Pharmaceutical Formulations: Tablet, Capsule

Brand Names: Diamox, Lopar, Avva SR, Acemax, Glumox, Gatomax, Actazid

8.3.2 Furosemide (Furesemide)

Structure:

Chemical (IUPAC) name: 4-Chloro-2-[(furan-2-ylmethyl)amino]-5-sulfamoylbenzoic acid.

It is a loop diuretic used for the treatment of hypertension and oedema. It is the first-line agent in patients with oedema due to congestive heart failure. It is also used for hepatic cirrhosis, renal impairment, nephrotic syndrome, in adjunct therapy for cerebral or pulmonary oedema where rapid diuresis is required, and in the management of severe hypercalcemia, it is used in combination with optimal rehydration therapy.

Molecular Weight: 330.74 g/mol

Physical Properties: Furosemide is an odourless white to slightly yellow crystalline powder, almost tasteless.

Storage: Stable under recommended storage conditions.

Uses: Furosemide is primarily used for the treatment of oedema but also in some cases of hypertension (where there is also kidney or heart impairment). It is often viewed as a first-line agent in most people with oedema caused by congestive heart failure because of its anti-vasoconstrictor and diuretic effects.

Adverse Effect:

Furosemide also can lead to gout caused by hyperuricemia. Hyperglycemia is also a common side effect. The tendency, as for all loop diuretics, to cause low serum potassium concentration (hypokalemia) has given rise to combination products, either with potassium or with the potassium-sparing diuretic amiloride (Co-amilofruse). Other electrolyte abnormalities resulting from furosemide use include hyponatremia, hypochloremia, hypomagnesemia, and hypocalcemia.

Dose:
1. Oral tablet- 20 mg, 40 mg, and 80 mg
2. Oral solution- 10 mg per 1 mL, 40 mg per 5 mL

The usual starting dose is 80 mg per day, taken as 40 mg twice each day.

Pharmaceutical Formulations: Tablet, Solution, Injection.

Brand Names: Discoid, Diural, Flusapex, Frudix, Frusetic, Frusid, Fulsix

8.3.3 Bumetanide

Structure:

Chemical (IUPAC) name: 3-butylamino-4-phenoxy-5-sulfamoyl-benzoic acid

It is sold under the brand name Bumex among others and is a medication used to treat swelling and high blood pressure. This includes swelling as a result of heart failure, liver failure, or kidney problems. It may work for swelling when other medications have not. For high blood pressure, it is not a preferred treatment. It is taken by mouth or by injection into a vein or muscle. Effects generally begin within an hour and last for about six hours.

Bumetanide was patented in 1968 and came into medical use in 1972. It is on the World Health Organization's List of Essential Medicines. It is available as a generic medication. In 2019, it was the 241st most commonly prescribed medication in the United States, with more than one million prescriptions.

Uses:

It is used to treat swelling and high blood pressure. This includes swelling as a result of heart failure, liver failure, or kidney problems. For high blood pressure, it is not a preferred treatment. It is taken by mouth or by injection into a vein or muscle.

Adverse Effect:

Common side effects include dizziness, low blood pressure, low blood potassium, muscle cramps, and kidney problems. Other serious side effects may include hearing loss and low blood platelets. Blood tests are recommended regularly for those on treatment. Safety during pregnancy and breastfeeding is unclear. Bumetanide is a loop diuretic and works by decreasing the reabsorption of sodium by the kidneys.

Dose:

1. Oral tablet- 0.5 mg, 1 mg and 2 mg
2. injectable solution- 0.25mg/mL

Pharmaceutical Formulations: Tablet, Injection.

Brand Names: Bumex, Burinex.

8.3.4 Chlorthalidone

Structure:

Chemical (IUPAC) name: (RS)-2-Chloro-5-(1-hydroxy-3-oxo-2,3-dihydro-1H-isoindol-1-yl)benzene-1-sulfonamide

Chlortalidone, also known as chlorthalidone, is a diuretic medication used to treat high blood pressure, swelling including that due to heart failure, liver failure, and nephrotic syndrome, diabetes insipidus, and renal tubular acidosis. In high blood pressure, it is a preferred initial treatment. It is also used to prevent calcium-based kidney stones. It is taken by mouth. Effects generally begin within three hours and last for up to three days. Chlortalidone is more effective than hydrochlorothiazide for the prevention of heart attack or stroke.

Uses:

It is used in hypertension. It treats oedema associated with heart, liver, kidney or lung diseases.

Adverse Effect:

Common side effects include low blood potassium, low blood sodium, high blood sugar, dizziness, and erectile dysfunction. Other side effects may include gout, low blood magnesium, high blood calcium, allergic reactions, and low blood pressure.

Dose:

Oral tablet- **Strengths:** 6.25 mg and 12 mg

Pharmaceutical Formulations: Tablet

Brand Names: Lupiclor, Thaloric, Aquiris, Chlorfid, Chhlohat, Hygroton, Thalitone, Chlorthalid.

8.3.5 Benzthiazide

Structure:

Chemical (IUPAC) name: 6-chloro-1,1-dioxo-3-(phenylmethylsulfanylmethyl)-4H-benzo[e][1,2,4]thiadiazine-7-sulfonamide

Benzthiazide is used to treat hypertension and oedema. Like other thiazides, benzthiazide promotes water loss from the body (diuretics). They inhibit Na^+/Cl^- reabsorption from the distal convoluted tubules in the kidneys. Thiazides also cause loss of potassium and an increase in serum uric acid. Thiazides are often used to treat hypertension, but their hypotensive effects are not necessarily due to their diuretic activity. Thiazides have been shown to prevent hypertension-related morbidity and mortality, although the mechanism is not fully understood. Thiazides cause vasodilation by activating calcium-activated potassium channels (large conductance) in vascular smooth muscles and inhibiting various carbonic anhydrases in vascular tissue.

Uses:

It is used in hypertension.

Adverse Effect:

Most common side effects includes gastrointestinal problems include gastric irritation, vomiting, cramping, nausea, anorexia, diarrhea, and constipation.

Dose:

Oral tablet- **Strengths:** 25 mg. It is prescribed in combination with Triamterene 50mg.

Pharmaceutical Formulations: Tablet

Brand Names: Ditide, Aquatag, Dihydrex, Diucen, Edemax, Exna, Foven

8.3.6 Metolazone

Structure:

Chemical (IUPAC) name: 7-chloro-2-methyl-4-oxo-3-o-tolyl-1,2,3,4-tetrahydroquinazoline-6-sulfonamide

Metolazone is used to reduce the swelling and fluid retention caused by heart failure or kidney disease. It also is used alone or with other medications to treat high blood pressure. Metolazone is in a class of medications called diuretics ('water pills').

Metolazone is a thiazide-like diuretic marketed under the brand names Zytanix, Metoz, Zaroxolyn, and Mykrox. It is primarily used to treat congestive heart failure and high blood pressure. Metolazone indirectly decreases the amount of water reabsorbed into the bloodstream so that blood volume decreases and urine volume increases. This lowers blood pressure and prevents excess fluid accumulation in heart failure. Metolazone is sometimes used with loop diuretics such as furosemide or bumetanide, but these highly effective combinations can lead to dehydration and electrolyte abnormalities.

It was patented in 1966 and approved for medical use in 1974.

Adverse Effect:

Common side effects of Metolazone includes dizziness, weakness, restlessness, headache, muscle cramps, joint pain or swelling, constipation, diarrhea etc.

Dose:

Oral tablet- **Strengths:** 1.25 mg, 2.5 mg, 5 mg

Pharmaceutical Formulations: Tablet

Brand Names: Zytanix, Metoral, Zalat, Metoz, Diurem, Rpmel, Metadure, Zalat

8.3.7 Xipamide

Structure:

Chemical (IUPAC) name: 4-chloro-N-(2,6-dimethylphenyl)-2-hydroxy-5-sulfamoylbenzamide

Xipamide is a sulfonamide diuretic drug marketed by Eli Lilly under the trade names Aquaphor (in Germany) and Aquaphoril (in Austria). It is used for the treatment of oedema and hypertension.

Uses:

It is use in hypertension. It treats oedema associated with heart, liver, kidney or lung diseases.

Adverse Effect:

Common side effects of Xipamide includes gastrointestinal disturbances, dizziness, decreased potassium level in blood, high level of uric acid in the blood and increased urination.

Dose:

Oral tablet- **Strengths:** 20 mg

Pharmaceutical Formulations: Tablet

Brand Names: Xipamid

8.3.8 Spironolactone

Structure:

Chemical (IUPAC) name: S-[(7R,8R,9S,10R,13S,14S,17R)-10,13-Dimethyl-3,5'-dioxospiro [2,6,7,8,9,11,12,14,15,16-decahydro-1H-cyclopenta[a]phenanthrene-17,2'-oxolane]-7-yl] ethanethioate.

Uses: Spironolactone is used to treat certain patients with hyperaldosteronism (the body produces too much aldosterone, a naturally occurring hormone); low potassium levels; heart failure; and in patients with oedema (fluid retention) caused by various conditions, including liver, or kidney disease.

Adverse Effect:

One of the most common side effects of spironolactone is frequent urination. Other general side effects include dehydration, hyponatremia (low sodium levels), mild hypotension (low blood pressure), ataxia (muscle incoordination), drowsiness, dizziness, dry skin, and rashes.

Because of its antiandrogenic activity, spironolactone can, in men, cause breast tenderness, gynecomastia (breast development), feminisation in general, and demasculinisation, as well as including loss of libido and erectile dysfunction. However, these side effects are usually confined to high doses of spironolactone.

Dose: Oral tablet- Strengths: 25 mg

Pharmaceutical Formulations: Tablet

Brand Names: Aldactone, Spirix, Spironot

QUESTION BANK

A. MULTIPLE CHOICE QUESTIONS

1. Diuretics are chemicals that _________ the rate of urine formation.
   A. Increases
   B. Decreases
   C. Stops
   D. None of the above

   Answer: A

2. Acetazolamide is
   A. Loop Diuretics
   B. Carbonic anhydrase inhibitors
   C. Potassium Sparing Diuretics
   D. Drugs acting on Distal Convoluted Tubule

   Answer: B

3. Following is used as a diuretic drug
   A. Furosemide
   B. Penicillin
   C. Dopamine
   D. Quinidine Sulphate

   Answer: A

4. Which of the following are Carbonic anhydrase inhibitors?
   A. Chlorthiazide
   B. Acetazolamide
   C. Furosemide
   D. Ethacrynic acid

   Answer: B

5. The parent heterocycle present in Metolazone is _________.
   A. Quinidone
   B. Furan

    C. Thiophene

    D. Quinazoline

Answer: C

6. Benzthiazide belongs to which class?
   A. Osmotic diuretics
   B. Thiazide diuretic
   C. Loop Diuretics
   D. Potassium Sparing Diuretics

Answer: B

7. Xipamide belong to which of the following category?
   A. Diuretic
   B. Antibiotic
   C. Antifungal
   D. Anti-pyretic

Answer: A

8. Lupiclor is brand name for _______________.
   A. Bumetanide
   B. Chlorthalidone
   C. Benzthiazide
   D. Metolazone

Answer: B

9. The parent heterocycle present in Furosemide is __________.
   A. Quinidone
   B. Furan
   C. Thiophene
   D. Quinazoline

Answer: B

10. Thiadiazol heterocycle present in which of the following diuretic?
    A. Acetazolamide
    B. Frusemide
    C. Bumetanide
    D. Chlorthalidone

Answer: A

B. SHORT ANSWER QUESTIONS

1. Define and classify Diuretics with examples.
2. Draw molecular formula, State properties and uses of Frusemide.
3. Give the physical properties and uses of Xipamide.
4. Give the physical properties and uses of Bumetanide.
5. Explain Metolazone as diuretic.
6. Give physical properties and structure of Acetazolamide.
7. Enlist names of Potassium Sparing Diuretics.
8. Enlist pharmaceutical formulations available for Frusemide.
9. Enlist different brand names for Frusemide.
10. Enlist pharmaceutical formulations available for Acetazolamide.

C. LONG ANSWER QUESTIONS

1. Give physical properties, pharmaceutical formulation and brand name of-
   1. Frusemide   2. Bumetanide   3. Benzthiazide,
2. Define and classify Diuretics with examples. Add a note on Furosemide.
3. Explain physical properties, pharmaceutical formulations, and brand names for Acetazolamide.

■■■

Chapter 9

HYPOGLYCEMIC AGENTS

CONTENTS

Hypoglycemic Agents:

- Insulin and Its Preparations
- Metformin*
- Glibenclamide*
- Glimepiride
- Pioglitazone
- Repaglinide
- Gliflozins
- Gliptins

♦ LEARNING OBJECTIVES ♦

After completing this chapter, the student should be able to understand:

1. Classification of hypoglycemic agents.
2. Chemical names for different compounds used as hypoglycemic agents.
3. Structures of hypoglycemic agents.
4. Uses, stability and storage conditions for hypoglycemic agents.
5. Different types of formulations and popular brand names for hypoglycemic agents.

9.1 INTRODUCTION

The pancreas secretes digestive enzymes, glucagon and insulin. An isolated group of cells within the pancreas is called as islets of Langerhans. These cells are divided into three types α cells (secretes glucagon), β cells (secretes insulin) and γ cells (secretes somatostatin). Insulin plays a vital role in the digestion and utilisation of food substances. It is essential for the phosphorylation of glucose to glucose-6-phosphate. Glucose-6-phosphate is further catabolised to give energy. In some individuals, the glucose level in the blood increases due to a lack of sufficient insulin. This condition is called hyperglycemia. The disease is known as diabetes mellitus.

Some of the more important symptoms associated with the disease are polydipsia, polyurea, ketonaemia, and ketonuria. Most patients can be classified clinically as having either insulin-dependent diabetes mellitus (Type-I diabetes) or non-insulin-dependent diabetes mellitus (Type-II diabetes).

Type-I diabetes is an auto-immune disease caused by the destruction of pancreatic islet cells. In Type-II diabetes, the cause of hyperglycemia is a combination of insulin resistance and a loss of secretory function by pancreatic β-cells. People who have diabetes are two to four times more likely to die from heart disease or have a stroke than people who do not have diabetes.

To date, there is no known cure for Type I (insulin-dependent) diabetes. Hypoglycemic agents lower blood sugar and are used to treat the symptoms of diabetes mellitus.

9.2 CLASSIFICATION OF HYPOGLYCEMIC AGENTS

Type II diabetes mellitus develops most often in people over 40 years of age, and obesity increases the risk of developing it. While diet and exercise are recommended as the first line of defence, many individuals are unable to control the disease through these means alone. For these people, one or more oral anti-diabetic agents may be used. Some people with type II diabetes require insulin to control the disease.

Oral anti-diabetic drugs act in one or more of the following:

1. Stimulating the release of insulin by the pancreas
2. Slowing the absorption of glucose from the intestines
3. Decreasing glucose synthesis and release by the liver
4. Making cells more sensitive to insulin (e.g., muscle and liver).

The hypoglycemic agents are classified into the following two main categories as follows:

A. Specific Hypo-glycemic Agents: *e.g.* Insulin and its preparation.

B. Synthetic compounds (Oral hypoglycemic agents):

The oral hypoglycemic agents are further classified as:

1. **Sulfonylurea derivatives:** *e.g.* Chlorpropamide, tolbutamide, glibenclamide, glipizide and dichlorphe-namide.
2. **Biguanide derivatives:** *e.g.* Phenformin, metformin.

9.3 SPECIFIC HYPOGLYCEMIC AGENT

Insulin

Insulin was the first hormone identified in the 1920s by Banting and Best. They discovered insulin by tying a string around the pancreatic duct of several dogs. When they examined the pancreas of these dogs several weeks later, all of the pancreatic digestive cells were died and were absorbed by the immune system, and the only thing left was thousands of pancreatic islets. They then isolated the protein from these islets and discovered insulin.

Sources of Insulin

The first successful insulin preparations came from cows (bovine) and later from pigs (porcine). Bovine and porcine insulin worked very well for the majority of patients, but some could develop an allergy or other types of reactions to the foreign protein (a foreign protein is a protein which is not native to humans). In the 1980s, technology had advanced to the point where we could make human insulin. The advantage would be that human insulin would have a much lower chance of inducing a reaction because it is not a foreign protein. The technology, which made this approach possible, was the development of recombinant DNA techniques. In simple terms, the human gene, which codes for the insulin protein, was cloned (copied) and then put inside of bacteria. A number of operations were performed on this gene to make the bacteria constantly make insulin. Big vats of bacteria now make tons of human insulin. From this, one can isolate pure human insulin.

Isolation of insulin

Isolation of insulin from animal pancreas involves the following steps:
   i.   Mince the pancreas of a slaughtered animal and extract with 80% ethanol containing small amounts of phosphoric acid (to adjust pH to 3)
   ii. Centrifuge the extract to separate proteins and fats
   iii. Raise the pH of the extract to 8.0 by adding ammonia solution and filter
   iv. Acidify and evaporate the filtrate to remove fatty material
   v.   Add picric acid to ethanolic solution to precipitate insulin as insulin picrate
   vi. Dissolve insulin picrate in acetone and re-precipitate as the hydrochloride salt
   vii. Insulin is further purified by chromatography.

Structure of Insulin

The minimum molecular weight of insulin is about 6000. Dinitro phenyl hydrazine (N-terminal amino acid determination method) showed the presence of two N-terminal amino acid residues, i.e. glycine and phenylalanine. Thus, insulin contains two peptide chains.

Insulin was oxidised with performic acid. This produced two peptides, which were separated by electrophoresis. The two peptide chains are referred to as the A chain and the B chain.

The peptide with N-terminal glycine residue was called the A-chain, and that with the N-terminal phenylalanine residue was called B-chain. Each chain was subjected to hydrolysis (with acids or enzymes). The products of hydrolysis were separated and examined by the DNP method.

Chain-A contains 21 amino acid residues, and the chain-B contains 30 amino acid residues. In chain-A, four cysteine acid residues, and in chain-B, two residues were present. Two disulphide bonds connect these two chains. Further, chain-A contains one intra-disulphide bond. These interactions have important clinical ramifications. Monomers and dimers readily diffuse into the blood, whereas hexamers diffuse very poorly. Hence, absorption of insulin preparations containing a high proportion of hexamers is delayed and slow. This problem, among others, has stimulated the development of a number of recombinant insulin analogues. The first of these molecules to be marketed is called insulin lispro—is engineered such that lysine and proline residues on the C-terminal end of the B chain are reversed; this modification does not alter receptor binding but minimises the tendency to form dimers and hexamers.

Although the amino acid sequence of insulin varies among species, certain segments of the molecule are highly conserved, including the positions of the three disulfide bonds, both ends of the A chain and the C-terminal residues of the B chain. These similarities in the amino acid sequence of insulin lead to a three-dimensional conformation of insulin that is very similar among species, and insulin from one animal is very likely biologically active in other species. Indeed, pig insulin has been widely used to treat human patients. The structure of insulin is as below:

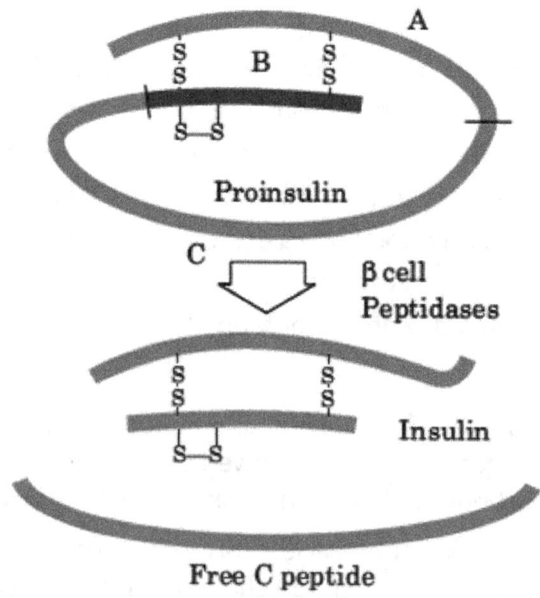

Fig. 9.1 Structure of Insulin

Biosynthesis of Insulin

Insulin is synthesised in significant quantities only in β cells in the pancreas. The insulin mRNA is translated as a single chain precursor called preproinsulin, and removal of its signal peptide during insertion into the endoplasmic reticulum generates proinsulin. Proinsulin consists of three domains: an amino-terminal B chain, a carboxy-terminal A chain and a connecting peptide in the middle known as the C peptide. Within the endoplasmic reticulum, proinsulin is exposed to several specific endo peptidases, which excise the C peptide, thereby generating the mature form of insulin. Insulin and free C peptide are packaged in the Golgi into secretory granules, which accumulate in the cytoplasm. When the β cell is appropriately stimulated, insulin is secreted from the cell by exocytosis and diffuses into islet capillary blood. C peptide is also secreted into the blood but has no known biological activity.

Insulin preparation

Human insulin is absorbed more quickly than beef or pork insulin. Thus the duration of action of human insulin is shorter. If insulin preparations were administered orally, it would be degraded in the gastrointestinal tract. Therefore, insulin must be administered by injection (IV or subcutaneous).

The following insulin preparations are used for the treatment of diabetes.

1. **Crystalline zinc insulin:** It is purified insulin crystallised as a zinc salt. It is administered subcutaneously and lowers the blood sugar within minutes. Hence it is known as rapid-acting insulin.

2. **Semilente insulin:** It is a suspension of amorphous insulin, which is also administered subcutaneously. It is another example of rapid action insulin.

3. **Isophane insulin:** It is a suspension of crystalline zinc insulin with a positively charged peptide mixture called protamine. Its duration of action is intermediate between crystalline zinc insulin and protamine zinc insulin. This is due to delayed absorption of insulin because of the conjugation of insulin with protamine to form a less soluble complex.

4. **Lente insulin:** It is a mixture of 30% of semilente insulin and 70% of ultralente insulin. It is intermediate action insulin that is administered subcutaneously.

5. **Protamine zinc insulin:** It is a prolonged action insulin preparation. It produces the maximum therapeutic effect in 24 hours. It is prepared by mixing crystalline zinc insulin with protamine.

6. **Extended insulin zinc suspension:** It is poorly soluble crystalline zinc insulin. This preparation has a delayed onset and prolonged duration of action.

Storage:

Insulin in powder form should be stored in airtight containers protected from light. The injections are required to be stored in a refrigerator at 2 to 8°C and not allowed to freeze.

The insulin receptor and mechanism of action:

Like the receptors for other protein hormones, the receptor for insulin is embedded in the plasma membrane. The insulin receptor is composed of two alpha subunits and two beta subunits linked by disulfide bonds. The alpha chains are entirely extracellular and house insulin binding domains, while the linked beta chains penetrate through the plasma membrane.

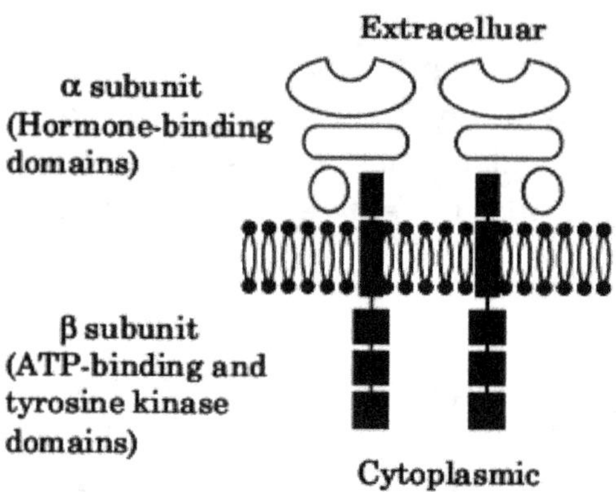

Fig. 9.2 The insulin receptor and mechanism of action

The insulin receptor is a tyrosine kinase. In other words, it functions as an enzyme that transfers phosphate groups from ATP to tyrosine residues on intracellular target proteins. Binding insulin to the alpha subunits causes the beta subunits to phosphorylate themselves (autophosphorylation), thus activating the catalytic activity of the receptor. The activated receptor then phosphorylates a number of intracellular proteins, which in turn alters their activity, thereby generating a biological response.

9.4 ORAL HYPOGLYCEMIC AGENT

Because of the ineffectiveness of insulin through the oral route in the treatment of diabetes mellitus, a search was made for compounds which could prove to be effective orally.

The following oral hypoglycemic agents are used to treat diabetes mellitus.

9.4.1 Metformin

Structure:

Chemical (IUPAC) name: 3-(diaminomethylidene)-1,1-dimethylguanidine;hydrochloride

Chemistry:

Metformin is another biguanide derivative. It is often prescribed for obese people with type 2 diabetes and does not cause weight gain. Chemically metformin is N, N-dimethyl biguanide. It is prepared by refluxing cyanoguanidine with dimethylamine hydrochloride at 135°C.

N . N Dimethylamine hydrochloride

Metformin hydrochloride

Properties:

It is a colourless crystalline powder. Metformin is freely soluble in water and is hygroscopic in nature.

Mechanism of action:

Metformin is an antihyperglycemic agent. The main causes for reduced glucose levels are:

i.   Increase in insulin action in peripheral tissues.

ii. Reduced hepatic glucose output due to inhibition of gluconeogenesis.

iii. Reduce the absorption of glucose from the intestine.

Uses:

Metformin is used as an anti-hyperglycemic agent. It is often given in combination with sulfonylureas.

Stability and storage conditions:

Keep this medication in the container it came in, tightly closed, and out of reach of children. Store it at room temperature and away from light, excess heat, and moisture.

Metformin from all trademarks was stable at all storage conditions for up to 30 days, retaining more than 90% of the initial amount of active drug, with a pH of 7.4 ± 0.3.

Dose:

Oral tablet- Strengths: 250 mg, 500 mg, 850 mg & 1gm

Pharmaceutical Formulations: Tablet

Brand Names: Glycomet, Cetapin, Melmet, Obimet, Obimet SR, Metadoze, Gluformin XL

9.4.2 Glibenclamide

Structure:

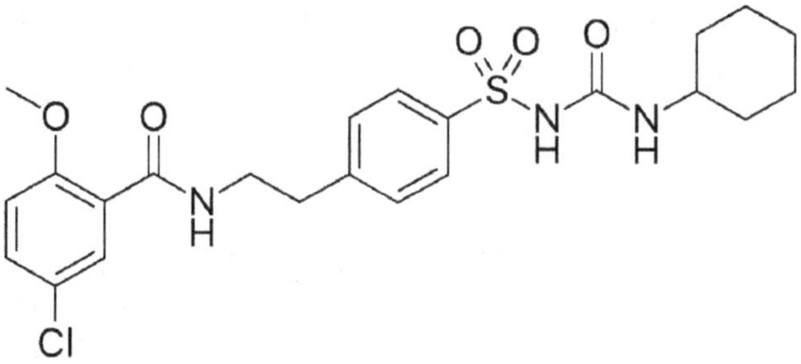

Chemical (IUPAC) name: 5-Chloro, N-[2-[4-[[[(cyclohexylamino) carbonyl]amino] sulfonyl] phenyl) ethyl], 2-methoxy benzamide.

Chemistry:

It is prepared by condensation of 5-chloro-2- methoxy benzoyl chloride with 4-amino ethyl benzene sulphonamide, followed by treatment with hexyl isocyanate.

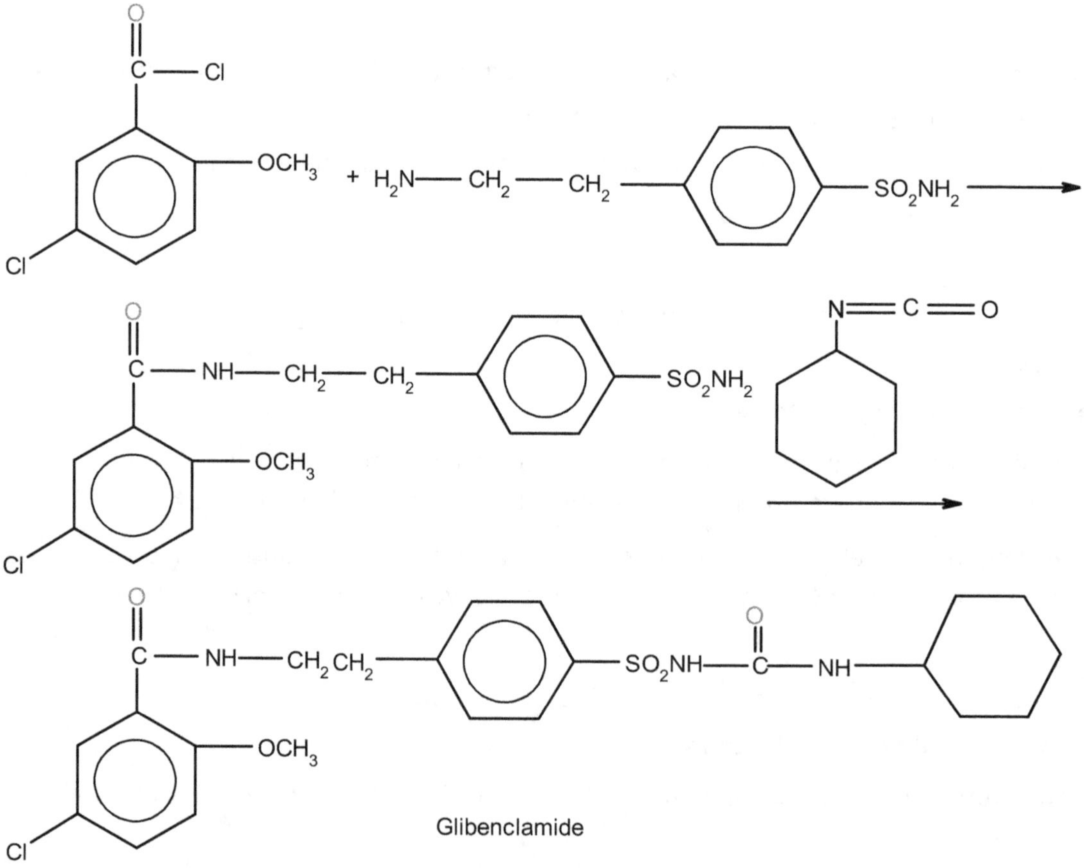

Glibenclamide

Properties:

It is a colourless crystalline powder. It is insoluble in water.

Mechanism of action:

It produces a hypoglycemic effect by stimulating the release of insulin from β-cells of the pancreas and by increasing the sensitivity of peripheral tissues to insulin. It is more potent than tolbutamide.

Uses:

It is used to control hyperglycemia in type-II diabetes mellitus.

Stability and storage conditions:

All formulations remained physicochemically and microbiologically stable at three different temperatures during the 90-day study. Therefore, glibenclamide formulations can be stored for at least 90 days at ≤40°C.

Should be stored at room temperature, keep it away from heat.

Dose:

Oral tablet- **Strengths:** 2.5 mg and 5 mg

Pharmaceutical Formulations: Tablet

Brand Names: Diaonil, Euglucon, Getrol, Glinil, Diabetnil, Aviglen, Glibamide

9.4.3 Glimepiride

Structure:

Chemical (IUPAC) name: 4-ethyl-3-methyl-N-[2-[4-[(4methylcyclohexyl)carbamoyl-sulfa-moyl]phenyl]ethyl]-5-oxo-2H-pyrrole-1-carboxamide

Mechanism of action:

Like all sulfonylureas, glimepiride acts as an insulin secretagogue. It lowers blood sugar by stimulating the release of insulin by pancreatic beta cells and by inducing increased activity of intracellular insulin receptors.

Uses:

Glimepiride is indicated to treat type 2 diabetes mellitus; its mode of action is to increase insulin secretion by the pancreas. However, it requires adequate insulin synthesis as a prerequisite to treatment appropriately. It is not used for type 1 diabetes because, in type 1 diabetes, the pancreas is not able to produce insulin.

Stability and storage conditions:

Store at room temperature, away from light and moisture.

Dose:

Oral tablet- **Strengths:** 0.5 mg, 1 mg 2 mg, 3 mg and 4 mg

Pharmaceutical Formulations: Tablet

Brand Names: Amaryl, Zoryl, Glimy, Azulix, Glimestar, Glimer, Diapriide, Glimisave

9.4.4 Pioglitazone

Structure:

Chemical (IUPAC) name: 5-(4-[2-(5-ethylpyridin-2-yl)ethoxy]benzyl)thiazolidine-2,4-dione,

Mechanism of action:

Pioglitazone selectively stimulates the nuclear receptor peroxisome proliferator-activated receptor gamma (PPAR-γ) and, to a lesser extent, PPAR-α. It modulates the transcription of the genes involved in controlling glucose and lipid metabolism in the muscle, adipose tissue, and the liver.

As a result, pioglitazone reduces insulin resistance in the liver and peripheral tissues, decreases gluconeogenesis in the liver, and reduces the quantity of glucose and glycated haemoglobin in the bloodstream.

Uses:

It is used to control hyperglycemia in type-II diabetes mellitus.

Stability and storage conditions:

Store at 20°C to 25°C (68°F to 77°F); excursions permitted to 15°C to 30°C (59°F to 86°F).

Dose:

Oral tablet- **Strengths:** 7.5 mg, 15 mg, 30 mg

Pharmaceutical Formulations: Tablet

Brand Names: Pioz, pioglar, Pionarm, Glitaris, Piokind, Piomed, Piopod

9.4.5 Repaglinide

Structure:

Chemical (IUPAC) name: (S)-(+)-2-ethoxy-4-[2-(3-methyl-1-[2-(piperidin-1-yl)phenyl] butylamino)-2-oxoethyl] benzoic acid

Mechanism of action:

Repaglinide lowers blood glucose levels by stimulating the release of insulin from the pancreas. This action depends upon functioning beta (ß) cells in the pancreatic islets. Insulin release is glucose-dependent and diminishes at low glucose concentrations.

Uses:

Repaglinide is used to treat type 2 diabetes (a condition in which the body does not use insulin normally and, therefore, cannot control the amount of sugar in the blood).

Stability and storage conditions:

Store at room temperature between $59-77^0$ F ($15-25^0$ C) away from light and moisture.

Dose:

Oral tablet- **Strengths:** 0.5 mg, 1 mg, 2 mg

Pharmaceutical Formulations: Tablet

Brand Names: Eurepa, Novonorm, Regan, Repide, Rapilin, Restrict.

9.4.6 Gliflozins

SGLT2 inhibitors, also called gliflozins, are a class of medications that alter the essential physiology of the nephron, unlike SGLT1 inhibitors, which modulate sodium/glucose channels in the intestinal mucosa. The foremost metabolic effect appears to show that this pharmaceutical class inhibits the reabsorption of glucose in the kidney and, therefore, lower blood sugar. They act by inhibiting sodium-glucose transport protein 2 (SGLT2). SGLT2 inhibitors are used in the treatment of type II diabetes mellitus (T2DM). Apart from blood sugar control, gliflozins have been shown to provide significant cardiovascular benefits in T2DM patients. Several medications in this class have been approved or are currently under development. In studies on canagliflozin, a member of this class, the medication was found to enhance blood sugar control and reduce body weight and systolic and diastolic blood pressure.

9.4.6.1 Canagliflozin

Structure:

Chemical (IUPAC) name: (2S,3R,4R,5S,6R)-2-[3-[[5-(4-fluorophenyl)thiophen-2-yl]methyl]-4-methylphenyl]-6-(hydroxymethyl)oxane-3,4,5-triol

It is a sodium-glucose co-transporter 2 (SGLT2) inhibitor used to manage hyperglycemia in type 2 diabetes mellitus (DM). Also used to reduce the risk of major cardiovascular events in patients with established cardiovascular disease and type 2 DM.

Canagliflozin, also known as Invokana, is a sodium-glucose cotransporter 2 (SGLT2) inhibitor used in the management of type 2 diabetes mellitus along with lifestyle changes, including diet and exercise Labels. Canagliflozin is the first oral antidiabetic drug approved for preventing cardiovascular events in patients with type 2 diabetes 8. Cardiovascular disease is the most common cause of death in these patients.

Mechanism of Action:

The sodium-glucose co-transporter-2 (SGLT-2) is found in the proximal tubules of the kidney and reabsorbs filtered glucose from the renal tubular lumen. Canagliflozin inhibits the SGLT-2 co-transporter. This inhibition leads to lower reabsorption of filtered glucose into the body and decreases the renal threshold for glucose (RTG), leading to increased glucose excretion in the urine.

Uses:

Canagliflozin is used with a proper diet and exercise program to control high blood sugar in people with type 2 diabetes. Controlling high blood sugar helps prevent kidney damage, blindness, nerve problems, loss of limbs, and sexual function problems.

Stability and storage conditions: It should be stored at 77°F (25°C).

Dose: 100 mg, 300 mg.

Pharmaceutical Formulation: Tablet.

Brand Names: Invokana

9.4.6.2 Dapagliflozin

Structure:

Chemical (IUPAC) name: *(2S,3R,4R,5S,6R)*-2-(4-Chloro-3-(4-ethoxybenzyl)phenyl)-6-(hydroxymethyl) tetrahydro -2H-pyran-3,4,5-triol

Dapagliflozin is a sodium-glucose co-transporter-2 inhibitor indicated for managing diabetes mellitus type 2. When combined with diet and exercise in adults, dapagliflozin helps to improve glycemic control by inhibiting glucose resorption in the proximal tubule of the nephron and causing glycosuria. Dapagliflozin was approved by the FDA on Jan 08, 2014

Mechanism of Action:

Dapagliflozin inhibits the sodium-glucose cotransporter-2 (SGLT2), which is primarily located in the proximal tubule of the nephron. SGLT-2 facilitates 90% of glucose resorption in the kidneys, so its inhibition allows glucose to be excreted in the urine. This excretion allows for better glycemic control and potentially weight loss in patients with type-2 diabetes mellitus

Uses:

Dapagliflozin is used along with diet and exercise to improve glycemic control in adults with type-2 diabetes and to reduce the risk of hospitalisation for heart failure among adults with type-2 diabetes and known cardiovascular disease or other risk factors. It appears more useful than Empagliflozin.

Stability and storage conditions:

Store at room temperature, away from light and moisture.

Dose: 05 mg, 10 mg.

Pharmaceutical Formulation: Tablet.

Brand Names: Forxiga

9.4.7 Gliptins

Gliptins represent a novel class of agents that improve beta cell health and suppress glucagon, resulting in improved post-prandial and fasting hyperglycemia. They function by augmenting the incretin system (GLP-1 and GIP), preventing their metabolism by dipeptidyl peptidase-4 (DPP-4).

9.4.7.1 Sitagliptin

Structure

Chemical (IUPAC) name: (R)-4-oxo-4-[3-(trifluoromethyl)-5,6-dihydro[1,2,4]triazolo[4,3-a]pyrazin-7(8H)-yl]-1-(2,4,5-trifluorophenyl)butan-2-amine

Sitagliptin was developed by Merck & Co. and approved for medical use in the United States in 2006. In 2019, it was the 88th most commonly prescribed medication in the United States, with more than 8 million prescriptions.

Mechanism of Action:

Sitagliptin works to competitively inhibit the enzyme dipeptidyl peptidase 4 (DPP-4). This enzyme breaks down the incretins GLP-1 and GIP, gastrointestinal hormones released in response to a meal. By preventing the breakdown of GLP-1 and GIP, they are able to increase the secretion of insulin and suppress the release of glucagon by the alpha cells of the pancreas. This drives blood glucose levels towards normal. As the blood glucose level approaches normal, the amounts of insulin released and glucagon suppressed diminishes, thus tending to prevent an "overshoot" and subsequent low blood sugar (hypoglycemia), which is seen with some other oral hypoglycemic agents.

Uses:

Sitagliptin is used to treat type 2 diabetes. It is generally less preferred than metformin or sulfonylureas.

Stability and storage conditions:

Sitagliptin should be stored at room temperature between 68°F and 77°F (20°C and 25°C). It can be stored briefly at a temperature between 59°F and 86°F (15°C and 30°C). Store this drug away from light.

Dose: 50 mg, 100 mg.
Pharmaceutical Formulations: Tablet
Brand names: Januvia, Ristaben, Tesavel

QUESTION BANK

A. MULTIPLE CHOICE QUESTIONS

1. Rosiglitazone mechanism of action is ________.
   A. Acts as PPAR gamma antagonist
   B. Inhibitor of alpha glucosidase
   C. Acts as an amylin analogue
   D. Acts as a dipeptidyl peptidase inhibitor

   Answer: A

2. Which of the following drug does not act by increasing insulin secretion?
   A. Exentide
   B. Sitagliptin
   C. Rosiglitazone
   D. Repaglinide

   Answer: C

3. Metformin acts by two mechanism ________.
   (P) Increasing insulin secretion
   (Q) Inhibiting α-glucosidase
   (R) Decreasing hepatic glucose production
   (S) Increasing insulin action in muscle and fat.
   (A) P, Q   (B) R, S
   (C) P, R   (D) Q, S

   Answer: B

4. Which of the following is rapid-acting insulin?
   A. Insulin Injection
   B. Insulin zinc suspension
   C. Globin zinc insulin injection
   D. Isophane insulin injection

   Answer: A

5. Metformin belongs to which of the chemical class of anti-diabetic drugs?
    A. Thiazolidinediones
    B. Sulfonylureas
    C. Biguanides
    D. Alpha-glucosidase inhibitor

Answer: C

6. First-generation sulfonylureas include all the following except ______.
    A. Acetohexamide
    B. Glipizide
    C. Tolazamide
    D. Tolbutamide

Answer: B

7. Which of the following phenylalanine derivative is used in type-2 diabetes?
    A. Miglitol
    B. Phenformin
    C. Repaglinide
    D. Rosiglitazone

Answer: C

8. Glycomet is the popular brand name for______.
    A. Metformin
    B. Glibenclamide
    C. Glimepiride
    D. Pioglitazone

Answer: A

9. Which of the following heterocyclic ring is present in Glimepiride?
    A. Thiphene
    B. Furan
    C. Pyrrole
    D. Quinidine.

Answer: C

10. Invokana is popular brand name for __________.
    A. Canagliflozin
    B. Dapagliflozin
    C. Sitagliptin
    D. Glimepiride

Answer: A

B. SHORT ANSWER QUESTIONS

1. Classify Hypoglycemic Agents with the structure of one drug from each class.
2. Enlist different Insulin preparation.
3. Give physical properties and structure of Metformin.
4. Enlist pharmaceutical formulations available for Glibenclamide.
5. Enlist different brand names for Metformin.
6. Enlist pharmaceutical formulations available for Pioglitazone.
7. Write a note on Gliflozins.
8. Write a note on Gliptins.
9. Enlist pharmaceutical formulations available for Glimepiride.
10. Explain Repaglinide as Hypoglycemic Agents.

C. LONG ANSWER QUESTIONS

1. Define Hypoglycemic Agents. Classify it in detail. Add a note on Metformin.
2. Define Hypoglycemic Agents. Classify it in detail. Add a note on Insulin and its preparations.
3. Define Hypoglycemic Agents. Classify it in detail. Explain Gliflozins as Hypoglycemic Agents.
4. Define Hypoglycemic Agents. Classify it in detail. Explain Gliptins as Hypoglycemic Agents.

■■■

Chapter 10

ANALGESIC AND ANTI-INFLAMMATORY AGENTS

CONTENTS

- A brief introduction to the pathophysiology and transmission of pain, its classification
- Classification of Analgesic and Anti-Inflammatory Agents
  - Narcotic Analgesics: Morphine Analogues & their classification
  - Nonsteroidal Anti-inflammatory Agents (NSAIDS) & their classification
- Some important Analgesic Agents and NSAIDs with respect to their chemical structure, name, mechanism of action, uses, side effects, incompatibilities, storage conditions, formulations and popular brands
  - o Morphine
  - o Codiene
  - o Meperidine
  - o Methadone
  - o Pentazocine
  - o Naloxone
  - o Aspirin
  - o Diclofenac
  - o Ibuprofen
  - o Piroxixam
  - o Celecoxib
  - o Mefenamic Acid
  - o Paracetamol
  - o Aceclofenac

♦ LEARNING OBJECTIVES ♦

After completing this chapter, the student should be able to understand:

1. The phenomenon of pain, its classification, transmission, mechanisms and receptors
2. The classification and mechanism of action of various narcotic analgesics.
3. The classification and mechanism of action of various non-steroidal anti-inflammatory agents (NSAIDS) narcotic analgesics.

4. Some selected narcotic analgesic agents and NSAIDS and information on their chemical structure, name, mechanism of action, uses and side effects, incompatibilities, storage conditions, formulations and popular brands.

10.1 INTRODUCTION

Pain is a reaction of the body to harmful stimuli and is, therefore, a protective early warning system. Pain is an ill-defined, unpleasant sensation evoked by an external or internal noxious stimulus.

Pain has been described by the International Association for the Study of Pain as an "*Unpleasant sensory and emotional experience associated with actual or potential tissue damage.*" In general, pain can be described as either acute or chronic.

Acute pain, which does not outlast the initiating painful stimulus, has three generally encountered origins.

1. Acute pain is of **superficial origin** from wounds, chemical irritants, and thermal stimuli, such as burns.
2. Acute pain of deep **somatic origin** arises from the injection of chemical irritants or from ischemia, such as with myocardial infarction.
3. Acute pain of **visceral origin** is most often associated with inflammation.

Chronic pain, by contrast, outlasts the initiating stimulus, which in many cases is of unknown origin. It is often associated with diseases such as cancer and arthritis. Treatment of chronic pain presents a challenge to the physician in that the underlying cause is often not readily apparent. Neuropathic pain, a type of chronic pain, responds poorly to opioids, and its causes include diabetic neuropathies, shingles (herpes zoster), ischemia following stroke, and phantom limb pain. Neuropathic pain responds well in many cases to therapies other than the use of opioids and nonsteroidal anti-inflammatory drugs (NSAIDs).

Types and reasons for chronic pain

- Neuropathic pain
- Inflammatory pain
- Bone cancer pain
- Fibromyalgia
- Migraine
- Psychogenic pain

Pain perception has 2 components;
- Nociceptive component
- Affective component

Pain sensation

- First pain : sharp, pricking, well defined, Aδ fibres
- Second pain : dull, aching, poorly localised, C fibres.

Transmission of Pain: transmission of nociceptive inputs from diverse nociceptors occurs *via* different fibre bundles.

Inhibitory and Facilitatory Mechanisms

- **Neurotransmitters**—chemicals that exert inhibitory or excitatory activity at post-synaptic nerve cell membranes. Examples include acetylcholine, norepinephrine, epinephrine, dopamine, and serotonin.
- **Neuromodulators**—endogenous opiates. Hormones in the brain. Alpha endorphins, beta-endorphins and enkephalins. Help to relieve pain.

Endogenous Opioid Peptides as Neuromodulators: A number of endogenous opioid peptides having morphine-like activity are found in the brain, pituitary, spinal cord, and GIT and act at specific receptors.

1. β-Endorphins-μ (beta-endorphins(opiopeptins)): in the anterior pituitary and hypothalamus. They have potent analgesic activity.
2. Enkephalins - μ & δ (*met-enkephalins & leu-enkephalins*)
3. Endogenous pentapeptides. May be involved in the modulation of the pain response.
4. Dynorphins- κ
5. Endomorphins- μ
6. Nociceptin- NOP receptor (nociceptin opioid peptide receptor)
7. Opioid receptors—binding sites not only for endogenous opiates but also for opioid analgesics to relieve pain. Several types of receptors: Mu, Kappa, Delta, Epsilon and Sigma.

Opioid Receptors where the opioid peptides act:

Receptor Type Action

μ: → Supraspinal analgesia, respiratory depression, euphoria, physical dependence.

μ1: → Analgesia.

μ2: → Respiratory depression

k: → Spinal analgesia, miosis and sedation.

σ: → Dysphoria, hallucination, respiratory & vasomotor stimulation.

RECEPTOR SITES:

1-Limbic system, Amygdaloid nucleus & hypothalamus.

2-Medial & lateral thalamus, *area postrema (nausea and vomiting)*.

3-The nucleus of the solitary tract (*cough*).

4-*Substantia gelatinosa* & some areas of the spinal cord.

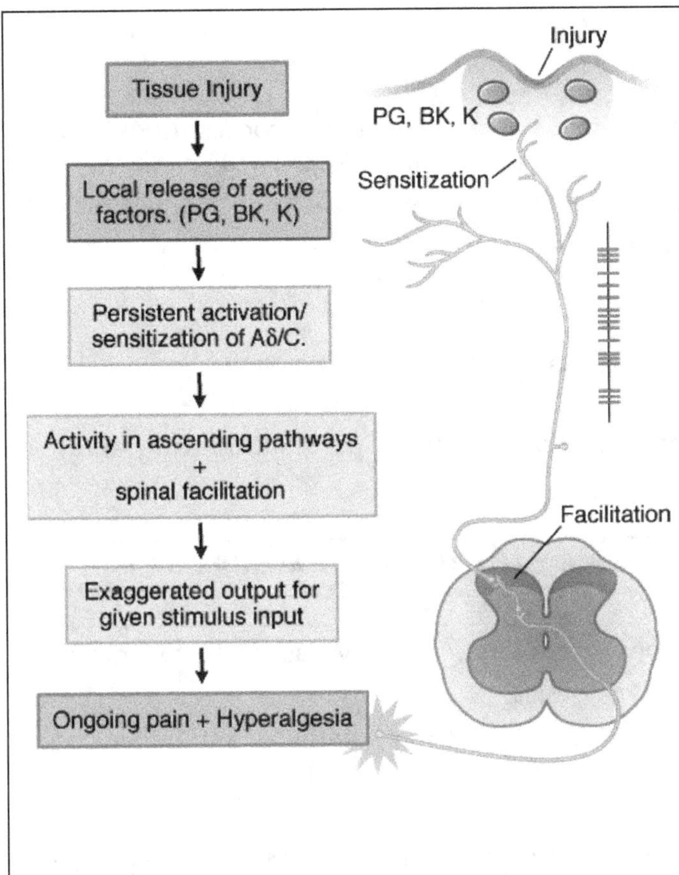

A6-fibres are the site for rapid transmission of sharp, painful stimuli. Such fibres are myelinated and enter the dorsal horn, from which point the ascending systems of the spinothalamic tract are activated.

C-fibers, which also enter the dorsal horn and synapse on spinothalamic tract neurons, are responsible for the slower transmission (fibers are not myelinated) of nociceptive impulses, resulting in a dull, aching sensation.

The A6-fibers and C-fibers are activated by mechano-receptors. A6-fibers and C-fibers are also activated by other types of nociceptors, such as those responding to heat and chemicals.

Fig. 10.1 Transmission of pain

Inflammation: Inflammation is a complex protective response of the organism to injury caused by damaging agents. It is aimed at inactivation or removal of these agents and promoting healing.

Mediators of Inflammation: The body defence system has many mediators of inflammation, as listed below;

- Prostaglandins
- Bradykinin
- Serotonin
- Histamine
- Interleukins-2 – 6, 10, 12,13
- Platelet-activating factor
- Gamma-Interferon
- Tumour Necrosis Factor

- Transforming Growth Factor
- Lymphotoxin

Analgesics are drugs which relieve pain without significantly altering consciousness. Such drugs eliminate pain by either decreasing the underlying cause of the pain (as do the nonopioid analgesics described later) or decreasing the transmission of nociceptive impulses and pain perception (as do the opioids).

NSAIDs: These are the agents which:
- Cause relief of pain -. Analgesic;
- Suppress the signs and symptoms of inflammation.
- Exert antipyretic action.
- Useful in pain related to inflammation, especially for superficial/integumental pain.

10.2 CLASSIFICATION OF ANALGESIC AND ANTI-INFLAMMATORY AGENTS

Thus, the analgesics and anti-inflammatory agents can be broadly classified into 2 groups;

A) Narcotic or Opioid or Centrally acting Analgesics

B) Non-Narcotic or Peripherally acting Analgesics and non-steroidal anti-inflammatory agents

A) Narcotic or Opioid Analgesics:

They bind to opioid receptors in the brain and spinal cord (SC) and even the periphery. They Relieve moderate to severe pain by inhibiting the release of Substance P in central and peripheral nerves, reducing the perception of pain sensation in the brain, producing sedation and decreasing emotional upsets associated with pain. (*Substance P (SP) is a neurotransmitter and a modulator of pain perception by altering cellular signalling pathways.*).

Classification of Opioid analgesic drugs

Type-I

Natural opium alkaloids:
- Morphine
- Codeine

Semisynthetic opiates:
- Diacetylmorphine (Heroin)
- Pholcodeine
- Oxycodone

Synthetic opioids:
- Pethidine (Meperidine)

- Fentanyl, Alfentanil, Sufentanil, Remifentanil
- Methadone
- Dextropropoxyphene
- Tramadol

Type-II

Narcotic Agonists include

- Natural Opium Alkaloids (e.g., **morphine, codeine**),
- Semisynthetic Analogs (e.g., hydromorphone, oxymorphone, oxycodone), and
- Synthetic Compounds (e.g., **meperidine**, levorphanol, **methadone**, sufentanil, alfentanil, fentanyl, remifentanil, and levomethadyl.

Mixed Agonist-antagonist Drugs (e.g., nalbuphine, **pentazocine)** have agonist activity at some receptors and antagonist activity at other receptors; also included are the partial agonists (e.g., butorphanol, buprenorphine).

Narcotic antagonists: Narcotic antagonists (e.g., **naloxone**) do not have agonist activity at any opioid receptor sites. Antagonists block the opiate receptor, inhibit the pharmacological activity of the agonist, and precipitate withdrawal in dependent patients.

B) Non-Narcotic or Peripherally acting Analgesics and non-steroidal anti-inflammatory agents

Classification of NSAIDs

i) Nonselective COX inhibitors

1. **Salicylates:** e.g. *Aspirin*, Sodium salicylate & diflunisal.
2. **Propionic acid derivatives**: e.g. *Ibuprofen*, Ketoprofen, Naproxen.
3. **Aryl acetic acid derivatives**: e.g. Diclofenac, Ketorolac, *Aceclofenac*
4. **Indole derivatives**: e.g. Indomethacin, Sulindac
5. **Alkanones:** e.g. Nabumetone.
6. **Oxicams:** e.g. *Piroxicam*, Tenoxicam
7. **Anthranilic acid derivatives (fenamates):** e.g. *Mefenamic acid* and Flufenamic acid.
8. **Pyrazolone derivatives:** e.g. Phenylbutazone, Oxyphenbutazone, Azapropazone (Apazone) & Dipyrone.
9. **Aniline derivatives (analgesic only):** e.g. *Paracetamol.*

ii) Selective COX inhibitors

1. **Preferential COX-2 inhibitors:** e.g. Nimesulide, Meloxicam, Nabumeton
2. **Selective COX-2 inhibitors:** e.g. *Celecoxib*, Parecoxib, Rofecoxib.

10.3 SOME IMPORTANT ANALGESIC DRUGS & NSAIDs

10.3.1 Morphine: (*Syn:* **Morphia. Morphium**)

Structure:

Systematic (IUPAC) Name: (4R,4aR,7S,7aR,12bS)-3-Methyl-2,3,4,4a,7,7a-hexahydro-1H-4,12-methano[1 benzofuro[3,2-e]isoquinoline-7,9-diol

Mechanism of Action: This drug produces most analgesic effects by binding to the mu-opioid receptor within the central nervous system (CNS) and the peripheral nervous system (PNS).

Uses: Morphine is used primarily to treat acute and chronic severe pain. Its duration of analgesia is about three to seven hours. It is also used to treat pain due to myocardial infarction and labour pains.[

Morphine is beneficial in reducing the symptom of shortness of breath due to both cancer and noncancer causes. In the setting of breathlessness at rest or minimal exertion from conditions such as advanced cancer or end-stage cardiorespiratory diseases.

Side Effects:

The most common adverse effects involve drowsiness. Stomach pain and cramps, dry mouth, headache, nervousness, mood changes, small pupils (black circles in the middle of the eyes, difficulty urinating or pain when urinating.

Other Side effects

Constipation: Like loperamide and other opioids, morphine acts on the myenteric plexus in the intestinal tract, reducing gut motility, and causing constipation.

Hormone imbalance: Clinical studies consistently conclude that morphine, like other opioids, often causes hypogonadism and hormone imbalances in chronic users of both sexes.

Effects on human performance: Prolonged use may cause impairment in CNS functioning and cognitive ability.

Stability and storage conditions: Morphine solutions should preferably be stored at room temperature in order to avoid precipitation at low temperatures and water evaporation at higher temperatures causing an increase in morphine concentration when stored in polymer reservoirs.

Incompatibility: Some drugs and formulations that may interact with morphine include: certain pain medications (mixed opioid agonist-antagonists such as butorphanol, nalbuphine, pentazocine), naltrexone, products that contain alcohol (such as cough-and-cold syrups), samidorphan.

Different types of formulations: Tablet, injection.

Popular brands: Some of the notable brand names in India are:

Sr. No.	Brand Name	Manufactures	Type
1	VERMOR-10	Verve Healthcare Ltd.	Tablet
2	MORPHIGESIC-10	West Coast Pharmaceuticals Ltd.	Tablet
3	RUMORF-10	Rusan Healthcare Ltd.	Tablet
4	VERMOR-10	Verve Healthcare Ltd.	Injection
5	DEMROL	WebMD Ltd.	Injection
6	QINARSOL-300	Cipla Limited	Injection
7	KWINIL	Intas Pharma. Pvt. Ltd.	Tablet

10.3.2 Codiene: (*Syn:* **3-Methylmorphine**)

Structure:

Systematic (IUPAC) Name: (5α,6α)-7,8-didehydro-4,5-epoxy-3-methoxy-17-methylmorphinan-6-ol

Uses:

1. Pain: Codeine is used to treat mild to moderate pain. It is commonly used to treat post-surgical dental pain.
2. Cough: Codeine is used to relieve coughing as a centrally acting anti-tussive agent, especially for chronic cough in adults.

3. Diarrhea: It is used to treat diarrhoea and diarrhoea-predominant irritable bowel syndrome, though better agents are now available.

Mechanism of Action: Codeine binds to mu-opioid receptors, which are involved in the transmission of pain throughout the body and central nervous system. The analgesic properties of codeine are thought to arise from its conversion to Morphine, although the exact mechanism of the analgesic action is unknown at this time. Label,

Side Effects: Common adverse effects associated with the use of codeine include drowsiness and constipation. Less common are itching, nausea, vomiting, dry mouth, miosis, orthostatic hypotension, urinary retention, euphoria, and dysphoria. Rare adverse effects include anaphylaxis, seizure, acute pancreatitis, and respiratory depression.

Stability and storage conditions: Codeine phosphate solution is stable in both amber polyethylene terephthalate bottles and amber polyethylene oral syringes for at least 98 days when stored at 22-25°C and protected from light

Incompatibility: Some products that may interact with this drug include: certain pain medications (mixed opioid agonists-antagonists such as butorphanol, nalbuphine, pentazocine), naltrexone, and samidorphan.

Different types of formulation: Tablet, Syrup.

Popular brands: It is manufactured by many companies. Some of the notable brand names in India are:

Sr. No.	Brand Name	Manufactures	Type
1	DURANZO [Codeine with Diclofenac]	Ajanta Pharma. Pvt. Ltd.	Tablet
2	ELUTREX [Codeine with Triprolidine]	Generic India	Syrup
3	COCOREX [Codeine with Chlorpheniramine maleate]	Taj Pharma Pvt. Ltd	Syrup
4	COREX [Codeine with Triprolidine]	Pfizer	Syrup

10.3.3 Meperidine: (*Syn:* **Pethidine**)

Structure:

Systematic (IUPAC) Name: Ethyl 1-methyl-4-phenylpiperidine-4-carboxylate

Uses: Pethidine is the most widely used opioid in labour and delivery.

Pethidine is the preferred painkiller for diverticulitis (a gastrointestinal disease characterised by inflammation of abnormal pouches—diverticula—which can develop in the wall of the large intestine.) because it decreases intestinal intraluminal pressure.

Mechanism of action: Like morphine, pethidine exerts its analgesic effects by acting as an agonist at the μ-opioid receptor

Side Effects: Common side effects of pethidine administration are primarily those of the opioids as a class: nausea, vomiting, dizziness, diaphoresis, urinary retention, and constipation.

Stability and storage conditions: Meperidine is stable at room temperature. Avoid freezing the injectable solution and protect it from light during storage.

Incompatibility: Pethidine has serious interactions that can be dangerous with monoamine oxidase inhibitors (e.g., furazolidone, isocarboxazid, moclobemide, phenelzine, procarbazine, selegiline, tranylcypromine). Such patients may suffer agitation, delirium, headache, convulsions, and/or hyperthermia. Fatal interactions have been reported, including death. Seizures may develop when tramadol is given intravenously following or with pethidine. It can interact as well with SSRIs and other antidepressants, antiparkinson agents, migraine therapy, stimulants and other agents causing serotonin syndrome. It is thought to be caused by an increase in cerebral serotonin concentrations. It is probable that pethidine can also interact with a number of other medications, including muscle relaxants, benzodiazepines, and ethanol.

Different types of formulation: Injection, Tablets.

Popular brands: It is manufactured by many companies. Some of the notable brand names in India are:

Sr. No.	Brand Name	Manufactures	Type
1	DEMEROL	Validus Pharmaceuticals Ltd.	Tablet
2	DEMEROL	Hospira Inc.	Injection
3	VERPAT-50	Verve Healthcare Pvt Ltd	Injection
4	VERMOR-SR	Verve Healthcare Pvt Ltd	Tablet

10.3.4 Methadone: (*Syn:* **Dolophine and Methadose**)

Structure:

Systematic (IUPAC) Name: (*RS*)-6-(Dimethylamino)-4,4-diphenylheptan-3-one

Uses: Pentazocine is used primarily to treat pain.

Mechanism of action: Methadone is a synthetic opioid analgesic with full agonist activity at the μ-opioid receptor. While agonism of the μ-opioid receptor is the primary mechanism of action for the treatment of pain, methadone also acts as an agonist of κ- and σ-opioid receptors within the central and peripheral nervous systems. Interestingly, methadone differs from morphine (which is considered the gold standard reference opioid) in its antagonism of the N-methyl-D-aspartate (NMDA) receptor and its strong inhibition of serotonin and norepinephrine uptake, which likely also contributes to its anti-nociceptive activity.

Side Effects: Side effects are similar to those of morphine, but pentazocine, due to its action at the kappa opioid receptor, is more likely to invoke psychotomimetic effects. High doses may cause high blood pressure or high heart rate. It may also increase cardiac work after myocardial infarction when given intravenously; hence, this use should be avoided where possible. Respiratory depression is a common side effect but is subject to a ceiling effect, such that at a certain dose, the degree of respiratory depression will no longer increase with dose increases. Likewise, rarely it has been associated with agranulocytosis, erythema multiforme and toxic epidermal necrolysis (hypoventilation), light-headedness, fainting, weight gain, Memory loss, Itching, Difficulty urinating, Swelling of the hands, arms, feet, and legs, Mood changes, euphoria, disorientation, Blurred vision, Decreased libido, difficulty in reaching orgasm, or impotence, Missed menstrual, periods, Skin rash, Central sleep apnea.

Withdrawal symptoms:

Physical symptoms: Lightheadedness, Tearing of the eyes, Mydriasis (dilated pupils), Photophobia (sensitivity to light), Hyperventilation syndrome (breathing that is too fast/deep), Runny nose, Fever, Sweating, Chills, Tremors, Akathisia (restlessness), Tachycardia (fast heartbeat), Aches and pains, often in the joints or legs

Cognitive symptoms: Suicidal ideation, Susceptibility to cravings, Depression, Spontaneous orgasm, Prolonged insomnia, Delirium, Auditory hallucinations, Visual hallucinations.

Stability and storage conditions: Methadone was stable (loss of potency, less than 5%) for up to 17, 11, 9, and 8 days when stored at 20-25 degrees C in Kool-Aid, Tang, apple juice, and Crystal Light, respectively, and for up to 29 days when stored at 20-25 degrees C in Crystal Light plus sodium benzoate.

Incompatibility: Anticonvulsants, such as phenobarbital, phenytoin, and carbamazepine. These drugs can cause methadone to stop working. HIV drugs such as abacavir, darunavir, efavirenz, nelfinavir, nevirapine, ritonavir, and telaprevir. Antibiotics, such as rifampin and rifabutin.

Different types of formulation: Tablet, Injection, Oral Concentrate.

Popular brands: Some of the notable brand names in India are:

Sr. No.	Brand Name	Manufactures	Type
1	DOLPHINE	Westward Pharmaceutical Corp. Ltd	Tablet
2	R-METH-5	Rusan Pharma Ltd.	Tablet
3	METHADONE HCl (ORAL CONCENTRATE)	Westward Pharmaceutical Corp. Ltd	Liquid
4	DOLPHINE	Westward Pharmaceutical Corp. Ltd	Injection

10.3.5 Pentazocine: (*Syn:* Talwin)

Structure:

(R)-Pentazocine

(S)-Pentazocine

Mechanism of action: Pentazocine is an agonist at the κ-receptors and produces analgesia. Higher doses of pentazocine elicit psychotomimetic and dysphoric effects. The exact mechanism for these effects is unknown but is thought to be caused by weak σ-receptor agonist activity and activation of supraspinal κ-receptors.

Side Effects: Side effects are similar to those of morphine, but pentazocine, due to its action at the kappa opioid receptor, is more likely to invoke psychotomimetic effects. High doses may cause high blood pressure or high heart rate. It may also increase cardiac work after myocardial infarction when given intravenously; hence, this use should be avoided where possible. Respiratory depression is a common side effect but is subject to a ceiling effect, such that at a certain dose, the degree of respiratory depression will no longer increase with dose increases. Likewise, rarely it has been associated with agranulocytosis, erythema multiforme and toxic epidermal necrolysis.

Stability and storage conditions: Store Pentazocine at room temperature, 15- and 30 degrees C. Store it in an airtight container. Prescription drugs - Drugs to be sold only under the prescription of a Registered Medical Practitioner.

Incompatibility: Avoid taking MAO inhibitors (isocarboxazid, linezolid, metaxalone, methylene blue, moclobemide, phenelzine, procarbazine, rasagiline, safinamide, selegiline, tranylcypromine) during treatment with this medication.

Different types of formulation: Tablet, Injection.

Popular brands: Some of the notable brand names in India are:

Sr. No.	Brand Name	Manufactures	Type
1	NAVCIN	Peark Pharmaceutical Pvt. Ltd	Injection
2	PANLAB	Peark Pharmaceutical Pvt. Ltd	Injection
3	VINZOCIN	Ruchem Pharmaceutical Pvt. Ltd	Injection
4	SOSEGON	Apothecar Pharmaceutical Ltd	Tablet

10.3.6 Naloxone: (*Syn:* **Narcan**)

It is a medication used to reverse the effects of opioids? It is commonly used to counter decreased breathing in opioid overdose.

Structure:

Systematic (IUPAC) Name: (4*R*,4a*S*,7a*R*,12b*S*)-4a,9-dihydroxy-3-(prop-2-en-1-yl)-2,3,4,4a,5,6-hexahydro-1*H*-4,12-methano[1]benzofurano[3,2-*e*]isoquinolin-7(7a*H*)-one

Mechanism of Action: Narcan (naloxone hydrochloride) is an opioid antagonist that antagonises opioid effects by competing for the same receptor sites. Naloxone hydrochloride reverses the effects of opioids, including respiratory depression, sedation, and hypotension.

Uses: Used as an antagonist to reverse effects of Opioid analgesic drugs

Opioid overdose: Naloxone is useful in treating both acute opioid overdose and respiratory or mental depression due to opioids.

Clonidine overdose: Naloxone can also be used as an antidote for overdose of clonidine, a medication that lowers blood pressure. Clonidine overdoses are of special relevance for children, in whom even small doses can cause significant harm.

Preventing recreational opioid use: Naloxone is poorly absorbed when taken by mouth, so it is commonly combined with a number of oral opioid preparations, including buprenorphine and pentazocine, so that when taken by mouth, only the opioid has an effect. However, if the opioid and naloxone combination is injected, the naloxone blocks the effect of the opioid. This combination is used in an effort to prevent non-medical use.

Other Uses: In people with shock, including septic, cardiogenic, hemorrhagic, or spinal shock, those who received naloxone had improved blood flow. The importance of this is unclear.

Naloxone is also experimentally used in the treatment of congenital insensitivity to pain with anhidrosis, an extremely rare disorder that renders one unable to feel pain or differentiate temperatures.

Naloxone can also be used to treat itchiness brought on by opioid use,[34] as well as opioid-induced constipation

Side Effects: Naloxone has little to no effect if opioids are not present. In people with opioids in their system, it may cause increased sweating, nausea, restlessness, trembling, vomiting, flushing, and headache, and has, in rare cases, been associated with heart rhythm changes, seizures, and oedema.

Stability and storage conditions: Naloxone is a fairly stable medication, with a shelf life between 18 months and two years. IN and IM naloxone should be stored between 59 and 86 degrees Fahrenheit and should be kept away from direct sunlight.

Incompatibility: Naloxone may also reduce or stop the effectiveness of these prescription medications: Some HIV/AIDS drugs like abacavir. Some cancer treatment medications, like acalabrutinib (Calquence) and abemaciclib (Verzenio) Herpes virus treatments like acyclovir.

Different types of formulations: Tablet.

Popular brands: Some of the notable brand names in India are

Sr.No.	Brand Name	Manufactures	Type
1	NARCOTAN	Avalon Pharma Ltd	Injection
2	N-XONE	Verve Pharma Ltd.	Injection
3	YASHINALOX	Yashica Pharmaceuticals Ltd	Injection
4	VEGNOK-LS (With Buprenorphine)	Vega Ltd.	Tablet

10.3.7 Aspirin: (*Syn:* **Acetylsalicylic Acid, ASA)**

Aspirin is known as a salicylate and a nonsteroidal anti-inflammatory drug (NSAID).

Structure:

Systematic (IUPAC) Name: 2-Acetoxybenzoic Acid

Mechanism of Action: Aspirin causes several different effects in the body, mainly the reduction of inflammation, analgesia (relief of pain), the prevention of clotting, and the reduction of fever. Much of this is believed to be due to decreased production of prostaglandins and TXA_2.

Uses:

1. Pain: Aspirin is an effective analgesic for acute pain, although it is generally considered inferior to ibuprofen because aspirin is more likely to cause gastrointestinal bleeding. Aspirin is generally ineffective for those pains caused by muscle cramps, bloating, gastric distension, or acute skin irritation.

2. Fever: Like its ability to control pain, aspirin's ability to control fever is due to its action on the prostaglandin system through its irreversible inhibition of COX.

3. Inflammation: Aspirin is used as an anti-inflammatory agent for both acute and long-term inflammation, as well as for the treatment of inflammatory diseases, such as rheumatoid arthritis

4. Heart attacks and strokes: Aspirin is an important part of the treatment of those who have had a heart attack. It is generally not recommended for routine use by people with no other health problems, including those over the age of 70.

5. Aspirin is a first-line treatment for the fever and joint-pain symptoms of acute rheumatic fever.

Side Effects: Common side effects may include: upset stomach, heartburn; drowsiness; or a mild headache, severe nausea, vomiting, or stomach pain; bloody or tarry stools, coughing up blood or vomit that looks like coffee grounds; fever lasting longer than 3 days; or swelling, or pain lasting longer than 10 days.

Stability and storage conditions: Aspirin should be stored at room temperature, 20°C to 25°C (68° F to 77°F), in a sealed container, avoiding moisture. Aspirin is an OTC drug and is available in generic form without a prescription from your doctor or other healthcare professional.

No change in appearance, decomposition, or dissolution rates was observed in unpackaged aspirin tablets stored at 40°C for 4 weeks. However, when stored at 60°C, tablets of 5 of the

6 brands showed whiskers on their surfaces along with an increase in salicylic acid content and a decrease in dissolution rate.

Incompatibility: Some products that may interact with this drug include: mifepristone, acetazolamide, "blood thinners" (such as warfarin and heparin), corticosteroids (such as prednisone), dichlorphenamide, methotrexate, valproic acid, herbal medications (such as ginkgo Biloba). etc.

Different types of formulations: Tablet, Minitablet.

Popular brands: Available as generic as well as a brand medicine, some of the notable brand names in India are

Sr.No.	Brand Name	Manufactures	Type
1	DISPIRIN	Reckitt Benckiser	Tablet
2	WELLSPRIN	Wellona Pharmaceuticals Pvt. Ltd.	Tablet
3	ECOSPRIN-75	USV Ltd	Minitablet
4	ASPIWELL-75 (Enteric Coated)	Wellona Pharmaceuticals Pvt. Ltd.	Minitablet

10.3.8 Diclofenac: (*Syn:* **Cataflam**)

Diclofenac is a nonsteroidal anti-inflammatory drug (NSAID) used to treat pain and inflammatory diseases such as gout.

Structure:

Systematic (IUPAC) Name: [2-(2,6-Dichloroanilino)phenyl]acetic acid

Mechanism of Action: Diclofenac inhibits cyclooxygenase-1 and -2, the enzymes responsible for the production of prostaglandin (PG) G_2, which is the precursor to other PGs.

Uses:

Diclofenac is used to treat inflammatory disorders that may include musculoskeletal complaints, especially arthritis, rheumatoid arthritis (R.A.), polymyositis, dermatomyositis, osteoarthritis, dental pain, temporomandibular joint (TMJ) pain, spondylarthritis, ankylosing spondylitis, attacks, and pain management in cases of kidney stones and gallstones. An additional indication is the treatment of acute migraines.

Diclofenac is used commonly to treat mild to moderate post-operative or post-traumatic pain, in particular when inflammation is also present and is effective against menstrual pain and endometriosis.

Diclofenac is also available in topical forms and has been found to be beneficial for osteoarthritis but not other types of long-term musculoskeletal pain.

Side Effects: Common side effects may include: Stomach ache, feeling or being sick (vomiting), diarrhoea, black poo or blood in your vomit – a sign of bleeding in your stomach, headaches, drowsiness, ringing in your ears (tinnitus).

Stability and storage conditions:

Store at a controlled room temperature between 20°C and 25°C.; Keep away from excess light and moisture. Keep out of the reach of children.

The diclofenac sodium suspensions were very stable, retaining at least 99.5% of the original concentration for up to 93 days, regardless of storage temperature. There were no apparent changes in the physical appearance of the suspensions, nor were there any substantial changes in odour or pH.

Incompatibility: Concomitant administration of diclofenac and aspirin or other NSAIDs is not advised because of the risk of increased adverse effects.

Drugs such as methotrexate and cyclosporine should not be taken with diclofenac to avoid the potential for kidney toxicity.

Diclofenac may decrease the antihypertensive effects of blood pressure lowering medications, including ACE inhibitors such as lisinopril, enalapril, perindopril, and ramipril, when taken together.

Diclofenac may reduce the natriuretic (sodium-losing) effects of frusemide and thiazide diuretics, and patients must be monitored for signs and symptoms of renal failure.

When taken with diclofenac, warfarin results in a severe risk of gastrointestinal bleeding. Lithium, when taken concomitantly with diclofenac, resulted in the elevation of plasma lithium levels and reduced renal lithium clearance.

Different types of formulations: Tablet, Injection.

Popular brands: Some of the notable brand names in India are

Sr. No.	Brand Name	Manufactures	Type
1	VOLTAC-50	Wellona Pharma. Pvt. Ltd.	Tablet
2	VOVERAN-D	Novartis. Ltd.	Tablet
3	COFENAC	Cipla Ltd	Tablet
4	SALINAC-SP (With Paracetamol & Serratiopeptidase)	Salvus Pharmaceuticals Pvt. Ltd.	Tablet
5	DICLONIES	Greystar Pharma Pvt. Ltd.	Injection
6	TOPAC- IM	Abbot Ltd.	Injection

10.3.9 Ibuprofen: (*Syn:* **Brufen**)

Ibuprofen is a medication in the nonsteroidal anti-inflammatory drug class that is used for treating pain, fever, and inflammation.

Structure:

Systematic (IUPAC) Name: (*RS*)-2-(4-(2-Methylpropyl)phenyl)propanoic acid

Mechanism of Action: Ibuprofen is a non-selective inhibitor of an enzyme called cyclooxygenase (COX), which is required to synthesise prostaglandins via the arachidonic acid pathway. COX is needed to convert arachidonic acid to prostaglandin H_2 (PGH_2) in the body. PGH_2 is then converted to prostaglandins.

Uses: Ibuprofen is used primarily to treat fever (including post-vaccination fever), mild to moderate pain (including pain relief after surgery), painful menstruation, osteoarthritis, dental pain, headaches, and pain from kidney stones. About 60% of people respond to any nonsteroidal anti-inflammatory drug; those who do not respond well to a particular one may respond to another.

It is used for inflammatory diseases such as juvenile idiopathic arthritis and rheumatoid arthritis. It is also used for pericarditis and patent ductus arteriosus.

Side Effects: Common side effects may include: Headaches, dizziness, Feeling sick (nausea), Being sick (vomiting), wind, and Indigestion.

Stability and storage conditions:

Store it at room temperature, away from excess heat and moisture; you should not keep it in the bathroom. Keep it away from children and pets.

Undiluted ibuprofen (5 mg/mL) stored in glass vials and ibuprofen diluted to 2.5 mg/mL with either NS or D5W and stored in polypropylene syringes will retain more than 92% of its initial concentration with storage for up to 14 days at 4°C.

Incompatibility: Drugs that may interact with ibuprofen include:

Lithium, warfarin, oral hypoglycemic, high dose methotrexate, medication for lowering blood pressure, viz., angiotensin-converting enzyme inhibitors, beta-blockers, diuretics.

Different types of formulations: Tablet, Suspension.

Popular brands: Some of the notable brand names in India are

Sr. No.	Brand Name	Manufacturer	Type
1	IMOL (with Paracetamol)	Zydus Cadila Ltd.	Suspension
2	IBUCARE	Advacare Ltd.	Powder for Suspension
3	BRUFEN (200mg, 400mg, 600mg)	Abbott Ltd	Tablet
4	COMBIFLAM (with Paracetamol)	Sanofi Ltd.	Tablet
5	COMBIFLAM (with Paracetamol)	Sanofi Ltd.	Syrup

10.3.10 Piroxicam: (*Syn:* Feldene)

Piroxicam is an NSAID that reduces pain, swelling, and joint stiffness from arthritis. Reducing these symptoms helps you do more of your normal daily activities.

Structure:

Systematic (IUPAC) Name: (4-Hydroxy-2-methyl-*N*-(2-pyridinyl)-2*H*-1,2-benzothiazine-3-carbox- amide 1,1-dioxide

Mechanism of Action: It acts as a non-selective COX inhibitor to exert both analgesic and antipyretic properties.

Uses: It is used in the treatment of certain inflammatory conditions like rheumatoid and osteoarthritis, primary dysmenorrhoea, and postoperative pain; and acts as an analgesic, especially where there is an inflammatory component.

Side Effects: Common side effects may include: diarrhoea, constipation, gas, headache, dizziness, and ringing in the ears.

Stability and storage conditions:

Store at room temperature 20°C to 25°C (68°F to 77°F); excursions permitted between 15°C to 30°C (59°F to 86°F).

Incompatibility: Drugs that may interact with Piroxicam include: Aliskiren, ACE inhibitors (such as captopril, lisinopril), angiotensin II receptor blockers (such as losartan, valsartan), cidofovir, corticosteroids (such as prednisone), lithium, "water pills" (diuretics such as furosemide).

Different types of formulations: Tablet, Capsule, Injection.

Popular brands: Some of the notable brand names in India are

Sr. No.	Brand Name	Manufacturer	Type
1	DOLONEX-DT	Pfizer Ltd.	Tablet
2	SALCAM-20	Sparc Remedies	Tablet
3	PIROWELL	Wellona Pharma Pvt. Ltd	Injection
4	PIROXICHEM	SKG Iternational Ltd.	Injection
5	FELDENE	Pfizer Ltd..	Capsule

10.3.11 Celecoxib: (*Syn:* **Celebrex)**

It is a COX-2 inhibitor and nonsteroidal anti-inflammatory drug.

Structure:

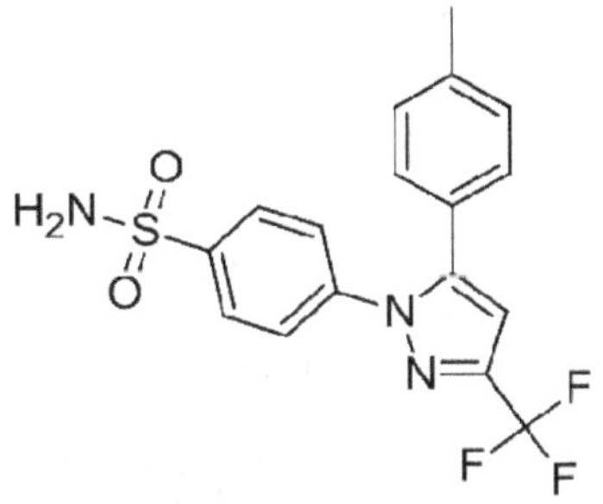

Systematic (IUPAC) Name: 4-[5-(4-Methylphenyl)-3-(trifluoromethyl)pyrazol-1-yl]ben-zenesulfon-amide

Mechanism of Action: It acts as a non-selective COX inhibitor to exert both analgesic and antipyretic properties.

Uses: Mefenamic acid is used to treat pain and inflammation in rheumatoid arthritis and osteoarthritis, postoperative pain, acute pain, including muscle and back pain, toothache and menstrual pain, as well as being prescribed for menorrhagia.

Side Effects: Common side effects may include: gas or bloating, sore throat, cold symptoms, constipation, dizziness, dysgeusia, and ringing in the ears.

Stability and storage conditions: Store at room temperature 20°C to 25°C (68°F to 77°F);

Incompatibility: Drugs that may interact with Celecoxib include:

Some products that may interact with this drug are aliskiren, ACE inhibitors (such as captopril, lisinopril), angiotensin II receptor blockers (such as valsartan, losartan), cidofovir, lithium, "water pills" (diuretics such as furosemide).

Different types of formulations: Tablet, Capsule, Injection.

Popular brands: Some of the notable brand names in India are

Sr. No.	Brand Name	Manufacturer	Type
1	COBIX-200	Cipla Ltd.	Capsule
2	CELEHEAL-200	Pharma India Pvt. Ltd	Capsule
3	WELLCOXIB-100	Wellona Pharma Pvt. Ltd	Capsule/Gelule
4	ELYXIB	Fierce Pharma Pvt Ltd.	Oral Solution

10.3.12 Mefenamic Acid: (*Syn:* Meftal)

Mefenamic acid is a member of the anthranilic acid derivatives class of nonsteroidal anti-inflammatory drugs and is used to treat mild to moderate pain.

Structure:

Systematic (IUPAC) Name: 2-(2,3-Dimethylphenyl)aminobenzoic acid

Mechanism of Action: The mechanism of action of mefenamic acid, like that of other NSAIDs, is not entirely understood but involves inhibition of cyclooxygenase (COX-1 and COX-2). Mefenamic acid is a potent inhibitor of prostaglandin synthesis *in vitro*.

Uses: Mefenamic acid is used to treat pain and inflammation in rheumatoid arthritis and osteoarthritis, postoperative pain, acute pain, including muscle and back pain, toothache and menstrual pain, as well as being prescribed for menorrhagia.

Side Effects: Common side effects may include: diarrhoea. Constipation, Gas or bloating, Headache, Dizziness, Nervousness, ringing in the ears.

Stability and storage conditions: Store at room temperature 15°C to 30°C (59°F to 68°F);

Incompatibility: Drugs that may interact with Celecoxib include:

Some products that may interact with this drug are aliskiren, ACE inhibitors (such as captopril, lisinopril), angiotensin II receptor blockers (such as valsartan, losartan), cidofovir, lithium, "water pills" (diuretics such as furosemide).

Different types of formulations: Tablet, Suspension

Popular brands: Some of the notable brand names in India are

Sr. No.	Brand Name	Manufacturer	Type
1	PONSTAN	Pfizer Ltd.	Tablets
2	MEFTAL-250	Blue Cross Labs Pvt. Ltd	Tablets
3	MEFSTAR FORTE (WITH PARACETAMOL)	Acurez Pharma Pvt. Ltd	Tablets

Table *Contd...*

Sr. No.	Brand Name	Manufacturer	Type
4	UNISPAS (WITH DICYCLOMIN)	Tanisq Lifecare Pvt Ltd.	Tablets
5	Celinyl-T (WITH TRANEXAMIC ACID)	Alpha Labs Pvt. Ltd	Tablets
6	MEFKID-PDS (WITH PARACETAMOL)	Hicure Biotech Pvt. Ltd.	Suspension
7	MEFSTAL FORTE (WITH PARACETAMOL)	Wellona Pharma Pvt. Ltd.	Suspension

10.3.13 Paracetamol: (*Syn:* Acetaminophen)

Paracetamol, also known as acetaminophen, is a medication used to treat fever and mild to moderate pain.

Structure:

Systematic (IUPAC) Name: *N*-(4-Hydroxyphenyl)acetamide

Mechanism of Action: Paracetamol appears to exert its effects through two mechanisms: the inhibition of cyclooxygenase and actions of its active metabolite arachidylaminophenol (AM404).

Uses: Paracetamol is a common painkiller used to treat aches and pain-Analgesic. It can also be used to reduce a high temperature-Antipyretic. It's available combined with other painkillers and anti-sickness medicines. It's also an ingredient in a wide range of cold and flu remedies.

Side Effects: Common side effects may include: skin rash that may include itchy, red, swollen, blistered or peeling skin, tightness in the chest or throat, trouble in breathing, swelling of mouth, lips, throat, low fever with nausea, stomach pain, and loss of appetite.

Stability and storage conditions: Store below 25°C in a dry place. Protect from light.

Incompatibility: There are no severe interactions with acetaminophen and other drugs. There are no serious interactions with acetaminophen and other drugs.

Different types of formulations: Tablet, Drops, Suspension, Injection.

Popular brands: Some of the notable brand names in India are

Sr. No.	Brand Name	Manufacturer	Type
1	PEACFMOL-I.V.	Asterisk Pvt. Ltd.	Injection-Infusion
2	CROCIN	GSK Ltd	Tablet, Drops, Suspension
3	DOLO-650	Micro Labs. Ltd	Tablets
4	CALPOL-500	GSK Ltd	Tablets

10.3.14 Aceclofenac: (*Syn:* **Aceclofar, Hifenac**)

Aceclofenac is a nonsteroidal anti-inflammatory drug analog of diclofenac. It is used for the relief of pain and inflammation.

Structure:

Systematic (IUPAC) Name: 2-[2-[2-[(2,6-Dichlorophenyl)amino]phenyl]acetyl] oxyacetic acid

Mechanism of Action: The mode of action of aceclofenac is largely based on the inhibition of prostaglandin synthesis. Aceclofenac is a potent inhibitor of the enzyme cyclo-oxygenase, which is involved in the production of prostaglandins.

Uses: It is used for the relief of pain and inflammation in rheumatoid arthritis, osteoarthritis and ankylosing spondylitis.

Side Effects: Common side effects may include - headache, nausea, vomiting, epigastric pain, gastrointestinal irritation, gastrointestinal bleeding, rarely diarrhoea, disorientation, excitation, coma, drowsiness, dizziness, tinnitus, hypotension, respiratory depression, fainting, occasionally convulsions.

Stability and storage conditions: Store it in an airtight container in a cool, dry place. Protect from light.

Incompatibility: Aceclofenac may decrease the antihypertensive activities of Amlodipine. The risk or severity of hyperkalemia can be increased when Aceclofenac is combined with Ammonium chloride. The risk or severity of gastrointestinal bleeding can be increased when Amoxapine is combined with Aceclofenac.

Different types of formulations: Tablet, Drops, Suspension, Injection.

Popular brands: Some of the notable brand names in India are

Sr. No.	Brand Name	Manufacturer	Type
1	HIFENAC	Intas Laboratories Ltd.	Tablet
2	ACECLO	Aristo Pharma Ltd	Tablets
3	MEDIFACE-SP (WITH PARACETAMOL & SERRATIOPEPIDASE)	Mediva Lifecare Ltd	Tablets
4	ACEFLOT-MR (WITH PARACETAMOL & CHLORZOXAZONE)	Sansavi Pharma Ltd.	Tablets

Table *Contd...*

Sr. No.	Brand Name	Manufacturer	Type
5	ACCLORACE	Saintroy Lifesciences	Injection
6	ACECLOWELL	Wellona Pharma Ltd.	Injection
7	ACEFLUR	Sencare LifeSciences Pvt. Ltd.	Suspension

QUESTION BANK

A. MULTIPLE CHOICE QUESTIONS

1. What are opioid analgesics?
   A. Render a specific portion of the body insensitive to pain
   B. Render the whole body of the patient insensitive to pain
   C. Render insensitive to pain by binding to opioid receptors
   D. Render a specific CNS of the body insensitive to pain

   Answer: C

   Explanation: *Analgesics, or pain killers, that bind to opioid receptors, which are found principally in the CNS and Gastrointestinal tract. Most of the time, the pain threshold level increases, thus decreasing the patient's perception of the pain occurring.*

2. Which of the following are natural opiates?
   A. Codeine
   B. Methadone
   C. Pentazocine
   D. Meperidine

   Answer: A

   Explanation: *Natural opiates are alkaloids contained in the resin of the opium poppy, including morphine and codeine. Semi-synthetic opiates are created from natural opiates such as hydromorphone, oxycodone, and diacetylmorphine. Fully synthetic opioids are such as fentanyl, methadone, etc.*

3. Which of the following is a semi-synthetic opiates?
   A. Codeine
   B. Oxycodone
   C. Morphine
   D. Pentazocine

   Answer: B

   Explanation: *Semi-synthetic opiates are created from natural opiates such as hydromorphone, oxycodone, and diacetylmorphine. These are made by simple modification of the existing natural chemical structure of the compound, making it more potent, less toxic etc. Natural opiates are alkaloids contained in the resin of the opium poppy, including morphine and codeine. Fully synthetic opioids are such as fentanyl, methadone, etc.*

4. Which of the following side effects of morphine is the most dangerous?
   A. Excitation
   B. Depression
   C. Constipation
   D. Tolerance and dependence

Answer: D

Explanation: *Dangerous side effects are tolerance and dependence, allied with the effects morphine can have on breathing. The most common cause of death from morphine overdose is suffocation. These side effects of one drug are particularly dangerous and lead to severe withdrawal symptoms when the drug is no longer taken.*

5. Which of the following opioid drug is used to reverse the effects of opioids as well as commonly used to counter decreased breathing in opioid overdose?
   A. Meprobamate
   B. Methadone
   C. Naloxone
   D. Mepiridine

Answer: C

Explanation: *Naloxone or N-Allylmorphine is the morphine antagonist and is used for deaddiction and detoxification of opioid effects.*

6. Which of the following NSAIDs is a better antipyretic drug?
   A. Aspirin
   B. Aceclofenac
   C. Acetaminophen
   D. Ibuprofen

Answer: C

7. Which of the following NSAIDs is a better anti-inflammatory drug?
   A. Aspirin
   B. Diclofenac
   C. Acetaminophen
   D. Ibuprofen

Answer: b

8. Which of the following NSAIDs is a selective COX-2 inhibitor?
   A. Aspirin
   B. Diclofenac
   C. Celecoxib
   D. Ibuprofen

Answer: C

9. Which of the following NSAIDs is an Aniline derivative?
    A. Aspirin
    B. Diclofenac
    C. Paracetamol
    D. Ibuprofen

Answer: C

10. Which of the following NSAIDs is an important part of hypertension and cardiovascular disease treatment?
    A. Aspirin
    B. Diclofenac
    C. Paracetamol
    D. Ibuprofen

Answer: A

11. Which of the following opioid drug is most widely used in labour and delivery?
    A. Meprobamate
    B. Methadone
    C. Naloxone
    D. Mepiridine

Answer: D

B. SHORT ANSWER QUESTIONS

1. Write the uses, side effects, stability and storage of the following

    a) Aspirin b) Ibuprofen c) Morphine d) Codiene e) Paracetamol f) Celecoxib g) Aceclofenac

2. Discuss the Mechanism of Action of Ibuprofen and Diclofenac.
3. Discuss the Mechanism of Action of Methadone and Mepiridine.
4. Discuss the Mechanism of Action of Piroxicam and Celecoxib.
5. Discuss the mechanism of Action of Mefenamic acid and Aceclofenac.
6. Write the structure and chemical names of the following compounds.

    a) Aspirin b) Ibuprofen c) Morphine d) Codiene e) Paracetamol f) Celecoxib g) Aceclofenac.

C. LONG ANSWER QUESTIONS

1. Define Analgesics and classify them and give one drug example of each class, its structure and mechanism of action.
2. Discuss any two opioid analgesics and two NSAIDs in detail.

ANTI-INFECTIVE AGENTS: ANTIFUNGAL AGENTS

CONTENTS

Antifungal Agents:

- Amphotericin-B
- Griseofulvin
- Miconazole
- Ketoconazole*
- Itraconazole
- Fluconazole*
- Naftifine Hydrochloride

♦ LEARNING OBJECTIVES ♦

After completing this chapter, the student should be able to understand:

1. Antifungal drugs
2. Fungal diseases
3. Classification of Antifungal drugs.
4. Chemical Structures of different Antifungal drugs.
5. Uses, Stability and storage conditions of Antifungal drugs.
6. Different types of formulations and popular brand names for Antifungal drugs.

11.1.1 INTRODUCTION

Until recently, chemotherapy for fungal infections has lagged far behind chemotherapy for bacterial infections. This lack of progress has resulted partly because the most common fungal infections in humans have been relatively superficial infections of the skin and mucosal membranes, and potentially lethal deep-seated infections have been quite rare. Because most humans with a normally functioning immune system are able to ward off invading fungal pathogens with little difficulty, the demand for improvements in antifungal therapy has been negligible. Immunocompromised patients, however, are very susceptible to invasive fungal infections. The onset of the AIDS epidemic, combined with the increased use of powerful immunosuppressive drugs for organ transplants and cancer chemotherapy, has resulted in a significantly increased incidence of life-threatening fungal infections and a corresponding increase in demand for new agents to treat these infections. The number of effective antifungal agents available is relatively small compared to those available to treat bacterial infections, but research in this area is quite active. Several new agents have been introduced in the last few years.

11.1.2 FUNGAL DISEASES

A fungus that invades the tissue can cause a disease confined to the skin that spreads into tissue, bones and organs or affects the whole body.

Symptoms depend on the area affected but can include skin rash or vaginal infection resulting in abnormal discharge.

Although most fungi are harmless to humans, some of them are capable of causing diseases under specific conditions. Fungi reproduce by releasing spores that can be picked up by direct contact or even inhaled. Fungal infections are most likely to affect the skin, nails, or lungs.

11.1.3 CLASSIFICATION OF ANTIFUNGAL AGENT

1. **Antibiotics**
    A. **Polyenes:** e.g. Amphotericin-B, nystatin, hamycin
    B. **Heterocyclic benzofuran derivative:** e.g. Griseofulvin
2. **Antimetabolite:** e.g. 5-Flucytocine
3. **Azole**
    A. **Imidazole**

       Systematic e.g. Econazole miconazole cotrimazole

       Topical e.g. Ketoconazole
    B. **Triazole:** e.g. Intraconazole
4. **Allylamines:** e.g. Terbinafine
5. **Other agents:** e.g. Benzoic acid, sodium thiosulfate, tolnaftae

11.1.4 SOME IMPORTANT ANTIFUNGAL DRUGS

11.1.4.1 Amphotericin-B

Structure:

Chemical (IUPAC) name:

(1R,3S,5R,6R,9R,11R,15S,16R,17R,18S,19E,21E,23E,25E,27E,29E,31E,33R, 35S,36R,37S)-33-[(3-amino-3,6-dideoxy-β-D-mannopyranosyl)oxy]-1,3,5,6,9,11,17,37-octahydroxy-15,16,18-trimethyl-13-oxo-14,39-dioxabicyclo [33.3.1] nonatriaconta-19,21,23,25,27,29,31-heptaene-36-carboxylic acid.

The isolation of amphotericin B (fungizone) was reported in 1956. The fermentation of beer of a soil culture of the actinomycete *streptomyces nodosus*, which was isolated in Venezuela. The first isolate from the streptomycete was a separable mixture of two compounds, designated amphotericins-A and B. In test cultures, compound B proved to be more active, and this is the one used clinically.

Mechanism of Action:

Amphotericin B is believed to interact with membrane sterols (ergosterol in fungi) to produce an aggregate that forms a transmembrane channel. Intermolecular hydrogen bonding interactions among hydroxyl, carboxyl and amino groups stabilise the channel in its open form, destroying symport activity and allowing the cytoplasmic contents to leak out. The effect is similar to cholesterol. This explains the toxicity in human patients. As the name implies, amphotericin B is an amphoteric substance, with a primary amino group attached to the mycosamine ring and a carboxyl group on the macrocycle.

The compound forms deep yellow crystals sparingly soluble in organic solvents but insoluble in water. Both acids and bases, the salts are only slightly soluble in water (0.1 mg/ml) and, hence, cannot be used systemically.

To create a parenteral dosage form, amphotericin B is stabilised as a buffered colloidal dispersion in micelles with sodium deoxycholate.

The barrel-like structure of the antibiotic develops interactive forces with the micellar components, creating a soluble dispersion. The preparation is light, heat, salt, and detergent sensitive. Parenteral amphotericin B is indicated for treating severe, potentially life-threatening fungal infections.

Side Effects:

The usefulness of amphotericin B is limited by a high prevalence of adverse reactions. Nearly 80% of patients treated with amphotericin B develop nephrotoxicity. Fever, headache, anorexia, gastrointestinal distress, malaise, and muscle and joint pain are common. Pain at the site of injection and thrombophlebitis are frequent complications of intravenous administration. The drug must never be administered intramuscularly. Clinical use of each of the approved lipid preparations has shown reduced renal toxicity.

Uses:

Liposomal amphotericin B has been explicitly approved for the treatment of pulmonary aspergillosis because of its demonstrated superiority to the sodium deoxycholate-stabilized suspension. Amphotericin B is used topically to treat cutaneous and mucocutaneous mycoses caused by *c. Albicans*. The drug is supplied in various topical forms, including a 3% cream, a 3% lotion, a 3% ointment, and a 100-mg/ml oral suspension. The oral suspension is intended for the treatment of oral and pharyngeal candidiasis. The patient should swish the suspension in his or her mouth and swallow it. The suspension has a very bad taste, so that compliance may be a problem.

Stability and storage conditions:

Amphotericin B 1 and 2 mg/mL was stable in 20% fat emulsion for four days at 20-25^{0}C exposed to fluorescent light and for seven days at 20-25^{0}C protected from light or at 4-8^{0}C; amphotericin B 0.5 mg/mL was stable in 20% fat emulsion for seven days under the three storage conditions.

Dose: The dose 50mg/vial, Abelcet Lipid Complex 5 mg/mL Concentrate for dispersion for infusion

Pharmaceutical Formulations: Suspension, Injection, powder, lyophilised, for solution

Brand Names: Abelcet, Ambisome, Amphotec, Fungizone

11.1.4.2 Griseofulvin

Structure:

Chemical (IUPAC) name: *(2S,5'R)*-7-chloro-3',4,6-trimethoxy-5'-methylspiro[1-benzofuran-2,4'-cyclohex-2-ene]-1',3-dione

Griseofulvin is an antifungal antibiotic produced by an unusual strain of Penicillium. It is used orally to treat superficial fungal infections, primarily fingernail and toenail infections, but it does not penetrate skin or nails if used topically. When given orally, however, plasma-borne griseofulvin becomes incorporated into keratin precursor cells and, ultimately, into keratin, which cannot then support fungal growth. The infection is cured when the diseased tissue is replaced by new, uninfected tissue, which can take months.

The mechanism of action of griseofulvin is through binding to the protein tubulin, which interferes with the function of the mitotic spindle and, thereby, inhibits cell division. Griseofulvin also may interfere directly with DNA replication. Griseofulvin is gradually being replaced by newer agents.

It is an antibiotic obtained from the fungus *penicillium griseofulvum*.

The compound is a white, bitter, heat-stable powder or crystalline solid that is sparingly soluble in water but soluble in alcohol and other nonpolar solvents. It is very stable when dry.

Griseofulvin has been used for a long time for the systemically delivered treatment of refractory ringworm infections of the body, hair, nails, and feet caused by species of dermatophytic fungi, including *trichophyton*, *microsporum*, and *epidermophyton*.

Stability and storage conditions: Griseofulvin preparation should generally be stored at a temperature less than 40⁰C, preferably between 15-30⁰C.

Dose: The dose is 500 mg, 125 mg/5 mL for suspension.

Pharmaceutical Formulations: Tablet, Suspension

Brand Names: Fulvicin P/G, Fulvicin U/F, Gris-PEG, Grisovin Fp, Griseofulvin

11.1.4.3 Miconazole

Structure:

Chemical (IUPAC) name: 1-[2-(2,4-dichlorophenyl)-2-[2,4-dichlorophenyl]-methoxy]ethyl]-1*h*-imidazole mononitrate (monistat, micatin) is a weak base with a pka of 6.65.

The nitric acid salt occurs as white crystals that are sparingly soluble in water and most organic solvents. The free base is available in an injectable form, solubilised with polyethylene glycol and castor oil, and intended for the treatment of severe systemic fungal infections, such as candidiasis, coccidioidomycosis, cryptococcosis, petriellidiosis, and paracoccidioidomycosis.

It may also be used for the treatment of chronic mucocutaneous candidiasis.

Although severe toxic effects from the systemic administration of miconazole are comparatively rare, thrombophlebitis, pruritus, fever, and gastrointestinal upset are relatively common.

Miconazole nitrate is supplied in various dosage forms (cream, lotion, powder, and spray) to treat tinea infections and cutaneous candidiasis.

Vaginal creams and suppositories are also available for the treatment of vaginal candidiasis. A concentration of 2% of the salt is used in most topical preparations.

Stability and storage conditions:

Store miconazole tablets at room temperature, 68°F to 77°F (20°C to 25°C). Protect the tablets from moisture. Store miconazole vaginal suppositories at 59°F to 86°F (15°C to 30°C).

Dose: The dose is 50 mg Tablet, 2 % Topical Cream, 200 Mg-2 % Suppository

Pharmaceutical Formulations: Suppository, Cream, Tablet

Brand Names: Monistat 3, Monistat-derm, Oravig.

11.1.4.4 Ketoconazole
Structure:

Chemical (IUPAC) name: 1-acetyl-4-[4-[[2-(2,4-dichlorophenyl)-2(1*h*-imidazole-1-yl-methyl)-1,3-dioxolan-4-yl]methoxy]phenyl]Piperazine.

It is a weakly basic compound that occurs as a white crystalline solid that is very slightly soluble in water.

Ketoconazole is a racemic compound consisting of the *cis-2S,4R* and *cis-2R,4S* isomers. An investigation of the relative potencies of the four possible diastereomers of ketoconazole against rat lanosterol 14-demethylase indicated that the *2S,4R* isomer was 2.5 times more active than its *2R,4S* enantiomer. The *trans*-isomers, *2S,4S* and *2R,4R*, are much less active.

Hepatotoxicity, primarily of the hepatocellular type, is the most severe adverse effect of ketoconazole.

Ketoconazole is known to inhibit cholesterol biosynthesis, suggesting that lanosterol 14-demethylase is inhibited in mammals as well as in fungi.

Ketoconazole is recommended for the treatment of the following systemic fungal infections: candidiasis (including oral thrush and the chronic mucocutaneous form), coccidioidomycosis, blastomycosis, histoplasmosis, chromomycosis, and paracoccidioido-mycosis.

It is also used orally to treat severe refractory cutaneous dermatophytic infections not responsive to topical therapy or oral griseofulvin.

Ketoconazole is also used topically in a 2% concentration in a cream and in a shampoo for the management of cutaneous candidiasis and tinea infections.

Stability and storage conditions: Store it at room temperature.

Dose: The dose is 200 mg Tablet, 2 % Topical Cream, 2 % Aerosol, foam Cream

Pharmaceutical Formulations: Aerosol, foam Cream, Shampoo, tablet, gel

Brand Names: Extina, Ketoderm Cream 2%, Nizoral, Xolegel.

11.1.4.5 Itraconazole

Structure:

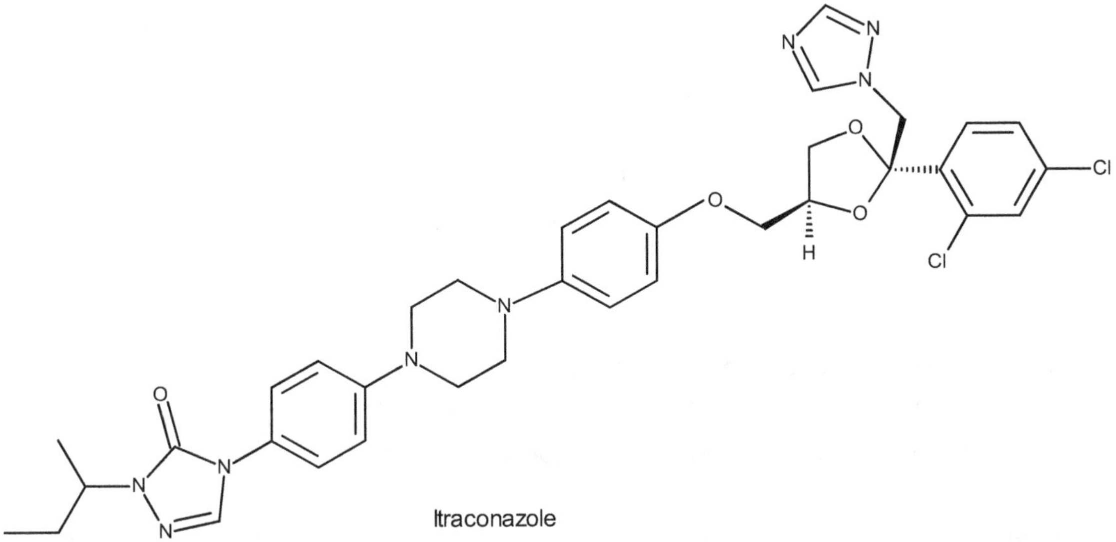

Itraconazole

Chemical (IUPAC) name: 4-[4-[4-[4-[[2-(2,4-dichlorophenyl)-2-1*h*-1,2,4-triazol-1-yl-methyl)-1,3-dioxolan-4-yl]methoxy]phenyl]-1-piperazinyl] phenyl]-2,4-dihydro-2-(1-methyl-propyl)-3*h*-1,2,4- triazol-3-one (sporanox) is a unique member of the azol class that contains two triazole moieties in its structure, a weakly basic 1,2,4-triazole and a nonbasic 1,2,4-triazol- 3-one.

Itraconazole is an orally active, broad-spectrum antifungal agent that has become an important alternative to ketoconazole.

It may also be effective in the treatment of pergellosis, disseminated and deep organ candidiasis, coccidioidal meningitis, and cryptococcosis.

Stability and storage conditions:

Capsules should be stored at room temperature, 15^0C - 25^0C (59^0F - 77^0F) and protected from light and moisture. Oral and injectable solutions should be stored below 25^0C (77^0F) but not frozen.

Dose: The dose of 200 mg

Pharmaceutical Formulations: Tablet, Capsule, injectable solutions

Brand Names: Onmel, Sporanox

11.1.4.6 Fluconazole

Structure:

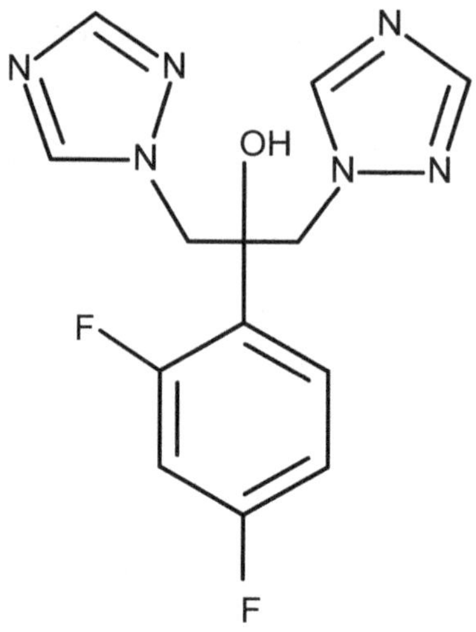

Chemical (IUPAC) name: α-(2,4-difluorophenyl)- α -(1*h*-1,2,4-triazol-1-ylmethyl)-1*h*-1,2,4-triazole-1-ethanol or 2,4-difluoro- α, α -bis(1*h*-1,2,4-triazol-1-ylmethyl)benzyl alcohol.

It is water soluble bis-triazole with broad-spectrum antifungal properties that is suitable for both oral and intravenous administration as the free base.

Intravenous solutions of fluconazole contain 2 mg of the free base in 1 ml of isotonic sodium chloride or 5% dextrose vehicle.

Fluconazole is an agent of choice for the treatment of cryptococcal meningitis and for prophylaxis against cryptococcosis in AIDS patients.

Although fluconazole is generally less effective than either ketoconazole or itraconazole against nonmeningeal coccidioidomycosis, it is the preferred therapy for coccidioidal meningitis. Fluconazole lends itself to one-dose therapies for vaginal candidiasis.

Stability and storage conditions:

Reconstituted fluconazole oral suspension is stable in both the product's original plastic bottles and amber polyethylene oral syringes for at least 70 days when stored at 22-25⁰C.

Dose: Tablet- 50mg, 100mg, 150mg

Pharmaceutical Formulations: Tablet

Brand Names: Flungi, Flumed, Fungout, Gusnil

11.1.4.7 Naftifine hydrochloride

Structure:

Chemical (IUPAC) name: *N*-methyl-*n*-(3-phenyl2-propenyl)-1-naphthalenemethanamine hydrochloride.

It is a white crystalline powder that is soluble in polar solvents such as ethanol and methylene chloride.

It is supplied in a 1% concentration in a cream and in a gel for the topical treatment of ringworm, athlete's foot, and jock itch. Although unapproved

For these uses, naftifine has shown efficacy for the treatment of ringworm of the beard, ringworm of the scalp, and tinea versicolor.

Stability and storage conditions:

Store in a cool, dry place out of direct sunlight. Store Naftifine Hydrochloride Cream USP, 2% at 20° to 25°C (68° to 77°F) [see USP Controlled Room Temperature

Dose: 2% Gel & Cream

Pharmaceutical Formulations: Gel, cream

Brand Names: Naftin (gel, cream)

QUESTION BANK

A. MULTIPLE CHOICE QUESTIONS

1. Which enzyme combination is involved in ergosterol biosynthesis?
    A. Lanosterol 14 alpha demethylase and squalene epoxidase
    B. Lanosterol epoxidase and squalene 16 alpha demethylase
    C. Lanosterol epoxidase and squalene 14 alpha demethylase
    D. Lanosterol 16 alpha demethylase and squalene epoxidase

Answer: A

2. Only topically used antifungal
   A. Clotrimazole
   B. Griseofulvin
   C. Ternbinafine
   D. Fluconazole

Answer: A

3. Which of the following is a Polyene antifungal agent?
   A. Griseofulvin
   B. Miconazole
   C. Ketoconazole*
   D. Amphotericin-B

Answer: D

4. Which of the following heterocyclic ring is present in Econazole, Miconazole, Cotrimazole & Ketoconazole?
   A. Pyrrol
   B. Imidazole
   C. Furan
   D. Benzfuran

Answer: B

5. In which of the following benzofuran ring is present?
   A. Ketoconazole
   B. Miconazole
   C. Griseofulvin
   D. Amphotericin-B

Answer: C

6. Mucormycosis, the drug of choice, is
   A. Amphotericin B
   B. Intraconazole
   C. Voriconazole
   D. Griseofulvin

Answer: A

7. The mode of action of polyenes involve
   A. Loss of membrane integrity and influx of ca^{+2}
   B. Inhibition of reproductive function
   C. Binds to ergosterol
   D. Destruction of pore-like molecular aggregates

Answer: C

8. Which of these is not a polyne?
   A. Terbinafine
   B. Nystatin
   C. Amphotericin B
   D. Hamaycin

   Answer: A

9. What adverse effects are associated with amphotericin B?
   A. Allergic reactions and reversible renal impairment on accumulation and hepatotoxicity
   B. Allergic reactions and nephrotoxicity
   C. Reversible renal impairment on accumulation nephrotoxicity and hepatotoxicity
   D. Allergic reactions hepatotoxicity and fungal infection, and superficial and invasive candidiasis

   Answer: B

10. What is an allylamines mode of action?
    A. Inhibit ergosterol synthesis *via* lanosterol epoxidase
    B. Inhibit ergosterol synthesis *via* squalene 14 alpha demethylase
    C. Inhibit ergosterol synthesis *via* synthesis of squalene epoxidase
    D. Inhibit ergosterol synthesis *via* lanosterol 14 alpha demethylase

    Answer: C

11. Terbinefine is an example often which class of antifungal
    A. Polyne
    B. Anti-metabolite
    C. Azole
    D. Allylamine

    Answer: D

12. Topical antifungal agents used for dermatophytes
    A. Hamycin
    B. Natamycin
    C. Nystatin
    D. Tolnaftate

    Answer: D

13. Drug not effective in candidial infection
    A. Fluconazle
    B. Ketoconazole
    C. Terbinafine
    D. Griseofulvin

    Answer: D

14. Extina is the brand name for which of the following drug?
    A. Fluconazle
    B. Terbinafine
    C. Griseofulvin
    D. Ketoconazole

Answer: D

15. Fulvicin P/G is the brand name for which of the following drug?
    A. Fluconazle
    B. Terbinafine
    C. Griseofulvin
    D. Ketoconazole

Answer: C

B. SHORT ANSWER QUESTIONS

1. Classify antifungal agents.
2. Enlist different pharmaceutical formulations available for Amphotericin-B.
3. Explain storage condition and stability for Amphotericin-B.
4. Enlist different pharmaceutical formulations available for Naftifine hydrochloride.
5. Enlist different brand names for Amphotericin-B.
6. Draw the structure for Fluconazole.
7. Enlist different brand names for Fluconazole.
8. Give the uses and different pharmaceutical formulations available for Ketoconazole.
9. Enlist different brand names for Ketoconazole.
10. Give the uses and different pharmaceutical formulations available for Miconazole.

C. LONG ANSWER QUESTIONS

1. Give the classification of antifungal drugs with examples. Discuss in details Amphotericin-B as an antifungal.
2. Give the classification of antifungal drugs with examples. Discuss in details Ketoconazole as an antifungal.
3. Explain Ketoconazole as an antifungal.
4. Give the classification of antifungal drugs, with examples. Discuss Stability and storage conditions, Dose, Pharmaceutical Formulations and Brand Names for Griseofulvin

ANTI-INFECTIVE AGENTS: URINARY TRACT ANTI-INFECTIVE AGENTS

CONTENTS

Urinary Tract Anti-Infective Agents:

- Norfloxacin,
- Ciprofloxacin
- Ofloxacin*
- Moxifloxacin

◆ LEARNING OBJECTIVES ◆

After completing this chapter, the student should be able to understand:

1. Anti-Infective Agents.
2. Quinolones as Anti-Infective Agents.
3. Classification of Anti-Infective Agents.
4. Chemical Names of Anti-Arrhythmic Drugs
5. Chemical Structures of different Anti-Infective Agents.
6. Uses, Stability and storage conditions of Anti-Infective Agents.
7. Different types of formulations and popular brand names Anti-Infective Agents.

11.2.1 INTRODUCTION

The quinolones are a family of synthetic broad-spectrum antibiotics. The term quinoles refers to potent synthetic chemotherapeutic antibacterial.

Quinolones

The quinolones comprise a series of **synthetic antibacterial agents** patterned after nalidixic acid, a naphthyridine derivative introduced for treating **urinary tract infections** in 1963.

Isosteric heterocyclic groupings in this class include the **quinolones** (e.g., norfloxacin, ciprofloxacin, lomefloxacin), the **naphthyridines** (e.g., nalidixic acid, enoxacin), and the **cinnolines** (e.g., cinoxacin).

Up to the present time, the clinical usefulness of the quinolones has been largely confined to the treatment of urinary tract infections.

For urinary tract infections, good oral absorption, activity against common Gram-negative urinary pathogens, and comparatively higher urinary (compared with plasma and tissue) concentrations are the key useful properties.

As a result of extensive structure-activity investigations leading to compounds with enhanced potency, an extended spectrum of activity, and improved absorption and distribution properties, the class has evolved to the point that certain newer members have.

11.2.2 CLASSIFICATION OF ANTIBACTERIAL QUINOLONES

The antibacterial quinolones can be divided into two classes based on their dissociation properties in physiologically relevant conditions.

The **first class**, represented by **nalidixic acid, oxolinic acid** (no longer marketed in the United States), and **cinoxacin**, possesses only the 3-carboxylic acid group as an ionisable functionality.

The **second class** of antibacterial quinolones embraces the **broad-spectrum fluoroquinolones** (namely, norfloxacin, enoxacin, ciprofloxacin, ofloxacin, lomefloxacin, and sparfloxacin), all of which possess, in addition to the 3-carboxylic acid group, a basic piperazine functionality at the 7- position and a 6-fluoro substituent.

11.2.3 MECHANISM OF ACTION

The bactericidal action of nalidixic acid and its congeners is known to result from the inhibition of DNA synthesis. This effect is believed to be caused by the inhibition of bacterial DNA gyrase (topoisomerase II), an enzyme responsible for introducing negative supercoils into circular duplex DNA. Negative supercoiling relieves the torsional stress of helical DNA, facilitates unwinding, and, thereby, allows transcription and replication to occur. Although nalidixic acid inhibits gyrase activity, it binds only to single-stranded DNA and not to the enzyme or double-helical DNA.

Bacterial DNA gyrase is a tetrameric enzyme consisting of two A and two B subunits encoded by the *gyrA* and *gyrB* genes. Bacterial strains resistant to the quinolones have been identified with decreased binding affinity to the enzyme because of amino acid substitution in either A or B subunits resulting from mutations in either *gyrA* or *gyrB* genes.

11.2.4 DIFFERENT DRUGS USED IN URINARY TRACT INFECTION

11.2.4.1 Norfloxacin

Structure:

Chemical (IUPAC) name: 1-Ethyl-6-fluoro-1,4-dihydro-4-oxo-7-(1-piperazinyl)-3-quino-linecarboxylic acid.

Properties:

It is a pale yellow crystalline powder that is sparingly soluble in water.

This quinoline has broad-spectrum activity against Gram-negative and Gram-positive aerobic bacteria.

Norfloxacin is indicated for the treatment of urinary tract infections caused by *E. coli, K. pneumonia, Enterobacter cloacae, Proteus mirabilis*, indole-positive *Proteus* spp., including *P. vulgaris, Providencia rettgeri, Morganella morganii, P. aeruginosa, S. aureus*, and *S. epidermidis*, and group-D streptococci.

It is generally not effective against obligately anaerobic bacteria.

Norfloxacin in a single 800-mg oral dose has also been approved to treat uncomplicated gonorrhoea.

Stability and storage conditions:

Store Norfloxacin tablets between temperatures of 15°C-30°C (59°F-86°F) in a tightly closed container.

Dose: The dose is 400

Pharmaceutical Formulations: Tablet

Brand Names: Noroxin, Norfloxacine-400

11.2.4.2 Ciprofloxacin

Structure:

Chemical (IUPAC) name: 1-Cyclopropyl-6-fluoro-1,4-dihydro-4-oxo-7-(1-piperazinyl)-3-quinoline carboxylic acid.

It is supplied in both oral and parenteral dosage forms.

The hydrochloride salt is available in 250-, 500-, and 750-mg tablets for oral administration. Intravenous solutions containing 200 and 400 mg are provided in concentrations of 0.2% in normal saline and 1% in 5% dextrose solutions.

Ciprofloxacin is an agent of choice for treating bacterial gastroenteritis caused by Gram-negative bacilli such as enteropathogenic *E. coli*, *Salmonella* spp. (including *S. typhi*), *Shigella* spp., *Vibrio* spp., and *Aeromonas hydrophilia*.

It is widely used for treating respiratory tract infections and is particularly effective for controlling bronchitis and pneumonia caused by Gram-negative bacteria.

Ciprofloxacin is also used for combating infections of the skin, soft tissues, bones, and joints. Both uncomplicated and complicated urinary tract infections caused by Gram-negative bacteria can be treated effectively with ciprofloxacin.

It is beneficial for controlling chronic infections characterised by renal tissue involvement.

The drug also has important applications in controlling venereal diseases.

A combination of ciprofloxacin with the cephalosporin antibiotic ceftriaxone is recommended as the treatment of choice for disseminated gonorrhoea, whereas a single-dose treatment with ciprofloxacin plus doxycycline, a tetracycline antibiotic, can usually eradicate gonococcal urethritis.

Stability and storage conditions:

Store the tablets and extended-release tablets at room temperature and away from excess heat and moisture (not in the bathroom). Store the suspension in the refrigerator or at room temperature, closed tightly, for up to 14 days. Do not freeze ciprofloxacin suspension.

Dose: The dose is 250 mg, 500 mg, and 750 mg for tablets, with 0.3 % eye drops

Pharmaceutical Formulations: Tablet, Solution, Drops, Ointment

Brand Names: Ciloxan, Cipro.

11.2.4.3 Ofloxacin

Structure:

Chemical (IUPAC) name: 9-Fluoro-2,3-dihydro-3-methyl-10(4-methyl-1-piperazin-yl)-7-oxo-7*H*-pyrido[1,2,3-de]-1,4,-benzoxazine-6-carboxylic acid.

It is a member of the quinolone class of antibacterial drugs wherein the 1- and 8- positions are joined in the form of a 1,4-oxazine ring.

Ofloxacin has been approved for treating infections of the lower respiratory tract, including chronic bronchitis and pneumonia, caused by Gram-negative bacilli.

It is also used for treating pelvic inflammatory disease and is highly active against gonococci and chlamydia.

In common with other fluoroquinolones, ofloxacin is not effective in the treatment of syphilis.

A single 400-mg oral dose of ofloxacin in combination with the tetracycline antibiotic doxycycline is recommended by the Centers for Disease Control and Prevention (CDC) for the outpatient treatment of acute gonococcal urethritis. Ofloxacin is also used for the treatment of urinary tract infections caused by

Gram-negative bacilli and for prostatitis caused by *E. coli*. Infections of the skin and soft tissues caused by staphylococci, streptococci, and Gram-negative bacilli may also be treated with ofloxacin.

Stability and storage conditions:

For the ophthalmic agents evaluated, gentamicin and ofloxacin, the inserts state that the drugs should be stored at 20 to 36^{0}C and 15^0 to 25^{0}C, respectively.

Dose: The dose is 20 mg tablet, 0.3 % ear drops

Pharmaceutical Formulations: Tablet, Solution

Brand Names: Floxin, Floxin Otic, Tritin 20

11.2.4.4 Moxifloxacin

It is a fluoroquinolone antibiotic used to treat various bacterial infections.

Chemical Name: 7-[(4aS,7aS)-1,2,3,4,4a,5,7,7a-octahydropyrrolo[3,4-b]pyridin-6-yl]-1-cyclopropyl-6-fluoro-8-methoxy-4-oxoquinoline-3-carboxylic acid; hydrochloride
Moxifloxacin is a synthetic fluoroquinolone antibiotic agent.

Mechanism of action

The bactericidal action of moxifloxacin results from the inhibition of the enzymes topoisomerase II (DNA gyrase) and topoisomerase IV. DNA gyrase is an essential enzyme that is involved in the replication, transcription and repair of bacterial DNA. Topoisomerase IV is an enzyme known to play a vital role in the partitioning of the chromosomal DNA during bacterial cell division.

Moxifloxacin is an antibiotic used to treat a number of bacterial infections. This includes pneumonia, conjunctivitis, endocarditis, tuberculosis, and sinusitis. It is used by mouth, by injection into a vein, or as an eye drop.

Common side effects include diarrhoea, dizziness, and headache. Severe side effects may include spontaneous tendon ruptures, nerve damage, and worsening of myasthenia gravis. The safety of use in pregnancy or breastfeeding is unclear. Moxifloxacin is in the fluoroquinolone family of medications. It usually results in bacterial death through blocking their ability to duplicate DNA. Moxifloxacin was patented in 1988 and approved for use in the United States in 1999. It is on the World Health Organization's List of Essential Medicines

Stability and storage conditions:

Store at 2°C- 25°C (36°F-77°F). Patients should be advised not to touch the dropper tip to any surface to avoid contaminating the contents. Patients should be advised not to wear contact lenses if they have signs and symptoms of bacterial conjunctivitis.

Dose: The dose is 400 mg tablet, 400 mg/250ml solution for infusion

Pharmaceutical Formulations: Tablet, Solution, Injection, Drops

Brand Names: Avelox, Moxeza, Vigamox

QUESTION BANK

A. MULTIPLE CHOICE QUESTIONS

1. Following drug is not used in Urinary tract infections
   A. Nalidixic acid
   B. Methanamine
   C. Rifampicin
   D. Gataifloxacin

   Answer: C

2. Which of the following is not effective in the treatment of syphilis?
   A. Ciprofloxacin
   B. Norfloxacin
   C. Ofloxacin
   D. Moxifloxacin

   Answer: C

3. Ciprofloxacin acts on _________ of bacteria
   A. DNA histone proteins
   B. DNA gyrase
   C. Cell wall
   D. mRNA polymerase

   Answer: B

4. Moxifloxacin is a ___________ fluoroquinolone antibiotic agent
   A. Synthetic
   B. Semisynthetic
   C. Natural
   D. None of the above

   Answer: A

5. Noroxin is the brand name for______
   A. Ciprofloxacin
   B. Norfloxacin
   C. Ofloxacin
   D. Moxifloxacin

   Answer: B

6. Ciprofloxacin is supplied in
   A. oral dosage form
   B. parenteral dosage forms
   C. Semisolid Dosage form
   D. Oral and parenteral dosage forms.

Answer: D

7. Ciloxan is the brand name for______
   A. Ciprofloxacin
   B. Norfloxacin
   C. Ofloxacin
   D. Moxifloxacin

Answer: A

8. Ofloxacin contains which of the following heterocyclic ring system?
   A. Quinolone
   B. Furan
   C. Pyrrole
   D. Pyrimidine

Answer: A

9. The bactericidal action of moxifloxacin results from inhibition of the enzymes
   A. Topoisomerase II (DNA gyrase).
   B. Topoisomerase IV
   C. Topoisomerase II (DNA gyrase) and topoisomerase IV.
   D. None of the above

Answer: C

10. Norfloxin is effective against
    A. Gram-negative aerobic bacteria.
    B. Gram-positive aerobic bacteria.
    C. Gram-negative and Gram-positive aerobic bacteria.
    D. None of the above.

Answer: C

B. SHORT ANSWER QUESTIONS

1. Classify Urinary Tract Anti-infective Agents.
2. Explain Norfloxacin as Urinary Tract Anti-infective Agent.
3. Enlist different pharmaceutical formulations available for Norfloxacin.
4. Explain storage condition and stability for Moxifloxacin.

5. Enlist different pharmaceutical formulations available for Moxifloxacin.

6. Enlist different brand names for Ciprofloxacin.

7. Enlist different pharmaceutical formulations available for Ofloxacin.

8. Enlist different brand names for Ofloxacin.

9. Draw the structure for Ofloxacin.

10. Give the storage condition for Ofloxacin.

C. LONG ANSWER QUESTIONS

1. Give the classification of Urinary Tract Anti-infective Agents with an example. Discuss in details Ciprofloxacin.

2. Give the classification of Urinary Tract Anti-infective Agents with an example. Discuss in details Ofloxacin.

3. Give storage condition, pharmaceutical formulation and brand names available for Ofloxacin.

■■■

ANTI-INFECTIVE AGENTS: ANTITUBERCULAR AGENTS

CONTENTS

- Brief introduction of Introduction-The organism, Pathogenesis of tuberculosis (TB), Signs and symptoms, Transmission of tuberculosis. Current drug therapy and its limitations.

- Classification of Antitubercular drugs -First line anti-tubercular drugs, Second line anti-tubercular drugs

- Brief introduction of Some selected Antitubercular drugs:
  - Isoniazid (INH),
  - Ethambutol
  - *Para*-Aminosalicylic acid
  - Pyrazinamide
  - Rifampicin
  - Bedaquine
  - Delamid
  - Pretomaid*

♦ LEARNING OBJECTIVES ♦

After completing this chapter, the student should be able to understand:

1. The causative organism and pathogenesis of tuberculosis, its signs and symptoms, its transmission,
2. Their classification and mechanism of action of various antitubercular drugs.
3. Important drugs used as antitubercular agents are their chemical structures, names, incompatibilities, stability and storage conditions and brands.
4. Resistance of *Mycobacterium tuberculosis* and types of TB.

11.3.1 INTRODUCTION

TB (Tuberculosis) - Infects one-fourth of the world population. New infections occur in about 1% of the population each year. Most hits are the underprivileged population – the third world and developing countries. This causes a great national economic burden. Five countries with the most significant numbers of cases include; China, India, South Africa,

Nigeria and Indonesia. More people in the developing world contract tuberculosis because of compromised immunity and metabolism, largely due to high rates of HIV infection.

The Organism: TB is usually caused by *Mycobacterium tuberculosis* [Mtb], a small, non-motile bacillus. Mtb is primarily a pathogen of the mammalian respiratory system, which attacks the lungs but can also affect other parts of the body. It is highly aerobic and requires high levels of oxygen. It has an unusual waxy coating of mycolic acid on its cell surface which makes it impervious to gram staining; therefore, it has to be detected by the acid-fast technique. The cell wall complex contains peptidoglycan, but otherwise, it is composed of complex lipids. The lipid fraction of MTB's cell wall consists of three major components, mycolic acids, cord factor, and wax-D.

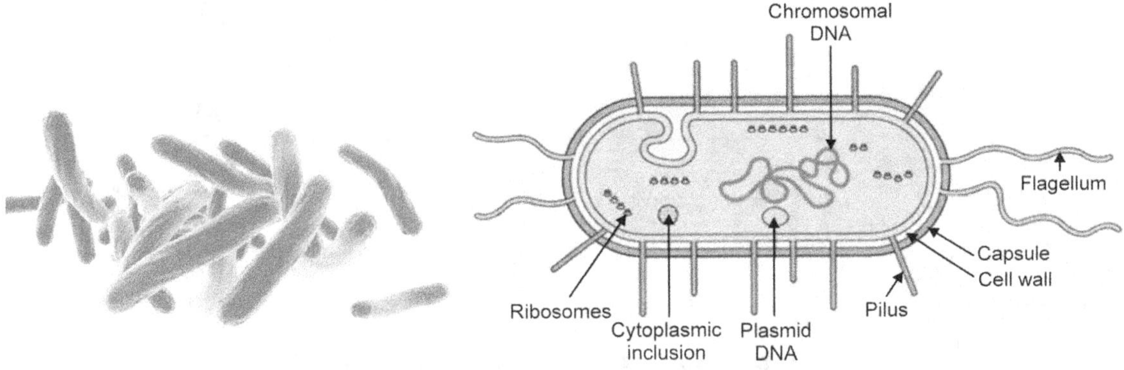

Mycobacterium tuberculosis [Mtb]   The general structure

Fig.11.3.1 Mtb (Mycobacterium tuberculosis) and its general structure

Pathogenesis of Tuberculosis:

Infection of a host by *M. tuberculosis* is initiated by inhalation of air droplets containing the bacilli from the cough or sneeze of infected persons. The organism then spreads from the site of infection to the lungs (primary infection site) and then through the lymphatic or blood systems to other parts of the body. Other infection sites are the Intestines, brain, bones, liver and kidney.

Once in the lung, bacilli are subjected to phagocytosis by the resident macrophages of the lung and phagocytosed by the destructive environment of lysosomes. However, some bacilli escape lysosomal delivery and survive. Infected macrophages can then either remain in the lung or are disseminated to other organs in the body.

Signs, symptoms & Diagnosis: Classic symptoms of active TB infection are as listed below; Early stages – free of symptoms, and many cases are found incidentally.

Systemic manifestations: Fatigue, malaise, anorexia, weight loss, low-grade fevers, night sweats, weight loss occurring late, characteristic cough which is frequent & produces mucoid or mucopurulent sputum, chronic cough with blood-tinged sputum, dull or tight chest pain, significant finger clubbing may also occur.

Some cases: acute high fever, chills, general flulike symptoms, pleuritic pain, productive cough

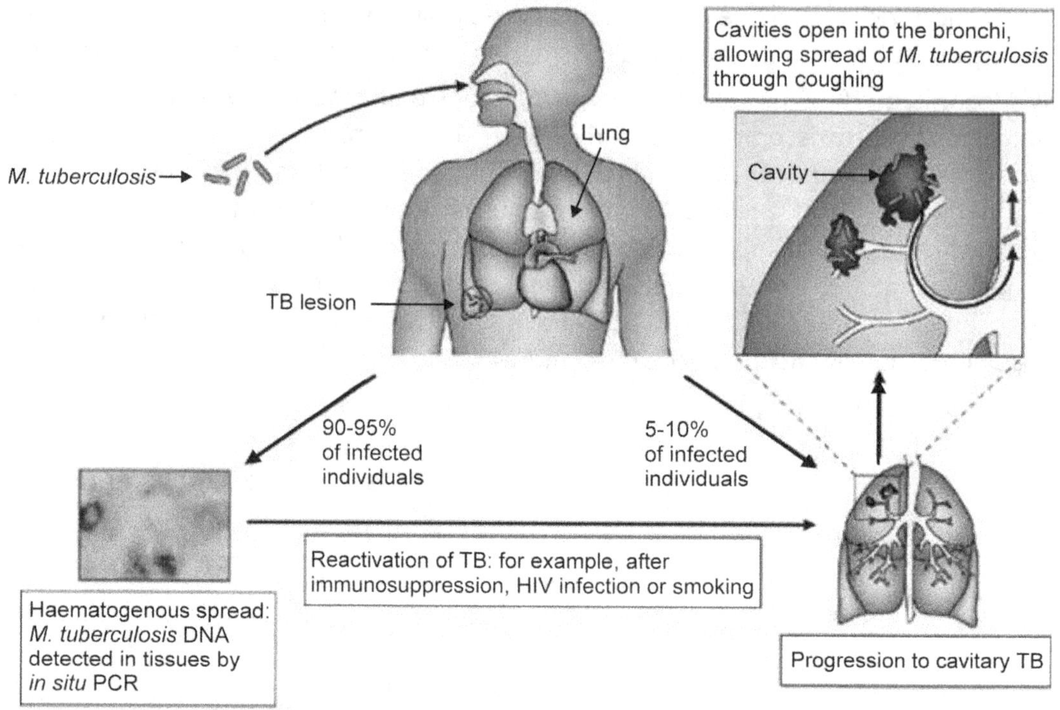

Fig. 11.3.2 The Pathogenesis of TB

Diagnosis: Diagnosis of active TB relies on radiology (chest X-rays), as well as microscopic examination and microbiological culture of body fluids. Diagnosis of latent TB relies on the tuberculin skin test (TST) and/or blood tests.

The most frequently used diagnostic methods for tuberculosis are the; *Tuberculin skin test, Chest X-ray* and *Bacteriologic Studies.*

SIGNS AND SYMPTOMS OF

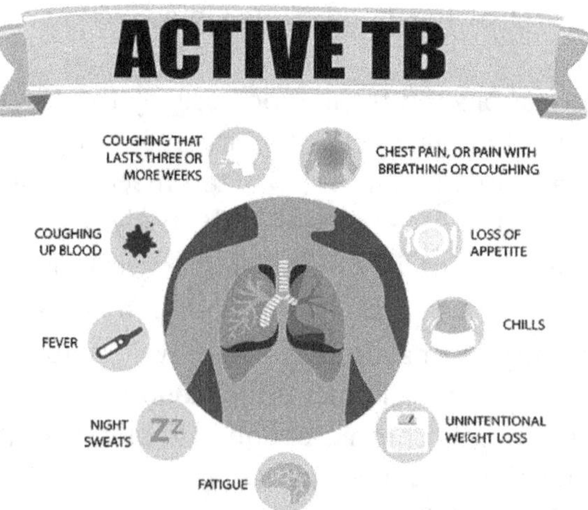

Fig. 11.3.3 The signs & symptoms of TB

Transmission of tuberculosis: When people with active pulmonary TB cough, sneeze, speak, sing, or spit, they expel infectious aerosol droplets 0.5 to 5.0 μm in diameter. A single sneeze can release up to 40,000 droplets. Each one of these droplets may transmit the disease since the infectious dose of tuberculosis is very low (the inhalation of fewer than 10 bacteria may cause an infection).

A person with active but untreated tuberculosis may infect 10–15 (or more) other people per year.

Transmission occurs from people with active TB, while those with latent infection are not thought to be contagious.

o Transmitted mainly by droplet infection and droplet nuclei – by sputum-positive patients with pulmonary TB

o Coughing generates the largest number of droplets of all sizes

o Frequency & vigour of cough & the ventilation of the enviroment influence transmission of infection

Fig. 11.3.4 Mode of transmission of TB

Current drug therapy and its limitations –

A long-term treatment with a combination of drugs is required to cure the disease without subsequent relapse, as well as for of death, prevention of transmission and resistance development.

Goals of anti-tubercular chemotherapy are 3 directional

1. *Kill dividing bacilli*: Patient is non-contagious: transmission of TB is interrupted.

2. *Kill persisting bacilli*: To effect a cure and prevent relapse.

3. *Prevent the emergence of resistance:* so that the bacilli remain susceptible to the drugs.

As the current treatment protocols are lengthy, usually 6–9 months, they are vulnerable to incidences of side effects, unsatisfactory patient compliance and slow improvement, thus, making the cure rates low. Inevitably as this complex therapy is administered under less than ideal conditions, drug-resistant forms of the disease have become increasingly common, leading the WHO to declare multi-drug resistant (MDR) TB a global threat and also strongly recommending DOTS (**D**irectly-**O**bserved **T**herapy).

Present efforts are on improving treatment by either shortening the length of treatment or using innovative drug delivery strategies in addition to alternative administration routes, or both. These efforts are likely to improve the efficacy of anti-tubercular chemotherapy, thereby enhancing patient compliance.

11.3.2 CLASSIFICATION OF ANTITUBERCULAR DRUGS

Antitubercular drugs are classified into two categories

a) First Line Antitubercular drugs

b) Second Line Antitubercular drugs

ANTI-TUBERCULAR DRUGS

First-line Drugs	Second-line Drugs
– Isoniazid	– Clarithromycin
– Rifampicin	– Ciprofloxacin or Levofloxacin
– Ethambutol	– Capreomycin
– Pyrazinamide	– Cycloserine
– Streptomycin (Aminoglycosides)	– Kanamycin
	– Amikasin

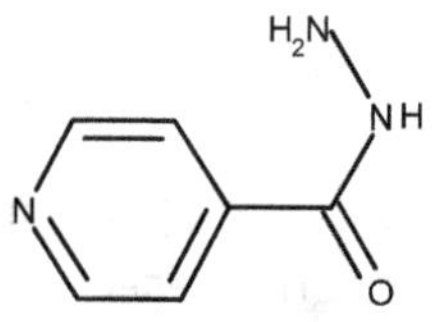

Isoniazid (NH)

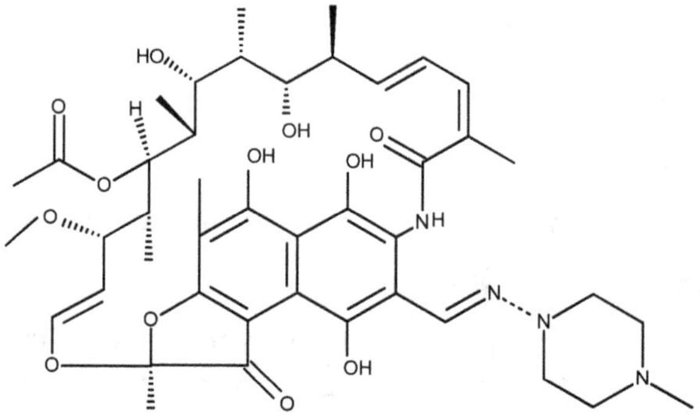

Rifampin (RIF)

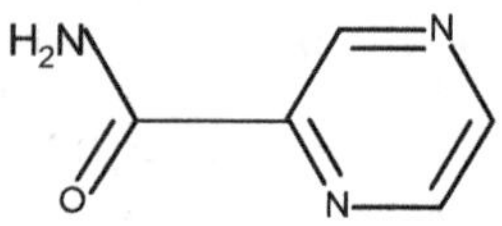

Pyrazinamide (PZA)

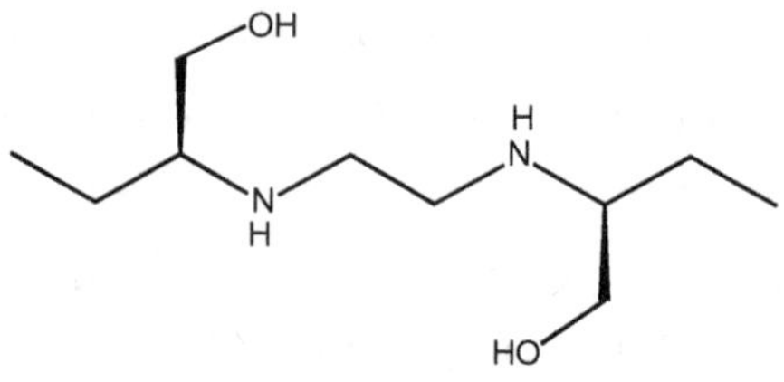

Ethambutol (ETB)

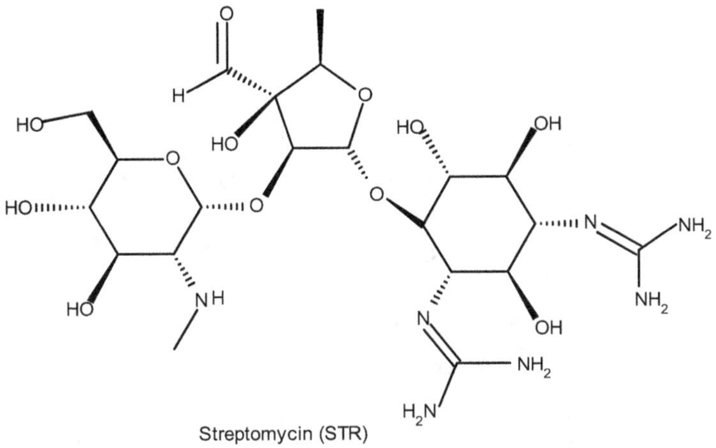

Fig. 11.3.5 First-line Anti-TB Drugs

Mechanism of Action (MoA) of First Line Anti-TB Drugs: The following table describes the MoA of First Line Anti-TB drugs-

Name of Drug	Mechanism of action
Rifampicin	It inhibits bacterial RNA synthesis by binding to the β-subunit of bacterial DNA-dependent RNA-polymerase which leads to blocking of the initiation of chain formation in RNA synthesis.
Isoniazid	It exerts its lethal effect through inhibition of synthesis of mycolic acids, through formation of a covalent complex with an acyl carrier protein and β-ketoacyl carrier protein synthetase.
Pyrazinamide	Its active metabolite pyrazinoic acid (PZA) inhibits growth of Mtb by lowering pH in the immediate surroundings, as well as by inhibition trans-translation, a key cellular process for managing damaged proteins and 'rescuing' nonfunctioning ribosomes in M. tuberculosis. It is also acts as an antimetabolite of nicotinamide thereby interfering NAD synthesis, inhibiting short-chain, fatty-acid precursors' synthesis.
Ethambutol	It inhibits mycobacterial arabinosyltransferases involved in the polymerization of D-arabinofuranose to arabinoglycan, an essential cell wall component.
Aminoglycoside antibiotics	They act by the irreversible inhibitors of protein synthesis through binding to specific 30-S-subunit ribosomal proteins. They inhibit bacterial topoisomerase II (DNA gyrase) and topoisomerase IV and thus inhibiting bacterial DNA synthesis.

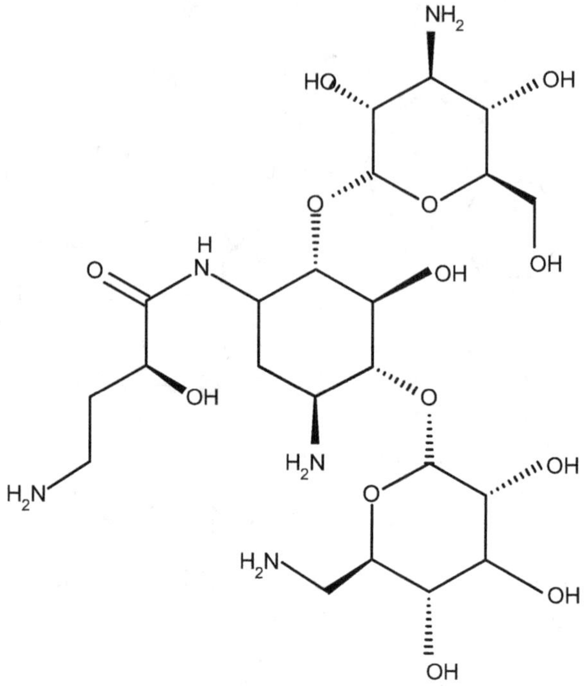

Amikacin

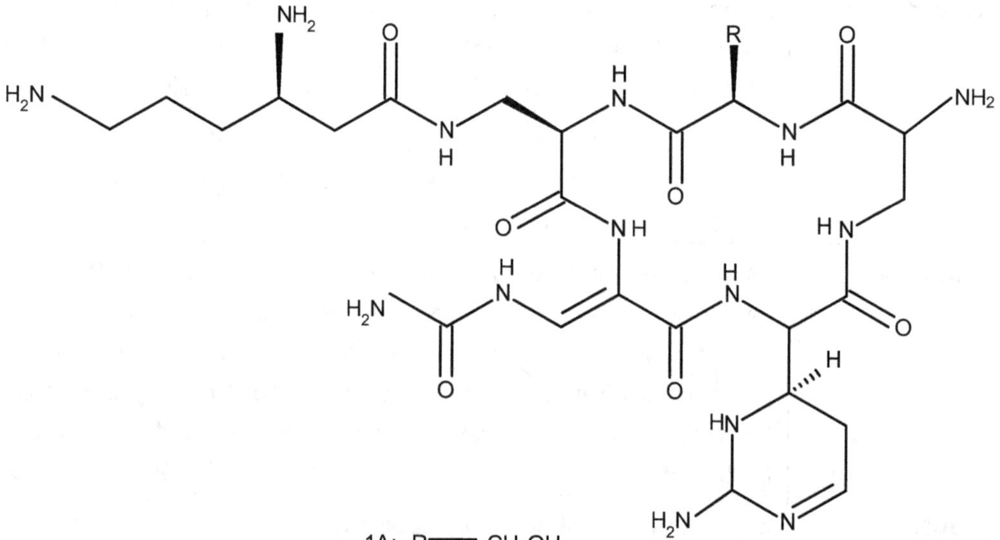

1A: R═══CH$_2$OH
1B: R═══CH$_3$

Capreomycin (1A and 1B)

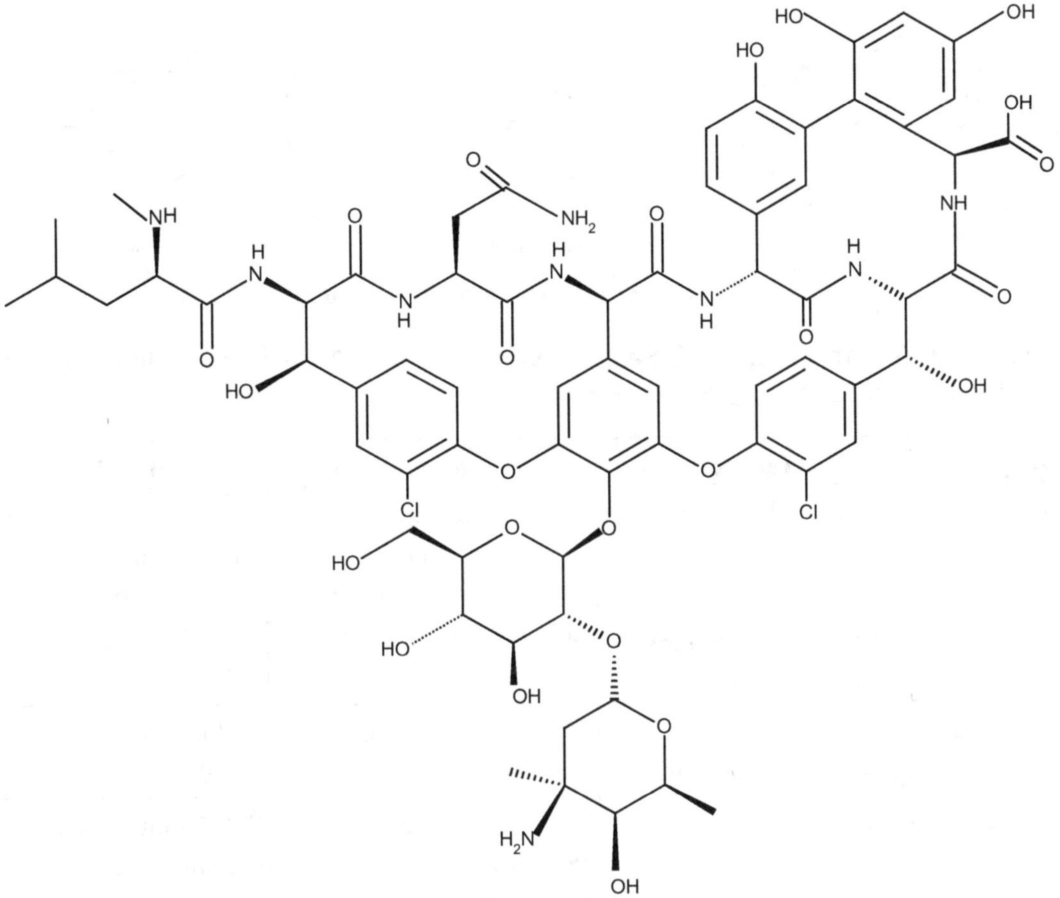

Vancomycin

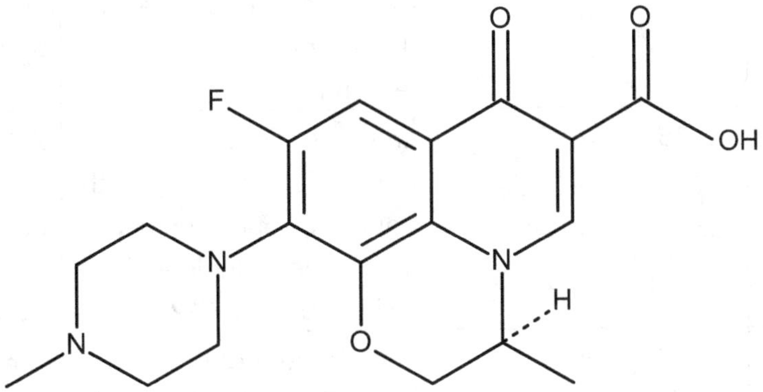

Levofloxacin

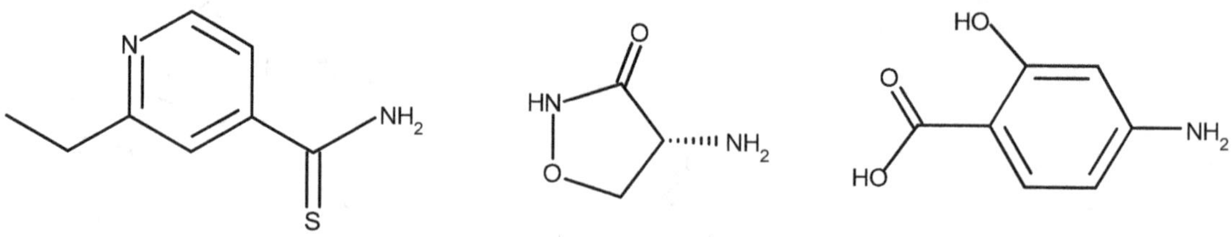

Ethionamide

Cyclosarine

p-Amino salicylic aicd

Fig. 11.3.6 Second-line Anti-TB Drugs

Mechanism of Action (MoA) of Second Line Anti-TB Drugs: The following table describes the MoA of Second Line Anti-TB drugs-

Drugs	Target	Mode of action	Mutation
Streptomycin	rpsL (S12 ribosomal protein) rrs (16S rRNA) gidB (7-methylguanosine methyltransferase)	Binds to the 16S rRNA, interferes with translation proofreading, and thereby inhibits protein synthesis	Mutation in codon 43 and 88 from lysine to arginine in rpsL
p-Aminosalicyclic acid	thyA (thymidylate synthase A)	Inhibit the folic acid synthesis	Mutation in thyA Thr202Ala
Ethionamide	inhA (enoyl reductase)	Inhibit the mycolic acid biosynthesis	C-15T mutation in the regulatory region S91A and 1194T
Fluoroquinolones	gyrA/gyrB (DNA gyrase)	Inhibiting the activity of both the DNA gyrase and the topoisomerase IV enzymes	Mutations in quinolone resistance determining region (QRDR) in gytA
Kanamycin	Rrs (16S rRNA) Eis (aminoglycoside acetyltransferase)	Inhibit the protein synthesis by binding to the four nucleotides of 16s rRNA and a single amino acid of protein S12	Mutations in the codon 1401, 1402 and 1484 in the rrs (16S rRNA)
Amikacin	rrs (16S rRNA)	Inhibit the protein synthesis by binding to the 30S ribosomal subunit	Same as kanamycin

11.3.3 SOME IMPORTANT ANTITUBERCULAR DRUGS

11.3.3.1 Isoniazid: (*Syn:* **Isonicotinic acid hydrazide, INH**)
Structure:

Systematic (IUPAC) Name: Pyridine-4-carbohydrazide

Uses:

1. It is used for the treatment of tuberculosis. For active tuberculosis as a first-line anti-tubercular drug.   It is often used together with rifampicin, pyrazinamide, and either streptomycin or ethambutol. For latent tuberculosis, it is often used by itself.

2. It may also be used for the prevention of atypical types of mycobacteria, such as *M. avium, M. kansasii,* and *M. xenopi.* 3.It is recommended that women with active tuberculosis who are pregnant or breastfeeding take isoniazid. Preventive therapy should be delayed until after giving birth

Side Effects:

- Hepatotoxicity -Elderly, slow acetylators are more prone
- Polyneuropathy-Prevented by concurrent pyridoxine administration.
- Rashes, acne
- Haematological – haemolytic anaemia in G6PD deficiency

– **Other Side effects** Common side effects of Isoniazid include increased blood levels of liver enzymes and numbness in the hands and feet.Serious side effects may include liver inflammation and acute liver failure.

Pyridoxine (Vit.B$_6$) may be given to reduce the risk of side effects

Stability and storage conditions: Store isoniazid oral solution at room temperature, 68°F to 77°F (20°C to 25°C). Store isoniazid tablets at 68°F to 77°F (20°C to 25°C) and protect them from moisture and light. Keep isoniazid in the container that it came in, and keep the container tightly closed.

Incompatibility: People taking isoniazid and acetaminophen are at risk of acetaminophen toxicity. Isoniazid is thought to induce a liver enzyme which causes a larger amount of acetaminophen to be metabolised to a toxic form.

Isoniazid decreases the metabolism of carbamazepine, thus slowing down its clearance from the body. People taking carbamazepine should have their carbamazepine levels monitored and, if necessary, have their dose adjusted accordingly.

It is possible that isoniazid may decrease the serum levels of ketoconazole after long-term treatment. This is seen with the simultaneous use of rifampin, isoniazid, and ketoconazole.

Isoniazid may increase the amount of phenytoin in the body. The doses of phenytoin may need to be adjusted when given with isoniazid.

Isoniazid may increase the plasma levels of theophylline. There are some cases of theophylline slowing down isoniazid elimination. Both theophylline and isoniazid levels should be monitored.

Valproate levels may increase when taken with isoniazid. Valproate levels should be monitored and its dose adjusted if necessary.

Different types of formulations: Tablet, injection drops

Popular brands: Some of the notable brand names in India are:

Sr. No	Brand Name	Manufactures	Type
1	ISO Sunibutol	Sunij Pharma Pvt. Ltd.	Tablet
2	Solonex	Mcleods Pharma Ltd.	Disintegrating Tablet
3	Tubernex Forte	Radicura Pharma Pvt Ltd	Tablet
4	3FD(INH-Rifampicin-Pyrazinamide)	Novartis India Ltd	Tablet
5	Isoniazid Injection USP	SG Pharma Pvt Ltd.	Injection

11.3.3.2 Ethambutol: (*Syn:*** Myambutol, Etibi, Servambutol)**

Structure:

Systematic (IUPAC) Name: (2S,2′S)-2,2′-(Ethane-1,2-diyldiimino)dibutan-1-ol

Uses: Ethambutol is used along with other medications to treat several infections, including tuberculosis, *Mycobacterium avium* complex, and *Mycobacterium kansasii*.

Side Effects: The common side effects of Ethambutol include problems with vision, joint pain, nausea, headaches, and feeling tired. Other side effects include liver problems and allergic reactions.

Stability and storage conditions: Ethambutol should be stored at room temperature (below 25°C) in the original packaging. PP bottle with a PP child-resistant closure.

Incompatibility: It is not recommended in people with optic neuritis, significant kidney problems, or under the age of five. Use during pregnancy or breastfeeding has not been found to cause harm.

Different types of formulation: Tablet

Popular brands: It is manufactured by many companies. Some of the notable brand names in India are:

Sr. No.	Brand Name	Manufactures	Type
1	Zytham	Zydus Cadila	Tablet
2	Combutol	Lupin Limited	Tablet
3	Mycobutol	Cadila Pharma	Tablet

11.3.3.3 *Para*-Aminosalicylic acid: (*Syn:* **PAS, 4-Aminosalicylic Acid, PASER**)

Structure:

Systematic (IUPAC) Name: 4-Amino-2-hydroxybenzoic acid

Uses: It is primarily used to treat tuberculosis. Specifically, it is used to treat active drug-resistant tuberculosis together with other antituberculosis medications.

It has also been used as a second-line agent to sulfasalazine in people with inflammatory bowel diseases such as ulcerative colitis and Crohn's disease. It is typically taken by mouth.

PAS has been shown to be a pro-drug, and it is incorporated into the folate pathway by dihydropteroate synthase (DHPS) and dihydrofolate synthase (DHFS) to generate a hydroxyl dihydrofolate antimetabolite, which in turn inhibits dihydrofolate reductase (DHFR) enzymatic activity.

Side Effects: The common side effects of *p*-aminosalicylic acid include nausea, abdominal pain, and diarrhoea. Other side effects may include liver inflammation and allergic reactions.

Stability and storage conditions: Store below 59°F (15°C) (in a refrigerator or freezer).

Incompatibility: It is not recommended in people with end-stage kidney disease. While there does not appear to be harm with use during pregnancy, it has not been well studied in this population. 4-Aminosalicylic acid is believed to work by blocking the ability of bacteria to make folic acid.

Different types of formulation: Tablets, Powder, Granules.

Popular brands: It is manufactured by many companies. Some of the notable brand names in India are

Sr. No.	Brand Name	Manufactures	Type
1	Q-PAS	Lupin Laboratories Ltd.	Granules
2	Mycopas	Macleods Pharmaceuticals Pvt. Ltd.	Granules
3	Monopas	Macleods Pharmaceuticals Pvt. Ltd.	Granules
4	INAPAS	USV-Corvette (A Div of USV Ltd.)	Powder
5	Mycopas	Macleods Pharmaceuticals Pvt. Ltd.	Tablet
6	Monopas	Macleods Pharmaceuticals Pvt. Ltd.	Tablet

11.3.3.4 Pyrazinamide: (*Syn:* **Rifater, Tebrazid**)

Structure:

Systematic (IUPAC) Name: pyrazine-2-carboxamide

Uses: This is a medication used to treat tuberculosis. For active tuberculosis, it is often used with rifampicin, isoniazid, and either streptomycin or ethambutol. It is generally not recommended for the treatment of latent tuberculosis. It is taken by mouth. It is a prodrug that stops the growth of M. tuberculosis. It diffuses into the granuloma of M. tuberculosis, where the tuberculosis enzyme pyrazinamide converts pyrazinamide to the active form pyrazinoic acid, the later makes pH acidic, which is inhibitory to the growth of *Mtb*.

Side Effects: The common side effects of Pyrazinamide include nausea, loss of appetite, muscle and joint pains, and rash. More severe side effects include gout, liver toxicity, and sensitivity to sunlight.

Stability and storage conditions: Store between 15°-25°C (59°-77°F).

Incompatibility: It may decrease the excretion of many drugs. It is not recommended in those with significant liver disease or porphyria. It is unclear if use during pregnancy is safe, but it is likely okay during breastfeeding.

Different types of formulation: Tablet.

Popular brands: Some of the notable brand names in India are:

Sr. No.	Brand Name	Manufactures	Type
1	PZA Ciba	Novartis India Ltd.	Tablet
2	Macrozide	Mcleods Ltd.	Tablet
3	Pyzina	Lupin Ltd	Tablet

11.3.3.5 Rifampicin: (*Syn:* Rifampin)

Structure:

It is a semisynthetic antibiotic produced from *Streptomyces mediterranei*. It has a broad antibacterial spectrum, including activity against several forms of Mycobacterium. In susceptible organisms, it inhibits DNA-dependent RNA polymerase activity by forming a stable complex with the enzyme. It thus suppresses the initiation of RNA synthesis, causing cell death. Rifampin is bactericidal and acts on both intracellular and extracellular organisms is a polyketide belonging to the chemical class of compounds termed ansamycins, so named because of their cyclic structure containing a napthoquinone core spanned by an aliphatic ansa chain. The napthoquinone chromophore gives rifampicin its characteristic red-orange crystalline colour. Rifampicin is made by the soil bacterium *Amycolatopsis rifamycinica*.

Systematic (IUPAC) Name: (7*S*,9*E*,11*S*,12*R*,13*S*,14*R*,15*R*,16*R*,17*S*,18*S*,19*E*,21*Z*)-2,15,17,27,29-pentahydroxy-11-methoxy-3,7,12,14,16,18,22-heptamethyl-26-{(*E*)-[(4-methylpiperazin-1-yl)imino] methyl} -6,23-dioxo-8,30-dioxa-24-azatetracyclo[23.3.1.1^{4,7}.0^{5,28}]triaconta-1(28),2,4,9,19,21,25(29),26-octaen-13-yl acetate.

Uses: it is an antibiotic used to treat several types of bacterial infections, including Tuberculosis, They are; *Mycobacterium avium* complex, leprosy, and Legionnaires' disease.

It is almost always used together with other antibiotics, except when given to prevent *Haemophilus influenzae* type b and meningococcal disease in people who have been exposed to those bacteria. Before treating a person for an extended period, measurements of liver enzymes and blood counts are recommended.

It is also used as a preventive treatment against *Neisseria meningitidis* (meningococcal) infections. Rifampicin is also recommended as an alternative treatment for infections by the tick-borne pathogens *Borrelia burgdorferi* and *Anaplasma phagocytophilum* when treatment with doxycycline is contraindicated, such as in pregnant women or in patients with a history of allergy to tetracycline antibiotics.

It is also sometimes used to treat infections by Listeria species, *Neisseria gonorrhoeae, Haemophilus influenza* and *Legionella pneumophila*. For these nonstandard indications, antimicrobial susceptibility testing should be done (if possible) before starting rifampicin therapy.

Side Effects:

The common side effects of Rifampicin include-

Liver toxicity—hepatitis, liver failure in severe cases; Respiratory—breathlessness, Cutaneous—flushing, pruritus, rash, hyperpigmentation, redness and watering of eyes, Abdominal — nausea, vomiting, abdominal cramps, diarrhoea, Flu-like symptoms—chills, fever, headache, arthralgia, and malaise. Rifampicin has good penetration into the brain, which may directly explain some malaise and dysphoria in a minority of users.

Allergic reaction—rashes, itching, swelling of the tongue or throat, severe dizziness, and trouble breathing.

Stability and storage conditions: Store at -20°C. Rifampicin is light-sensitive, so keep it in the dark or wrap it in a foil. It also degrades in solution reasonably quickly, so make stocks and replace them every couple of months, store stock solution, plates and media in the dark and wrap in foil. Such solutions are stable for 3 months if stored below 4°C.

Incompatibility: Studies and case reports have demonstrated that rifampin accelerates the metabolism of several drugs, including oral anticoagulants, oral contraceptives, glucocorticoids, digitoxin, quinidine, methadone, hypoglycemics, and barbiturates.

When rifampin is given concomitantly with other hepatotoxic medications such as halothane or isoniazid, the potential for hepatotoxicity is increased. The concomitant use of rifampin and halothane should be avoided.

Different types of formulations: Capsule and Tablet.

Popular brands: Some of the notable brand names in India are:

Sr. No.	Brand Name	Manufactures	Type
1	Macox	Macleods Pharmaceuticals Pvt. Ltd.	Dispersible Tablet
2	Cavind, Cavikid	Merind Ltd.	Tablet
3	Ticin	Themis Pharmaceuticals Ltd	Tablet
4	R-Cin	Lupin, Ltd	Syrup, Suspension
5	Gocox	IPCA Labs Ltd.	Capsule
6	Ripharmed	Pharmed Labs Pvt Ltd	Capsule

New Antitubercular Drugs

11.3.3.6 Bedaquiline: (*Syn:* **Sirturo**)

It is a medication used to treat active tuberculosis. Specifically, it is used to treat multi-drug-resistant tuberculosis (MDR-TB) along with other medications for tuberculosis. It is used by mouth.

Structure:

Systematic (IUPAC) Name: (1R,2S)-1-(6-Bromo-2-methoxy-3-quinolyl)-4-dimethylamino-2-(1-naph -thyl) 1-phenylbutan-2-ol

Mechanism of Action: its mechanism of action is adenosine triphosphate synthase inhibition. It has potent activity against M. Tuberculosis.

Uses: Its use was approved in December 2012 by the U.S. Food and Drug Administration (FDA) for use in tuberculosis (TB) treatment, as part of a Fast-Track accelerated approval, for use only in cases of and the more resistant extensively drug-resistant tuberculosis.

As of 2013, both the World Health Organization (WHO) and US Centers for Disease Control (CDC) have recommended (provisionally) that bedaquiline be reserved for people with multidrug-resistant tuberculosis when an otherwise recommended regimen cannot be designed.

Side Effects: The most common side effects of bedaquiline in studies were nausea, joint and chest pain, and headache. The drug also has a black-box warning in its packaging inserts for increased risk of death and arrhythmias, as it may prolong the QT interval by blocking the hERG channel.

Stability and storage conditions: Tablets dispensed outside the original container should be stored in a tight light-resistant container with an expiration date not to exceed 3 months. Store at 25 °C (77 °F); excursions permitted to 15-30 °C (59-86 °F).

Incompatibility: It is contraindicated in persons with a low amount of magnesium, calcium and/or potassium in the blood; those with havingtorsades de points, a type of abnormal heart rhythm, slow heartbeat, prolonged QT interval on EKG.

Following drugs to be avoided for concomitant use with bedaquiline: carbamazepine; certain medications for human immunodeficiency virus (HIV) infection including efavirenz, indinavir, lopinavir, nelfinavir.

Different types of formulations: Tablet.

Popular brands: Presently, only SIRTURO tablets by Johnson & Johnson are available in India, in 2023 onwards, other companies shall be permitted.

11.3.3.7 Delamanid: (*Syn:* **Deltyba**)

It is a medication used to treat active tuberculosis. Specifically, it is used to treat multi-drug-resistant tuberculosis (MDR-TB) along with other medications for tuberculosis. It is used by mouth.

Structure:

Systematic (IUPAC) Name: (2*R*)-2-Methyl-6-nitro-2-[(4-{4-[4-(trifluoromethoxy)phenoxy]-1-piperidinyl}phenoxy)methyl]-2,3-dihydroimidazo[2,1-*b*][1,3]oxazole.

Mechanism of Action: Delamanid is a dihydro-nitroimidazooxazole derivative. It acts by inhibiting the synthesis of mycobacterial cell wall components, methoxy mycolic acid and ketomycolic acid. Delamanid is a pro-drug which gets activated by the enzyme deazaflavin dependent nitroreductase (Rv3547)

Uses: Delamanid is used, along with other antituberculosis medications, for active multidrug-resistant tuberculosis.

Side Effects: Common side effects of Delamanid include headache, dizziness, and nausea. Other side effects include QT prolongation

Stability and storage conditions: Store in the original container at room temperature below 25° C. Protect from excess heat and moisture. Keep the tablets out of reach of children.

Incompatibility: Delamanid is metabolised by the liver enzyme CYP3A4; therefore, strong inducers of this enzyme can reduce its effectiveness.

E.g.; Carbamazepine, Phenytoin, Rifampin, Rifabutin, Nevirapine, Rifapentine, Primidone

Different types of formulations: Tablet.

Popular brands: Presently, only DELTYBA tablets by Mylan Labs are available in India under the agreement with Otsuka Pharmaceutical Co., Ltd., Japan.

11.3.3.8 Pretomaid: (*Syn:* **Dovprela**)

It is a medication used to treat active tuberculosis. Specifically, it is used to treat multi-drug-resistant tuberculosis (MDR-TB) along with other medications for tuberculosis. It is used by mouth.

Structure:

Systematic (IUPAC) Name: (6*S*)-2-Nitro-6-{[4-(trifluoromethoxy)benzyl]oxy}-6,7-dihydro-5*H*-imidazo[2,1-*b*][1,3]oxazine

Mechanism of Action: Pretomanid kills actively replicating M. tuberculosis by inhibiting mycolic acid biosynthesis, thereby blocking cell wall production. Under anaerobic conditions, against non-replicating bacteria, pretomanid acts as a respiratory poison following nitric oxide release.

Uses: Approved in 2019 by US FDA, Pretomanid is indicated in combination with bedaquiline and linezolid, in adults, for the treatment of pulmonary extensively drug-resistant (XDR) or treatment-intolerant or nonresponsive multidrug-resistant (MDR) tuberculosis (TB)

Side Effects: Common side effects of Pretomaid include nerve damage, acne, vomiting, headache, low blood sugar, diarrhoea, and liver inflammation.

Stability and storage conditions:

Do not store above 30°C. The shelf-life at this storage condition is - 24 months for the bottle packs and - 18 months for the blister packs.

Incompatibility: Co-administration of pretomanid with rifampin and efavirenz resulted in a decrease in pretomanid plasma concentrations. Avoid co-administration of the combination regimen of Pretomanid Tablets, bedaquiline, and linezolid with rifampin, efavirenz, or other strong or moderate CYP3A4 inducers.

Pretomanid Tablets used in combination with bedaquiline and linezolid are contraindicated in patients for whom bedaquiline and/or linezolid is contraindicated

Different types of formulations: Tablet.

Popular brands: Presently, only Dovprela tablets by Mylan Labs are available in India,

Resistance to Drugs - MDR, XDR & TDR-TB

1. Multidrug Resistant Tuberculosis (MDR-TB)

Multi drugs resistance (MDR) TB refers to simultaneous resistance to at least two or more of the five first-line anti-TB drugs (isoniazid, rifampicin, pyrazinamide, ethambutol, and streptomycin). Treatment for MDR-TB is long-lasting, less effective, costly, and poorly tolerated.

Two reasons or types of resistance

- Primary resistance: As the patient is infected and exposed to resistant strain right in the beginning itself
- Secondary resistance: due to inadequate or faulty treatment regimen, drug-drug interactions, malabsorption, etc.

2. Extensively drug-resistant tuberculosis (XDR-TB)

The advent of XDR-TB (Extensively drug-resistant TB), which is commonly defined as strains resistant to all the current first-line drugs, as well as any of fluoroquinolones and, at least, to one of the three injectable second-line drugs (capreomycin, kanamycin or amikacin) is a major disturbing development in the resurgence of TB, worldwide.

3. Totally drug-resistant tuberculosis (TDR-TB)

This term has been nowadays commonly referred to the tuberculosis, caused by the strains resistant to all available first-line as well as second-line TB drugs. This is the most severe form of TB.

This highlights the urgent need for the discovery and development of new drugs.

QUESTION BANK

A. MULTIPLE CHOICE QUESTIONS

1. Which of the following ring system is present in Pyrazinamide?
   A. Pyridine
   B. Piperidine

   C.   Pyrazine

   D.   Pyridazine

Answer: C

2. Which of the following ring system is present in INH?

   A.   Pyridine

   B.   Piperidine

   C.   Pyrazine

   D.   Pyridazine

Answer: A

3. Ethambutol is basically a substituted _______________?

   A.   Ethanol

   B.   Propanol

   C.   Butanol

   D.   Pentanol

Answer: C

4. Synonyms for PAS _____________________.

   A.   4-Amino-2-hydroxybenzoic acid

   B.   *p*-Aminosalicylic acid

   C.   PASER

   D.   All of the above

Answer: D

5. Which of the following antitubercular drug is an antibiotic?

   A.   Rifampicin

   B.   Pyrazinamide

   C.   Ethambutol

   D.   PAS

Answer: A

6. Which of the following drug/drugs are recommended for multidrug-resistant TB treatment?

   A.   Bedaquine

   B.   Delamid

   C.   Pretomaid

   D.   All of the above

Answer: D

7. Which of the following is not a First Line Antitubercular drug?
   A.   Ethambutal
   B.   Rifamopicin
   C.   Isoniazid
   D.   PAS

Answer: D

8. Which is the most severe infection among the following?
   A.   Lung TB
   B.   MDR TB
   C.   XDR TB
   D.   TDR TB

Answer: D

9. The long form of recommended abbreviation by WHO for TB treatment- DOT is ______.
   A.   Directly-Observed Therapy
   B.   Directly-Observed Tuberculosis
   C.   Do-Observation and Treat
   D.   Drug &-Operation Therapy

Answer: A Directly-Observed Therapy

10. The pathogenic organism for tuberculosis is a____________.
   A.   Bacteria
   B.   Virus
   C.   Fungi
   D.   Protozoa

Answer: A

B. SHORT ANSWER QUESTIONS

1. Write the uses, side effects, stability and storage of the following
   a) INH, b) Ethambutol, c) Pyrazinamide, d) PAS, e) Bedaquiline, f) Delamanid
   g) Pretomaid, h) Rifampicin

2. Discuss the Mechanism of Action of INH and Rifampicin.

3. Discuss the Mechanism of Action of PAS and Pyrazinamide.

4. Discuss the Mechanism of Action of Ethambutol and Bedaquiline

5. Discuss the mechanism of Action of Delamanid and Pretomaid

6. Discuss the current antitubercular drug therapy and its limitations

7. Write the structure and chemical names of the following compounds.
   a) INH, b) Ethambutol, c) Pyrazinamide, d) PAS, e) Bedaquiline, f) Delamanid
   g) Pretomaid

C. LONG ANSWER QUESTIONS

1. Discuss the organism, Pathogenesis of tuberculosis (TB), Signs and symptoms, Transmission of tuberculosis

2. Classify various conventional antitubercular drugs and discuss their use, mechanism of action as well as storage conditions.

■■■

ANTI-INFECTIVE AGENTS: ANTIVIRAL AGENTS

CONTENTS

Antiviral Agents:
- Amantadine Hydrochloride
- Idoxuridine
- Acyclovir*
- Foscarnet
- Zidovudine,
- Ribavirin,
- Remdesivir
- Favipiravir

♦ LEARNING OBJECTIVES ♦

After completing this chapter, the student should be able to understand:

1. Classification of Antiviral Agents
2. Chemical Names of Antiviral Agents
3. Chemical Structures of different Antiviral Agents
4. Uses, Stability and storage conditions of Antiviral Agents
5. Different types of formulations and popular brand names Anti- Antiviral Agents

11.4.1 INTRODUCTION

Viruses are the smallest of the human infectious agents and range in size from about 20 nm to about 300 nm in diameter. They contain one kind of nucleic acid, either RNA or DNA, as their entire genome, which codes for a variety of enzymes and other proteins used in the replication and transmission of the organism. It can be argued that a virus does not qualify as a true-life form since it is nothing more than a nucleic acid strand with associated proteins and cannot move on its own power. However, when it attaches itself to a host cell, it internalises itself and forces the host to make additional copies of the virus, demonstrating a clear reproductive plan. During replication, it uses host cellular and biochemical processes and, thus, in a sense, takes in "nutrients" in order to survive and multiply.

In some cases, viruses respond to external conditions and escape the immune response by integrating into the host DNA, demonstrating the ability to respond to external stimuli. Although viruses are simple organisms, they are a significant causative agent for numerous human diseases and, as such, represent one of the significant challenges in the area of drug discovery. Agents that are used clinically for a variety of viral diseases act by targeting processes that are specific to the virus, such as a unique viral enzyme or a necessary process, such as transcription. However, to date, no drug has been discovered that is truly curative for viral infection. In addition, because viruses have the ability to undergo mutations, resistance to existing therapies can develop. The discovery of new antiviral agents is thus an important ongoing effort in medicinal chemistry.

11.4.2 CLASSIFICATION OF ANTIVIRAL AGENTS

1. Anti-herpes agents: e.g. Velcyclovir, Idoxuridine, Pamacyclovir, Acyclovir, Famcyclovir.

2. Anti-retroviral agents (Anti HIV Drugs)
  A. Nucleoside Reverse Transcriptase Inhibitors (NRTI)
     e.g. Abacavir, Zidovivodine, Stavivodine, Didanosine, Zalcitabine, Lamivudine
  B. Non-Nucleoside Reverse Transciptase Inhibitors (NNRTI)
     e.g. Efaviranz, Navirabine, Delaviridine Loviride
  C. Protease Inhibitors
     e.g. Saquinavir, Indonavir, Faminavir
  D. Fusion Inhibitors
     e.g. Infuviritile
  E. Integrase Inhibitors
     e.g. Reltegravir

3. Anti-influenza Agents: e.g. Anatadine, Rimantidine

4. Nuraminidase inhibitors: e.g. Oseltamavir, Zanamavir

5. Other: e.g. Interferons

11.4.3 SOME IMPORTANT ANTIVIRAL AGENTS

11.4.3.1 Amantadine Hydrochloride
Structure:

Chemical (IUPAC) Name: 1-Adamantanamine hydrochloride

Uses:

Amantadine is clinically effective in preventing and treating all A strains of influenza, particularly A2 strains of Asian influenza virus and, to a lesser extent, German measles (rubella) or togavirus. It also shows in vitro activity against influenza B, parainfluenza (paramyxovirus), respiratory syncytial virus (RSV), and some RNA viruses (murine, Rous, and Eshsarcoma viruses). Many prototype influenza A viruses of different human subtypes (H1N1, Fort Dix, H2N2, Asian type, and H3N2, Hong Kong type) are also inhibited by amantadine hydrochloride in vitro and in animal model systems. If given within the first 48 hours of the onset of symptoms, amantadine hydrochloride is effective in respiratory tract illness resulting from influenza A but not influenza B virus infection, adenoviruses, and RSV.

Side Effects:

Generally, the drug has low toxicity at therapeutic levels but may cause severe CNS symptoms such as nervousness, confusion, headache, drowsiness, insomnia, depression, and hallucinations. Gastrointestinal (GI) side effects include nausea, diarrhoea, constipation, and anorexia. Convulsions and coma occur with high doses and in patients with cerebral arteriosclerosis and convulsive disorders.

Storage conditions: Store amantadine at room temperature between 68°F and 77°F (20°C and 25°C). It can be temporarily stored in temperatures from 59°F to 86°F (15°C to 30°C). Don't store this medication in moist or damp areas.

Dose: The dose of 100mg

Pharmaceutical Formulations: Tablet, Capsule

Brand Names: Amantrel, Comantrel, Neaman, Amantral

11.4.3.2 Idoxuridine Trifluoride

Structure:

Chemical (IUPAC) Name: 5-Iodo-2-deoxyuridine

It was introduced in 1963 for the treatment of herpes simplex keratitis. The drug is an iodinated analogue of thymidine that inhibits the replication of several DNA viruses in vitro. The susceptible viruses include the herpesviruses and poxviruses (vaccinia).

Uses:

In the United States, idoxuridine is approved only for the topical treatment of herpes simplex virus (HSV) keratitis; although outside the United States, a solution of idoxuridine in dimethyl sulfoxide is available for the treatment of herpes labialis, genitals, and zoster.

The use of idoxuridine is limited because the drug lacks selectivity; low subtherapeutic concentrations inhibit the growth of uninfected host cells.

The effective concentration of idoxuridine is at least 10 times greater than that of acyclovir.

Idoxuridine occurs as a pale yellow, crystalline solid that is soluble in water and alcohol but poorly soluble in most organic solvents. The compound is a weak acid, with a pKa of 8.25. Aqueous solutions are slightly acidic, yielding a pH of about 6.0. Idoxuridine is light and heat sensitive. It is supplied as a 0.1% ophthalmic solution and a 0.5% ophthalmic ointment.

Storage conditions: Store it in the refrigerator but do not freeze it.

Dose: The dose 0.1%/ml eye drops, 1gm/1 vial

Pharmaceutical Formulations: Drops, injection, Ointment.

Brand Names: Ridinox, Idurin Eye, Xurin Eye, Toxil Ointment, Ifex -M, Toxil.

11.4.3.3 Acyclovir

Structure:

Chemical (IUPAC) Name: 9-[2-(Hydroxyethoxy)methyl]-9H-guanine

It is the most effective of a series of acyclic nucleosides that possess antiviral activity.

Uses:

Two dosage forms of acyclovir are available for systemic use: oral and parenteral.

Oral acyclovir is used in the initial treatment of genital herpes and to control mild recurrent episodes. It has been approved for short-term treatment of shingles and chickenpox caused by VZV.

Intravenous administration is indicated for initial and recurrent infections in immunocompromised patients and the prevention and treatment of severe episodes.

The injectable form is a sodium salt, which is supplied as a lyophilised powder, equivalent to 50 mg/mL of active acyclovir dissolved in sterile water for injection. As the solution is strongly alkaline (pH 11), it must be administered by slow and constant intravenous infusion to avoid irritation and thrombophlebitis at the injection site.

Adverse reactions:

Adverse reactions are few. Some patients experience occasional GI upset, dizziness, headache, lethargy, and joint pain.

An ointment composed of 5% acyclovir in a polyethylene glycol base is available for the treatment of initial, mild episodes of herpes genitalis. The ointment is not an effective prevention of recurrent episodes.

Storage conditions: Store acyclovir suspension at 59°F to 77°F (15°C to 25°C) and protect it from light. Store acyclovir capsules at room temperature, 68°F to 77°F (20°C to 25°C), and protect them from moisture.

Dose: The dose Injectable solution 50 mg/mL, Oral suspension-200 mg/5mL, Powder for injection 500 mg/vial, 1 g/vial, Tablet-400 mg, 800 mg, Capsule- 200 mg

Pharmaceutical Formulations: Tablet, Capsule, Injectable solution, Suspension

Brand Names: Zovirax

11.4.3.4 Foscarnet

Structure:

HO–P(=O)(OH)–C(=O)–OH

Chemical (IUPAC) Name: Phosphonoformic acid (phosphonomethanoic acid), known by its brand name Foscavir. It is an antiviral medication which is primarily used to treat viral infections involving the Herpesviridae family. It is classified as a pyrophosphate analogue DNA polymerase inhibitor. Foscarnet is the conjugate base of a chemical compound with the formula $HO_2CPO_3H_2$ (Trisodium phosphonoformate).

Uses:

This phosphonic acid derivative (marketed by Clinigen as foscarnet sodium under the trade name Foscavir) is an antiviral medication used to treat herpes viruses, including drug-resistant cytomegalovirus (CMV) and herpes simplex viruses types 1 and 2 (HSV-1 and HSV-2). It is particularly used to treat CMV retinitis. Foscarnet can be used to treat highly treatment-experienced patients with HIV as part of salvage therapy.

Side effects:

Nephrotoxicity- an increase in serum creatinine levels and renal injury can occur in patients receiving foscarnet. Other nephrotoxic drugs should be avoided. Nephrotoxicity is usually reversible and can be reduced by dosage adjustment and adequate hydration.

Electrolyte disturbances-hypocalcemia and hypomagnesemia can occur, and regular monitoring of electrolytes is necessary to avoid clinical toxicity.

Genital ulceration-a less common reported side effect which occurs more in men and usually during induction use of foscarnet. It is most likely a contact dermatitis due to high concentrations of foscarnet in urine. It usually resolves rapidly following the discontinuation of the drug.

CNS-less common side effects of perioral paresthesia, irritability and altered mental states.

Storage conditions: Foscarnet sodium 12 mg/mL in 0.9% sodium chloride injection was stable for up to 30 days when stored at 25^0C and exposed to light, 25^0C and protected from light, or 5^0C and protected from light.

Dose: The dose is 2.4g/100mL.

Pharmaceutical Formulations: Injection

Brand Names: Foscavir

11.4.3.5 Zidovudine

Structure:

Chemical (IUPAC) Name: 3' Deoxy-3'-azido-thymidine 1-[(2R,4S,5S)-4-Azido-5-(hydroxy-methyl) oxolan-2-yl]-5-methylpyrimidine-2,4-dione

Zidovudine is recommended for managing adult patients with symptomatic HIV infection who have a history of confirmed *Pneumocystis carinii* pneumonia or an absolute CD4 (T4 or TH cell) lymphocyte count below 200/mm3 before therapy.

Anaemia and granulocytopenia are the most common toxic effects associated with Zidovudine.

For oral administration, Zidovudine is supplied as 100-mg capsules and as a syrup containing 10 mg of Zidovudine per mL. The injectable form of Zidovudine contains 10 mg/mL and is injected intravenously.

Storage conditions: Store zidovudine tablets at room temperature, 68°F to 77°F (20°C to 25°C). Store zidovudine capsules and oral solution between 59°F and 77°F (15°C to 25°C).

Dose: The dose 100 mg, 150 mg, 300 mg 600 mg

Pharmaceutical Formulations: Tablet, Capsule, Suspension, Injection.

Brand Names: Lazid, RAZIV Lamda-Z, Lamda Z, Cytocom -E, Zivudin, Duovir, Zidomax, Lamda-Z, Zidolam, Duovir E Kit, Zidolam, Lamuzid, Virocomb, Viro-Z

11.4.3.6 Ribavirin

Structure:

Chemical (IUPAC) Name: 1-β-D-Ribofuranosyl-1,2,4-thiazole-3-carboxamide.

The compound is a purine nucleoside analogue with a modified base and a D-ribose sugar moiety.

Ribavirin inhibits the replication of a wide variety of RNA and DNA viruses, including orthomyxoviruses, paramyxoviruses, arenaviruses, bunyaviruses, herpesviruses, adenoviruses, poxvirus, vaccinia, influenza virus (types A and B), parainfluenza virus, and rhinovirus.

Despite the broad spectrum of activity of ribavirin, the drug has been approved for only one therapeutic indication—the treatment of severe lower respiratory infections caused by RSV in carefully selected hospitalised infants and young children.

The mechanism of action of ribavirin is not known.

However, the broad antiviral spectrum of ribavirin suggests multiple modes of action.

Ribavirin occurs as a white, crystalline, polymorphic solid that is soluble in water and chemically stable.

It is supplied as a powder to be reconstituted in an aqueous aerosol containing 20 mg/mL of sterile water. The aerosol is administered with a small-particle aerosol generator (SPAG).

The role of ribavirin in these events has not been determined. Anaemia, headache, abdominal pain, and lethargy have been reported in patients receiving oral ribavirin.

Storage conditions: Store ribavirin tablets and capsules at room temperature, between 68°F and 77°F (20°C to 25°C). Store ribavirin oral solution at room temperature, between 68°F and 77°F (20°C to 25°C), or in the refrigerator, between 36°F and 46°F (2°C to 8°C).

Dose: The dose is 800 to 1400 mg

Pharmaceutical Formulations: Tablet, Capsule

Brand Names: Copegus Rebetol, RibaPak, Ribasphere, Ribapak, RibaTab

11.4.3.7 Remdesivir

Structure:

Chemical (IUPAC) name:- (2S)-2-{(2R,3S,4R,5R)-[5-(4-Aminopyrrolo[2,1-f] [1,2,4]triazin-7-yl)-5-cyano-3,4-dihydroxy-tetrahydro-furan-2-ylmethoxy]phenoxy-(S)-phosphorylamino} propionic acid 2-ethyl-butyl ester

Remdesivir, sold under the brand name Veklury, is a broad-spectrum antiviral medication developed by the biopharmaceutical company Gilead Sciences. It is administered via injection into a vein. During the COVID-19 pandemic, remdesivir was approved or authorised for emergency use to treat COVID-19 in numerous countries.

Remdesivir was originally developed to treat hepatitis C and was subsequently investigated for Ebola virus disease and Marburg virus infections before being studied as a post-infection treatment for COVID-19.

Side Effects

The most common adverse effects in people treated with remdesivir were respiratory failure and blood biomarkers of organ impairment, including low albumin, low potassium, low count of red blood cells, low count of thrombocytes, and elevated bilirubin (jaundice). Other reported adverse effects include gastrointestinal distress, elevated transaminase levels in the blood (liver enzymes), infusion site reactions, and electrocardiogram abnormalities.[14] Remdesivir may cause infusion-related reactions, including low blood pressure, nausea, vomiting, sweating or shivering.

Storage conditions: Do not reuse or save unused remdesivir lyophilised powder for infusion for future use. This product doesn't contain preservatives.

Lyophilised Powder: Store redeliver for injection, 100 mg, vials below 30°C (below 86°F) until required for use.

After reconstitution, vials can be stored up to 4 hours at room temperature (20°C to 25°C [68°F to 77°F]) before administration or 24 hours at refrigerated temperature (2°C to 8°C [36°F to 46°F]). Dilute within the same day as administration.

Dose: injection, lyophilized powder for reconstitution- 100mg/vial, injection, concentrated solution 100mg/20mL (5mg/mL)

Pharmaceutical Formulations: IV Injection

Brand Names: Veklury, Remdac, Redyx Cipremi

11.4.3.8 Favipiravir

Structure:

Chemical (IUPAC) name: - 6-Fluoro-3-hydroxypyrazine-2-carboxamide

Favipiravir, sold under the brand name Avigan among others, is an antiviral medication used to treat influenza in Japan. It is also being studied to treat a number of other viral infections, including SARS-CoV-2.

Uses:

Favipiravir has been approved to treat influenza in Japan. It is, however, only indicated for novel influenza (strains that cause more severe disease) rather than seasonal influenza. As of 2020, the probability of resistance developing appears low.

Side Effects:

There is evidence that uses during pregnancy may result in harm to the baby. Teratogenic and embryotoxic effects were shown on four animal species.

Storage conditions: Short Term Storage: +4⁰C. Long-Term Storage: -20⁰C. Keep at a cool and dry place.

Dose: The dose of 200 mg

Pharmaceutical Formulations: Tablet

Brand Names: Avigan, Avifavir, Avipiravir, Areplivir, FabiFlu, Favipira, Reeqonus, Qifenda

QUESTION BANK

A. MULTIPLE CHOICE QUESTIONS

1. Which of the following is an Anti-herpes agent?
   A. Acyclovir
   B. Foscarnet
   C. Zidovudine
   D. Ribavirin

   Answer: A

2. Tick the drug, a derivative of pyrophosphate
   A. Foscarnet
   B. Zidovudine
   C. Vidarabine
   D. Acyclovir

   Answer: A

3. Which of the following is Anti-influenza Agent?
   A. Idoxuridine
   B. Acyclovir

    C. Foscarnet
    D. Amantadine Hydrochloride

Answer: A

4. Which of the following is Nucleoside Reverse Transcriptase Inhibitor (NRTI)?
    A. Acyclovir
    B. Foscarnet
    C. Zidovudine
    D. Ribavirin

Answer: C

5. Remdac is popular brand name for__________.
    A. Zidovudine
    B. Ribavirin
    C. Favipiravir
    D. Remdesivir

Answer: D

6. During the COVID-19 pandemic, which of the following medicine was approved or authorised for emergency use?
    A. Ribavirin
    B. Zidovudine
    C. Remdesivir
    D. Favipiravir

Answer: C

7. Which of the following heterocycle is present in Favipiravir?
    A. Pyrrole
    B. Pyrazine
    C. Furan
    D. Benzene

Answer: B

8. Which of the following heterocycle is present in Ribavirin?
    A. Thiazole
    B. Pyrazine
    C. Furan
    D. Benzene

Answer: A

9. Zovirax is popular brand name for__________.
    A. Idoxuridine
    B. Acyclovir

   C. Foscarnet

   D. Zidovudine

Answer: B

10. Toxil is popular brand name for__________.

   A. Idoxuridine

   B. Acyclovir

   C. Foscarnet

   D. Zidovudine

Answer: A

B. SHORT ANSWER QUESTIONS

1. Classify antiviral agents.

2. Enlist names of Anti-herpes agents.

3. Enlist pharmaceutical formulations available for Acyclovir.

4. Enlist different brand names for Idoxuridine.

5. Enlist pharmaceutical formulations available for Remdesivir.

6. Give the uses of Remdesivir.

7. Explain Remdesivir as an Anti-viral agent.

8. Explain Amantadine as Anti-influenza Agent.

9. Draw structure for Foscarnet.

10. Enlist different brand names for Favipiravir.

C. LONG ANSWER QUESTIONS

1. Give physical properties, pharmaceutical formulation and brand name of-

   1. Idoxuridine   2. Acyclovir

2. Define and classify Anti-viral agents with examples. Add a note on Acyclovir.

3. Explain physical properties, pharmaceutical formulations, and brand names for Remdesivir.

ANTI-INFECTIVE AGENTS: ANTIMALARIALS

CONTENTS

- Brief introduction of the infection, causative organism, its vector,
- The life cycle of malarial parasite & some important terminologies.
- Classification of Antimalarial drugs- different types.
- Malarial chemotherapy
- Some important antimalarial drugs, their chemical name, chemical structure (compounds marked with*) uses, stability and storage conditions, different types of formulations and their popular brand names
  - Quinine Sulphate,
  - Chloroquine Phosphate*,
  - Primaquine Phosphate,
  - Mefloquine*,
  - Cycloguanil,
  - Pyrimethamine,
  - Artemisinin

◆ LEARNING OBJECTIVES ◆

After completing this chapter, the student should be able to understand:

1. The causative organism and pathogenesis of malaria, its types and lifecycle. The disease signs and symptoms, its transmission,
2. The classification and mechanism of action of various antimalarial drugs.
3. Important drugs used as antitubercular agents are their chemical structures, names used, incompatibilities, stability and storage conditions and brands.

11.5.1 INTRODUCTION

Malaria is one of the most important of transmittable parasitic diseases. It still remains a major threat to life, especially in the IIIrd World countries. It is endemic in more than 100 countries in Asia, Africa and South & Central America. It is estimated that over 80 million are afflicted by this every year. Its near eradication in post-World War II led to euphoria, which has now turned to severe pessimism. One million die from the disease each year.

Etiology and Symptoms: Malaria is caused by four species of plasmodium protozoa; *P.falciparum, P.malariae, P.vivax, and P.ovale.* The plasmodium spends half of its life cycle in the vector; female *Anopheles* mosquitoes.

Signs and symptoms: Malaria is characterised by attacks of high fever, chills, nausea and vomiting. Initial manifestation of malaria are non-specific and resembles to flu-like symptoms. It includes headache, fever, shivering, arthralgia (joint pain), and myalgia (muscle pain).

The paroxysm (outburst), which includes fever spikes, chills and rigours, is classical for malaria.

The typical paroxysmal attack comprises three distinct stages:

a) **Cold stage-_**The onset is with lassitude (weakness), headache, nausea and chilly sensation followed by rigours. The stage lasts for ¼ - 1 hour

b) **Hot stage-_**The patient feels burning hot, and the skin is hot and dry to the touch. The headache is intense. The pulse rate is high. The stage lasts for 2-6 hours

c) **Sweating stage-** Fever comes down with profuse sweating. The pulse rate gets slower, and the patient feels relieved. The stage lasts 2-4 hours.

These paroxysms have different frequencies in different species of malarial parasites.

- In *P. vivax* and *P. ovale* after every two days- Tertian fever
- In *P. malariae* after every 3 days- Quartan fever
- While in *P. falciparum*, it recurs every 36-48 hours

These paroxysmal attacks coincide with the release of successive broods of merozites into the bloodstream.

Relapse *Vs* Recrudesence

Depending upon the cause, **recurrence** can be classified either as recrudescence or relapse.

Recrudescence is when symptoms return after a symptoms-free period. It is due to parasites surviving in the blood as a result of inadequate or ineffective treatment.

Relapse is when symptoms reappear after the parasites have been eliminated from blood but persist as dormant hypnozites in liver cells.

Relapse is common in *P.ovale* and *P.vivax* infection

Recrudescence is commonly seen in *P.falciparum*

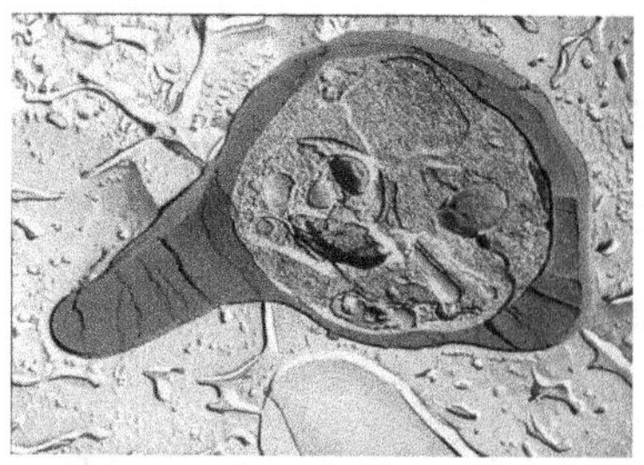 

Fig. 11.5.1 The causative parasite- Plasmodium & the vector- female Anopheles mosquitoes

11.5.2 THE LIFE CYCLE OF MALARIAL PARASITE

The life cycle of the malarial parasite has two interdependent life cycles- it is important to understand a few facts given below before the actual life cycle.

1. Sexual cycle, which occurs in the mosquito &

2. Asexual cycle which occurs in the human

The knowledge of the life cycles is essential in understanding antimalarial drug treatment. Drugs are effective only during the asexual cycle, which has two phases or stages:

a. Exoerythrocytic phase/stage; which occurs "outside" the erythrocyte and is also known as the tissue phase/stage and

b. Erythrocytic phase/stage, which occurs "inside" the erythrocyte and is also known as the blood phase/stage.

In short, the life cycle of the malarial parasite can be narrated as below;

1. The plasmodium species afflicting men live in female anopheles mosquito during the sexual phase of their LC and are injected into men as **sporozoites.**

2. These sporozites rapidly disappear from human circulation and invade the liver parenchymatous cells, where they undergo an asexual division called **schizogony.** Each sporozoite grows within the liver cell and undergoes repeated nuclear division to form a **schizont (merozite).** This process is called as a **pre-erythrocytic stage** or **1° tissue stage** and takes 6-12 days (6-8 days for *p.vivax;* 5.5 -7 days for *p.falciparum;* 9 days for *p.ovale* and 13-15 days for *p.malariae.*). No symptoms of the disease are seen during this stage.

3. The liver cells finally burst open, and merozoites are released into the blood. Some of the merozoites reinvade liver cells and undergo further asexual division. This is called the **exo-erythrocytic stage** or the **2° tissue stage**. These forms are called exo-erythrocytic forms and are responsible for the relapses of malaria even after the blood cycle is suppressed.

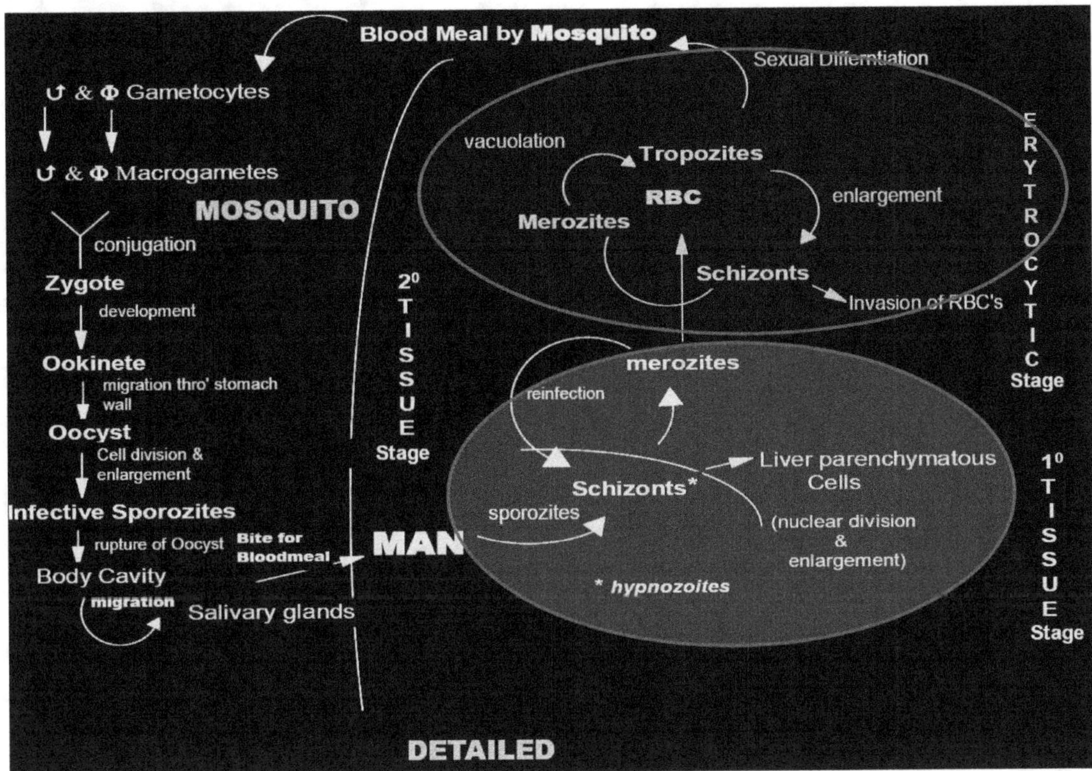

Fig.11.5.2 The life cycle of the malarial parasite-Schematic

4. The other merozoites invade RBC's and initiate a cycle of **erythrocytic schizogony**. The young merozites grow within the RBCs to form vacuolated cells or **tropozites,** which further grow and form **erythrocytic schizonts,** which rupture the RBCs. Malaria pigment (**hemozoin**) is produced by the breakdown of haemoglobin and is liberated in the bloodstream to enter fresh RBCs. The production and release of asexual erythrocytic forms into the bloodstream are responsible for fever and other symptoms of malaria (rigours). Fresh invasion of new RBCs and asexual development of new merozites continues.

5. Some of the daughter merozites fail to reproduce asexually and differentiate into male and female forms known as **gametocytes,** which are ingested by the mosquitoes when they suck fresh blood from the host. Any asexual forms ingested by a mosquito are destroyed in its stomach. But the gametocytes survive and change rapidly.

6. The male and female gametocytes undergo nuclear division to form **male macrogametes** and **female macrogametes**, respectively.

7. These male and female macrogametes conjugate and the **zygote** is formed. This zygote alters its shape and becomes worm-like and is called an **ookinete**.

8. The ookinete enters the stomach wall of the mosquito to form the rounded **oocyst,** which undergoes nuclear division and enlargement and finally ruptures to release

thousands of mobile infective sporozoites, which enter the body cavity of the mosquito and migrate to its salivary glands.

9. The infective sporozoites then enter the vertebrate host, and the whole cycle begins when the mosquito takes the host's blood meal.

11.5.3 CLASSIFICATION OF ANTIMALARIAL DRUGS

Based on their action at a particular stage of the Plasmodial lifecycle as well as their chemical structure as well as their mechanism of action, various antimalarial drugs are classified into 3 categories of classifications.

1. Chemical Classification
2. Biological Classification
3. Mixed Biological & Chemical Classification

1. Chemical Classification:

Based on the chemical classes (structures), antimalarials are classified into various major categories as follows in the below figure.

4-Aminoquinolines

Table *Contd...*

Quinoline methanols

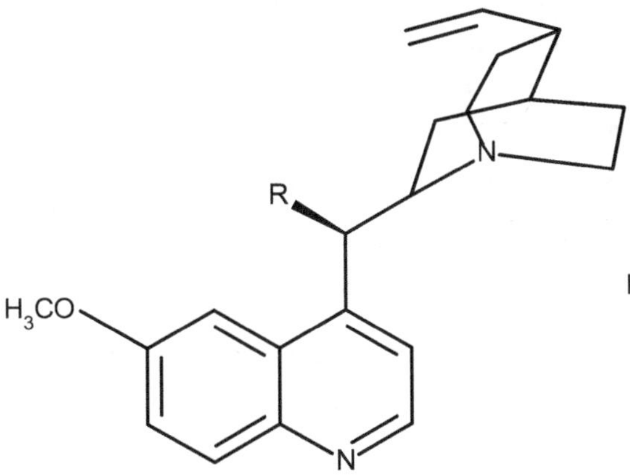

Quinine

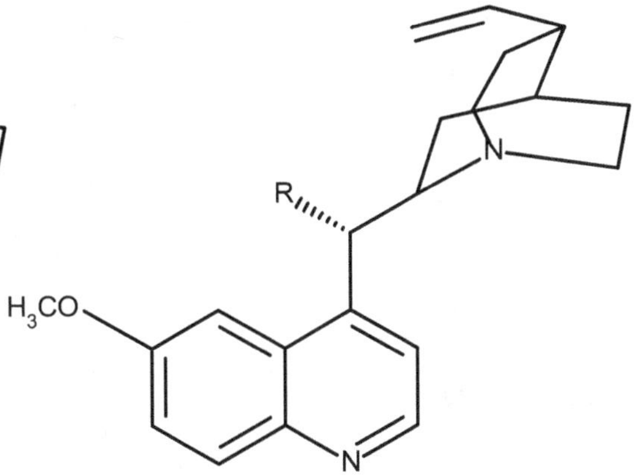

Quinidine

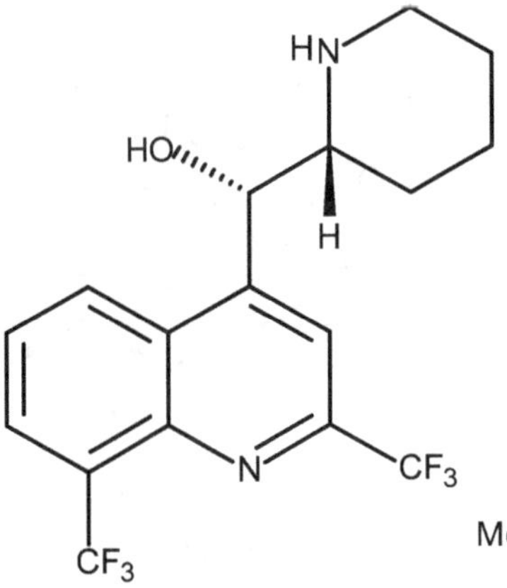

Mefloquine

Biguanides

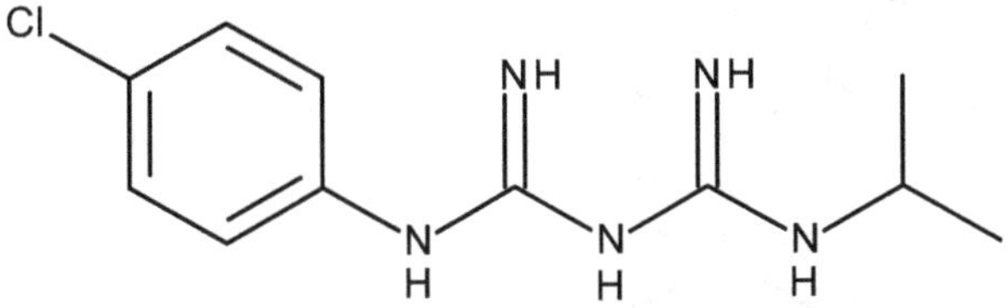

Proguanil (Chloroguanide)

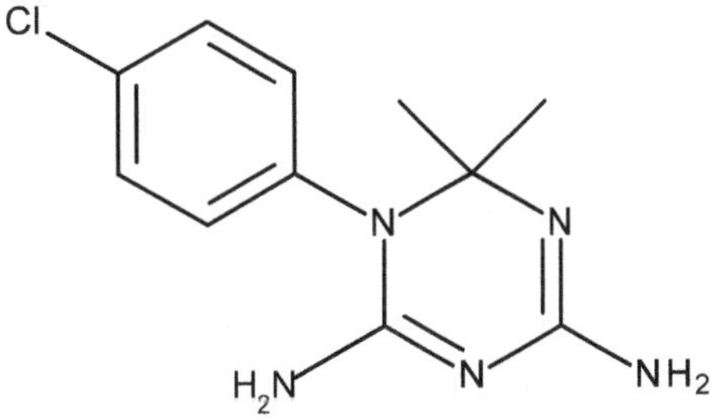

Cycloguanil

Table *Contd...*

Diaminopyrimidines (DHFR inhibitors)
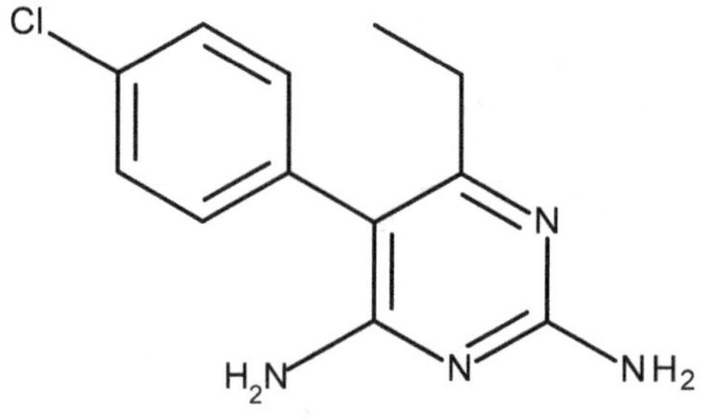 Pyrimethamine
Sesquiterpine lactones
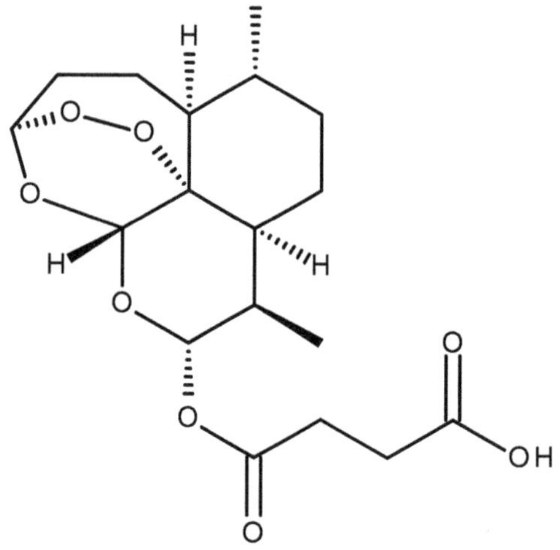 Artesunate 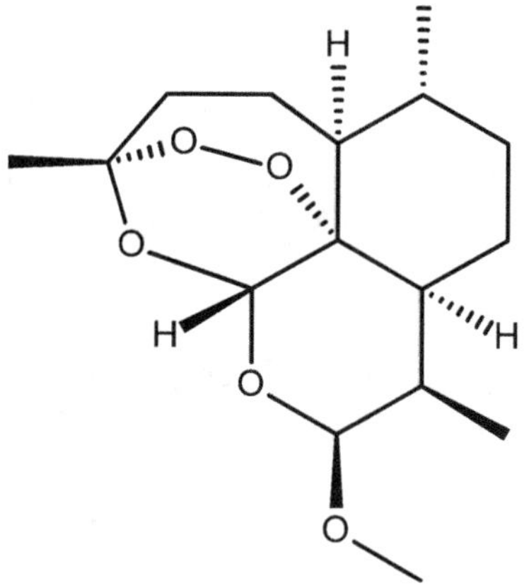 Artemether 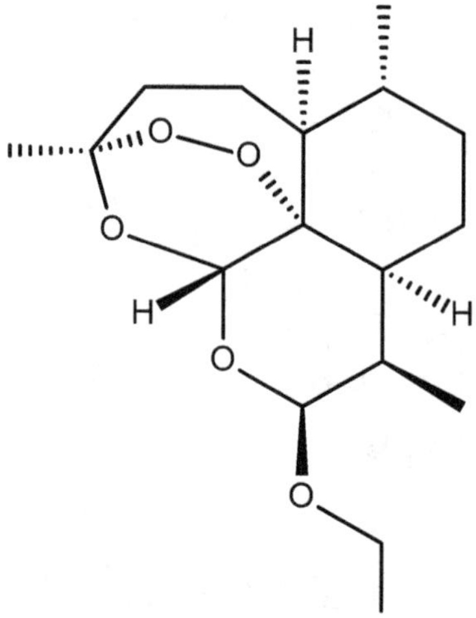 Arteether

Table *Contd...*

Naphthyridines

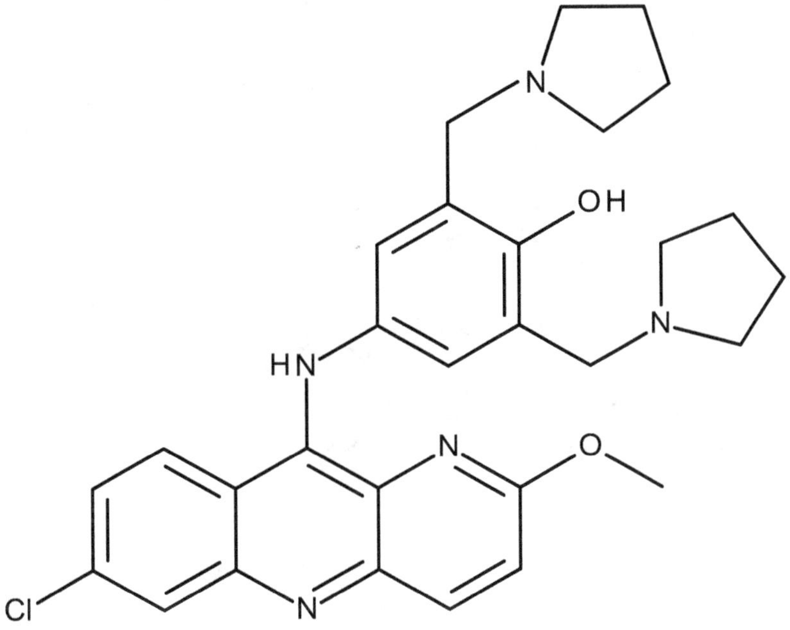

Pyronaridine

Amino alcohols

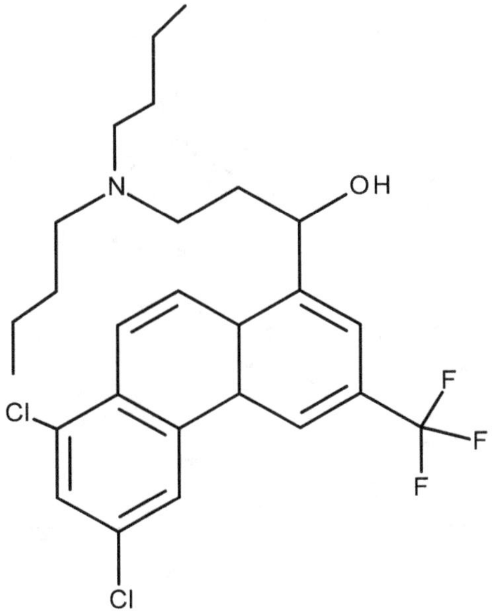

Halofantrine

Table *Contd...*

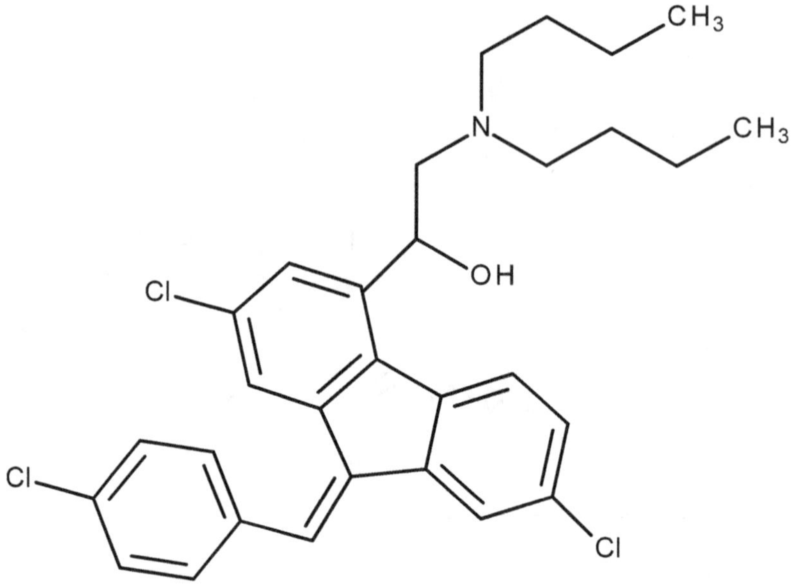

Lumefantrine

Naphthoquinones

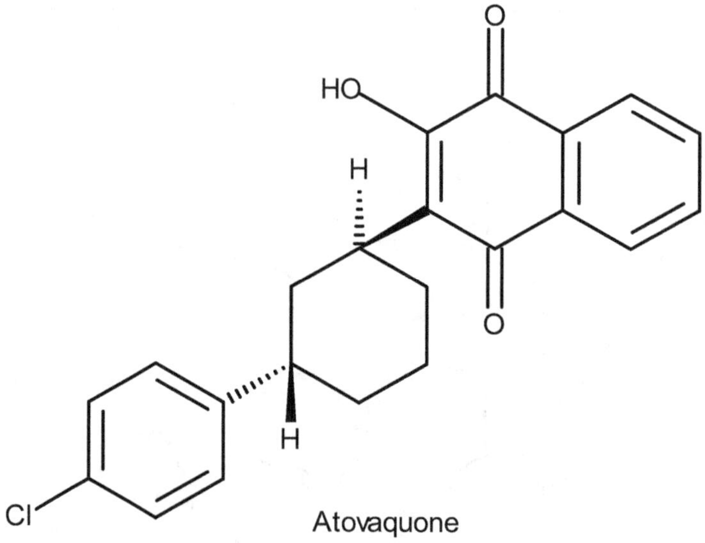

Atovaquone

Table *Contd...*

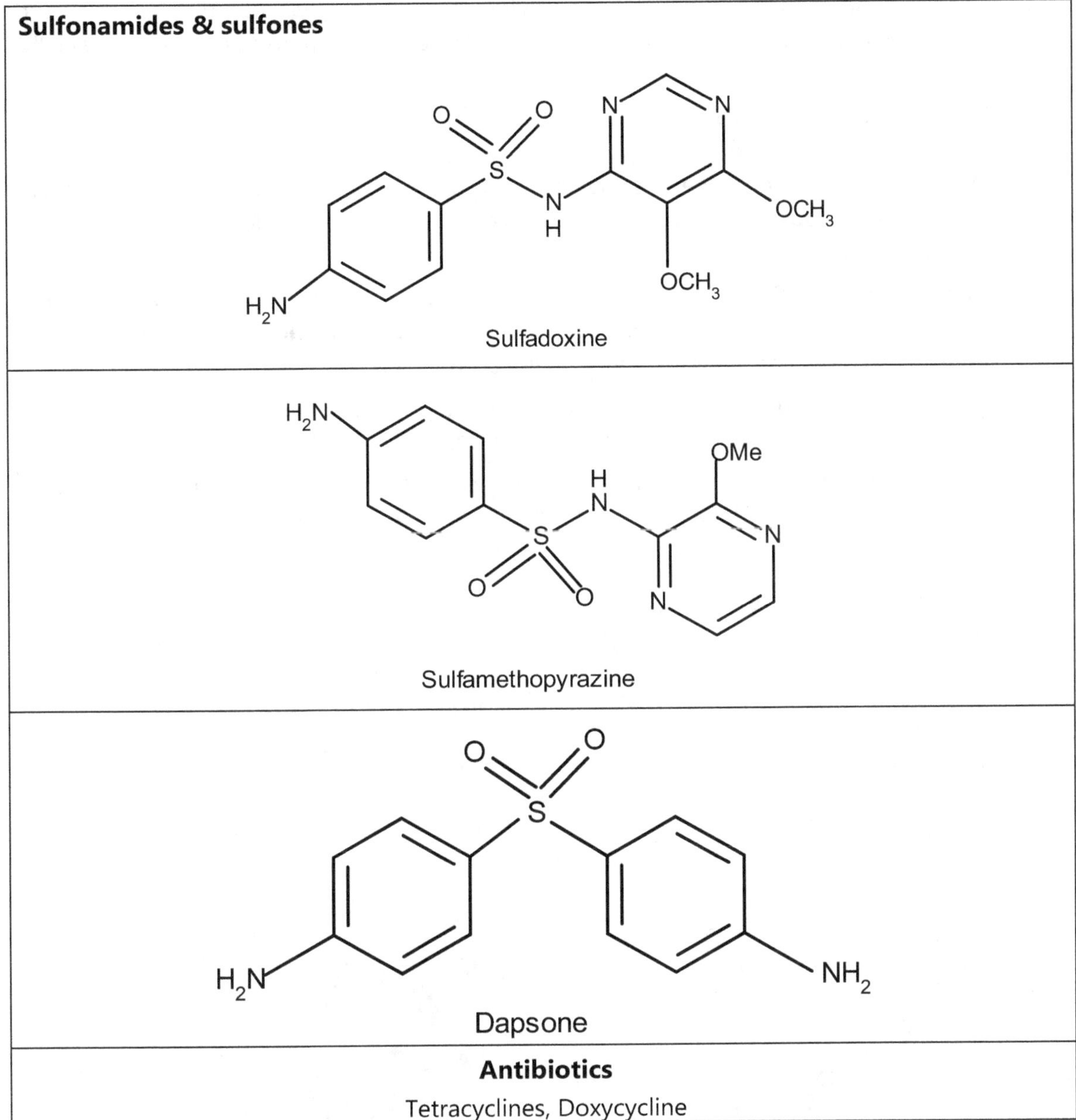

Fig. 11.5.3 Antimalarial Drugs: Chemical Classification

2. Biological Classification:

1. Tissue Schizonticides: (Primary Causal Prophylactics): Inhibit the growth of the pre-erythrocytic stages of the parasites in the liver. E.g., Proguanil and Pyrimethamine are used alone or in combination or sulfonamides as Casual prophylactics to prevent the infection. These drugs eradicate the exoerythrocytic form (liver-tissue stages), preventing the parasite's entry into the blood. Useful for prophylaxis.

2. Hypnozoiticides: Kill the dormant long living liver stages (hypnozoites) and are used as antirelapse drugs in *P.vivax* and *P.ovale*. E.g., 8-Aminoquinolines especially, primaquine.

3. Blood Schizonticides: (Suppressives): Act only in the erythrocytic stages and are generally used in therapy. E.g., Chloroquine and Quinine. Chloroquine is also employed as a suppressive prophylactic. These drugs can cure cases of falciparum malaria or suppress relapses. Drug delivery into the bloodstream can be accomplished rapidly.

4. Gametocytocides: Destroy the sexual stages of the parasites in the blood, including mature gametocytes of *P. falciparum*. Thereby **preventing** transmission of the disease. E.g., 8-Aminoquinolines especially, primaquine.

5. Sporontocides: Inhibitory action on developing oocysts and sporozoites in the mosquitoes. E.g., Pyrimethamine and cycloguanil.

   Radical Curatives: Drugs that destroy both exoerythrocytic (tissue) and erythrocytic (blood) forms. E.g., Combination of chloroquine + primaquine.

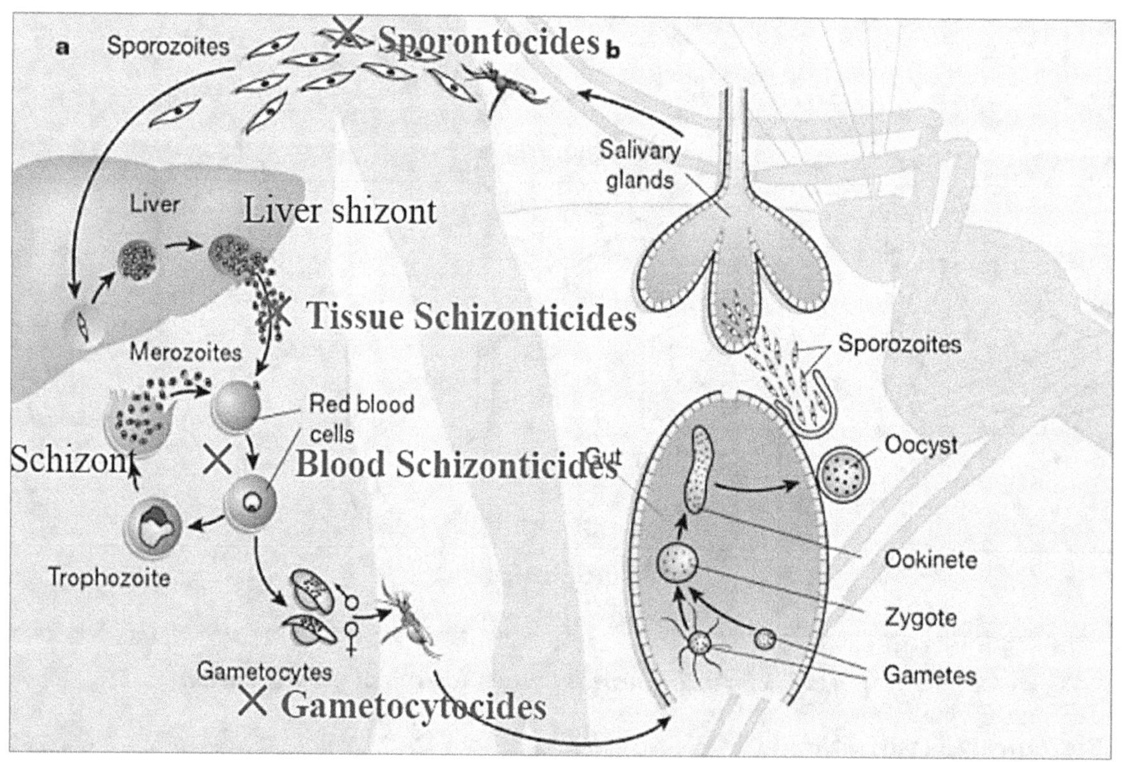

Fig. 11.5.4 Classification of Antimalarial agents based on the parasite life cycle, [i.e., depending on which stage of the life cycle of the organism is affected.]

3. Mixed Chemical / Biological Classification:

Alternatively, the drugs are classified into 3 main categories.

1. 8- Aminoquinolines; 2. Antimetabolites and 3. Blood schizonticides.
   a) 8-Aminoquinolines are particularly effective on dormant parasite stages such as gametocytes and hypnozoites, although they have some effect at toxic concentrations on the asexual stages in the blood, *e.g.*, primaquine.
   b) Antimetabolites active in all parasite growing stages, e.g., pyrimethamine, cycloguanil, sulfonamides and sulfones.
   c) Blood Schizonticides active against only blood stages of malarial parasite *e.g.*, quinine, quinidine, chloroquine, amodiaquine, mefloquine, 9-aminoacridines; mepacrine, quinacrine).

11.5.4 SOME IMPORTANT ANTIMALARIAL DRUGS

11.5.4.1 Quinine Sulphate: (*Syn:* Qualaquin, Quinbisul)

Structure:

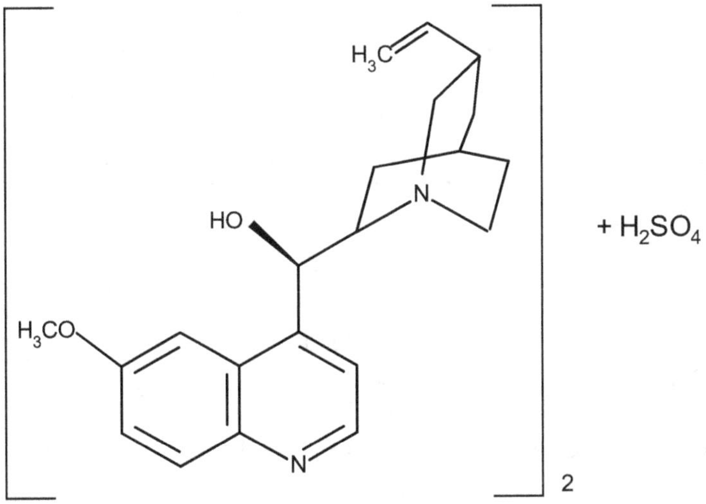

Systematic (IUPAC) Name: (*R*)-(6-Methoxyquinolin-4-yl)[(1*S*,2*S*,4*S*,5*R*)-5-vinylquinuclidin-2-yl]methanol

Mechanism of Action: Quinine inhibits nucleic acid synthesis, protein synthesis, and glycolysis in Plasmodium falciparum and can bind with hemazoin in parasitised erythrocytes. However, the precise mechanism of the antimalarial activity of quinine sulfate is not completely understood.

Uses: Quinine is an erythrocytic schizontocide for all species of plasmodia. It is, however, inactive against pre-erythrocytic stages. It is gametocidal against *P. vivax* & *P. malaria* when given primaquine for *vivax* malaria.

It is less effective and more toxic than chloroquine, therefore, the latter is preferred over it.

Quinine is also used to treat lupus and arthritis.

Side Effects:

The most common adverse effects involve a group of symptoms called cinchonism, which can include headache, vasodilation and sweating, nausea, tinnitus, hearing impairment, vertigo or dizziness, blurred vision, and disturbance in colour perception.

Other Side effects

Quinine can cause unpredictable severe, life-threatening blood and cardiovascular reactions, including hemolytic-uremic syndrome/thrombotic thrombocytopenic purpura Pyridoxine (Vit.B6), which may be given to reduce the risk of side effects.

Stability and storage conditions: Store in cool, dry conditions and well-sealed containers

Incompatibility: Quinine has diverse unwanted interactions with numerous prescription drugs, such as potentiating the anticoagulant effects of warfarin

Different types of formulations: Tablet, injection, suspension.

Popular brands: Some of the notable brand names in India are:

Sr. No	Brand Name	Manufactures	Type
1	CINKONA	IPCA Laboratories Ltd.	Tablet
2	REZ-Q	Plethico Pharmaceuticals	Tablet
3	CINKONA	IPCA Laboratories Ltd.	Suspension
4	CINKONA	IPCA Laboratories Ltd.	Injection
5	Isoniazid Injection USP	SG Pharma Pvt Ltd.	Injection
6	QINARSOL-300	Cipla Limited	Injection
7	KWINIL	Intas Pharma. Pvt. Ltd.	Tablet

11.5.4.2 Chloroquine Phosphate: (*Syn:* **Lariago, Aralen**)

Structure:

Systematic (IUPAC) Name: (*RS*)-*N*'-(7-chloroquinolin-4-yl)-*N*,*N*-diethyl-pentane-1,4-diamine

Uses:

1. Chloroquine has been used in the treatment and prevention of malaria from *Plasmodium vivax, P. ovale,* and *P. malariae.* It is generally not used for Plasmodium falciparum as there is widespread resistance to it.

2. In the treatment of amoebic liver abscess, chloroquine may be used instead of or in addition to other medications in the event of failure of improvement with metronidazole or another nitroimidazole or intolerance to metronidazole or a nitroimidazole.

3. As it mildly suppresses the immune system, chloroquine is used in some autoimmune disorders, such as rheumatoid arthritis and has an off-label indication for lupus erythematosus.

Mechanism of Action: As a blood schizonticide. It actively concentrates in the sensitive intra-erythrocytic plasmodia by accumulating in the acidic vesicles of the parasite. By its weakly essential nature, it raises the vesicular pH and thereby interferes with the degradation of haemoglobin by parasitic lysosomes.

The polymerisation of toxic haeme to nontoxic parasite pigment hemozoin is inhibited by the formation of chloroquine-heme complex, damaging the plasmodial membranes. Clumping of pigment and changes in parasite membranes follows death.

Side Effects: include blurred vision, nausea, vomiting, abdominal cramps, headache, diarrhoea, swelling legs/ankles, shortness of breath, pale lips/nails/skin, muscle weakness, easy bruising/bleeding, hearing and mental problems.

Stability and storage conditions: Store at room temperature (15-25°C). Protect from light. Product in powder form is stable for 6 months at room temperature when properly stored.

Incompatibility: It is not recommended in people with optic neuritis, significant kidney problems, or under the age of five. Use during pregnancy or breastfeeding has not been found to cause harm in patients with known hypersensitivity to 4-aminoquinoline compounds and for long-term therapy in children.

Different types of formulation: Tablet

Popular brands: It is manufactured by many companies. Some of the notable brand names in India are:

Sr. No.	Brand Name	Manufactures	Type
1	Quinodev	Dev Life Corpn. Pvt. Ltd.	Tablet
2	Lariago	IPCA Labs Limited	Tablet/Injection
3	Torris Chloroquine Phosphate	Pharma Pluss Distributors	Injection

11.5.4.3 Primaquine Phosphate: (*Syn:* **Malirid, Neo-Quipenyl, Pimaquin, Pmq)**

Structure:

Systematic (IUPAC) Name: (*RS*)-*N*-(6-methoxyquinolin-8-yl)pentane-1,4-diamine

Uses: Primaquine is primarily used to prevent relapse of malaria due to *Plasmodium vivax* and *Plasmodium ovale.* It eliminates hypnozoites, the dormant liver form of the parasite,[10] after the organisms have been cleared from the bloodstream.

Primaquine is also used in the treatment of *Pneumocystis pneumonia* (PCP), a fungal infection commonly occurring in people with AIDS and, more rarely, in those taking immunosuppressive drugs. To treat PCP effectively, it is usually combined with clindamycin.

Mechanism of action: Primaquine is lethal to *P. vivax* and *P. ovale* in the liver stage and to *P. vivax* in the blood stage through its ability to do oxidative damage to the cell. However, the exact mechanism of action is not fully understood.

Side Effects: Common side effects of primaquine administration include nausea, vomiting, and stomach cramps.

Stability and storage conditions: Primaquine phosphate tablets should be stored in well-closed, light-resistant containers at a temperature less than 40 °C, preferably between 15-30 °C.

Incompatibility: Some products that may interact with this drug include: penicillamine, quinacrine, and drugs that may cause decreased blood cells (such as trimethoprim, zidovudine, pyrimethamine, azathioprine).

Different types of formulation: Tablets.

Popular brands: It is manufactured by many companies. Some of the notable brand names in India are:

Sr. No.	Brand Name	Manufactures	Type
1	Malirid.	Ipca Laboratories Ltd.	Tablet
2	Primarid	Themis Medicare Ltd	Tablet
3	R C Vax	Unicure India Pvt Ltd	Tablet
4	Primelife	Leo Pharmaceuticals	Tablet

11.5.4.4 Mefloquine: (*Syn:* Lariam)

Structure:

Systematic (IUPAC) Name: [(R*,S*)-2,8-Bis(trifluoromethyl)quinolin-4-yl]-(2-piperidyl) methanol

Uses: Mefloquine is used as a treatment for chloroquine-sensitive or resistant Plasmodium falciparum malaria and is deemed a reasonable alternative for uncomplicated chloroquine-resistant Plasmodium vivax malaria.

Mechanism of action: It's mechanism of action is not completely understood. Some studies suggest that mefloquine specifically targets the 80S ribosome of the Plasmodium falciparum, inhibiting protein synthesis and causing subsequent schizonticidal effects.

Side Effects: Common side effects of Mefloquine include vomiting, diarrhoea, headaches, and a rash. Severe side effects requiring hospitalisation are rare but include mental health problems such as depression, hallucinations, and anxiety and neurological side effects such as poor balance, seizures, and ringing in the ears.

Stability and storage conditions: Mefloquine hydrochloride is photolabile, and tablets of the drug should be stored at 25 °C but may be exposed to 15-30 °C.

Incompatibility: It is known to interact with drugs; artemether-lumefantrine, beta-blockers (such as atenolol, propranolol), chloroquine, halofantrine, ketoconazole, quinidine, quinine, drugs for seizures (such as phenytoin, valproic acid), ziprasidone.

Different types of formulation: Tablet.

Popular brands: Some of the notable brand names in India are:

Sr.No.	Brand Name	Manufactures	Type
1	Meflotas-250	Intas Laboratories Pvt. Ltd	Tablet
2	Mefloc	Aristo Pharma Ltd.	Tablet
3	Mefliam Plus	Cipla Ltd	Tablet
4	Larimef	IPCA Labs Ltd.	Tablet

11.5.4.5 Cycloguanil

Structure:

Systematic (IUPAC) Name: 1-(4-chlorophenyl)-6,6-dimethyl-1,3,5-triazine-2,4-diamine

Uses: As an antimalarial, blood schizonticide. The continuing development of resistance to current antimalarial drugs has led to renewed interest in studying the use of cycloguanil in combination with other drugs.

Mechanism of Action: Cycloguanil is a dihydrofolate reductase inhibitor, so it acts as an antimetabolite for the plasmodium and is a metabolite of the antimalarial drug proguanil.

Side Effects: The common side effects of Cycloguanil include anorexia, dizziness, vomiting, diarrhoea, mouth ulcers, a low white cell count, and a low platelet count. The other developed extensive purpura, epistaxis, and vomiting.

Stability and storage conditions: Store at - Dry, dark and at 0-4^{0}C for short term (days to weeks) or -20^{0}C for long term (months to years).

Incompatibility: The therapeutic efficacy of Cycloguanil can be decreased when used in combination with Folic acid and Leucovorin.

Different types of formulations: Tablet.

Popular brands: Only available as the generic medicine

11.5.4.6 Pyrimethamine: (*Syn:* Daraprim)

It is a medication used to treat active tuberculosis. Specifically, it is used to treat multi-drug-resistant tuberculosis (MDR-TB) along with other medications for tuberculosis. It is used by mouth.

Structure:

Systematic (IUPAC) Name: 5-(4-chlorophenyl)-6-ethyl- 2,4-pyrimidinediamine

Mechanism of Action: its mechanism of action is a selective plasmodial DHFR inhibitor (Antimetabolite).

Uses: It is primarily active against *Plasmodium falciparum* but also against *Plasmodium vivax*. Due to the emergence of pyrimethamine-resistant strains of *P. falciparum*, pyrimethamine alone is seldom used now, but in combination with a long-acting sulfonamide such as sulfadiazine,

Other Uses: Pyrimethamine is also used in combination with sulfadiazine to treat active toxoplasmosis

Side Effects: When higher doses are used, as in the treatment of toxoplasmosis, pyrimethamine can cause gastrointestinal symptoms such as nausea, vomiting, glossitis, anorexia, and diarrhoea. A rash, which can be indicative of a hypersensitivity reaction, is also seen, particularly in combination with sulfonamides. Central nervous system effects include ataxia, tremors, and seizures. Hematologic side effects such as thrombocytopenia, leukopenia, and anaemia can also occur.

Stability and storage conditions: Pyrimethamine is stable for at least 91 days in an oral suspension stored in plastic or glass prescription bottles at 4 or 25 degrees C

Incompatibility: Some products that may interact with this drug include: lorazepam, penicillamine, sulfa drugs (such as sulfamethoxazole), drugs that can lower folate levels (such as phenytoin, trimethoprim), drugs that can lower blood counts (such as proguanil, zidovudine, chemotherapy including methotrexate, daunorubicin, cytosine).

Different types of formulations: Tablet.

Popular brands: Mostly available in combination with Sulfadoxine (25 mg: 500mg).

Sr. No.	Brand Name	Manufactures	Type
1	Malocide	Torrent Pharma Ltd	Tablet
2	Crydoxin FM	GSK Pharma Ltd.	Tablet
3	Laridox	IPCA Labs Ltd	Tablet
4	Pyralfin	Lupin Ltd.	Tablet

11.5.4.7 Artemisinin: (*Syn:* **Qinghaosu**)

It is a medication used to treat active tuberculosis. Specifically, it is used to treat multi-drug-resistant tuberculosis (MDR-TB) along with other medications for tuberculosis. It is used by mouth.

Structure:

Systematic (IUPAC) Name: (3*R*,5a*S*,6*R*,8a*S*,9*R*,12*S*,12a*R*)-Octahydro-3,6,9-trimethyl-3,12-epoxy-12*H*-pyrano[4,3-*j*]-1,2-benzodio- xepin-10(3*H*)-one

Mechanism of Action: Artemisinin is believed to act via a two-step mechanism. Artemisinin is first activated by intraparasitic heme-iron, which catalyses the cleavage of this endoperoxide. A resulting free radical intermediate may then kill the parasite by alkylating and poisoning one or more essential malarial protein(s).

Uses: The World Health Organization (WHO) recommends artemisinin or one of its derivatives — typically in combination with a longer-lasting partner drug — as frontline therapy for all cases of malaria. Artemisinins are not used for malaria prevention because of the extremely short activity (half-life) of the drug. To be effective, it would have to be administered multiple times each day.

Side Effects: The side effects of the artemisinin class of medications are similar to the symptoms of malaria: nausea, vomiting, loss of appetite, and dizziness. Mild blood abnormalities have also been noted. A rare but serious adverse effect is an allergic reaction.

Stability and storage conditions: An artemisinin derivatives AS, AM, and DHA degrade extensively when stored at 60°C for up to 21 days.

Incompatibility: The metabolism of Artemisinin can be increased when combined with drugs like Fluvastatin, Fluvoxamine, Fosphenytoin, Golimumab, etc.

Different types of formulations: Tablet, Capsule.

Popular brands: Only available as the generic medicine

QUESTION BANK

A. MULTIPLE CHOICE QUESTIONS

1. Which of the following ring system is present in Pyrimethamine?
   A. Pyrimidine
   B. Quinoline

    C. Benzopyrimidine

    D. Triazine

Answer: A

2. Which of the following ring system is present in Cycloguanil?

    A. Pyrimidine

    B. Quinoline

    C. Benzopyrimidine

    D. Triazine

Answer: D

3. Chloroquine is basically a _______________?

    A. 8-Aminoquinoline

    B. 6-Aminoquinoline

    C. 2-Aminoquinoline

    D. 4-Aminoquinoline

Answer: D

4. Primaquine is basically a ___________________

    A. 8-Aminoquinoline

    B. 6-Aminoquinoline

    C. 2-Aminoquinoline

    D. 4-Aminoquinoline

Answer: A

5. _________________ is used against chloroquine-resistant *P.falciparum* malaria

    A. Mefloquine

    B. Artemisinin

    C. Arteether

    D. Any of above

Answer: D

6. _________________ is a DHFR inhibitor drug

    A. Chloroquine

    B. Pyrimethamine

    C. Quinine

    D. Pyrimethamine

Answer: D

7. Which of the following is not a blood schizonticide drug?
    A. Chloroquine
    B. Pyrimethamine
    C. Quinine
    D. Primaquine

Answer: D

8. Quartan fever is observed in infection by ___________________?
    A. *P. falciparum*
    B. *P. malariae*
    C. *P. vivax*
    D. *P. ovale*

Answer: B

9. Recrudescence is commonly seen in the malaria caused by_______________.
    A. *P. falciparum*
    B. *P. malariae*
    C. *P. vivax*
    D. *P. ovale*

Answer: A

10. ______________________ is a hypnozoiticide.
    A. Chloroquin
    B. Mefloquin
    C. Primaquin
    D. Quinine

Answer: C

B. SHORT ANSWER QUESTIONS

1. Write the uses, side effects, stability and storage of the following.
    a) Chloroquin, b) Primaquin, c) Mefloquin, d) Quinine, e) Pyrimethamine, f) Cycloguanil g) Artemisinin

2. Discuss the Mechanism of Action of Chloroquine and Primaquine.

3. Discuss the Mechanism of Action of Mefloquine and Quinine

4. Discuss the Mechanism of Action of Ethambutol and Bedaquiline

5. Discuss the mechanism of Action of Pyrimethamine and Cycloguanil

6. Write the structure and chemical names of the following compounds.)
    a) Chloroquin, b) Primaquin, c) Mefloquin, d) Quinine, e) Pyrimethamine, f) Cycloguanil g) Artemisinin

C. LONG ANSWER QUESTIONS

1. Discuss the malarial parasite, its life cycle, Signs and symptoms, and Transmission of malaria

2. Classify various antimalarial drugs and discuss their use, mechanism of action as well as storage conditions.

■■■

ANTI-INFECTIVE AGENTS: SULFONAMIDES

CONTENTS

- Sulfanilamide
- Sulfadiazine
- Sulfamethoxazole
- Sulfacetamide*
- Mafenide Acetate
- Cotrimoxazole [Trimethoprim with Sulfamethoxazole]
- Dapsone*

♦ LEARNING OBJECTIVES ♦

After completing this chapter, the student should be able to understand:

1. Historical development of Sulphonamide.
2. Chemistry of Sulphonamide.
3. Classification of Sulphonamide.
4. SAR of Sulphonamide.
5. Different medicinal agents from sulphonamide and sulfone class.
6. Mechanism of action of Sulphonamide.
7. Chemical Names of Sulphonamide.
8. Chemical Structures of different Sulphonamide.
9. Uses, Stability and storage conditions of Sulphonamide.
10. Different types of formulations and popular brand names for Sulphonamide.

11.6.1 INTRODUCTION

The antibacterial properties of the sulphonamides were discovered in the mid-1930s after an incorrect hypothesis, but after observing the results carefully and drawing correct conclusions. Prontosil rubrum, a red dye, was one of a series of dyes examined by Gerhard Domagk of Bayer of Germany in the belief that it might be taken up selectively by certain pathogenic bacteria and not by human cells, in a manner analogous to the way that the Gram stain works, and thus serve as a selective poison to kill these cells.

The dye, indeed, proved active in vivo against streptococcal infections in mice. Curiously, it was not active in vitro. Trefouel and others soon showed that the urine of prontosil rubrum–treated animals were bioactive in vitro. Fractionation led to the identification of the active substance as *p*-aminobenzenesulfonic acid amide (sulfanilamide), a colourless cleavage product formed by reductive liver metabolism of the administered dye. Today, we would call prontosil rubrum a prodrug.

Prontosil Rubrum   Sulfanilamide

The discovery of sulfanilamide's in vivo antibacterial properties ushered in the modern anti-infective era, and Domagk was awarded a Nobel Prize for medicine in 1939. Once mainstays of antimicrobial chemotherapy, the sulfonamides have decreased enormously in popularity and are now comparatively minor drugs. The relative inexpensiveness of the sulfonamides is one of their most attractive features and accounts for much of their persistence on the market.

11.6.2 CHEMISTRY OF SULPHONAMIDE

Aniline has a pKa of 4.6 pKa is nothing but its salt acting as conjugated acid. A negative charge on a nitrogen atom is unstable unless it can be delocalized by resonance. This happens in sulphonamides. In the case of sulphonamides, there occurs loss of amine H^+, as it has pKa 10.4, *i.e.* single pKa usually given for sulphonamides refers to loss of amine proton. The resulting negative charge is resonance stabilized.

Sulphonamides are almost excreted in the urine. Sulphanilamides are not so soluble in water unless the pH is above the pKa; little water-soluble salt is present. Because the urine usually has about pH 6.0, all sulphonamides are in relative insoluble, non–ionized form in the kidneys. Thus attempts are made to make the sulphanilamide more soluble in urine.

11.6.3 CLASSIFICATION OF SULPHONAMIDE

A. On the basis of its chemical nature

1. **N_1- Substituted Sulphonamides**

   a) **Acyclic N_1 Substituents:** e.g. Sulphacetamide, Sulphaguanidine

   b) **Heterocyclic N_1 Substituents:** e.g. Sulphadiazine, Sulphathiazole

2. **N₄- Substituted Sulphonamides:** e.g. Sulphasalazine

3. **N₁ & N₄ - Substituted Sulphonamides:** e.g. Phthalylsulpharthiazole and Succinylsulphathiazole

B. On the basis of duration of action -

1. **Short-acting - (half-life 4-8 hours):** e.g. Sulphadiazine

2. **Intermediate-acting - (half-life 08-12 hours):** e.g. Sulphamoxazole, Sulphamethoxazole

3. **Long-acting - (half-life 7 days):** e.g. sulphadoxine, sulphamethapyrazone

4. **Special purpose sulfonamide:** e.g. Sulphacetamide, Sulphasalazine, Silver sulphadiazine

C. On the basis of therapeutic uses

1. **Sulphonamides used for systemic infection**

   a) **Used for bacterial infections:** e.g. Sulphadiazine

   b) **Used for urinary tract infection:** e.g. Cotrimozazole

   c) **Used for respiratory tract infection:** e.g. Sulphadiazine

2. **Sulphonamides used for local infection**

   a) **Used for burn therapy:** e.g. Silver sulphadiazine

   b) **Used for ophthalmic infection therapy:** e.g. Sulphacetamide

3. **Mixed Sulphonamide**

   a) **Trisulfapyrimidenes (oral suspension):** Sulfadiazine + Sulfamerazine + Sulfamethazine

   b) **Trisulfadiazine (Tablet):** Sulfadiazine + Sulfamerazine + Sulfamethazine

   c) **Sulfadoxine and pyrimethamine- prophylaxis of malaria**

   d) **Sulphamethoxazole + Trimethoprim (Cotrimoxazole)**

11.6.4 THERAPEUTIC APPLICATION

Of the thousands of sulfonamides that have been evaluated, only a few are still available and are often used in combination with other agents. The surviving sulfonamides include sulfisoxazole, which is used in combination with erythromycin. It has a comparatively broad antimicrobial spectrum in vitro, especially against gram-negative organisms, but clinical use is generally restricted due to the development of bacterial resistance. Susceptible organisms may include Enterobacteriaceae (*E. coli*, *Klebsiella*, and *Proteus*) and *Streptococcus pyogenes*, *Streptococcus pneumonia*, and *Haemophilus*.

The remaining sulfonamides are not used systemically. Sulfadiazine, in the form of its silver salt, is used topically for the treatment of burns and is effective against a range of bacteria and fungus, whereas sulfacetamide is used ophthalmically for the treatment of eye infections caused by susceptible organisms. Sulfasalazine stands out from the typical sulfonamide because, although it is administered orally, the drug is not absorbed in the gut, so the majority of the dose is delivered to the distal bowel. In addition, the drug is a prodrug that undergoes reductive metabolism by gut bacteria converting the drug into sulfapyridine and 5-aminosalicyclic acid, the active component **(Fig. 11.6.1)**.

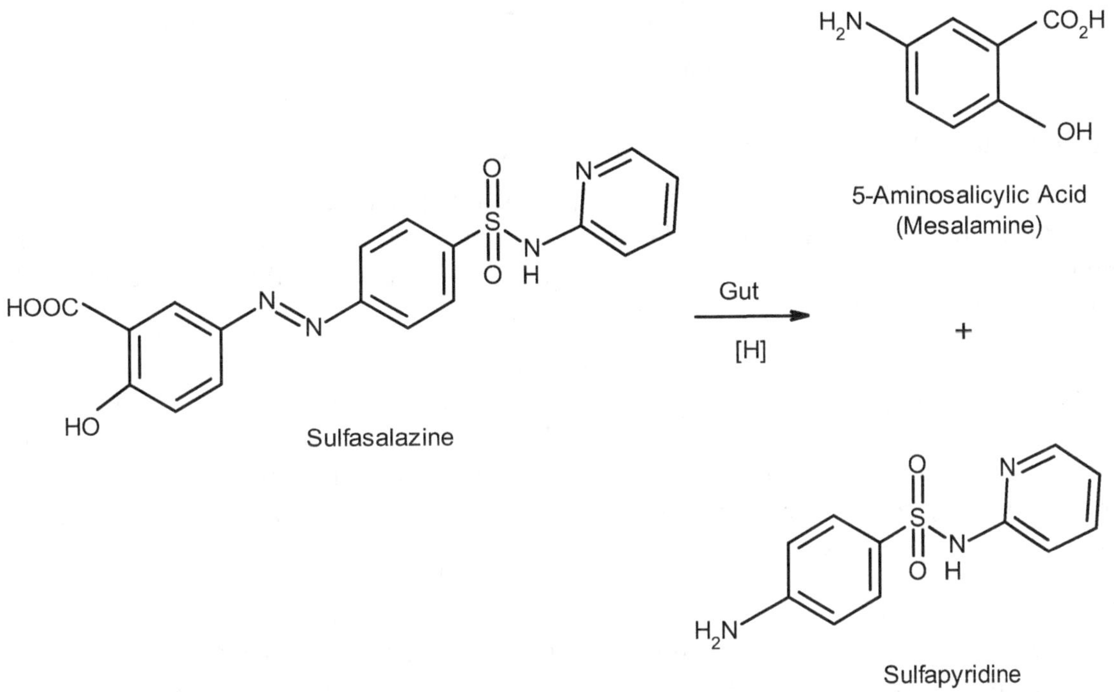

Fig. 11.6.1 Activation of sulfasalazine to 5-aminosalicylic acid

The liberation of 5-aminosalicylic acid (mesalamine), an anti-inflammatory agent, is the purpose for administering this drug. This agent is used to treat ulcerative colitis and Crohn's disease. Direct administration of salicylates is otherwise irritating to the gastric mucosa.

11.6.5 ADVERSE EFFECTS

Allergic reactions are the most common and take the form of a rash, photosensitivity, and drug fever. Less common problems are kidney and liver damage, hemolytic anaemia, and other blood problems. The most severe adverse effect is the Stevens-Johnson syndrome, characterised by sometimes fatal erythema multiforme and ulceration of mucous membranes of the eye, mouth and urethra. Fortunately, these effects are comparatively rare.

11.6.6 SOME IMPORTANT SULPHONAMIDE

11.6.6.1 Sulfanilamide

Structure:

Chemical (IUPAC) name: (4-aminophenyl)sulfonylazanide

It is an organic sulfur compound structurally similar to p-aminobenzoic acid (PABA) with antibacterial properties. Sulfanilamide competes with PABA for the bacterial enzyme dihydropteroate synthase, thereby preventing the incorporation of PABA into dihydrofolic acid, the immediate precursor of folic acid.

Sulfanilamide is a sulfonamide antibacterial drug. Chemically, it is an organic compound consisting of aniline derivatised with a sulfonamide group. Powdered sulfanilamide was used by the Allies in World War II to reduce infection rates and contributed to a dramatic reduction in mortality rates compared to previous wars. Sulfanilamide is rarely if ever, used systemically due to toxicity and because more effective sulfonamides are available for this purpose.

Mechanism of Action

As a sulfonamide antibiotic, sulfanilamide functions by competitively inhibiting enzymatic reactions involving para-aminobenzoic acid (PABA). Specifically, it competitively inhibits the enzyme dihydropteroate synthase. PABA is needed in enzymatic reactions that produce folic acid, which acts as a coenzyme in synthesizing purines and pyrimidines. Mammals do not synthesise their folic acid, so they are unaffected by PABA inhibitors, which selectively kill bacteria.

However, this effect can be reversed by adding the end products of one-carbon transfer reactions, such as thymidine, purines, methionine, and serine. PABA can also reverse the effects of sulfonamides.

Storage conditions: Store at room temperature below $86^{0}F$ ($30^{0}C$) away from light and moisture.

Dose: The dose of 50mg

Pharmaceutical Formulations: Tablet, cream.

Brand Names: Similia Sulphanilamide Trituration Tablet 3X, AVC

11.6.6.2 Sulphadiazine

Structure:

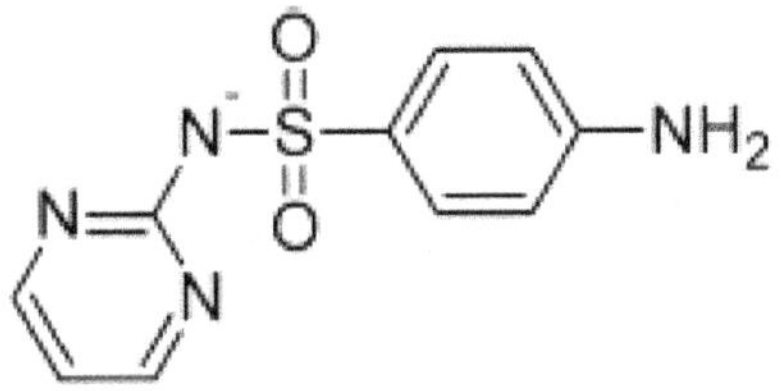

Chemical (IUPAC) name: (4-aminophenyl)-sulphonyl-pyridine-2-yalazanide

Soluble sulfadiazine is an anhydrous, white, colourless, crystalline powder soluble in water (1:2) and slightly soluble in alcohol.

Its water solutions are alkaline (pH 9–10) and absorb carbon dioxide from the air, with precipitation of sulfadiazine. It is administered as a 5% solution in sterile water intravenously for patients requiring an immediately high blood level of sulfonamide.

Storage conditions: Store sulfadiazine tablets at room temperature, 68°F to 77°F (20°C to 25°C). Protect sulfadiazine tablets from light. Keep sulfadiazine in the container that it came in and keep the container tightly closed.

Dose: The dose 500 mg, 2.5%/5%/15gm

Pharmaceutical Formulations: Tablet, cream.

Brand Names: Salphaz Tablet 500mg), Zad G Cream.

11.6.6.3 Sulfamethoxaole

Structure:

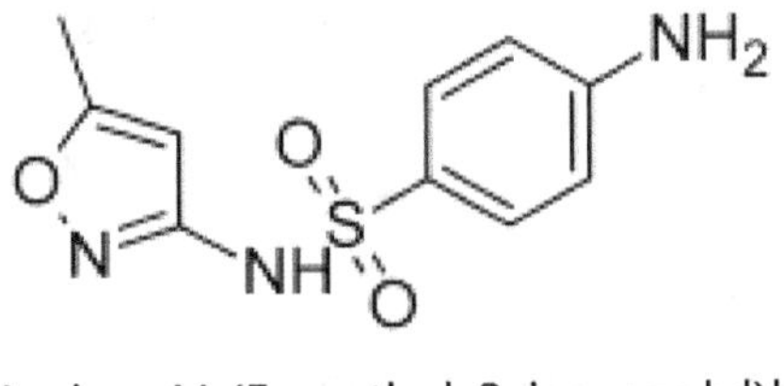

Chemical (IUPAC) name: 4-Amino-*N*-(5-methyl-3-isoxazolyl)benzene sulfonamide or *N*1-(5-methyl-3-isoxazolyl) sulfanilamide

Sulfamethoxazole's plasma half-life is 11 hours. Sulfamethoxazole is a sulfonamide drug closely related to sulfisoxazole in chemical structure and antimicrobial activity. It occurs as a tasteless, odourless, almost white crystalline powder. The solubility of sulfamethoxazole in the pH range of 5.5 to 7.4 is slightly lower than that of sulfisoxazole but higher than that of sulfadiazine, sulfamerazine, or sulfamethazine.

Following oral administration, sulfamethoxazole is not absorbed as completely or as rapidly as sulfisoxazole, and its peak blood level is only about 50% as high.

Commercially available sulfamethoxazole tablets and oral suspension should be protected from light and stored at a temperature less than 40 °C

Storage conditions: Store at a controlled room temperature of 15°-30°C (59°-86°F).

Dose: The dose 200 mg, 800 mg,

Pharmaceutical Formulations: Tablet, Injection Suspension

Brand Names: Bactrim, Septra, Sulfatrim

11.6.6.4 Sulfacetamide

Structure:

Chemical (IUPAC) name: *N*-[(4-Aminophenyl)sulfonyl]-acetamide or *N*-sulfanilylacetamide or *N*1-acetylsulfanilamide.

Sulfacetamide's plasma half-life is 7 hours. This compound is a white crystalline powder, soluble in water (1:62.5 at 37°C) and in alcohol. It is very soluble in hot water, and its water solution is acidic.

It has a pKa of 5.4.

Sulfacetamide is slightly irritant when UV-A light is present. In the presence of light, sulfacetamide gets sensitised and degraded, which might cause irritation which will lead to toxicity when it is used continuously. In the dark only slight irritation has been shown. Therefore it should be stored in the dark.

Storage conditions: It should be stored in the dark

Dose: The dose is 50mg, 10 % eye drops

Pharmaceutical Formulations: Ointment, lotion, cream

Brand Names: Cetamide, Klaron, Ovace, sulfacetamide sodium 10 % eye drops

11.6.6.5 Mefenide acetate

Structure:

Chemical (IUPAC) name: 4-(Aminomethyl) benzenesulfonamide acetate

It is a homologue of the sulfanilamide molecule.

It is a topical non absorbable Sulfonamides used for burn therapy. It is not a true sulfanilamide-type compound, as it is not inhibited by PABA. Its antibacterial action involves a mechanism that differs from that of true sulfanilamide-type compounds.

This compound is particularly effective against *Clostridium welchii* in the topical application and was used during World War II by the German army for the prophylaxis of wounds.

It is not effective orally. It is currently used alone or with antibiotics in the treatment of slow-healing, infected wounds.

Some patients treated for burns with large quantities of this drug have developed metabolic acidosis. To overcome this adverse effect, a series of new organic salts were prepared.

The acetate in an ointment base proved to be the most efficacious.

Storage conditions: Store the reconstituted solution at 20° to 25°C (68° to 77°F). Limited storage periods at 15° to 30°C (59° to 86°F) are acceptable.

Dose: The 5% topical solution

Pharmaceutical Formulations: Solution

Brand Names: Sulfamylon

11.6.6.6 Trimethoprim

Structure:

Chemical (IUPAC) name: Trimethoprim (5-[(3,4,5-trimethoxyphenyl)methyl]-2,4-pyrimidinediamine or 2,4-diamino-5-(3,4,5-trimethoxybenzyl) pyrimidine)

It is closely related to several antimalarials but does not have good antimalarial activity by itself; it is, however, a potent antibacterial.

Originally introduced in combination with sulfamethoxazole, it is now available as a single agent. Approved by the FDA in 1980, trimethoprim as a single agent is used only for the treatment of uncomplicated urinary tract infections. The concern is that when used as a single agent, bacteria now susceptible to trimethoprim will rapidly develop resistance.

In combination with a sulfonamide, however, the bacteria will be less likely to do so. They will not survive long enough to easily develop resistance to both drugs.

It is given in combination with sulphamethoxazole which shows the synergetic action. This combination is commonly used for anti-bacterial action, which is called as **co-trimoxazole** in the proportion of 5:1

Mechanism of Action

A further step in the pathway leading from the pteroates to folic acid and on to DNA bases requires the enzyme dihydrofolate reductase. Exogenous folic acid must be reduced step-wise to dihydrofolic acid and then to tetrahydrofolic acid, an important cofactor essential for supplying a one-carbon unit in thymidine biosynthesis and, ultimately, for DNA synthesis (Fig. 11.6.2).

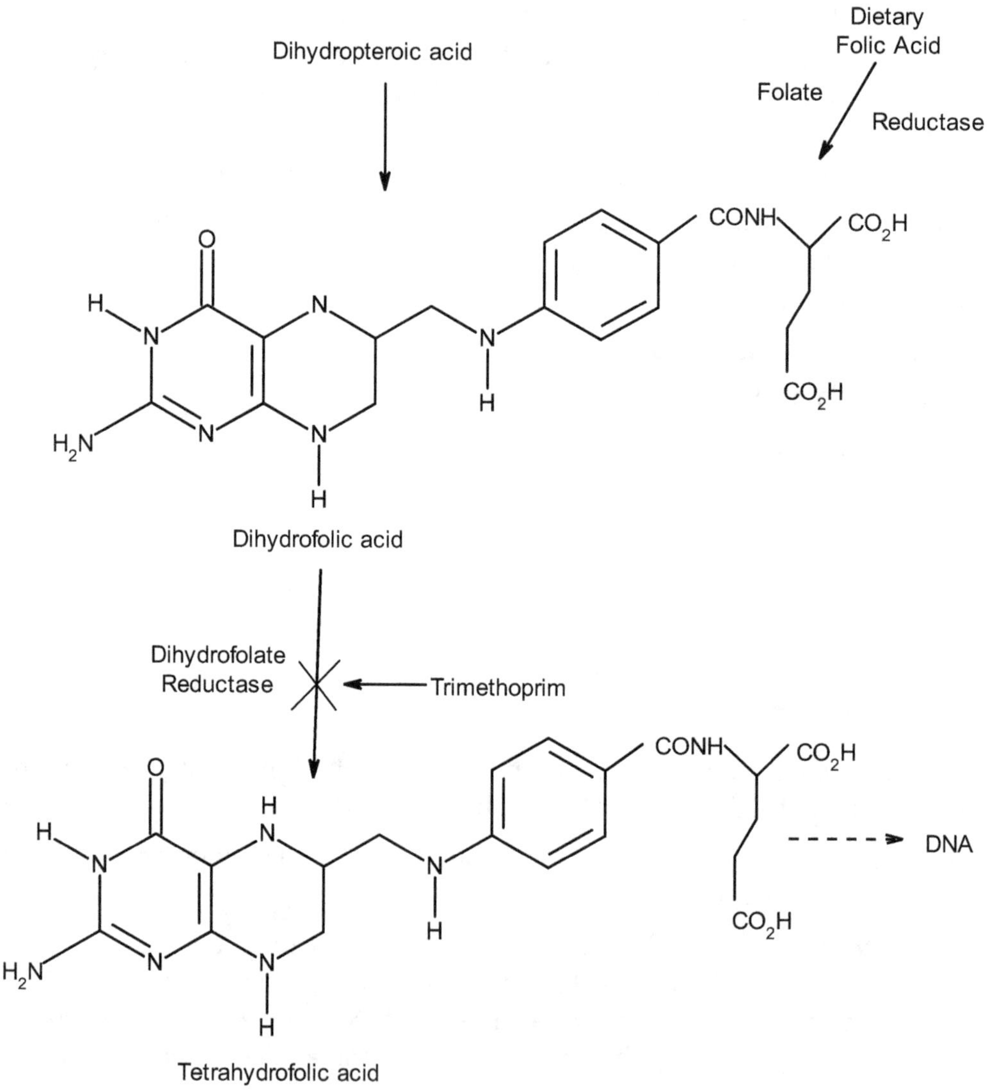

Fig. 11.6.2 Site of action of trimethoprim.

The same enzyme must also reduce endogenously produced dihydrofolate. Inhibition of this crucial enzyme had been widely studied in attempts to find anticancer agents by starving rapidly dividing cancer cells of needed DNA precursors. Antifolates such as methotrexate arose from such studies. Methotrexate, however, is much too toxic to be used as an antibiotic. Subsequently, trimethoprim was developed in 1969 by George Hitchings and Gertrude Elion (who shared a Nobel Prize for this and other contributions to chemotherapy in 1988). This inhibitor prevents tetrahydrofolic acid biosynthesis and results in bacteriostasis. Trimethoprim selectivity between bacterial and mammalian dihydrofolate reductases results from the subtle but significant architectural differences between these enzyme systems. Whereas the bacterial enzyme and the mammalian enzyme both efficiently catalyse the conversion of dihydrofolic acid to tetrahydrofolic acid, the bacterial enzyme is sensitive to inhibition by trimethoprim by up to 100,000 times lower concentrations than is a vertebrate enzyme. This difference explains the useful selective toxicity of trimethoprim.

Therapeutic Application:

Trimethoprim can be used as a single agent clinically for the oral treatment of uncomplicated urinary tract infections caused by susceptible bacteria (predominantly community-acquired *E. coli* and other Gram-negative rods). It is, however, most commonly used in a 1:5 fixed concentration ratio with the sulfonamide sulfamethoxazole (Bactrim, Septra). This combination is not only synergistic in vitro but is less likely to induce bacterial resistance than either agent alone. It is rationalised that microorganisms not completely inhibited by sulfamethoxazole at the pteroate condensation step will not likely be able to push the lessened amount of substrates that leak through past a subsequent blockade of dihydrofolate reductase. Thus these agents block sequentially at two different steps in the same essential pathway, and this combination is extremely difficult for a naive microorganism to survive. It is also comparatively uncommon that a microorganism will successfully mutate to resist both enzymes during therapy. However, if the organism is already resistant to either drug at the outset of therapy, much of the advantage of the combination is lost.

Pairing these two particular antibacterial agents was based on pharmacokinetic factors. For such a combination to be useful in vivo, the two agents must arrive at the necessary tissue compartment where the infection is at the correct time and in the proper ratio. In this context, the optimum ratio of these two agents in vitro is 1:20. Administration of the 1:5 combinations of the two drugs orally produces the desired 1:20 ratio in the body once a steady state is reached. The combination is used for oral treatment of urinary tract infections, shigellosis, otitis media, traveler's diarrhoea, community-acquired methicillin-resistant *Staphylococcus aureus* (MRSA), acute exacerbations of chronic bronchitis, and pneumocystis pneumonia. The pneumonia-causing fungus *Pneumocystis jiroveci* (previously classified as *Pneumocystis carinii*) is an opportunistic pathogen for immunocompromised individuals.

Side effects:

The most frequent side effects of trimethoprim-sulfamethoxazole are rash, nausea, and vomiting. Blood dyscrasias are less common, as is pseudomembranous colitis (caused by non–antibiotic-sensitive opportunistic gut anaerobes, often *Clostridium difficile*). Despite a significant effort, no structurally related analogue has emerged to compete with trimethoprim.

Resistance:

Bacterial resistance to trimethoprim is increasingly common. In pneumococcal infections, it can result from a single amino acid mutation (Ile-100 to Leu) in the dihydrofolate reductase enzyme. Overexpression of dihydrofolate reductase by *Staphylococcus aureus* has also been reported in resistant strains.

Storage conditions: Store the reconstituted solution at 20° to 25°C (68° to 77°F). Limited storage periods at 15° to 30°C (59° to 86°F) are acceptable.

Dose: 480 mg

Pharmaceutical Formulations: Solution, Suspension, Drops

Brand Names: Co-Trimoxazole Tablets 480mg, Bactrim, Intrim DS, Bactrim Suspension, Larprim Drops, Septram P Susp, Colizol DS, Colizol Susp.

11.6.6.7 Sulfones

The **sulfones** are primarily of interest as antibacterial agents, though there are some reports of their use in the treatment of malarial and rickettsial infections. They are less effective than sulfonamides. PABA partially antagonises the action of many of the sulfones, suggesting that the mechanism of action is similar to that of the sulfonamides. Further, infections that arise in patients being treated with sulfones are cross-resistant to sulfonamides. Several sulfones have proved useful in the treatment of leprosy, but among them, only dapsone is clinically used today.

It has been estimated that there are about 11 million cases of leprosy in the world, of which about 60% are in Asia (with 3.5 million in India alone). The first reports of dapsone resistance prompted the use of multidrug therapy with dapsone, rifampin, and clofazimine combinations in some geographic areas. The search for antileprotic drugs has been hampered by the inability to cultivate *M. leprae* in artificial media and by the lack of experimental animals susceptible to human leprosy.

A method of isolating and growing *M. leprae* in the footpads of mice and in armadillos has been reported and has permitted a much more comprehensive range of research.

Sulfones were introduced into the treatment of leprosy after it was found that sodium glucosulfone was effective in experimental tuberculosis in guinea pigs. The parent sulfone, dapsone (4,4-sulfonyldianiline), is the prototype for various analogues that have been widely studied.

11.6.6.7.1 Dapsone

Structure:

$$H_2N-\underset{\displaystyle O}{\overset{\displaystyle O}{\underset{\|}{\overset{\|}{S}}}}-NH_2$$

Chemical (IUPAC) name: Dapsone (4,4-sulfonylbisbenzeneamine or 4,4-sulfonyldianiline or *p,p*-diaminodiphenylsulfone

It occurs as an odourless, white crystalline powder that is very slightly soluble in water and sparingly soluble in alcohol. The pure compound is light stable, but traces of impurities, including water, make it photosensitive and thus susceptible to discolouration in light. Although no chemical change is detectable following discolouration, the drug should be protected from light.

Uses:

Dapsone is used in the treatment of both lepromatous and tuberculoid types of leprosy.

Dapsone is used widely for all forms of leprosy, often in combination with clofazimine and rifampin. Initial treatment often includes rifampin with dapsone, followed by dapsone alone. It is also used to prevent the occurrence of multibacillary leprosy when given prophylactically. Dapsone is also the drug of choice for dermatitis herpetiformis and is sometimes used with pyrimethamine for the treatment of malaria and with trimethoprim for PCP.

Side effects:

Serious side effects can include hemolytic anaemia, methemoglobinemia, and toxic hepatic effects. Hemolytic effects can be pronounced in patients with glucose-6-phosphate dehydrogenase deficiency. During therapy, all patients require frequent blood counts.

Storage conditions: The drug should be kept at room temperature between 68°F and 77°F (20°C and 25°C).

Dose: The dose is 25 mg, 100mg- tablet.

Pharmaceutical Formulations: Tablet, Gel

Brand Names: Dapsone 25, Aczone Gel, acnesone Gel.

QUESTION BANK

A. MULTIPLE CHOICE QUESTIONS

1. Combination of sulfonamides with trimethoprim-
   A. Decreases the unwanted effects of sulfonamide
   B. Increases the antimicrobial activity
   C. Decreases the antimicrobial activity
   D. Increases the elimination of sulfonamides

   Answer: B

2. Trimethoprim acts by _______ in bacteria.
   A. Inhibiting DHFRase
   B. Inhibiting protein
   C. Inhibiting cell wall
   D. Inhibiting cell membrane

   Answer: A

3. The following measures are necessary for the prevention of sulfonamide precipitation and crystalluria-
   A. Taking of drinks with acid pH
   B. Taking of drinks with alkaline pH
   C. Taking of saline drinks
   D. Restriction of drinking

   Answer: B

4. Which of the following sulfonamide is used topically?
   A. Silver sulfadiazine
   B. Sulphadoxine
   C. Sulphamethoxazole
   D. Dapsone

   Answer: A

5. Sulphamethoxazole is combined with-
   A. Clotrimazole
   B. Trimethoprim
   C. Cotrimoxazole
   D. Nitrofurantoin

   Answer: B

6. Sulphonamide injection causes a decrease in folic acid by-
    A. Competitive inhibition
    B. Noncompetitive inhibition
    C. Uncompetitive inhibition
    D. Allosteric inhibition

    Answer: A

7. Mechanism of action of sulphonamide antibacterial effect is-
    A. Inhibition of dihydropteroate reductase
    B. Inhibition of dihydropteroate synthase
    C. Inhibition of cyclooxygenase
    D. Activation of DNA gyrase

    Answer: B

8. Bactrim is brand name for which of the following sulphonamide
    A. Sulfamethoxaole
    B. Sulfanilamide
    C. Sulfadiazine
    D. Cotrimoxazole

    Answer: B

9. Which of the following is short-acting sulphonamide?
    A. Sulphadiazine
    B. Sulphamoxazole,
    C. Sulphamethoxazole
    D. Sulphadoxine

    Answer: A

10. Which of the following is long-acting sulphonamide?
    A. Sulphadiazine
    B. Sulphamoxazole,
    C. Sulphamethoxazole
    D. Sulphadoxine

    Answer: D

B. SHORT ANSWER QUESTIONS

1. Classify Sulphonamide.
2. Enlist different brand names for Sulfanilamide
3. Give the uses and different pharmaceutical formulations available for Sulfadiazine.

4. Enlist different brand names for Sulfacetamide.
5. Give the uses and different pharmaceutical formulations available for Cotrimoxazole.
6. Give the uses and different pharmaceutical formulations available for Dapsone.
7. Enlist different brand names for Dapsone.
8. Give storage conditions for Trimethoprim.
9. Explain in short Mafenide Acetate.
10. Explain the mechanism of action of Sulphonamide.

C. LONG ANSWER QUESTIONS

1. Discuss SAR, the Mode of action of Sulphonamides.
2. What are sulphonamides? Classify it. Explain in detail Sulfacetamide.
3. What are sulphonamides? Classify it. Write a note on Cotrimoxazole.
4. Classify Sulphonamide On the basis of therapeutic uses.
5. What are sulphonamides? Classify it. Explain its therapeutic applications of it.

■■■

Chapter 12

ANTIBIOTICS

CONTENTS

Antibiotics:
Penicillin G, Amoxicillin*, Cloxacillin, Streptomycin,
Tetracyclines:
Doxycycline, Minocycline,
Macrolides:
Erythromycin, Azithromycin,
Miscellaneous:
Chloramphenicol* Clindamycin *

♦ LEARNING OBJECTIVES ♦

After completing this chapter, the student should be able to understand:

1. Classification of Antibiotics.
2. Chemical names for different antibiotics.
3. Structures of different Antibiotics, tetracyclines, and macrolides.
4. Uses, stability and storage conditions for different Antibiotics, tetracyclines, and macrolides.
5. Different types of formulations and popular brand names of different Antibiotics, tetracyclines, and macrolides.

12.1 INTRODUCTION

Since Alexander Fleming accidentally discovered penicillin in 1929, the number of antibiotics that have been added to our therapeutic armamentarium has grown tremendously. Along with immunising biologicals, antibiotics have turned the tide in terms of the treatment of infectious diseases. They are truly medical miracles. Yet, because of the overuse of many of these agents and the biochemical fickleness of many bacteria, resistance to antibiotics has become a severe problem in the 21st century. Indeed, there are now organisms that cannot be arrested or killed by any common antibiotics.

Clearly, new approaches are needed. This chapter will present each of the antibiotics as chemical classes, such as tetracyclines, aminoglycosides, macrolides, and _-lactam antibiotics. In each class, it is important to note the different ranges of bacteria that each will treat. Notably, it is difficult to define an absolute range for any antibiotic. Populations of bacteria differ in their characteristics among regions and hospitals in those regions. So, sometimes, an antibiotic is listed as effective against a particular bacterial strain, but the concentration needed to kill the organism is too high. The bottom line is that when describing the efficacy of an antibiotic against a particular organism, it is necessary to look at clinical results. This chapter will describe ranges of activity in a general sense. The history of antibiotics is an interesting one and is well worth considering. Hence, the chapter will begin with a brief historical overview.

12.2 HISTORICAL BACKGROUND

Sir Alexander Fleming's accidental discovery of the antibacterial properties of penicillin in 1929[1] is largely credited with initiating the modern antibiotic era. Not until 1938, however, when Florey and Chain introduced penicillin into therapy, did practical medical exploitation of this important discovery begin to be realised. Centuries earlier, humans had learned to use crude preparations empirically for the topical treatment of infections, which we now assume to be effective because of the antibiotic substances present. As early as 500 to 600 BC, moulded curd of soybean was used in Chinese folk medicine to treat boils and carbuncles. Mouldy cheese had also been used for centuries by Chinese and Ukrainian peasants to treat infected wounds. The discovery by Pasteur and Joubert in 1877 that anthrax bacilli were killed when grown in culture in the presence of certain bacteria, along with similar observations by other microbiologists, led Vuillemin[2] to define antibiosis (literally "against life") as the biological concept of survival of the fittest, in which one organism destroys another to preserve itself. The word antibiotic was derived from this root. The use of the term by the lay public, as well as the medical and scientific communities, has become so widespread that its original meaning has become obscured. In 1942, Waksman[3] proposed the widely cited definition that "an antibiotic or antibiotic substance is a substance produced by microorganisms, which has the capacity of inhibiting the growth and even of destroying other microorganisms." Later proposals[4–6] have sought both to expand and to restrict the definition to include any substance produced by a living organism that is capable of inhibiting the growth or survival of one or more species of microorganisms in low concentrations. The advances made by medicinal chemists to modify naturally occurring antibiotics and prepare synthetic analogues necessitated the inclusion of semisynthetic and synthetic derivatives in the definition. Therefore, a substance is classified as an antibiotic if the following conditions are met:

1. It is a product of metabolism (although it may be duplicated or even have been anticipated by chemical synthesis).

2. It is a synthetic product produced as a structural analogue of a naturally occurring antibiotic.

3. It antagonises the growth or survival of one or more species of microorganisms.

4. It is effective in low concentrations.

The isolation of the antibacterial antibiotic tyrocidin from the soil bacterium Bacillus brevis by Dubois suggested the probable existence of many antibiotic substances in nature. It provided the impetus for the search for them. An organised search of the order Actinomycetales led Waksman and associates to isolate streptomycin from *Streptomyces griseus*. The discovery that this antibiotic possessed in vivo activity against Mycobacterium tuberculosis in addition to numerous species of Gram-negative bacilli was electrifying. It was now evident that soil microorganisms would provide a rich source of antibiotics. Broad screening programs were instituted to find antibiotics that might be effective in the treatment of infections hitherto resistant to existing chemotherapeutic agents, as well as to provide safer and more effective chemotherapy. The discoveries of broad-spectrum antibacterial antibiotics such as chloramphenicol and the tetracyclines, antifungal antibiotics such as nystatin and griseofulvin, and the ever-increasing number of antibiotics that may be used to treat infectious agents that have developed resistance to some of the older antibiotics, attest to the spectacular success of this approach as it has been applied in research programs throughout the world.

12.3 CLASSIFICATION OF ANTIBIOTICS

I] **Depending upon clinical effectiveness, the spectrum of activity and degree of selectivity,** those inhibiting only one group of microorganisms are called '**narrow spectrum antibiotics**

e.g. Nystatin and bacitracin.

These antibiotics exhibit a high degree of selectivity. A few antibiotics exhibit a high degree of selectivity.

A few antibiotics that inhibit both gram-positive and gram-negative bacteria and/or other intracellular organism may be termed broad-spectrum antibiotics, e.g. Chloramphenicol and tetracycline.

II] **Depending upon the source from which antibiotics are obtained, they can be classified as follows:**

A. **Natural:** these antibiotics are obtained from the large-scale fermentation of microorganisms. e.g. Bacitracin and polymixin are obtained from some bacilli while streptomycin, tetracycline etc., from streptomyces species.

B. **Semi-synthetic:** The observation that 6-aminopenicillanic acid can be obtained from cultures of *P. chrysogenum* that were depleted of side chain precursors led to the development of this class. For example, during the commercial production of benzylpenicillin (penicillin G), phenylacetic acid is added to the medium in order to achieve the predominance of the product.

C. **Synthetic:** this class includes antibiotics which are having purely synthetic origin. For example, chloramphenicol, a broad spectrum antibiotic initially isolated from a fermented media in 1947 and later was produced synthetically on a commercial basis.

III] The third basis of classification involves the difference in mechanism of action; accordingly, these agents can be divided as:

A. **Drugs that interfere with the biosynthesis of bacterial cell-wall**

   e.g. Penicillins, cephalosporins, cycloserine, Bacitracin and Vancomycin

B. **Drugs that interfere in the functioning cytoplasmic membrane**

   e.g. Polymixins, amphotericin B and Nystatin.

C. **Drugs that interfere with the protein biosynthesis**

   e.g. Erythromycin, lincomycins, Tetracyclines and chloramphenicol.

D. **Drugs that interfere with the nucleic acid biosynthesis**

   e.g. Actinimycin, Griseofulvin and Rifampin.

IV] Antibiotics can be, in general, classified as:

A. **β- lactam antibiotics**
B. **Aminoglycosides antibiotics**
C. **Tetracycline antibiotics**
D. **Peptide antibiotics**
E. **Macrolide antibiotics**
F. **Lincomycins antibiotics**
G. **Unclassified antibiotics**

12.4 MECHANISM OF ACTION

The manner in which antibiotics exert their actions against susceptible organisms varies. The mechanisms of action of some of the more common antibiotics are summarised in the following table.

Site of action	Antibiotics	Process Interrupted	Type of activity
Cell wall	Bacitracin	Mucopeptide synthesis	Bactericidal
	Cephalosporin	Cell wall cross linking	Bactericidal
	Cycloserine	Synthesis of cell wall peptides	Bactericidal
	Penicillines	Cell wall cross linking	Bactericidal
	Vancomycin	Mucopeptide synthesis	Bactericidal
Cell membrane	Amphotericin B	Membrane Function	Fungicidal
	Nystatin	Membrane Function	Fungicidal
	Polymixins	Membrane Integrity	Bactericidal

Table *Contd...*

Site of action	Antibiotics	Process Interrupted	Type of activity
Ribosomes 50S subunit	Chloramphenicol	Protein Synthesis	Bacteriostatic
	Erythromycin	Protein Synthesis	Bacteriostatic
	Lincomycin	Protein Synthesis	Bacteriostatic
Ribosomes 30S subunit	Aminoglycosides	Protein Synthesis and fidelity	Bactericidal
	Tetracyclines	Protein Synthesis	Bacteriostatic
Nucleic acids	Actinomycin	DNA and mRNA synthesis	Pancidal
	Griseofulvin	Cell division, microtubule assembly	Fungistatic
DNA and/or mRNA	Mitomycin C	DNA synthesis	Pancidal
	Rifampin	mRNA Synthesis	Bactericidal

In many instances, the mechanism of action is not fully known; for a few (e.g., penicillins), the site of action is known, but precise details of the mechanism are still under investigation. The biochemical processes of microorganisms are a lively subject for research, for an understanding of those mechanisms that are peculiar to the metabolic systems of infectious organisms is the basis for the future development of modern chemotherapeutic agents.

Antibiotics that **interfere with the metabolic systems found in microorganisms** and not in mammalian cells are the most successful anti-infective agents.

For example, antibiotics that **interfere with the synthesis of bacterial cell walls** have a high potential for selective toxicity.

Some antibiotics structurally resemble some essential metabolites of microorganisms, which suggests that **competitive antagonism** may be the mechanism by which they exert their effects.

Thus, cycloserine is believed to be an anti-metabolite for D-alanine, a constituent of bacterial cell walls.

Many antibiotics selectively interfere with **microbial protein synthesis** (e.g., the aminoglycosides, tetracyclines, macrolides, chloramphenicol, and lincomycin) or **nucleic acid synthesis** (e.g., rifampin).

Others, such as polymyxins and polyenes, are believed to **interfere with the integrity and function of microbial cell membranes.**

The mechanism of action of an antibiotic determines, in general, whether the agent exerts a bactericidal or bacteriostatic action. The distinction may be necessary for the treatment of severe, life-threatening infections, particularly if the natural defence mechanisms of the host are either deficient or overwhelmed by the infection.

In such situations, a bactericidal agent is obviously indicated. Much work remains to be done in this area, and as mechanisms of action are revealed, the development of improved structural analogues of effective antibiotics probably will continue to increase.

12.5 BETA- LACTAM ANTIBIOTICS

Antibiotics that possess the β -lactam (a four-membered cyclic amide) ring structure are the dominant class of agents currently used for the chemotherapy of bacterial infections.

β -lactam

The first antibiotic to be used in therapy, penicillin (penicillin G or benzylpenicillin), and a close biosynthetic relative, phenoxymethylpenicillin (penicillin V), remain the agents of choice for the treatment of infections caused by most species of Gram-positive bacteria.

The discovery of a second major group of β-lactam antibiotics, the cephalosporins, and chemical modifications of naturally occurring penicillins and cephalosporins have provided semisynthetic derivatives that are variously effective against bacterial species known to be resistant to penicillin, in particular, penicillinase-producing staphylococci and Gram-negative bacilli.

Thus, apart from a few strains that have either inherent or acquired resistance, almost all bacterial species are sensitive to one or more of the available β -lactum antibiotics.

12.5.1 Penicillin G

Structure:

Chemical (IUPAC) Name: (2S,5R,6R)-3,3-Dimethyl-7-oxo-6-(2-phenylacetamido)-4-thia-1-azabicyclo[3.2.0]heptane-2-carboxylic acid

For years, the most popular penicillin has been penicillin G or benzylpenicillin. In fact, with the exception of patients allergic to it, penicillin G remains the agent of choice for treating more different kinds of bacterial infections than any other antibiotic.

It was first made available as the water-soluble salts of potassium, sodium, and calcium. These salts of penicillin are inactivated by the gastric juice and are ineffective when administered orally unless antacids, such as calcium carbonate, aluminium hydroxide, and magnesium trisilicate, or a substantial buffer, such as sodium citrate, are added.

Also, because penicillin is absorbed poorly from the intestinal tract, oral doses must be very large, about five times the amount necessary with parenteral administration.

Only after the production of penicillin had increased enough to make low-priced penicillin available did the oral dosage forms become popular. The water-soluble potassium and sodium salts are used orally and parenterally to achieve high plasma concentrations of penicillin G rapidly.

The more water-soluble potassium salt usually is preferred when large doses are required.

Situations in which hyperkalemia is a danger, however, as in renal failure, require the use of the sodium salt; potassium salt is preferred for patients on salt-free diets or with congestive heart conditions.

The rapid elimination of penicillin from the bloodstream through the kidneys by active tubular secretion and the need to maintain an effective concentration in blood has led to the development of "repository" forms of this drug.

Suspensions of penicillin in peanut oil or sesame oil with white beeswax added were first used to prolong the duration of injected forms of penicillin. This dosage form was replaced by a suspension in vegetable oil, to which aluminium monostearate or aluminium distearate was added. Today, most repository forms are suspensions of high molecular weight amine salts of penicillin in a similar base.

Storage condition: Store in a refrigerator, 2° to 8° (36° to 46°F).

Dose: Oral tablet- Strengths: Pediatric patients-900,000 units to 1,200,000 units, Adult patients-2,400,000 units.

Pharmaceutical Formulations: Injection.

Brand Names: Bicillin C-R, Bicillin C-R 900/300

12.5.2 Amoxicillin

Structure:

Chemical (IUPAC) Name:, 6-[D-(α)--amino-*p*- *hydroxyphenylacetamido]* penicillanic acid.

It is semisynthetic penicillin introduced in 1974 and is simply the *p-hydroxy* analogue of ampicillin, prepared by acylation of 6-APA with *phydroxyphenylglycine.*

Early clinical reports indicated that orally administered amoxicillin possesses significant advantages over ampicillin, including more complete GI absorption to give higher plasma and urine levels, less diarrhoea, and little or no effect of food on absorption. Thus, amoxicillin has largely replaced ampicillin for the treatment of specific systemic and urinary tract infections for which oral administration is desirable.

Amoxicillin is a fine, white to, off-white, crystalline powder that is sparingly soluble in water. It is available in various oral dosage forms.

Aqueous suspensions are stable for 1 week at room temperature.

Storage Condition: Store the capsules and tablets at room temperature and away from excess heat and moisture. The liquid medication should preferably be kept in the refrigerator, but it may be stored at room temperature. Do not freeze. Dispose of any unused liquid medication after 14 days.

Dose: Oral Solution-50 mg/5mL, 125 mg/5mL, 200 mg/5mL, 250 mg/5mL, 400 mg/5mL
    Capsule- 250 mg, 500 mg
    Tablet- 500 mg, 875 mg
    Chewable Tablet- 125 mg, 250 mg
    Extended-release (Moxatag)- 775 mg

Pharmaceutical Formulations: Tablet, Capsule, Oral Solution

Brand Names: Amoxil, Moxatag, and Trimox.

12.5.3 Cloxacillin

Structure:

Chemical (IUPAC) Name: (2S,5R,6R)- 6- ({[3- (2- chlorophenyl)- 5- methylisoxazol- 4- yl] carbonyl}amino)- 3,3- dimethyl- 7- oxo- 4- thia- 1- azabicyclo[3.2.0]heptane- 2- carboxylic acid

A semi-synthetic penicillin antibiotic which is a chlorinated derivative of oxacillin.

Mechanism of Action

By binding to specific penicillin-binding proteins (PBPs) located inside the bacterial cell wall, cloxacillin inhibits the third and last stage of bacterial cell wall synthesis. Cell lysis is then mediated by bacterial cell wall autolytic enzymes such as autolysins; it is possible that cloxacillin interferes with an autolysin inhibitor.

Storage condition: The capsules should be stored at room temperature below 25°C. The reconstituted solutions should be refrigerated.

Dose: Strengths: injection- 500mg/500mg/5ml, 62.5mg, 62.5mg 125mg, 250mg 500mg.

Pharmaceutical Formulations: Tablet Injection, Dry Syrup, Capsule

Brand Names: Duoclox, Numox, Baxin, Duoclox

12.6 AMINOGLYCOSIDE ANTIBIOTICS

12.6.1 Streptomycin sulfate

Structure:

NH
HN   NH$_2$
HO   OH
NH
HO
H
N   NH$_2$
H
O
O
O
OH
O   H
H
N
HO
OH
OH
1.5H$_2$SO$_4$

Chemical (IUPAC) Name: 2-[(1R,2R,3S,4R,5R,6S)-3-(diaminomethylideneamino)-4-[(2R,3R, 4R,5S)-3-[(2S,3S,4S,5R,6S)-4,5-dihydroxy-6-(hydroxymethyl)-3-(methylamino)oxan-2-yl]oxy-

4-formyl-4-hydroxy-5-methyloxolan-2-yl]oxy-2,5,6-trihydroxycyclohexyl]guanidine;sulfuric acid

Streptomycin sulfate is a white, odorless powder that is hygroscopic but stable toward light and air.

It is freely soluble in water, forming solutions that are slightly acidic or nearly neutral.

It is very slightly soluble in alcohol and is insoluble in most other organic solvents.

Aqueous solutions may be stored at room temperature for 1 week without any loss of potency, but they are most stable if the pH is between 4.5 and 7.0.

The solutions decompose if sterilised by heating, so sterile solutions are prepared by adding sterile distilled water to the sterile powder.

Clinically, a problem that sometimes occurs with the use of streptomycin is the early development of resistant strains of bacteria, necessitating a change in therapy. Other factors that limit the therapeutic use of streptomycin are chronic toxicities. Neurotoxic reactions have been observed after the use of streptomycin.

These are characterised by vertigo, disturbance of equilibrium, and diminished auditory perception.

Additionally, nephrotoxicity occurs with some frequency. Patients undergoing therapy with streptomycin should have frequent checks of renal monitoring parameters. Chronic toxicity reactions may or may not be reversible.

Minor toxic effects include rashes, mild malaise, muscular pains, and drug fever.

Storage condition: Store in-between 2 to 8° C.

Dose: Powder for injection- 1g

Pharmaceutical Formulations: Injection

Brand Names: Isos, Ambistryn S, Cipstryn, Merstrep.

12.7 TETRACYCLINE

Chemistry:

Among the most important broad-spectrum antibiotics are members of the tetracycline family.

Nine such compounds-tetracycline, rolitetracycline, oxytetracycline, chlortetracycline, demeclocycline, meclocycline, methacycline, doxycycline, and minocycline have been introduced into medical use.

Several others possess antibiotic activity. The tetracyclines are obtained by fermentation procedures from *Streptomyces* spp. or by chemical transformations of the natural products.

Their chemical identities have been established by degradation studies and confirmed by the synthesis of three group members, oxytetracycline, 6-demethyl-6-deoxytetracycline, and anhydrochlortetracycline, in their (alpha) forms. The essential members of the group are derivatives of an octahydronaphthacene, a hydrocarbon system that comprises four annulated six-membered rings. The group name is derived from this tetracyclic system. The antibiotic spectra and chemical properties of these compounds are very similar but not identical. The stereochemistry of the tetracyclines is very complex.

Carbon atoms 4, 4a, 5, 5a, 6, and 12a are potentially chiral, depending on substitution. Oxytetracycline and doxycycline, each with a 5-hydroxyl substituent, have six asymmetric centres; the others, lacking chirality at C-5, have only five.

Structure:

Tetracycline is amphoteric in nature. Its acid salts like hydrochloride and sulfosalicycylates are water soluble, and essential salts like NaOH and KOH are not stable. It forms complexes with divalent poly metals, which are water insoluble. At intermediate pH-epimerization at C-4 occurs.

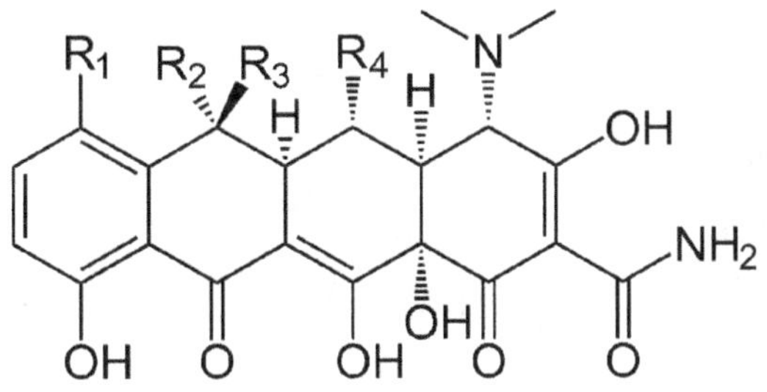

12.7.1 Doxycyclinec

Structure:

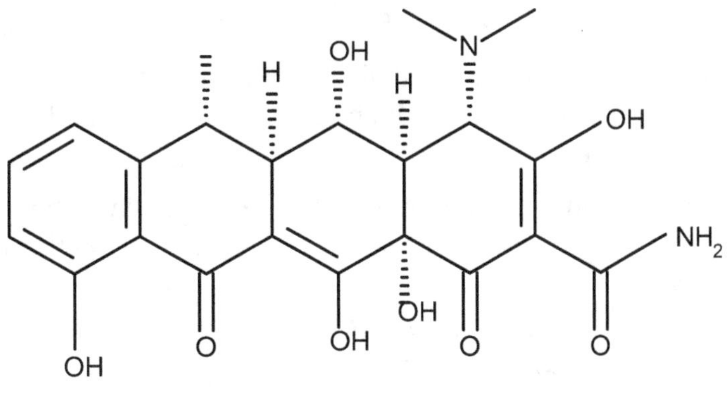

Chemical (IUPAC) Name: (4S,4aR,5S,5aR,6R,12aS)-4-(Dimethylamino)-3,5,10,12,12a-penta-hydroxy-6-methyl-1,11-dioxo-1,4,4a,5,5a,6,11,12a-octahydrotetracene-2-carboxamide

It was obtained first in small yields by a chemical transformation of oxytetracycline, but it is now produced by catalytic hydrogenation of methacycline or by reduction of a benzyl mercaptan derivative of methacycline with Raney nickel.

The latter process produces a nearly pure form of the 6 α -methyl epimer. The 6 α - methyl epimer is more than 3 times as active as its β- epimer.

Apparently, the difference in orientation of the methyl groups, which slightly affects the shapes of the molecules, causes a substantial difference in biological effect. Also, the absence of the 6-hydroxyl group produces a compound that is very stable in acids and bases and that has a long biological half-life.

In addition, it is absorbed very well from the GI tract, thus allowing a smaller dose to be administered. Doxycycline is available as a hydrated salt, a hydrochloride salt solvated as the hemiethanolate hemihydrate, and a monohydrate. The hydrate form is sparingly soluble in water and is used in a capsule; the monohydrate is water insoluble and is used for aqueous suspensions, which are stable for up to 2 weeks when kept in a cool place.

Storage Condition: Tablets, capsules, and syrup should be kept at room temperature 15^0C to 30^0C (59^0F to 86^0F) in tight, light-resistant containers. Powder for injection should be stored at or below 25^0C (77^0F) and protected from light.

Dose: Strengths: 100 mg, 200 mg

Pharmaceutical Formulations: Tablet, Capsule.

Brand Names: Vibramycin, Doryx, Doryx MPC, Oracea, Acticlate, Atridox, Doxy 100, and Doxy 200

12.7.2 Minocycline

Structure:

Chemical (IUPAC) Name: (2E,4S,4aR,5aS,12aR)-2-(Amino-hydroxy-methylidene)-4,7-bis (dimethylamino)-10,11,12a-trihydroxy-4a,5,5a,6- tetrahydro-4H-tetracene-1,3,12-trione

Minocycline, 7-dimethylamino-6-dimethyl-6-deoxytetracycline, the most potent tetra-cycline currently used in therapy, is obtained by reductive methylation of 7-nitro-6-dimethyl-6-deoxytetracycline.

Minocycline, 7-dimethylamino-6-dimethyl-6-deoxytetracycline, the most potent tetra-cycline currently used in therapy, is obtained by reductive methylation of 7-nitro-6-dimethyl-6-deoxytetracycline.

Minocycline has been recommended for the treatment of chronic bronchitis and other upper respiratory tract infections.

Despite its relatively low renal clearance, partially compensated for by high serum and tissue levels, it has been recommended for the treatment of urinary tract infections.

It has been effective in the eradication of N. meningitidis in asymptomatic carriers.

Dose: Store it at room temperature and away from excess heat and moisture—store minocycline pellet-filled capsules and extended-release tablets away from light.

Strengths: 50 mg, 45 mg, 65 mg, 100 mg

Pharmaceutical Formulations: Tablet, capsule

Brand Names: Minoz ER 45 Tablet, Minoz OD 100 Capsule MR, Minoz 100 Tablet, Minoz 50 Tablet, Minoz ER 65 Tablet, Cynomycin 50 Capsule, Cynomycin 100 Capsule

12.8 MACROLIDE ANTIBIOTICS

Among the many antibiotics isolated from the actinomycetes is the group of chemically related compounds called the *macrolides*. In 1950, picromycin, the first of this group to be identified as a macrolide compound, was first reported. In 1952, erythromycin and carbomycin were reported as new antibiotics, and they were followed in subsequent years by other macrolides.

Currently, more than 40 such compounds are known, and new ones are likely to appear in the future. Of all these, only two, erythromycin and oleandomycin, have been available consistently for medical use in the United States.

In recent years, interest has shifted away from novel macrolides isolated from soil samples (e.g., spiramycin, josamycin, and rosamicin), all of which thus far have proved to be clinically inferior to erythromycin and semisynthetic derivatives of erythromycin (e.g., clarithromycin and azithromycin), which have superior pharmacokinetic properties because of their enhanced acid stability and improved distribution properties.

Chemistry:

The macrolide antibiotics have three common chemical characteristics:

1. a large lactone ring (which prompted the name *macrolide*),
2. a ketone group, and
3. a glycosidically linked amino sugar.

Usually, the lactone ring has 12, 14, or 16 atoms in it, and it is often unsaturated, with an olefinic group conjugated with the ketone function. (The polyene macrocyclic lactones, such as natamycin and amphotericin B; the ansamycins, such as rifampin; and the polypeptide lactones generally are not included among the macrolide antibiotics.)

They may have, in addition to the amino sugar, a neutral sugar that is linked glycosidically to the lactone ring. Because of the dimethyl amino group on the sugar moiety, the macrolides are bases that form salts with pKa values between 6.0 and 9.0.

This feature has been used to make clinically useful salts.

The free bases are only slightly soluble in water but dissolve in somewhat polar organic solvents.

They are stable in aqueous solutions at or below room temperature but are inactivated by acids, bases, and heat.

Mechanism of action and resistance for macrolide antibiotics:

Some details of the mechanism of antibacterial action of erythromycin are known. It binds selectively to a specific site on the 50S ribosomal subunit to prevent the translocation step of bacterial protein synthesis.

It does not bind to mammalian ribosomes. Broadly based, nonspecific resistance to the antibacterial action of erythromycin among many species of Gram-negative bacilli appears to be largely related to the inability of the antibiotic to penetrate the cell walls of these organisms. In fact, the sensitivities of members of the Enterobacteriaceae family are pH-dependent, with MICs decreasing as a function of increasing pH.

Furthermore, protoplasts from Gram-negative bacilli, which lack cell walls, are sensitive to erythromycin. A highly specific resistance mechanism to the macrolide antibiotics occurs in erythromycin-resistant strains of S. *aureus*.

Such strains produce an enzyme that methylates a specific adenine residue at the erythromycin-binding site of the bacterial 50S ribosomal subunit. The methylated ribosomal RNA remains active in protein synthesis but no longer binds erythromycin. Bacterial resistance to the lincomycins also occurs by this mechanism.

12.8.1 Erythromycin

Structure:

Chemical (IUPAC) Name: (3R,4S,5S,6R,7R,9R,11R,12R,13S,14R)-6-{[(2S,3R,4S,6R)-4-(Dimethylamino)-3-hydroxy-6-methyloxan-2-yl]oxy}-14-ethyl-7,12,13-trihydroxy-4-{[(2R,4R,5S,6S)-5-hydroxy-4-methoxy-4,6-dimethyloxan-2-yl]oxy}-3,5,7,9,11,13-hexamethyl-1-oxacyclotetradecane-2,10-dione

Early in 1952, the isolation of erythromycin from *Streptomyces erythraeus*. It achieved rapid early acceptance as a well-tolerated antibiotic of value for the treatment of various upper respiratory and soft-tissue infections caused by Gram-positive bacteria.

It is also effective against many venereal diseases, including gonorrhoea and syphilis, and provides a valuable alternative for treating many infections in patients allergic to penicillins.

More recently, erythromycin was shown to be an effective therapy for Eaton agent pneumonia (*Mycoplasma pneumoniae*), venereal diseases caused by *Chlamydia*, bacterial enteritis caused by *Campylobacter jejuni*, and Legionnaires disease.

Erythromycin is a very bitter, white or yellow-white, crystalline powder. It is soluble in alcohol and the other common organic solvents but only slightly soluble in water.

Storage Condition: Commercially available erythromycin topical solutions and gels should be stored at 15- 30°C; exposure to heat or open flames should be avoided. The topical ointment should be stored at a temperature less than 27°C.

Dose: Strengths: 100 mg, 125 mg, 250 mg, 500 mg

Pharmaceutical Formulations: Drop, Tablet, Liquid.

Brand Names: Althrocin, Erythro, Eromycin KID Emcin, Emgel

12.8.2 Azithromycin

Structure:

Chemical (IUPAC) Name: (2R,3S,4R,5R,8R,10R,11R,12S,13S,14R)-2-ethyl-3,4,10-trihydroxy-3,5,6,8,10,12,14-heptamethyl-15-oxo-11-{[3,4,6-trideoxy-3-(dimethylamino)-β-D-xylo-hexopyranosyl] oxy}-1-oxa-6-azacyclopentadec-13-yl 2,6-dideoxy-3C-methyl-3-O-methyl-α-L-ribo-hexopyranoside

Azithromycin is a semisynthetic derivative of erythromycin, prepared by Beckman rearrangement of the corresponding 6-oxime, followed by N-methylation and reduction of the resulting ring-expanded lactam. It is a prototype of a series of nitrogen-containing, 15-membered ring macrolides known as azalides.

It is more active against Gram-negative bacteria and less active against Gram-positive bacteria than its close relatives.

The greater activity of azithromycin against *H. influenzae*, *M. catarrhalis*, and *M. pneumoniae*, coupled with its extended half-life, permits a 5-day dosing schedule for treating respiratory tract infections caused by these pathogens.

The clinical efficacy of azithromycin in the treatment of urogenital and other sexually transmitted infections caused by Chlamydia trachomatis, *N. gonorrhoeae*, *H. ducreyi*, and Ureaplasma urealyticum suggests that single-dose therapy with it for uncomplicated urethritis or cervicitis may have advantages over the use of other antibiotics.

Storage Condition: Store azithromycin tablets, suspension, and extended-release suspension at room temperature and away from excess heat and moisture (not in the bathroom). Do not refrigerate or freeze the extended-release suspension. Discard any azithromycin suspension that is left over after 10 days or is no longer needed.

Dose: Strengths: 200 mg, 500 mg

Pharmaceutical Formulations: Tablet, Capsule, Dry Syrup, Oral Solution

Brand Names: Azithral, Azee, Azicip, Azithral, Azee, Azithral XL

12.9 MISCELLANEOUS ANTIBIOTICS

12.9.1 Chloramphenicol

Structure:

Chemical (IUPAC) Name: 2,2-dichloro-N-[(1R,2R)-1,3-dihydroxy-1-(4-nitrophenyl)propan-2-yl]acetamide

The first of the widely used broad-spectrum antibiotics, chloramphenicol, was isolated by Ehrlich et al. in 1947. They obtained it from *Streptomyces venezuelae*; an organism found in a sample of soil collected in Venezuela. Since then, chloramphenicol has been isolated as a product of several organisms found in soil samples from widely separated places.

It was the first and still is the only therapeutically important antibiotic to be so produced in competition with microbiological processes. Diverse synthetic procedures have been developed for chloramphenicol. The commercial process generally used starts with p-nitro acetophenone.

Chloramphenicol is a white, crystalline compound that is very stable. It is very soluble in alcohol and other polar organic solvents but only slightly soluble in water. It has no odour but has a very bitter taste.

Chloramphenicol is very stable in the bulk state and in solid dosage forms. In solution, however, it slowly undergoes various hydrolytic and light-induced reactions.

Chloramphenicol is recommended specifically for the treatment of serious infections caused by strains of Gram-positive and Gram-negative bacteria that have developed resistance to penicillin G and ampicillin, such as *H. influenzae, Salmonella typhi, S. pneumonia, B. fragilis,* and *N. meningitidis.*

Because of its penetration into the central nervous system, chloramphenicol is a particularly important alternative therapy for meningitis.

It is not recommended for the treatment of urinary tract infections because 5% to 10% of the unconjugated form is excreted in the urine. Chloramphenicol is also used for the treatment of rickettsial infections, such as Rocky Mountain spotted fever. Because it is bitter, this antibiotic is administered orally either in capsules or as the palmitate ester.

Chloramphenicol palmitate is insoluble in water and may be suspended in aqueous vehicles for liquid dosage forms.

Chloramphenicol is administered parenterally as an aqueous suspension of wonderful crystals or as a solution of the sodium salt of the succinate ester of chloramphenicol.

Sterile chloramphenicol sodium succinate has been used to prepare aqueous solutions for intravenous injection.

Storage Condition: Chloramphenicol in eye drops will be most stable if stored at refrigerator temperature

Dose: Strengths: 250mg, 500 mg 1% w/w), 125mg/ 5mL

Pharmaceutical Formulations: Capsule, Eye Ointment, eye drops, Oral Suspension, Injections.

Brand Names: Paraxin, Paraxin, Kemnicol, Chloromycetin Aplicap, Chlorocol, Paraxin

12.9.2 Clindamycin

Structure:

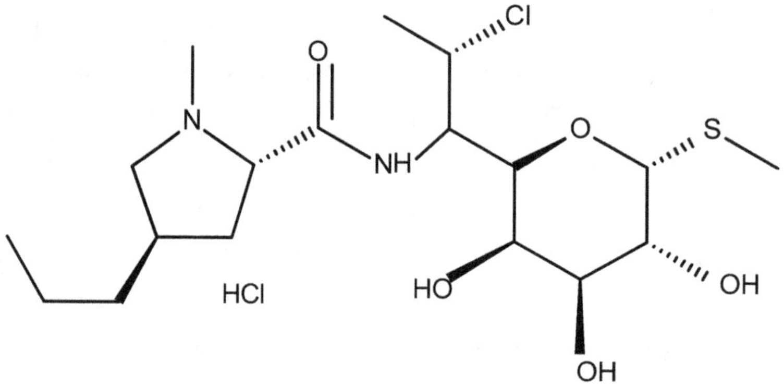

Chemical (IUPAC) Name:

In 1967, Magerlein et al. reported that replacing the 7(*R*)-hydroxy group of lincomycin with chlorine with inversion of configuration resulted in a compound with enhanced antibacterial activity in vitro.

Clinical experience with this semisynthetic derivative, clindamycin, 7(*S*)-chloro-7-deoxylincomycin (Cleocin), released in 1970, has established its superiority over lincomycin is even more significant in vivo.

Improved absorption and higher tissue levels of clindamycin and its greater penetration into bacteria have been attributed to a higher partition coefficient than that of lincomycin.

Structural modifications at C-7 (e.g., 7(*S*)-chloro and 7(*R*)-OCH_3) and of the C-4 alkyl groups of the hygric acid moiety appear to influence the activity of congeners more through an effect on the partition coefficient of the molecule than through a stereospecific binding role.

Changes in the α-thiolincosamide portion of the molecule seem to decrease activity markedly, however, as evidenced by the marginal activity of 2-deoxylincomycin, its anomer, and 2- *O*-methylincomycin.

Exceptions to this are fatty acid and phosphate esters of the 2-hydroxyl group of lincomycin and clindamycin, which are hydrolysed rapidly in vivo to the parent antibiotics. Clindamycin is recommended for the treatment of a wide variety of upper respiratory, skin, and tissue infections caused by susceptible bacteria. Its activity against streptococci, staphylococci, and pneumococci is indisputably high. It is one of the most potent agents available against some non–spore-forming anaerobic bacteria, the *Bacteroides* spp. in particular.

An increasing number of reports of clindamycin-associated GI toxicity, which range in severity from diarrhoea to an occasionally severe pseudomembranous colitis, have, however, caused some clinical experts to call for a reappraisal of the role of this antibiotic in therapy. Clindamycin- (or lincomycin)-associated colitis may be particularly dangerous in elderly or debilitated patients and has caused deaths in such individuals.

The colitis, which is usually reversible when the drug is discontinued, is now believed to result from an overgrowth of a clindamycin-resistant strain of the anaerobic intestinal bacterium *Clostridium difficile*. The intestinal lining is damaged by a glycoprotein endotoxin released by the lysis of this organism.

Storage Condition: Keep this medication in the container it came in, tightly closed, and out of reach of children. Store it at room temperature and away from excess heat and moisture. Do not refrigerate clindamycin liquid because it may thicken and become hard.

Dose: Strengths: 300 mg 1% w/w gel

Pharmaceutical Formulations: Capsule, Injection, Gel

Brand Names: Clid-Injection, Clid-Capsule, Clid Gel, Clindagel, Erytop Gel.

QUESTION BANK

A. MULTIPLE CHOICE QUESTIONS

1. Penicillin was accidentally discovered in
    A. 1922
    B. 1924
    C. 1921
    D. 1928

Answer: D

2. The antimicrobial activity of penicillin is due to ____________.
    A. Thiazolidine ring
    B. Beta-lactam ring
    C. 6-amino penicillanic acid
    D. None of the above

Answer: C

3. The penicillin G preparation with the longest duration of action is _________
    A. Benzathine penicillin
    B. Sodium penicillin

   C. Potassium penicillin

   D. Procaine penicillin

Answer: A

4. A drug which interferes with bacterial cell wall synthesis is?
   A. Chloramphenicol
   B. Tetracycline
   C. Colistin
   D. Penicillin & cephalosporin

Answer: D

5. The action of penicillin requires the presence of a cell wall that contains
   A. Proteoglycans
   B. Peptidoglycans
   C. n-acetyl glucosamine
   D. n-acetyl muramic acid

Answer: B

6. Which of the following is broad-spectrum penicillin?
   A. Oxacillin
   B. Methicillin
   C. Ampicillin
   D. Azlocillin

Answer: C

7. Erythromycin is an antibiotic. Belong to the class __________.
   A. Beta-lactam
   B. Aminoglycoside
   C. Macrolide
   D. Peptide

Answer: C

8. __________________ is an organism from which streptomycin is isolated.
   A. *Streptomyces griseus*
   B. *Streptomyces fradiae*
   C. *Streptomyces alboniger*
   D. *Streptomyces rimosus*

Answer: A

9. Which of the following is not present in macrolide?
   A. A large lactone ring
   B. A glycosidically linked amino sugar

    C. A spiroketal group

    D. A ketone group

Answer: B

10. Azee is the popular brand name for _______________.

    A. Doxycycline

    B. Minocycline

    C. Erythromycin

    D. Azithromycin

Answer: D

B. SHORT ANSWER QUESTIONS

1. Classify Antibiotics.
2. Write a note on β–lactam antibiotics.
3. Write a note on Streptomycin as Aminoglycoside antibiotics
4. Enlist different kinds of formulations available for Doxycycline along with brand names.
5. Write a note on Chloramphenicol.
6. Enlist different kinds of formulations available for Amoxicillin along with brand names.
7. Give Storage conditions, different kinds of formulations available for Clindamycin, and brand names.
8. Give the structure for Penicillin G.
9. Enlist different kinds of formulations available for Azithromycin along with brand names.
10. Write a note on Tetracycline as an antibiotic.

C. LONG ANSWER QUESTIONS

1. Define Antibiotics. Classify it based on Chemistry. Write a note on β–lactam antibiotics.
2. Define Antibiotics. Give classification, Give Storage conditions, different kinds of formulations available for Clindamycin, and brand names.
3. Define Antibiotics. Classify it. Add a note on Aminoglycosides as antibiotics.
4. Define Antibiotics. Classify it. Add a note on Tetracyclines.
5. Define Antibiotics. Classify it. Add a note on Chloramphenicol.

■■■

Chapter 13

ANTI-NEOPLASTIC AGENTS

CONTENTS

- Cyclophosphamide
- Busulfan
- Mercaptopurine
- Fluorouracil
- Methotrexate
- Dactinomycin
- Doxorubicin Hydrochloride
- Vinblastine Sulphate
- Cisplatin
- Dromostanolone Propionate

♦ LEARNING OBJECTIVES ♦

After completing this chapter, the student should be able to understand:

1. Classification of Anti-Neoplastic Agents.
2. Chemical names for different compounds used as Anti-Neoplastic Agents.
3. Structures of Anti-Neoplastic Agents.
4. Uses, stability and storage conditions Anti-Neoplastic Agents.
5. Different types of formulations and popular brand names for Anti-Neoplastic Agents.

13.1 INTRODUCTION

The American Cancer Society defines cancer as a group of diseases characterised by uncontrolled growth and the spread of abnormal cells that, left untreated, may lead to death. Related to this definition is the term neoplasia, which is the uncontrolled growth of new tissue, the product of which is known as a tumour, and these tumours may be either malignant or benign. Malignant tumours have the capability of invading surrounding tissues and moving to distant locations in the body in a process known as metastasis, characteristics that benign tumours do not possess.

Treatment of malignant tumours or cancer has generally involved initially surgical removal followed by radiation and/or chemotherapy, if necessary. In those cases where complete surgical removal is not feasible, radiation and chemotherapy become the only available options.

The term chemotherapy, in the strictest sense, refers to drugs that are used to kill cells and includes both antibiotics and agents used in the treatment of cancer, but it is often used to refer exclusively to anticancer agents, also known as antineoplastics.

Traditional chemotherapy has been based on the principle of selective toxicity; however, this has been difficult to achieve in the case of cancer cells because these cells utilise the biochemical pathways used by normal cells. In many cases, the agents have attempted to exploit the increased proliferative rates of cancer cells compared with normal cells. This has been difficult to achieve even in a relative sense because, in part, of the fact that not all normal tissue is slowly proliferative and, conversely, not all cancer cells are highly proliferative.

Increased knowledge of intercellular and intracellular communication has led to the development of several newer agents that have shown some effectiveness in treating several cancers, especially when used in combination with more traditional agents. These have included several monoclonal antibodies that target the overproduction of growth factor receptors and TK inhibitors that target the transduction process involved in growth factor stimulation.

13.2 CLASSIFICATION OF ANTINEOPLASTIC AGENTS

I. Cytotoxic drugs (directly act on cells)

    a) Alkylating agents

        i. Nitrogen mustards

        ii. Ethyleneimine

        iii. Alkyl sulfonate

        iv. Nitrosoureas

        v. Triazine

    b) Antimetabolites (act on the metabolic pathway involved in DNA synthesis)

        i. Folate antagonist

        ii. Purine antagonist

        iii. Pyrimidine antagonist

c) Plant derivatives
- i. Vinca alkaloids
- ii. Taxanes
- iii. Epipodophyllotoxin

d) Antibiotics

II. Hormones (mainly steroids which suppress hormone secretion or antagonise hormone action)
- **a) Glucocorticoids**
- **b) Estrogen**
- **c) Progestins**
- **d) Antiandrogens**

III. Miscellaneous (include Hydroxyurea, Cisplatin, Monoclonal antibodies and *L. Asparginase*)

13.3 SOME IMPORTANT ANTINEOPLASTIC AGENTS

13.3.1 Cyclophosphamide

Structure:

Chemical (IUPAC) name: (RS)-N,N-bis(2-chloroethyl)-1,3,2-oxazaphosphinan-2-amine 2-oxide

Cyclophosphamide (CP), also known as cytophosphane, among other names, is a medication used for chemotherapy and to suppress the immune system. As chemotherapy, it is used to treat lymphoma, multiple myeloma, leukaemia, ovarian cancer, breast cancer, small cell lung cancer, neuroblastoma, and sarcoma. As an immune suppressor, it is used in nephrotic syndrome, granulomatosis with polyangiitis, and following an organ transplant, among other conditions. It is taken by mouth or injection into a vein.

Uses:

Cyclophosphamide is used to treat cancers and autoimmune diseases. It is used to quickly control the disease. Due to its toxicity, it is replaced as soon as possible by less toxic drugs.

Regular and frequent laboratory evaluations are required to monitor kidney function, avoid drug-induced bladder complications and screen for bone marrow toxicity.

Cancer: The main use of cyclophosphamide is with other chemotherapy agents in treating lymphomas, some forms of brain cancer, neuroblastoma, leukaemia and some solid tumours.

Autoimmune diseases: Cyclophosphamide. Although concerns about toxicity restrict its use in patients with severe disease, it remains an essential treatment for life-threatening autoimmune diseases where disease-modifying antirheumatic drugs (DMARDs) have been ineffective. For example, systemic lupus erythematosus with severe lupus nephritis may respond to pulsed cyclophosphamide. Cyclophosphamide is also used to treat minimal change disease, severe rheumatoid arthritis, granulomatosis with polyangiitis, Goodpasture syndrome and multiple sclerosis.

AL amyloidosis: Cyclophosphamide, used in combination with or, has documented efficacy as an off-label treatment of AL amyloidosis. It appears to be an alternative to the more traditional treatment with melphalan in people who are ill-suited for autologous stem cell transplants.

Graft-versus-host disease: Graft-versus-host disease (GVHD) is a major barrier to allogeneic stem cell transplant because of the immune reactions of donor cells against the person receiving them. GVHD can often be avoided by T-cell depletion of the graft. The use of a high-dose cyclophosphamide post-transplant in a half-matched or haploidentical donor hematopoietic stem cell transplantation reduces GVHD, even after using a reduced conditioning regimen.

Side Effects: Common side effects include low white blood cell counts, loss of appetite, vomiting, hair loss, and bleeding from the bladder. Other severe side effects include an increased future risk of cancer, infertility, allergic reactions, and pulmonary fibrosis. Cyclophosphamide is in the alkylating agent and nitrogen mustard family of medications. It is believed to work by interfering with the duplication of DNA and the creation of RNA.

Storage conditions: Preparations should be stored under refrigeration in glass containers and used within 14 days. Cyclophosphamide for Injection, USP contains cyclophosphamide monohydrate and is supplied in vials for single-dose use. Store vials at or below 25°C (77°F).

Dose: The dose is 50 mg, 200 mg, 500 mg- tablet.

Pharmaceutical Formulations: Injection, Tablet

Brand Names: Oncoxan, Endoxan-Asta, Cycrame, CTX-GLS, Cyclomet, Endoxan-N

13.3.2 Busulfan

Structure:

Chemical (IUPAC) name: Butane-1,4-diyl dimethanesulfonate

Busulfan is a chemotherapy drug in use since 1959. It is a cell cycle non-specific in the class of alkyl sulfonates. Its chemical designation is 1,4-butanediol dimethanesulfonate.

Uses: Busulfan is used to treat a certain type of chronic myelogenous leukaemia (CML; a type of cancer of the white blood cells). Busulfan is in a class of medications called alkylating agents. It works by slowing or stopping the growth of cancer cells in your body.

Side effects: The common side effect observed with administration of busulfan are nausea, vomiting, diarrhoea, constipation, loss of appetite, weight loss, mouth sores, upset stomach, abdominal pain, dizziness, weakness, swelling of the ankles/feet/hand, flushing (warmth, redness, or tingly feeling), headache, trouble sleeping, swelling or irritation around the IV needle, missed menstrual periods, hair loss, darkened skin colour, and fatigue etc.

Side Effects: Common side effects observed with the administration of busulfan are nausea, diarrhoea, loss of appetite or weight, constipation, sores in the mouth and throat, dry mouth, headache, difficulty falling asleep or staying asleep, feeling unusually anxious or worried, dizziness etc.

Storage conditions: Unopened vials of Busulfan Injection must be stored under refrigerated conditions between 2°C to 8°C (36°F to 46°F).

Dose: The dose is 2mg- tablet.

Pharmaceutical Formulations: Injection, Tablet

Brand Names: Mylefan, Myleran, Busulfex

13.3.3 Mercaptopurine

Structure:

Chemical (IUPAC) name: Mercaptopurine (6-MP)

It is taken by mouth. Mercaptopurine was approved for medical use in the United States in 1953. It is on the World Health Organization's List of Essential Medicines.

Uses: It is a medication used for cancer and autoimmune diseases. Specifically, it is used to treat acute lymphocytic leukaemia (ALL), chronic myeloid leukaemia (CML), Crohn's disease, and ulcerative colitis. For acute lymphocytic leukaemia, it is generally used with methotrexate.

Side Effects: Common side effects include bone marrow suppression, liver toxicity, vomiting, and loss of appetite. Other serious side effects include an increased risk of future cancer and pancreatitis. Those with a genetic deficiency in thiopurine S-methyltransferase are at higher risk of side effects. Use in pregnancy may harm the baby. Mercaptopurine is in the thiopurine and antimetabolite family of medications.

Storage conditions: Store it at room temperature and away from excess heat and moisture

Dose: 50 mg, 20 mg/ml

Pharmaceutical Formulations: Oral suspension, Tablet

Brand Names: Mercapto, Empurine, Purinethol, Puri-Nethol, Purinetone.

13.3.4 Fluorouracil

Structure:

Chemical (IUPAC) name: 5-fluorouracil

The drug is available in a 500-mg or 10-mL vial for IV use and as a 1% and 5% topical cream.

The mechanism of action includes inhibition of the enzyme TS by the deoxyribose monophosphate metabolite, 5-FdUMP. The triphosphate metabolite is incorporated into DNA, and the ribose triphosphate into RNA. These incorporations into growing chains result in inhibition of the synthesis and function of DNA and RNA. Resistance can occur as a result of increased expression of TS, decreased levels of reduced folate substrate 5,10-methylenetetrahydrofolate, or increased levels of dihydropyrimidine dehydrogenase. Dihydropyrimidine dehydrogenase is the main enzyme responsible for 5-FU catabolism.

Uses: 5-FU is used in the treatment of several carcinoma types, including breast cancer, colorectal cancer, stomach cancer, pancreatic cancer, and topical use in basal cell cancer of the skin.

Side Effects: Toxicities include dose-limiting myelosuppression, mucositis, diarrhoea, and hand-foot syndrome (numbness, pain, erythema, dryness, rash, swelling, increased pigmentation, nail changes, pruritus of the hands and feet).

Storage conditions: Store at room temperature 15°to 30°C (59° to 86°F). Protect from light. Retain in carton until time of use.

Dose: 250mg/5ml, 500mg/1Vial, 500 mg

Pharmaceutical Formulations: Injection, Cream

Brand Names: Adrucil Efudex, Carac, Tolak, Fluoroplex

13.3.5 Methotrexate

Structure:

Chemical (IUPAC) name: (2S)-2-[(4-{[[(2,4-Diaminopteridin-6-yl)methyl](methyl)amino} benzoyl) amino]pentanedioic acid

The drug is available in 50-, 100-, 200-, and 1,000-mg vials for IV use. The mechanism of action of methotrexate involves inhibition of DHFR, leading to a depletion of critically reduced folates. The reduced folates are necessary for the biosynthesis of several purines and pyrimidines.

Uses: Methotrexate is used to treat several cancer types, including breast cancer, bladder cancer, colorectal cancer, and head and neck cancer.

Side Effects: Methotrexate toxicity includes myelosuppression, mucositis, nausea, vomiting, severe headaches, renal toxicity, acute cerebral dysfunction, skin rash, and hyper-pigmentation.

Storage conditions: Store at 20°C to 25°C (68°F to 77°F); excursions permitted to 15°C to 30°C (59°F to 86°F). Protect from light. After the first puncture, store multiple-dose vials at 2°C to 8°C and use within 30 days.

Dose: 5 mg, 7.5 mg, 10 mg, 15 mg, 50 mg

Pharmaceutical Formulations: Solution, Tablet,

Brand Names: Reditrex, Trexall, Xatmep, Otrexup, Rasuvo

13.3.6 Dactinomycin

Structure:

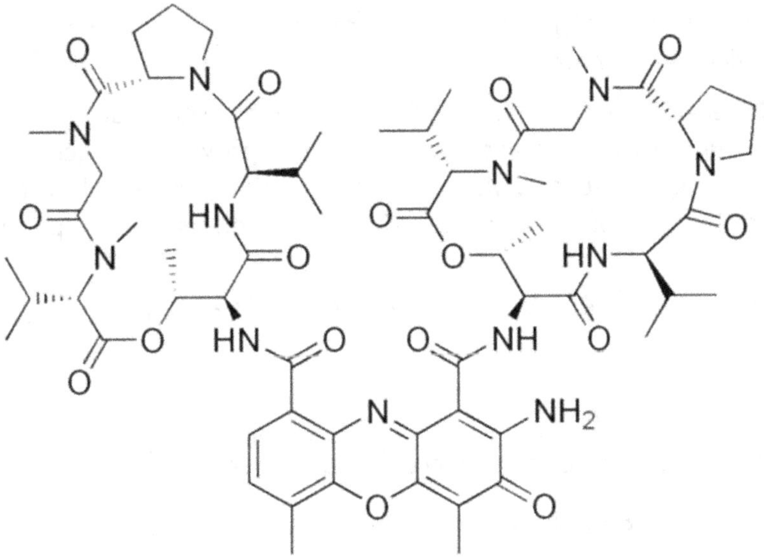

or

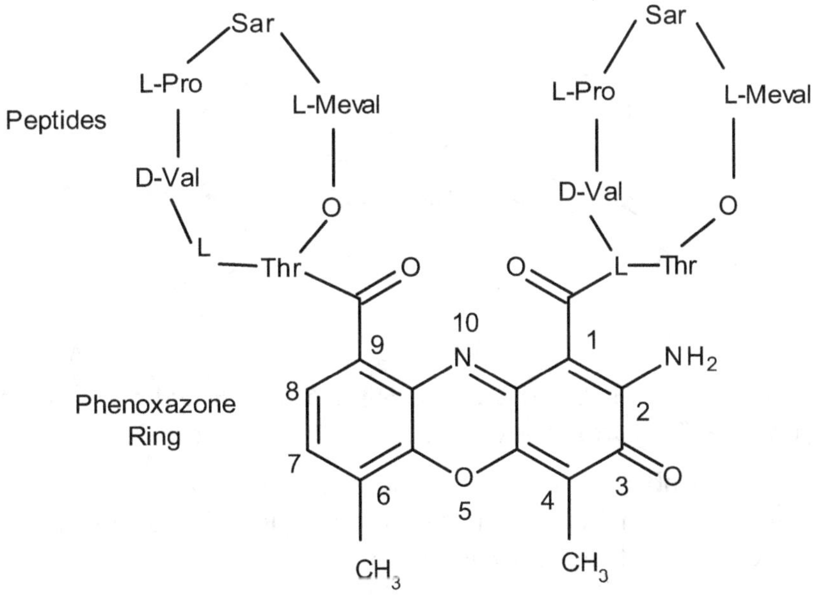

Chemical (IUPAC) name: 2-Amino-N,N'-bis[(6S,9R,10S,13R,18aS)-6,13-diisopropyl-2,5,9-trimethyl-1,4,7,11,14-pentaoxohexadecahydro-1H-pyrrolo[2,1-i][1,4,7,10,13]oxatetraazacyclohexadecin-10-yl]-4,6-dimethyl-3-oxo-3H-phenoxazine-1,9-dicarboxamide

Dactinomycin is available in vials containing 0.5 mg of the drug for reconstitution in sterile water for IV administration.

Uses: This antibiotic is most effective in the treatment of rhabdomyosarcoma and Wilms tumour in children as well as in the treatment of choriocarcinoma, Ewing sarcoma, Kaposi sarcoma, and testicular carcinoma.

Side Effects: Myelosuppression is dose-limiting, with both leucopenia and thrombocytopenia being the most likely presentation. Nausea and vomiting occur shortly (2 hours) after treatment and may be severe. Mucositis and diarrhoea also result from irritation of the GI tract. Hair loss is commonly associated with the agent, as is hyperpigmentation of the skin and erythema.

Storage conditions: Store at 20-25°C (68-77°F). Protect from light and humidity.

Dose: 500mcg

Pharmaceutical Formulations: Injection

Brand Names: Dactino, Dactinoget, Dacmozen, Cosmegen

13.3.7 Doxorubicin Hydrochloride

Structure:

Chemical (IUPAC) name: (7S,9S)-7-[(2R,4S,5S,6S)-4-Amino-5-hydroxy-6-methyloxan-2-yl]oxy-6,9,11-trihydroxy-9-(2-hydroxyacetyl)-4-methoxy-8,10-dihydro-7H-tetracene-5,12-dione

Doxorubicin is available as both the conventional dosage form and a liposomal preparation, both of which are administered by infusion. Doxorubicin HCl powder is available in 10-, 20-, 50-, and 150-mg vials and

Uses: It is widely used in treating various cancers, including leukaemias, soft and bone tissue sarcomas, Wilms tumour, neuroblastoma, small cell lung cancer, and ovarian and testicular cancer.

Side effects: More common side effects observed with administration of Doxorubicin Hydrochloride are Hair loss, thinning of hair, nausea and vomiting and sores in the mouth and on the lips.

Storage conditions: Store at 20-25°C (68-77°F). Protect from light and humidity.

Dose: 10 mg, 50 mg, 2mg/mL

Pharmaceutical Formulations: Injection

Brand Names: Zodox, Cardia, Adrosal, Dobixin, Oncordia, Adrim, Docemax

13.3.8 Vinblastine Sulphate

Structure:

Chemical (IUPAC) name: dimethyl (2β,3β,4β,5α,12β,19α)-15-[(5S,9S)-5-ethyl-5-hydroxy-9-(methoxycarbonyl)-1,4,5,6,7,8,9,10-octahydro-2H-3,7-methanoazacycloundecino[5,4-b] indol-9-yl]-3-hydroxy-16-methoxy-1-methyl-6,7-didehydroaspidospermidine-3,4-dicarboxylate

Uses: Vinblastine Sulphate is available as a powder in 10-mg vials and as a solution in 10- and 25-mL vials for IV administration in the treatment of various cancers, including Hodgkin's disease, lymphocytic lymphoma, histiocytic lymphoma, advanced mycosis fungoides, advanced testicular carcinoma, and Kaposi sarcoma. It has also been used in treating choriocarcinoma and breast cancer when other therapies have failed.

Side Effects: Myelosuppression is commonly seen with vinblastine and is dose-limiting. Inflammation of the GI tract is more commonly seen with vinblastine than vincristine. Nausea and vomiting may also occur. Other adverse effects include alopecia, secretion of antidiuretic hormone, headache, and depression. Neurotoxicity is mild compared with vincristine, but peripheral neuropathy may be seen. An additional manifestation of this neurotoxicity is hypertension related to disruption of autonomic function.

Storage conditions: Store in a refrigerator (2°C - 8°C). Keep the vial in the outer carton to protect it from light.

Dose: 1ml

Pharmaceutical Formulations: Injection

Brand Names: Vincristine Sulphate, Alcrist, VCR, Oncocristin-AQ, Biocristin

13.3.9 Cisplatin

Structure:

$$Cl_2Pt(NH_3)_2$$

Chemical (IUPAC) name: (SP-4-2)-diamminedichloridoplatinum(II)

Cisplatin was discovered in 1845 and licensed for medical use in 1978 and 1979. It is on the World Health Organization's List of Essential Medicines.

Uses: Cisplatin is a chemotherapy medication used to treat a number of cancers. These include testicular cancer, ovarian cancer, cervical cancer, breast cancer, bladder cancer, head and neck cancer, oesophagal cancer, lung cancer, mesothelioma, brain tumours and neuroblastoma. It is given by injection into a vein.

Side Effects: Common side effects include bone marrow suppression, hearing problems, kidney damage, and vomiting. Other serious side effects include numbness, trouble walking, allergic reactions, electrolyte problems, and heart disease. Use during pregnancy can cause harm to the developing fetus.

Storage conditions: Cisplatin, both undiluted in glass containers and diluted with NaCl 0.9% in PE bags, remains stable (< 10% degradation) for at least 30 days at room temperature when protected from light.

Dose: 50 mg, 1 mg, 0.5mg/mL

Pharmaceutical Formulations: Injection, Tablet

Brand Names: Cisplatinum Korea, Plationco, Cisteen, Duplat, Platicis, Kemoplat, Platinex, Cisplan, Platicis

13.3.10 Dromostanolone Propionate

Structure:

Chemical (IUPAC) name: (2R,5S,8R,9S,10S,13S,14S,17S)-17-hydroxy-2,10,13-trimethyl-1,2,4,5,6,7,8,9, 11,12,14,15,16,17-tetradecahydrocyclopenta[a]phenanthren-3-one

Uses: Drostanolone propionate or dromostanolone propionate is an androgen and anabolic steroid (AAS) medication used to treat women's breast can but is now no longer marketed. It is given by injection into a muscle. Drostanolone propionate was first described in 1959 and was introduced for medical use in 1961.

In addition to its medical use, drostanolone propionate is used to improve physique and performance. The drug is a controlled substance in many countries, and so non-medical use is generally illicit.

Side Effects: Side effects of drostanolone propionate include symptoms of masculinisation like acne, increased hair growth, voice changes, and increased sexual desire. It has no risk of liver damage.

Storage conditions: Dry, dark and at 0^0- 4^0C for short term (days to weeks) or -20^0C for long term (months to years).

Dose: 40mg/ml, 100mg/ml, 125 mg/ml, 200mg/ml, 500 mg/ml, 1000 mg/ml,

Pharmaceutical Formulations: Injection

Brand Names: Drolban, Masterid, Masteril, Masteron, Masterone, Mastisol, Metormon, Permastril, Rometholone.

QUESTION BANK

A. MULTIPLE CHOICE QUESTIONS

1. Oncoxan is popular brand name for ______________.
   A. Cyclophosphamide
   B. Busulfan

C. Mercaptopurine

D. Fluorouracil

Answer: A

2. Which of the following heterocyclic ring is present in Dromostanolone Propionate?

   A. Naphthalene

   B. Phenanthrene

   C. Anthracene

   D. Quinidine

Answer: B

3. Which of the following drug is available as both the conventional dosage form and a liposomal preparation?

   A. Doxorubicin Hydrochloride

   B. Vinblastine Sulphate

   C. Cisplatin

   D. Dromostanolone Propionate

Answer: A

4. Which of the following heterocyclic ring is present in Methotrexate?

   A. Naphthalene

   B. Phenanthrene

   C. Anthracene

   D. Pteridin

Answer: D

5. In which of the following drugs have the Phenoxazone ring?

   A. Doxorubicin Hydrochloride

   B. Vinblastine Sulphate

   C. Dactinomycin

   D. Cisplatin

Answer: C

B. SHORT ANSWER QUESTIONS

1. Classify Anti-Neoplastic Agents.

2. Write a short note on Anti-Neoplastic Agents.

3. Give physical properties and structure of Cyclophosphamide.

4. Enlist pharmaceutical formulations available for Busulfan.

5. Enlist different brand names for Fluorouracil.

6. Enlist pharmaceutical formulations available for Dactinomycin

7. Write a note on Vinblastine Sulphate

8. Write a note on Cisplatin.

9. Enlist pharmaceutical formulations available for Dromostanolone Propionate

10. Explain Methotrexate as Anti-neoplastic Agents.

C. LONG ANSWER QUESTIONS

1. Define Anti-neoplastic Agents. Classify it in detail. Add a note on Cyclophosphamide.

2. Define Anti-neoplastic Agents. Classify it in detail. Add a note on Methotrexate as Anti-neoplastic Agents.

■■■